TRANSFORMING SUSTAINABILITY STRATEGY INTO ACTION

TRANSFORMING SUSTAINABILITY STRATEGY INTO ACTION

The Chemical Industry

Edited by

Beth Beloff
Marianne Lines
Dicksen Tanzil
BRIDGES to Sustainability
Houston, TX

A JOHN WILEY & SONS, INC., PUBLICATION

For general information on our other products and services or for technical support, please contact our
Customer Care Department within the United States at (800) 762-2974, outside the United States at (317)
572-3993 or fax (317) 572-4002.

Wiley also publishes its books in a variety of electronic formats. Some content that appears in print may
not be available in electronic formats. For more information about Wiley products, visit our web site at
www.wiley.com.

Library of Congress Cataloging-in-Publication Data:

Transforming sustainability strategy into action : the chemical industry / edited by Beth
Beloff, Marianne Lines, Dicksen Tanzil.
  p. cm.
 ISBN-13: 978-0-471-64445-3 (cloth)
 ISBN-10: 0-471-64445-5 (cloth)
 1. Chemical industry--Environmental aspects--Management. 2. Sustainable development.
3. Industrial management--Environmental aspects. I. Beloff, Beth, 1947- II. Lines,
Marianne, 1954- III. Tanzil, Dicksen, 1973-

 HD9650.5.T73 2005
 660'.068'2--dc22
                                                            2005043920
Printed in the United States of America

10 9 8 7 6 5 4 3 2 1

To Mark for his love, support, and encouragement; to my children – Justin, Carrie, Aaron and Darcy – who inspire my passion to preserve and enhance future opportunities for their generation and their children's. Justin, your new journey to understand sustainability is truly breathtaking! May you help drive the change we all seek!

Beth Beloff

To Brian and my family, who have encouraged me to follow my passions, to grow with change and to make a difference.

Marianne Lines

To Yanhui for her heartfelt trust and support; to many who have taught me so much.

Dicksen Tanzil

CONTENTS

FOREWORD

Dawn Rittenhouse

A sustainable chemical industry. To many that sounds like an oxymoron. How can the industry that produced the materials that caused Love Canal, Bhopal, and the ozone hole ever be sustainable?

However, I would argue that the question cannot be "Can the chemical industry become sustainable?" The questions must be "How does the chemical industry become sustainable?" Today, synthetic chemicals are part of every product we use, from cell phones and computers, to our clothes, our automobiles, and our housing. BASF got it perfect with their tag line – "We don't make the products you use, we make the products you use better." If we do not figure out how to make the chemical industry sustainable, we will not, by definition, develop a sustainable world.

Right now, sustainability is a journey . . . maybe someday, if we are successful, it will be an endpoint. The chemical industry took the first step of the journey in 1985 with Responsible Care®. Responsible Care® was fundamentally about taking responsibility for our operations, and through product stewardship, our products throughout the lifecycle. Not a bad start; in fact, for the time, an amazingly proactive initiative.

Despite the effort that has gone into implementing Responsible Care® globally, public perception of the industry is still very low. Probably not surprising. The World Wildlife Fund (WWF) tested the blood of 14 European Union Environmental Ministers to determine levels of persistent chemicals (they all had man-made chemicals in their blood) and recent studies have indicated that Arctic animals have high levels of the brominated compounds that are typically used as fire retardants. With that kind of news in the media, it is hard to imagine a positive industry image.

So, 20 years on from the introduction of Responsible Care®, what is next? Many chemical companies are starting to ask the hard questions about what a sustainable chemical industry looks like ... what's in, what's out and, more importantly, what does the path look like, and how are we going to get from here to there?

Fortunately, we are not alone. Many other sectors are also struggling with the question and are developing ways to approach sustainability that in some cases are driving demand for more sustainable solutions from the chemical industry. HP, for example, has an extensive supplier survey to understand how their suppliers operate and how they manage their operations to enhance safety, health, environmental, and social performance. SC Johnson has a process called the "Greenlist" to help their product designers focus on raw materials that are more sustainable in the long term. The chemical industry is also reaching out to a broad group of stakeholders to help with what sustainability really means. Dow has a Corporate Environmental Advisory Committee, and DuPont has a Biotechnology Advisory Panel.

Beth, Marianne, and Dicksen have gathered a lot of information on tools like metrics and lifecycle analysis. They have also gathered information on what a number of companies have done and why. There are no silver bullets or quick fixes in the book. None of the companies is close to operating in a truly sustainable way. They have taken on different approaches so may be ahead in one area while not engaged in another area at all. But credit is due; they are leaders out there trying to forge a path.

Ideally, the companies who are leaders will be successful financially because nothing encourages imitation like the opportunity to gain a competitive advantage, or the fear of losing market position. If the leadership companies outperform their peers, then the business case will be evident and the transition to more sustainable actions will be quick. More than likely though, there will be successes that encourage imitation and there will be failures that become the reasons for some to maintain the status quo rather than seeing them as learning opportunities.

Transforming Sustainability Strategy into Action: The Chemical Industry provides an understanding of what the many facets of sustainability mean to the chemical industry and how one may take on the challenge of building an integrated sustainability approach. Anybody interested in sustainability can learn from this book, whether you have been engaged in the process for a while or are just beginning to think about sustainability as a strategy. Enjoy the journey.

BRAD ALLENBY

Underlying the interest in "green chemistry," "sustainable engineering," "sustainability," and similar formulations is the beginning of what for many people is a frightening realization: we now live in an anthropogenic world. The Industrial Revolution was in many ways a revolution in material sciences and chemistry, and even today the implications of nanotechnology – chemistry at small scale – continue that revolution. The history of the human species, in fact, is in large part a history of learning to manipulate the material world, and our success in that endeavor is a significant reason why we now live in what the journal *Nature* calls "the

Anthropocene" – the Age of Humans (Editorial, *Nature*, 2003). Humans increasingly dominate the dynamics of many natural systems and material cycles; given the coming convergence of foundational technologies, especially nanotechnology, biotechnology, information and communication technology, and cognitive sciences (dubbed "NBIC"), this trend will not only continue, but intensify. And our use and management of materials will remain a central feature of our continuing terraforming of this planet.

The evolution of this human world raises a number of difficult questions. For one thing, many people would rather it did not exist, and so retreat to simplistic formulations that may be psychologically satisfying, but are increasingly unrealistic, even dysfunctional. So it becomes a major challenge just to recognize what it is we have already wrought, and what is coming down the road, quite rapidly, directly at us. Beyond that, of course, even if we perceive the difficulties, we lack tools and methodologies with which to deal with them. In particular, we are institutionally and intellectually constrained in our ability to deal rationally with complex adaptive systems, such as complicated material and technological patterns in a rapidly evolving and increasingly global economy.

But this should not be taken pessimistically to mean that we are powerless; rather, it should be encouragement to continue to research and learn, and not to let the best become the enemy of the good. In this regard, the history of chlorofluorocarbons (CFCs) offers some useful insights. First, it is important to remember than when CFCs were introduced, they are adopted because they were that era's equivalent of "green chemistry." They were much safer than the chemicals they replaced (ammonia in refrigeration uses, for example, or toxic chlorinated solvents in cleaning applications); they were stable; they were energy efficient – in other words, they were a perfect example of "green chemistry." Only later did it turn out that they had the unfortunate and unanticipated ability to migrate to the upper atmosphere and there break down to reduce stratospheric ozone concentrations. This threatened the ozone layer, which absorbs high-energy ultraviolet radiation from the sun before it gets to the Earth's surface and thereby protects biological systems from its detrimental effects. As this effect was validated, a number of initiatives, including international legal agreements and technological innovation, reduced CFC emissions with the result that, at this point, we have both reasonably safe industrial processes and products, with much less potential for stratospheric ozone depletion.

What are the lessons from this experience? First, that "green chemistry" and other reductionist approaches, while often quite valuable, must be understood at the context of the modern globalized economy, where chemical and material science innovations can rapidly expand to a scale where fundamental natural systems can be impacted. In other words, it is inadequate to consider systemic effects only at a small scale; rather, we must learn to explore potential impacts at all scales, including emergent behaviors at very large scales that may be completely unanticipated. We must also learn to consider the relevant systems as a whole, not jsut as particular interests or academic disciplines dictate. Bluntly put, "it's the system, stupid." Secondly, the story does not end with the successful commercial implementation of a material or chemical innovation. Rather, we must learn to dialog with these systems on an ongoing basis, so that subtle but important effects are not missed, or ignored until they become irreversible except at great cost. This capability, which right now is not the responsibility of any institution, should be a priority for all concerned. But – and this is the good news – the CFC story, for a number of reasons, is a success story. It was complicated chemistry, complicated politics, emergent and unanticipated behaviors – but in a very short time, relatively speaking, our human systems responded. We are not powerless, unless we choose to render ourselves powerless.

We do so in two important ways. First, blinded by ideology or simply afraid of the magnitude of the challenge posed to us by the anthropogenic Earth, we may choose not to perceive what is actually going on around us. Secondly, even if we do realize what is going on, we can choose not to act, either because inaction is more comfortable, or because we cannot at this point know what might be best, rather than simply an improvement. Both options are arguably unethical.

Transforming Sustainability Strategy into Action: The Chemical Industry takes another path. Eschewing simplistic silver bullets, it pushes us forward both conceptually and operationally into new approaches, new ways of thinking, that in turn enable us to begin a rational, and ethical, interaction with this unique and strange new age that we have done so much to bring about. Forget Mars: humans have terraformed a planet already, and this book is another small step in responsibly responding to that reality. Certainly we do not yet understand the complex systems within which we are already operating, nor do we know what the best actions might always be. As this book illustrates, however, we have both the will and the capability to begin building a better world. And so the Anthropocene begins.

REFERENCE

Editorial, "Welcome to the Anthropocene" *Nature*, 424; 709 (2003).

PREFACE

When I was approached by Wiley to write a book on sustainable development in the chemical industry, my first concern was that, by addressing only the chemical industry, the view on sustainability would be too limited. Upon greater reflection, this industry represents an opportune lens through which to fashion a book on sustainable development for numerous reasons. Chemicals have been linked to the first trumpeting of alarm that heralded in the environmental movement, with the writing of *Silent Spring*. The chemical industry was among the first to organize on a significant scale a voluntary program with hopes of transforming itself and the perception of others into a responsible environmental actor (Responsible Care®). And lastly, the industry's products are ubiquitous in commerce and are found in the supply chains, if not lifecycles, of virtually all products and services. Correspondingly, they are also found in all ecosystems around the globe. So, utilizing the issues of sustainability to be representative of the broader set confronting business in general makes sense.

The second concern I pondered was how to present a book that is balanced in representing the perspectives of multiple stakeholder groups of importance, broad in the handling of an array of topics, yet pragmatic in its quest to be of value to those most interested in operationalizing sustainability in their daily activities, on "Monday morning." It was clear that I needed to work on a team with others representing areas of knowledge deeper than my own, and further, that we needed to maintain, wherever possible, the voices of others as co-authors.

And so I asked Marianne Lines, my dear friend and colleague, to become co-editor. Marianne's exceptional organizational skills and sense of enjoyment from details, well-honed experience in pulling together expert stakeholders, and fundamental knowledge of sustainability issues were my primary reasons for inviting her to participate, and she has surpassed all of my expectations.

Further, Marianne and I wanted a third team member imbued with specific knowledge of the chemical industry and with technical expertise in the area of chemical products and processes. Dicksen Tanzil kindly accepted our invitation to fill that spot. His dedication, sense of purpose, intelligence, and willingness to take on any task required, made him an invaluable member of the team.

This rounded out our editorial team. And so BRIDGES to Sustainability, our non-profit organization housing the three of us in our quest to bridges between technical and management issues, between short- and long-term values, between companies and their stakeholders, and between strategies and actions, became again a bridge between thought and words.

Finally, we wanted to create a comprehensive resource on the topic of operationalizing sustainability, which could be used as a reference on a variety of sustainability how-to topics. As a result, we developed a compendium drawing upon conceptual, technical, and practical perspectives of sustainability. Rather than being comprehensive, we offer representative approaches in short form, with references as to where to find more information.

Our main goal of this book is to provide a framework for companies to adopt sustainable business practices; to demonstrate how businesses can translate sustainability strategies into action. Our focus is how to operationalize sustainability, and we seek to be relevant to not only managers in industry, but also to future managers in business and engineering educational programs.

We are motivated by a desire to be better: better corporate players, more socially responsible, creating fewer negative impacts and more positive benefits from our actions as well as those of others constituting our "value chain", improving the quality of life in our communities, protecting our children's quality of life in the future. If this book offers insights into how to move closer to any of these, we have accomplished our mission. Please take as many of these ideas as you wish, expand upon them, and use them. That will be the ultimate test of the success of this endeavor.

Beth Beloff
Marianne Lines
Dicksen Tanzil

ACKNOWLEDGMENTS

Assembling and writing portions of this book have led us on an incredible journey. We never envisioned how much effort this would entail, or how much give-and-take and rethinking of the final product would be involved with every interaction with each contributor, author, or reviewer. We have had the kind support of professional colleagues, friends, and family, all of whom have helped to broaden our understanding of our mission and keep us on target. And to the many agents of change who have inspired us by their own thoughts, actions, and written words, we say thank you. You know who you are.

Brad Allenby, Chuck Bennett, Karen Coyne, Pogo Davis, Dawn Rittenhouse, and Darlene Schuster all took time out of their busy schedules to comment among the way about the direction we were taking and gave valuable insights as to how to correct course.

The earliest reviewers of our book concept gave us much appreciated feedback and guidance. They include Tarcisio Alvarez-Rivero, Karen Coyne, Pogo Davis, Tom Gladwin, Dawn Rittenhouse, and Darlene Schuster. Further guidance on overall concept came from Chuck Bennett and Frank Dixon. Both wrote various pieces that fundamentally inspired our compendium, even though their contributions are not directly found in the book. Brad Allenby, John Carberry, and Dave Constable provided significant critique to the sections on Sustainability Planning and Design. Beverley Thorpe provided invaluable insight into public perceptions of the chemical industry. Ken Geiser and Joel Tickner kept us abreast of emerging international policies on the management of chemicals. Karen Coyne jumped in without hesitation to organize the auditing section when we were confronted with a vacuum there, and Art Gillen filled the same role in the management systems area when we were confronted with similar issues.

Joanna Underwood reconnected as a friend and colleague, and provided moral support as well as professional wisdom. Mitch Mathis and Marilu Hastings, in spite of experiencing a time of great personal angst, came through with their personal and professional support. Ann Goodman shared her writings to inspire our case study development. Priscilla Johnson shared invaluable insights into the business case for sustainability. Dave Taschler and Tim Donnelly provided excellent insights and critique of our business case section, as did Gautham Parthasarathy from Solutia.

The development of the Chemical Industry Sustainability Survey was more of a challenge than expected. The PricewaterhouseCoopers team, comprised of Andy Savitz, Doug Hileman, and Michael Besly, was extraordinary in their dedication to an excellent product. When funding for the effort fell through, they graciously contributed their time and effort, for which we are extremely grateful. Through industry focus groups we vetted the survey findings to both clarify and validate interpretations. Darlene Schuster, as leader in AIChE's Institute for Sustainability, was a dedicated partner in the review of the survey instrument, marketing of the survey, and in the organization of the focus groups. Charlene Wall (BASF), through AIChE's Center for Sustainable Technology Practices, was also instrumental in organizing an excellent group of industry leaders to participate in the focus groups. Those participants include Emanuel Baba (FMC), Earl Beaver (IfS), Tim Donnelly (Rohm & Haas), Karen Koster (Cytec), Joe Machado (Shell Chemicals), Ada Nelson (BP), Dawn Rittenhouse (DuPont), Dave Taschler (Air Products), Joe Zola (Eastman Chemicals), and Elaine Zoeller (Eastman Chemicals). Terry Yosie (ACC) was also generous with his time in participating in a focus group.

Finally, we would like to thank the Board and Advisory Council, as well as the extended family, of BRIDGES to Sustainability for their dedication to the mission of the organization and their crafting a more sustainable future with us.

1

INTRODUCTION

BETH BELOFF
MARIANNE LINES

BRIDGES to Sustainability

In recent years, many top leaders have committed their companies to the concept of sustainable development. The practical implications of such a commitment for a corporation is that performance is considered along three dimensions – economic, environmental, and social – rather than a single-minded drive for immediate economic value. While this growing commitment to sustainability has been driven by many complex factors, such as changing societal expectations, stakeholder activism, and regulatory activity, for reasons regarding competitive advantage, business leaders are considering the possibility of long-term economic performance being enhanced by a commitment to strong environmental and social performance. This means that in the next decade or so, competitive advantage may migrate to those firms that learn to create customer and shareholder value in ways that do not harm the environment and benefit a broader spectrum of society.

For those of us who have been working in the field encompassing sustainable development, sustainability, or corporate social responsibility – among an array of other terms-of-art – it is heartening to observe the recent explosion of venues and opportunities featuring this topic. In the United States, these expanding venues include publication of books and journal articles, conferences, technical society and industry association initiatives, and university initiatives in the form of new centers and institutes, course additions, or campus physical site planning and management. Recent initiatives among technical and industry associations include the new Institute for Sustainability (IfS) at the American Institute of Chemical Engineers (AIChE), Green Chemistry Institute at the American Chemical

Transforming Sustainability Strategy into Action: The Chemical Industry, Edited by B. Beloff, M. Lines, and D. Tanzil
Copyright © 2005 John Wiley & Sons, Inc.

Society (ACS), US Green Building Council, US Business Council for Sustainable Development (USBCSD), SD Planner and the newly formed Metrics task forces at the Global Environmental Management Initiative (GEMI), as well as the Auditing Roundtable's new initiative to include sustainability auditing.

In addition, there is a proliferation of global agreements, standards, guidance documents, and frameworks related to sustainability practices, and increased stakeholder pressure to subscribe to them. They include the United Nations' Global Compact, Caux Roundtable Principles for Business, CERES Principles, Sustainability Reporting Guidelines of the Global Reporting Initiative (GRI), the Global Sullivan Principles, Organisation for Economic Development and Cooperation's Guidelines for Multinational Enterprises, Social Accountability SA 8000, and Dow Jones Sustainability Index (refer to Sections 4.7 and Appendix 2 of this book). The investment community is beginning to pay attention to sustainability, as sustainability ratings with respect to corporate performance are drawing the attention of Socially Responsible Investing (SRI) funds as the number of rating organizations is growing. The evidence that sustainability is becoming a more accepted concept to business and is emerging as a driving force is a subtext of this book.

While many companies embrace the concept of sustainability, few know how to make it operational. Further, there is a lack of accepted methods to assess an organization's progress toward this important goal. If the goal of sustainability is to be incorporated into management decision-making, indicators of progress are needed that take into consideration the impacts of the organization, its processes and products, in terms of resource depletion and pollution emissions, corporate and societal costs and value-added, and impacts along the lifecycle that the processes, products, and services generate.

The principal strength of this book is that it provides a framework to enable companies to adopt sustainable business practices. The book is intended to:

- provide managers with a practical framework to identify and assess options for improving the sustainability of their companies' and supply chains' current and future business practices, products, and manufacturing or production methods;
- demonstrate how businesses in the chemical sector can translate sustainability strategies into action;
- focus on operationalizing the environmental, economic, and social value of sustainable development for chemical industries; and
- contribute to the body of evidence regarding the business case for sustainability.

Building on current initiatives and leveraging partnerships, this unique work provides practical mechanisms to help managers understand how to recognize and monitor the value of sustainability in their decisions. The work focuses on operational aspects, decision support, practical tools for measuring progress, and case studies of barriers and opportunities associated with the pursuit of corporate sustainability – all with a look to the chemical industry.

The focus on the chemical industry is made because of the prominence of its role in the global economy and in society. The impact of the chemical industry is felt in every area of commerce, as chemicals are ubiquitous in all value chains and affect all ecosystems, no matter how seemingly pristine, on the planet. The lessons presented here, then, can extend to other business sectors.

The book is intended for a primary audience comprised of professionals in the chemical industries, including those designated managers in business, research, and development, operations, health, safety and environment, security, and sustainable development. Yet, managers from other industries in the value chain of chemical companies will benefit from this text. Future decision-makers, such as students, are provided with a framework with which to lead in a world where social and environmental stewardship skills are just as important as technical, financial or marketing skills.

As this is a book for industry and its current and future practitioners of sustainability, the intent has been to present a balanced perspective, one from multiple points of reference. This book represents the views of many different experts on the subject of sustainability. It is written in their voices. Lessons can be learned not only from what companies feel they do right but also by what they either have done or have been perceived as doing wrong.

While some of the issues dealt with in the book are contentious, they have served to galvanize efforts by the public and forward-looking governments and companies to reduce the impacts of chemicals harmful to ecosystems and human health. It is in that spirit that we present the nature of the controversy. Where companies are spotlighted regarding their practices, opinions are those of the individual authors and are *not* necessarily reflections of how the editors feel. They have been included in the interest of presenting an array of stakeholder positions for purposes primarily of advancing the broader discussion and ultimately encouraging action.

1.1 ORGANIZATION OF THE BOOK

The book has been organized into the following major sections.

Chapter 2: Addressing Sustainability in the Chemical Industry
This chapter looks at the scope and scale of the chemical industry; the industry's response to formative developments and drivers; and the evolution of its signature program, Responsible Care®, with a look at its current role in advancing sustainability as well as future positioning.

Chapter 3: Views on Key Issues Facing the Chemical Industry
This chapter looks at the public's perception of the chemical industry. The numbers of chemicals produced and the range of products that contain the chemicals have raised serious concerns about the impacts of chemicals on human health and ecosystems, which often boil down to the public's mistrust of the industry as a whole. This chapter highlights key issues, challenges and opportunities for the chemical industry, including the complex process of quantitative risk assessment

and the limits of this approach; the limits of the current risk management approach and emerging policy directions that the industry must face; impacts to human health and ecosystems through the chemical life cycle of manufacture, use and disposal; security vulnerabilities; and balancing the implementation of sustainability programs with the sector's competitiveness.

Chapter 4: Planning Frameworks for Sustainable Development
The planning section of this book offers a range of planning frameworks, from an elaboration on the elements to consider in planning for sustainability and steps to take, to broader conceptual frameworks regarding the systems in which business operates, what contributes to their *un*sustainability, and how to make the systems as well as the companies operating within them more sustainable. This chapter attempts to capture the essence of these "frameworks" as a way of demonstrating the breadth of ways to move sustainable development into management practice. It is complemented by a more in-depth discussion of the application of management systems and auditing protocols to sustainability presented in Chapter 6.

Chapter 5: Designing for Sustainable Development
Design is a critical element in the implementation of sustainable development. Meeting the needs of the growing global community while minimizing negative impacts to the environment and societal well-being requires us to come up with alternative patterns of resource utilization, production, and consumption. This chapter provides an overview of approaches to designing for sustainability, details cradle-to-cradle materials assessment and product design, and highlights aspects of more sustainable process design strategies.

Chapter 6: Implementing Sustainable Development; Decision-Support Approaches and Tools
This chapter describes some of the approaches and tools that companies use to demonstrate their commitment to sustainability and support decision-making. While the approaches and tools are demonstrated throughout the examples and case studies elsewhere in the book, a few are covered here in more depth. This section looks at management systems, auditing, indicators and metrics, assessing value, reporting, security, and corporate social responsibility. In particular, the authors identify opportunities for leveraging existing efforts and look at the role of each of the tools in advancing sustainability.

Chapter 7: Future Directions for the Chemical Industry
The transition to a sustainable chemicals industry requires a thorough reconceptualization of the industry and its products. New directions for the industry are emerging. This chapter takes a broad, futuristic, and macro-view of the chemical industry. We find here ideas that link chemical production to biological processes, the phase out and substitution of the most dangerous chemicals, and the dawn of new means of transforming chemicals at the molecular and submolecular level. Chemical stewardship, biomimicry, chemical substitution, and nanotechnology all provide approaches that are potentially cleaner, greener, and more productive.

Chapter 8: The Business Case for Sustainable Development

The key question asked by companies considering pursuit of sustainability practices is what is the business case – the business value proposition – that can create justification for allocating resources and the attention of corporate boards and executives to this practice area. As was written in *Walking the Talk* (Holliday *et al.*, 2002), companies tend to get involved in activities long before they can prove the business case for doing so. However, the case is emerging and aspects of it are presented in this chapter. The chapter is divided into four parts: (1) results from The 2004 Chemical Industry Sustainability Survey and related Focus Groups, developed in a collaboration between BRIDGES to Sustainability, PricewaterhouseCooper (PwC), and AIChE, and conducted by PwC for this book; (2) an overview of sustainability and performance, linking the intangibles of sustainability to market performance; (3) five business cases presented by sustainability managers at companies with significant chemical operations and a major customer of chemicals; and (4) various other provocative perspectives on the business case for the industry.

The editors recognize that the book's objective, of providing a framework for companies to adopt sustainable business practices, is ambitious. If we have achieved but a fraction of this goal, then the book is of great value to industry. However, collectively as a society, the journey towards sustainable development is in the early stages and our knowledge continues to strengthen. We cannot promise that a company will become sustainable just by adopting the actions outlined in this book, but we can assure you that the company will have begun one of the most important journeys of its evolution.

What does sustainable development/sustainability mean?

- In 1987, the World Commission on Environment and Development (Brundtland Commission) published a report entitled "Our Common Future," in which it defined Sustainable Development as "Development that meets the needs of the present without compromising the ability of future generations to meet their own needs."

- More recently, the concept of sustainability has been captured within the framework of the "Triple Bottom Line," which companies seek to address in order to minimize harm resulting from their activities and to create economic, social, and environmental value. The Global Reporting Initiative (GRI) is one effort to develop a triple-bottom-line reporting framework.

- The Dow Jones Sustainability Group Index defines "corporate sustainability" as "a business approach to create long-term shareholder value by embracing opportunities and managing risks deriving from economic, environmental and social developments."

- A Native American sentiment, oft quoted, opines that we do not inherit the Earth from our ancestors; we borrow it from our children.

REFERENCE

C. O. Holliday, Jr., S. Schmidheiny and P. Watts. *Walking the Talk. The Business Case for Sustainable Development.* Greenleaf Publishing, San Francisco, 2002.

2

ADDRESSING SUSTAINABILITY IN THE CHEMICAL INDUSTRY

Marianne Lines

BRIDGES to Sustainability

2.1 INTRODUCTION

This book's focus on the chemical industry reflects the industry's prominent role in the global economy and society. The chemical industry is very diverse, producing thousands of substances that are present in countless products. No longer associated with just commodity chemicals, plastics, and additives, chemical companies have made significant forays into consumer care products, pharmaceuticals, biotechnology, agribusiness, and many other sectors that touch upon almost every aspect of our lives. Chemical companies provide a wide range of products and solutions, from automobile components to construction materials to consumer goods (Innovest, 2002). Because of its size and diversity, the industry has had a significant impact on our global economy, the environment, and society.

The chemical industry has a long history of product and process innovation and has provided enormous benefits to society. Since the late 19th century the industry has continually developed new products and processes, improved its functionality and cost-effectiveness, and displaced many traditional materials. Its products are now integrated into many economic sectors and it has a history of collaborating with its customers to develop innovative solutions. As a result there has been a dramatic increase in the use of chemicals and plastics, and these have provided tremendous cost savings as well as improved functionality (Sherman, Chapter 4, Section 4.6 of this volume). Safer and more efficient automobiles, and large-scale

Transforming Sustainability Strategy into Action: The Chemical Industry, Edited by B. Beloff, M. Lines, and D. Tanzil
Copyright © 2005 John Wiley & Sons, Inc.

production of life-saving drugs, to name a few, also bear witness to the societal benefits produced by the industry.

Nevertheless, events during the past 35 years have directed public and government scrutiny on the chemical industry and its unintended impacts on the environment and society. For instance, despite evidence of health effects linked to chemical exposure, a great majority of the chemicals found in our air, water, food, and everyday products lack basic safety data for human and ecosystem health, particularly that of developing organisms. Of the 100,000 chemicals in common use, 30,000 are used in volumes of one ton or more. Of these, 95 percent have little or no environmental or human health data, simply because prior to 1980 no data were required before marketing in the United States or Europe (Thorpe, Chapter 3, Section 3.2 of this volume).

This chapter will look at the scope and scale of the chemical industry; the industry's response to formative developments and drivers; and the evolution of its signature program, Responsible Care®, with a look at its role in advancing sustainability.

2.2 UNDERSTANDING THE CHEMICAL INDUSTRY

2.2.1 Sector Profile

There is no single definition of the chemical industry or a universal categorization of the industry's subsectors. There are, in fact, numerous ways of identifying subsectors of the chemical industry, including, according to the Standard Industrial Classification (SIC) codes (US OMB, 1987), the North American Industry Classification System (NAICS) (US OMB, 2000), the European Union NACE Codes for the nomenclature of economic activities,[1] the Japanese standard industrial classification,[2] and according to S&P definitions (S&P, 2004), to name a few.

The Organisation for Economic Co-Operation and Development's (OECD) categorization, demonstrating how diverse the industry is, divides the chemical industry into four groupings (OECD, 2001):

- *Basic chemicals* (or commodity chemicals).   Markets for basic chemicals are primarily in other basic chemicals, specialty chemicals, and other chemical products, as well as in other manufactured goods (textiles, automobiles, appliances, furniture, and so on) or in the processing applications (pulp and paper, oil refining, aluminum processing, and so on). They include bulk inorganics and organics (e.g., ammonia, gases, acids, salts), petrochemicals (e.g., benzene, ethylene, xylene, toluene, butadiene, methane), fertilizers, industrial chemicals, plastics, resins, elastomers, fibers, dyestuffs, and so on.

- *Specialty chemicals.*   These chemical substances (e.g., adhesives and sealants, catalysts, coatings, electronic chemicals, plastic additives), which are derived from basic chemicals, are more technologically advanced products

[1]See http://www/top500.de/nace4-e.htm. Accessed October 2004.
[2]See http://www.stat.go.jp/english/index/seido/sangyo/3.htm. Accessed October 2004.

than basic chemicals. They are manufactured in lower volumes than basic chemicals and have a higher value-added because they cannot easily be duplicated by other producers or are shielded from competition by patents.

- *Life science products.* These include pharmaceuticals, products for crop protection and products of modern biotechnology. Technological advantages are extremely important and R&D spending for this sector is the highest among all industries.

- *Consumer care products.* This includes soap, detergents, bleaches, laundry aids, hair care products, skin care products, fragrances, and so on. These products are formulated products, employing what is often simple chemistry but featuring a high degree of differentiation along branding lines. Research and development expenses are rising and many of these products are becoming high-tech in nature.

2.2.2 Contribution to the Economy

The global chemical industry today produces tens of thousands of substances. The substances can be mixed by the chemical industry and sold and used in this form, or they can be mixed by downstream customers of the chemical industry (e.g., retail stores that sell paint). It is important to note that most of the output from chemical companies is used by other chemical companies or other industries (e.g., metal, glass, electronics), and chemicals produced by the chemical industry are present in countless products used by consumers (e.g., automobiles, toys, paper, clothing) (OECD, 2001).

The chemical industry has a prominent role in the global economy, accounting for about 13 percent of world trade (Krueger and Sein, 2002). Almost every country has a chemical industry, yet almost 80 percent of the world's total output is currently being produced by only 16 countries: the United States, Japan, Germany, China, France, the United Kingdom, Italy, Korea, Brazil, Belgium/Luxembourg, Spain, the Netherlands, Taiwan, Switzerland, and Russia.

In 2001, the value of worldwide chemical output was reported to be $1.67 trillion, generated from the regions or countries shown in Figure 2.1. More recent data reported in *Chemical & Engineering News* (McCoy *et al.*, 2004) suggests small fluctuations on these data over the past few years.

In the European Union, some 25,000 chemical companies employ 1.7 million, or 7 percent of the overall workforce in the manufacturing industry (CEFIC, 2004). While the chemical industry employs only about 1 million people in the United States, it is one of the largest industries within the U.S. economy and contributed $460 billion worth of shipments in 2000, accounting for more than 1 percent of total U.S. GDP and nearly 12 percent of the manufacturing GDP (Innovest, 2002).

In the United States alone, the industry produces more than 70,000 different chemical substances from 12,000 operating chemical plants. The industry concentrates 63 percent of its production in 10 U.S. states: Texas, New Jersey, Louisiana, Illinois, North Carolina, California, Ohio, Pennsylvania, New York, and South Carolina.

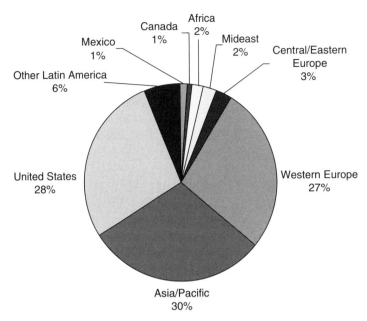

Figure 2.1. Percent of total worldwide chemical production in 2001. (*Source*: American Chemistry Council, 2002.)

2.3 DRIVERS OF SUSTAINABILITY

Major events and developments during the past 35 years have caused the chemical industry to examine ways to improve its environmental, social and economic performance. Drivers encouraging industry's action include expanded regulation of industry processes and products; increased public right-to-know initiatives and the public reporting of environmental, health and safety (EHS) performance; greater stakeholder participation in government and corporate decision-making; the trend of restructuring, mergers, and the globalization of business activities; global environmental management systems and standards; and the re-examining of security practices (Yosie, 2003). Some of the many milestones that have marked society's journey toward sustainable development and those developments that have influenced the chemical industry are highlighted in the Sustainable Development Timeline produced by the International Institute for Sustainable Development (IISD, 2002) (Table 2.1).

2.3.1 Chemical Sector Survey

To corroborate and explore the importance of the drivers commonly cited in the literature, in 2004, BRIDGES to Sustainability (BRIDGES), in collaboration with PricewaterhouseCoopers LLC (PwC), developed a survey addressing the chemical industry's view of sustainability and its response to sustainability issues. BRIDGES,

TABLE 2.1.   The Sustainable Development Timeline

The Sustainable Development Timeline captures some of the key events. The original version of **The Sustainable Development Timeline** was published in 1998 with the support of the IDRC. The second edition was published in 1999. **IISD** has prepared this third edition of the Timeline in advance of the **World Summit on Sustainable Development.**

Silent Spring was published in 1962. The book's release was considered by many to be a turning point in our understanding of the interconnections among the environment, the economy, and social well-being. Since then, many milestones have marked the journey toward sustainable development.

1960s

1962 – *Silent Spring* by Rachel Carson brought together research on toxicology, ecology, and epidemic to suggest that agricultural pesticides were building to catastrophic levels. This was linked to damage to animal species and to human health.

1967 – **EDF Environmental Defense Fund** formed to pursue legal solutions to environmental damage. EDF goes to court to stop the Suffolk County Mosquito Control Commission from spraying DDT on the marshes of Long Island. http://www.environmentaldefense.org/

1968 – **Biosphere**. Intergovernmental Conference for Rational Use and Conservation of Biosphere (UNESCO) is held; early discussions of the concept of ecologically sustainable development. http://www.unesco.org/

1968 – **Paul Ehrlich publishes** *Population Bomb* on the connection between human population, resource exploitation, and the environment. http://www.pbs.org/population_bomb/

1969 – *Friends of the Earth* forms as a nonprofit advocacy organization dedicated to protecting the planet from environmental degradation; preserving biological, cultural, and ethnic diversity; and empowering citizens to have a voice in decision-making. http://www.foe.org/

1969 – *National Environmental Policy Act* is passed in the United States, creating the Council on Environmental Quality and establishing a national policy for the environment. http://es.epa.gov/oeca/ofa/nepa.html

1969 – **Partners in Development/**

1970s

1970 – **IDRC** Report of the Commission on International Development, chaired by former Prime Minister of Canada, Lester B. Pearson. The first of the international commissions to consider new approach to development focused on research and knowledge in the South. Led to the formation of the IDRC. http://www.idrc.ca/

1970 – **First Earth Day** held as a national teach-in on the environment. An estimated 20 million people participated in peaceful demonstrations across the United States. http://earthday.envirolink.org/history.html

1970 – **Natural Resources Defense Council** forms with a professional staff of lawyers and scientists to push for comprehensive U.S. environmental policy. http://www.nrdc.org/

(continued)

TABLE 2.1. *Continued*

1971 – Greenpeace starts up in Canada and launches an aggressive agenda to stop environmental damage through civil protests and nonviolent interference. http://www.greenpeace.org/

1971 – Founex Report is prepared by a panel of experts meeting in Founex, Switzerland, in June 1971. It calls for the integration of environment and development strategies.

1971 – Polluter Pays Principle. OECD Council says that those causing pollution should pay the costs.

1971 – International Institute for Environment & Development (IIED) established in Britain with a mandate to seek ways to make economic progress without destroying the environmental resource base. http://www.iied.org/

1971 – Rene Dubos and Barbara Ward write *Only One Earth*. The book sounds an urgent alarm about the impact of human activity on the biosphere but also expresses optimism that a shared concern for the future of the planet could lead humankind to create a common future.

1972 – UN Conference on Human Environment/UNEP held in Stockholm under the leadership of Maurice Strong. The conference is rooted in the regional pollution and acid rain problems of northern Europe. The conference leads to establishing many national environmental protection agencies and the United Nations Environment Programme (UNEP). http://www.unep.org/

1972 – Environnement et Développement du Tiers-Monde (ENDA) is established to provide courses and training about environment and development in Africa. In 1978 it refocuses, becoming an international voluntary nonprofit organization concerned with empowering local peoples, eliminating poverty, and research and training for sustainable development at all levels. http://www.enda.sn/

1972 – Club of Rome publishes Limits to Growth. The report is extremely controversial because it predicts dire consequences if growth is not slowed. Northern countries criticize the report for not including technological solutions while Southern countries are incensed because it advocates abandonment of economic development. http://www.clubofrome.org/

1973 – United States enacts Endangered Species Act to better safeguard, for the benefit of all citizens, the nation's heritage in fish, wildlife, and plants. http://www.audubon.org/campaign/esa/esa.html

1973 – Chipko movement born in India in response to deforestation and environmental degradation. The actions of the women of the community influenced both forestry and women's participation in environmental issues. http://www.rightlivelihood.se/recip1987_2.html

1973 – OPEC oil crisis fuels limits to growth debate.

1974 – Rowland and Molina release CFCs work in the scientific journal, *Nature*, calculating that continued use of CFC gases at an unaltered rate would critically deplete the ozone layer. http://www.ourplanet.com/imgversn/92/rowland.html

(*continued*)

TABLE 2.1. *Continued*

1974 – **Latin American World Model developed** by the Fundacio Bariloche; presented at the Second IIASA Symposium on Global Modelling. It is the South's response to Limits to Growth and calls for growth and equity for the Third World.

1975 – **CITES**. Convention on International Trade in Endangered Species of Flora and Fauna comes into effect. http://www.cites.org

1975 – **Worldwatch Institute** established in the United States to raise public awareness of global environmental threats and catalyze effective policy responses. Begins publishing annual State of the World in 1984. http://www.worldwatch.org/

1976 – **Habitat**. First global meeting to link environment and human settlement.

1977 – **Greenbelt Movement** starts in Kenya. It is based on community tree-planting to prevent desertification.

1977 – **UN Conference on Desertification** is held. http://infoserver.ciesin.org/docs/002-478/002-478.html

1978 – **Amoco Cadiz oil spill** off the coast of Brittany. http://www.mairiebrest.fr/amocosymposium/

1978 – **OECD Directorate of the Environment** relaunches research on environment and economic linkages. http://www.oecd.org/

1979 – **Convention on Long-Range Transboundary Air Pollution** is adopted. http://sedac.ciesin.org/pidb/texts/transboundary.air.pollution.1979.html

1979 – **Banking on the Biosphere**. IIED report on practices of nine multilateral development agencies, including the World Bank, sets the stage for reforms, which are still under way.

1979 – **Three Mile Island nuclear accident** occurs in Pennsylvania, USA. http://www.libraries.psu.edu/crsweb/tmi/tmi.htm

1980s

1980 – **World Conservation Strategy** released by IUCN. The section "Towards Sustainable Development" identifies the main agents of habitat destruction as poverty, population pressure, social inequity, and the terms of trade. It calls for a new international development strategy with the aims of redressing inequities, achieving a more dynamic and stable world economy, stimulating economic growth, and countering the worst impacts of poverty. http://www.iucn.org/

1980 – **Independent Commission on International Development Issues** publishes North–South, A Programme for Survival (Brandt Report). It asks for a reassessment of the notion of development and calls for a new economic relationship between North and South.

1980 – **U.S. President Jimmy Carter** authorizes study leading to the Global 2000 report. This report recognizes biodiversity for the first time as a critical characteristic in the proper functioning of the planetary ecosystem. It asserts that the robust nature of ecosystems is weakened by species extinction.

1981 – **World Health Assembly** unanimously adopts a Global Strategy for Health for All by the year 2000. Affirmed that the major social goal of governments and WHO should be the attainment of a level of health by all people of the world that would permit them to lead socially and economically productive lives. http://www.who.org/

(continued)

TABLE 2.1. *Continued*

1982 – **World Resources Institute** established in the United States begins publishing annual assessments of World Resources in 1986. http://www.wri.org/

1982 – **UN Convention on the Law of the Sea** is adopted. It establishes material rules concerning environmental standards as well as enforcement provisions dealing with pollution of the marine environment. http://www.un.org/Depts/los/index.htm

1982 – **International debt crisis** erupts and threatens the world financial system. It turns the 1980s into a lost decade for Latin America and other developing regions.

1982 – **The United Nations World Charter for Nature** published. It adopts the principle that every form of life is unique and should be respected regardless of its value to humankind. It also calls for an understanding of our dependence on natural resources and the need to control our exploitation of them.
http://sedac.ciesin.org/pidb/texts/world.charter.for.nature.1982.html

1983 – **Development Alternatives** established in India. It fosters a new relationship among people, technology and the environment in the South in order to attain sustainable development. http://www.devalt.org/

1984 – **Toxic chemical leak** leaves 10,000 dead and 300,000 injured in Bhopal, India. http://www.bhopal.net/

1984 – **Drought in Ethiopia.** Between 250,000 and 1 million people die from starvation.

1984 – **Third World Network** is founded during an international conference, "The Third World: Development or Crisis?" which was organized by the Consumers Association of Penang. TWN's role is to be the activist voice of the South on issues of economics, development, and environment. http://www.twnside.org.sg/twnintro.htm

1984 – **International Conference on Environment and Economics (OECD).** Concludes that the environment and economics should be mutually reinforcing. Helped to shape Our Common Future.

1985 – **Responsible Care**® An initiative of the Canadian Chemical Producers. Provides a code of conduct for chemical producers, which is now adopted in many countries. http://www.ccpa.ca/

1985 – **Climate change.** Austria meeting of World Meteorological Society, UNEP, and the International Council of Scientific Unions reports on the build-up of CO_2 and other "greenhouse gases" in the atmosphere. They predict global warming. http://www.unep.ch/iuc/submenu/infokit/factcont.htm

1985 – **Antarctic ozone hole** discovered by British and American scientists.

1986 – **Accident at nuclear station** in Chernobyl generates a massive toxic radioactive explosion. http://www-bcf.usc.edu/~meshkati/chernobyl.html

1987 – **Our Common Future Brundtland Report.** Report of the World Commission on Environment and Development weaves together social, economic, cultural, and environmental issues and global solutions. Chaired by Norwegian Prime Minister Gro Harlem Brundtland. Popularizes term "sustainable development."

1987 – **Development Advisory Committee.** DAC members of OECD evolve guidelines for environment and development in bilateral aid policies. http://www.oecd.org/home/

(continued)

TABLE 2.1. *Continued*

1987 – **Montreal Protocol** on Substances that Deplete the Ozone Layer is adopted.
http://www.unep.ch/ozone/home.htm

1988 – **Chico Mendes**. Brazilian rubber tapper fighting the destruction of the Amazon
rainforest is assassinated. Scientists use satellite photos to document what the Amazon fires
are doing to the rainforest. http://www.chicomendes.com/

1988 – **Intergovernmental Panel on Climate Change** established to assess the most up-to-
date scientific, technical, and socioeconomic research in the field. http://www.ipcc.ch/

1989 – **Exxon Valdez tanker runs aground** dumping 11 million gallons of oil into Alaska's
Prince William Sound. http://www.oilspill.state.ak.us/

1989 – **Stockholm Environment Institute** established as an independent institute for
carrying out global and regional environmental research. http://www.sei.se/

1990s

1990 – **Regional Environmental Centre for Central and Eastern Europe** established as an
independent, nonprofit organization to assist environmental nongovernmental
organizations, governments, businesses, and other environmental stakeholders to fulfill
their role in a democratic, sustainable society. http://www.rec.org/

1990 – **UN Summit for Children**. Important recognition of the impact of the environment on
future generations. http://www.unicef.org/wsc/

1990 – **International Institute for Sustainable Development (IISD)** established in Canada.
Begins publishing the Earth Negotiations Bulletin in 1992. http://www.iisd.org/

1991 – **The Canadian east coast cod fishery collapses** when only 2700 tonnes of spawning
biomass are left after a harvest of 190,000 tonnes.
http://www.greenpeace.org/~comms/cbio/cancod.html

1991 – **Hundreds of oil fires burn** out of control in Kuwait for months following the Persian
Gulf War.

1992 – **The Business Council for Sustainable Development** publishes Changing Course.
Establishes business interests in promoting SD practices. http://www.wbcsd.ch/

1992 – **Earth Summit**. UN Conference on Environment and Development (UNCED) held in
Rio de Janeiro, under the leadership of Maurice Strong. Agreements reached on Agenda 21,
the Convention on Biological Diversity, the Framework Convention on Climate Change,
the Rio Declaration, and nonbinding Forest Principles. Concurrent NGO Global Forum
publishes alternative treaties. http://www.unep.org/unep/partners/un/unced/home.htm

1992 – **The Earth Council** is established in Costa Rica as a focal point for facilitating follow-
up and implementation of the agreements reached at the Earth Summit, and linking national
SD councils. http://www.ecouncil.ac.cr/

1993 – **President's Council for Sustainable Development** in United States announced by
President Bill Clinton. They publish Sustainable America: A New Consensus for
Prosperity, Opportunity, and a Healthy Environment for the Future in 1996.
http://clinton2.nara.gov/PCSD/

1993 – **First meeting of the UN Commission on Sustainable Development** established to
ensure effective follow-up to UNCED, enhance international cooperation, and rationalize
intergovernmental decision-making capacity. http://www.un.org/esa/sustdev/

(*continued*)

TABLE 2.1. *Continued*

1993 – World Conference on Human Rights. Governments reaffirmed their international commitments to all human rights. Appointment of the first UN High Commissioner for Human Rights. http://www.unhchr.ch/

1994 – Global Environment Facility. Billions of aid dollars restructured to give more decision-making power to developing countries. The GEF affirms its commitment to fund projects that are country-driven, based on national priorities, and reflect the incremental costs of meeting international commitments that achieve global environmental benefits. http://www.gefweb.org/

1994 – North American Free Trade Agreement (NAFTA) enters into force. The side agreement – the North American Agreement on Environmental Cooperation – establishes the Commission for Environmental Cooperation (CEC). http://www.cec.org/

1995 – The execution of Ken Saro-Wiwa in Nigeria brings international attention to the linkages between human rights, environmental justice, security, and economic growth. http://www.mosopcanada.org/text/ken.html

1995 – World Trade Organization established. Formal recognition of trade, environment, and development linkages. http://www.wto.org/

1995 – World Summit for Social Development held in Copenhagen, Denmark. First time that the international community has expressed a clear commitment to eradicate absolute poverty. http://www.un.org/esa/socdev/wssd/index.html

1995 – Fourth World Conference on Women held in Beijing, China. Delegates adopt the Beijing Declaration and Platform for Action. These documents recognize that the status of women has advanced but obstacles still remain to the realization of women's rights as human rights. http://www.undp.org/fwcw/daw1.htm

1996 – The Summit of the Americas on Sustainable Development held in Santa Cruz, Bolivia. This Summit identified the joint efforts needed to reach SD in the hemisphere. http://www.summit-americas.org/boliviaplan.htm

1996 – ISO 14001 formally adopted as a voluntary international standard for corporate environmental management systems.

1997 – Asian ecological and financial chaos. Landclearing fires intensified by an El Niño induced drought result in haze blanketing the region and causing over US$1.4 billion in health costs and at least that amount in direct fire-related damage. Concurrently, the market crashes, raising questions about currency speculation and need for government economic reforms. http://www.ciaonet.org/isa/dap01

1997 – Signing of the Kyoto Protocol. Delegates to the UN Framework Convention on Climate Change Third Conference of the Parties (COP-3) sign the Kyoto Protocol. This document sets goals for greenhouse gas emission reduction and establishes emissions trading in developed countries and the clean development mechanism for developing countries. http://unfccc.int/resource/convkp.html

1997 – UN General Assembly review of Earth Summit progress. Special session acts as a sober reminder that little progress has been made in implementing the Earth Summit's Agenda 21 and ends without significant new commitments. http://www.iisd.ca/linkages/csd/ungass.html

(*continued*)

TABLE 2.1. *Continued*

1998 – Controversy over genetically modified organisms. Global environmental and food security concerns raised over genetically modified (GM) food products. The EU blocks import of GM crops from North America and farmers in developing countries rebel against "terminator technology," seed that will only germinate once. http://scope.educ.washington.edu/gmfood/

1998 – Unusually severe weather. China experiences worst floods in decades; two-thirds of Bangladesh under water for several months from torrential monsoons; Hurricane Mitch destroys parts of Central America; 54 countries hit by floods and 45 by drought; Earth hits highest global temperature ever recorded. http://lwf.ncdc.noaa.gov/oa/climate/research/1998/ann/extremes98.html

1998 – MAI. Environmental groups, social activists and concerned citizens effectively lobby against the multilateral agreement on investment (MAI). That, along with disagreement by governments over the scope of the exceptions being sought, led to the demise of the negotiations. http://www.citizen.org/trade/issues/mai/articles.cfm?ID = 1021

1999 – The World Commission on Forests and Sustainable Development releases its report Our Forests ... Our Future. This independent Commission, after extensive hearings with stakeholders worldwide, concluded that the world's material needs from forests can be satisfied without jeopardizing them by changing the way we value and manage forests. http://www.iisd.org/wcfsd/default.htm

1999 – Launch of the first global sustainability index tracking leading corporate sustainability practices worldwide. Called the Dow Jones Sustainability Group Indexes, this tool provides a bridge between those companies implementing sustainability principles and investors looking for trustworthy information to guide sustainability focused investment decisions. http://www.sustainabilityindex.com/

1999 – Third World Trade Organization Ministerial Conference held in Seattle, Washington, United States. Thousands of demonstrators take to the streets to protest the negative effects of globalization and growth of global corporations, and along with deep conflicts among delegates inside, scuttle the negotiations. The first of many such antiglobalization protests, they signal a new era of confrontation between disaffected stakeholders and those in power. http://www.iisd.org/trade/wto/seattleandsd.htm

2000s

2000 – Increasing urbanization. Almost half of the world's population now lives in cities that occupy less than two percent of the Earth's land surface, but use 75 percent of Earth's resources. http://www.aaas.org/international/atlas/contents/pages/population06.html

2000 – The Second World Water Forum and Ministerial held in The Netherlands and attended by 5700 participants from all parts of the world and 120 ministers. It results in the Declaration of The Hague on Water Security in the 21st Century. The World Commission on Water for the 21st Century also releases its "World Water Vision" for the sustainable use and management of water resources. http://www.worldwaterforum.net/

2000 – United Nations Millennium Summit. This largest ever gathering of world leaders adopted the United Nations World Summit Declaration, which spells out values and principles, as well as goals in key priority areas. World leaders agreed that the UN's first priority was the eradication of extreme poverty and highlighted the importance of a fairer world economy in an era of globalization. http://www.un.org/millennium/summit.htm

(*continued*)

TABLE 2.1. *Continued*

2000 – Miss Waldron's red colobus monkey declared extinct. It is the first extinction in
several centuries of a member of the Primate Order, to which human beings belong.
According to the IUCN Red Book, 11,046 species are threatened with extinction.
http://wcs.org/news/wcsreports/6989/- story4

2001 – Terrorists representing anti-Western, nonstate interests and ideologies, bomb the
World Trade Center and Pentagon, the first serious attack on U.S. soil since 1814, thus
marking the end of an era of unhindered economic expansion. Repercussions are felt
throughout the world, as stockmarkets and economies stumble and the United States gears
up for a war on terrorism with its first target being terrorists' networks in Afghanistan.
http://www.globalpolicy.org/wtc/wtc/index.htm

2001 – Fourth Ministerial Conference of the World Trade Organization held in Doha,
Qatar, sets the stage for the next round of world trade talks and features environment and
development issues in several sections of The Doha Declaration. The WTO Committee on
Trade and Environment is given a watching brief on the entire negotiation agenda to ensure
that environmental issues are adequately addressed. NGOs and the WTO agree to
reinterpret the Agreement on Intellectual Property Rights regarding access to medicines
and public health. http://www.ictsd.org/ministerial/index.htm

2001 – The Marrakech Accords, agreed to at COP-7, finalize how the Kyoto Protocol
should work, lay the groundwork for ratification by different countries and mark the
conclusion of a process begun at COP-4 in Buenos Aires.
http://www.iisd.ca/linkages/download/pdf/enb12189e.pdf

2002 – World Summit on Sustainable Development held in Johannesburg, South Africa.
World governments, concerned citizens, UN agencies, multilateral financial institutions,
and other major groups participate and assess global change since the United Nations
Conference on Environment and Development (UNCED) in 1992.
http://www.johannesburgsummit.org/

Reprinted with permission of the **(IISD) International Institute for Sustainable Development**, http://
www.iisd.org/ © IISD, 3rd edition, 2002.

PwC, and the American Institute of Chemical Engineers (AIChE) organized two
focus groups of industry sustainability managers to discuss the survey findings in
greater depth. The following summarizes key findings from both the survey and
the focus group discussions; results of both efforts are detailed in Chapter 8.

2.3.2 Survey and Focus Groups Key Findings

2.3.2.1 Public Reporting Demands. Environmental mishaps of the past have
galvanized the chemical industry to respond to the public's concerns and ensure
compliance with environmental regulations, public right-to-know initiatives, and
public reporting requirements related to EHS performance. These requirements
reflect a growing demand on the part of all stakeholders for greater participation
in decisions of government and industry, and for greater transparency. This is
reflected in the growth of sustainability reporting; most survey respondents issue

or plan to issue a sustainability report in the near future (see Section 6.5 for a more detailed discussion on sustainability reporting).

2.3.2.2 Transparency. Transparency has become a critical business issue. The Sarbanes–Oxley Act of 2002 is the legislative incarnation of the spotlight that investors, consumers, and employees now shine on the financial statements of a company (see Funk, Chapter 8, Section 8.3 of this volume, for an in-depth look at the business case). While reducing environmental impacts, engaging stakeholders more effectively and responding to their demands for transparency are seen, at a minimum, as providing license to operate in the community; they are also seen as leading to benefits, both "hard" and "soft," including enhanced reputation, cost savings, and innovation.

2.3.2.3 Regulatory Requirements. From a global perspective, the changing regulatory environment, with REACH considerations in the European Union and compliance with the Kyoto Protocol, is influencing the strategies of U.S.-based companies operating globally. On 14 April 2003, the *New York Times* reported that "the European Union is adopting environmental and consumer protection legislation that will go further in regulating corporate behavior than almost anything the United States government has enacted in decades." Chemical companies are therefore being driven toward sustainability by changing regulatory requirements that will affect their global operations.

2.3.2.4 Globalization. Globalization is another prominent driver of sustainability, as indicated in the survey and focus groups. Companies with global operations are seeking to work toward global best practices and are pressured increasingly by customers, investors, and NGOs to be responsible for the impacts from not only their operations, but also those of their suppliers and customers, regardless of where in the world they are located.

2.3.2.5 Competitive Advantage. There is great concurrence among sustainability managers, from both the survey and the focus groups, as well as from company case studies found in this volume, that even more important drivers of sustainability considerations come from the desire to gain competitive advantage and preserve the long-term viability of the company. When asked what are the greatest business risks associated with (not considering) sustainability, the top responses – increased costs, loss of competitive advantage, and damaged reputation – can all be interpreted as linked to financial performance and competitive advantage. Further, when asked about the greatest opportunities derived from the pursuit of sustainability, the top responses – reduced costs, enhanced reputation, and innovation – are also all related to competitive advantage. To companies more mature in the practice of sustainability, it is seen as a portal for developing innovations – new products, services, business alliances, and markets – and therefore as a growth engine (see DuPont case study, Section 8.3, and Future Directions for the Chemicals Industry, Section 7.1). Sustainability can be a way to view where the market may be going in the future and to predict the need for disruptive technologies and discontinuities with respect to corporate strategies. As such, it has the potential for positioning the company to do business quite differently and to be differentiated from the pack.

Managers are discovering also that the intangible indicators that gauge sustainability can also be indicators of efficacy – that is, how well a company is *run*. From the management of corporate liabilities to new market ventures, a sustainable business strategy can improve all segments of corporate activity (Funk, Section 8.3; also see Section 8.9 business case).

2.3.2.6 Market Value. Investors are increasingly valuing sustainability practices and SRI (Socially Responsible Investing) funds are found to outperform the S&P 500 and Dow Jones Global Index by significant amounts. In the chemicals industry, a Value Creation Index (VCI) model reveals that the top four indicators for the chemical industry that are highly correlated with the market value of equity of these companies are related to innovativeness, strategic alliances, leadership, and environmental and social responsibility (see Section 8.9, business case).

2.3.3   Current Status of Sustainability Programs

While the vast majority of survey respondents believe sustainability to be of the highest importance and that it will continue to be so in the next five years, sustainability is not yet fully integrated into their core business strategy; less than half have written sustainability policies, and only a few have established a sustainability management system. Around a third have what they refer to as triple-bottom-line metrics to track performance in this area. Only about a half of survey respondents have a senior-level executive or executive committee with dedicated responsibilities to coordinate and promote sustainability. Further, management of sustainability is fragmented across multiple organizational and functional boundaries and not organized within one function or group.

2.4   THE ROLE OF RESPONSIBLE CARE® IN ADVANCING SUSTAINABLE DEVELOPMENT

Few will disagree that the Responsible Care® program is one of the most recognized voluntary environmental initiatives launched by an industry sector. Many proponents acknowledge that Responsible Care® has reduced releases to air, land, and water, and improved worker and community safety. While the social dimension was core to the original vision of Responsible Care®, the program's impact on public trust has met with mixed reviews. On the economic front, Responsible Care® is yet to be recognized as a business-driven initiative.

So what role will Responsible Care® play in advancing sustainability? There are clearly two views of that potential role:

On one hand, Responsible Care® is seen as "compliance focused" and, while considered a component of sustainability, does not currently offer the platform for embracing all of the concepts of sustainability (focus groups above; SustainAbility, 2004). Responsible Care® is viewed as a sector-specific collaborative initiative, whereas sustainability broadly encompasses issues of competitive positioning and business growth, which involve full engagement of internal and external stakeholders. "Even as Responsible Care® expands focus to include more sustainability parameters, it is not (seen as) sufficient to drive an overall sustainability program, performance, and the business value derived from sustainable development" (Savitz, Chapter 8, Section 8.2).

On the other hand, one needs only to examine the original intent of Responsible Care® (see Box 2.1) and a recent commitment to revitalize Responsible Care® globally (see Box 2.2), to recognize that it could be the chemical industry's approach to meet the sustainability expectations of the industry and its stakeholders.

This section reviews the evolution of Responsible Care®, with a look at its plans to become a broader business-driven initiative and its role in advancing sustainability.

BOX 2.1   THE EVOLUTION OF AN ETHIC AND A COMMITMENT

JEAN BELANGER

To tell the story of Responsible Care is to tell the story of how an industry has evolved over a quarter century. And trust is what it is all about – the key driver.

The initial concepts of Responsible Care, in Canada, grew out of a growing concern, in the early 1980s, that the chemical industry risked losing its mandate to produce in Canada. In that time frame, pressures to regulate the chemical industry were building up and were particularly exacerbated by a major hazardous material train derailment in late 1979 that resulted in the evacuation of the fifth largest city in Canada. The leaders of the chemical industry, through their membership on the Board of Directors of the Canadian Chemical Producers' Association (CCPA), had the foresight to understand that the crux of the problem was one of trust. The public, in fact, believed that the industry knew about the dangers associated with its products, but it did not tell anyone and even worse, it just did not care.

The industry leaders deliberately chose the path that would be anchored on gaining the trust of the community, both near the plant and, further, to all society. Trust, therefore, became the key driver. Building trust required the application of three fundamental principles, which formed the cornerstones of Responsible Care:

- Doing the right thing;
- Caring about products from cradle to grave;
- Being open and responsive to public concerns.

Concerned about its eroded credibility, not just with the public but also with government decision-makers, the CCPA developed a Statement of Policy on Responsible Care® in 1979. By 1985, a once simple one-page statement of principles became the Responsible Care® program endorsed by the CCPA, and adopted in 1988 by the American Chemistry Council (then the Chemical Manufacturers Association). What was initiated in the early part of the 1980s as a simple one-page statement of principles, touching the singular aspect of environmental responsibility, has since evolved into a set of codes, verification processes, visible performance measurement and advisory panels. There has, however, been one constant, namely, that the whole exercise of Responsible Care is about building trust through ethical behavior and listening attentively to the evolving concerns of the public.

Today Responsible Care® exists in 52 countries worldwide and has become the industry's initiative to address environmental, health, and safety (EHS), security, outreach, process safety and "cradle to grave" product management.

Jean Belanger's more indepth look at the evolution of Responsible Care® can be found in *Chemistry International.* Mar/Apr 2005, pp. 4–9.

2.4.1 Public Trust and Environmental Performance

The main objectives of Responsible Care® were to regain public trust and demonstrate leadership beyond compliance by improving the environmental performance of the industry as a whole and by improving community relations.

In the 1980s, Responsible Care's six Codes of Management Practices served as a framework for the chemical companies to use internal processes to improve performance. Overall, environmental performance did improve because of a commitment to the codes and expanding regulatory requirements (Yosie, 2003). Members of the CCPA steadily improved their environmental and workplace health and safety records throughout the 1990s. Their records indicate a steady decline in workplace injuries and a marked reduction in the frequency and severity of transportation incidents (CCPA, 1999).

Evaluating the program's impact on public trust is more complicated. Although Responsible Care® may have helped arrest the precipitous decline in trust that marked the early to mid-1980s, industry polls continue to reveal low overall levels of public confidence in the industry. To address the concerns of its diverse stakeholders, the chemical industry has formed hundreds of community advisory councils and partnerships with community and environmental activists in an attempt to hear their points of view (see Section 6.7). While individual companies report improved community relations and while the industry is much more open than it has ever been, polls indicate that it remains low in the public's esteem (Moffet, 1999).

A survey completed in 1999 indicates that although the public believes the chemical industry provides valuable products, creates employment, and contributes to positive economic growth, it "does no better than a fair job when it comes to minimizing risks to health and the environment, considering the future effects of chemicals, and assuming responsibility for their activities" (Earnscliffe Research and Communications, 1999). In fact, despite the work, the industry still faces huge hurdles in terms of public acceptance and confidence. These concerns and the public's unfavorable perception of the industry are covered in Chapter 3.

BOX 2.2 TAKING PERFORMANCE TO A NEW LEVEL: THE RESPONSIBLE CARE® GLOBAL CHARTER

TERRY F YOSIE

Recently (2003), the global chemical industry, acting through the International Council of Chemical Associations (ICCA), has undertaken a strategic re-examination of Responsible Care and committed to comprehensive and significant performance commitments in the 52 nations that currently implement the program. Working through a CEO-level task force comprised of business leaders from Asia, Europe, and North and South America, this reassessment aimed to:

- Review the current status of Responsible Care implementation worldwide;
- Obtain external stakeholder expectations of chemical industry performance;
- Develop consistent core commitments for the industry worldwide;
- Achieve continued performance improvements;
- Deliver significant business value and reputation benefits;
- Enable the industry to speak with single, stronger voice on performance issues.

The result of this review is the Responsible Care Global Charter (2004; Appendix 1), a document that goes beyond the original elements of the program since its inception in 1985 and focuses on new and important challenges facing both the chemical industry and global society, including: the growing public dialog over sustainable development; public health issues related to the use of chemical products; and the need for greater industry transparency.

The Responsible Care® Global Charter contains nine key elements. They commit the global chemical industry to:

- Adopt Responsible Care core principles;
- Implement the program based upon a set of eight fundamental features;
- Advance sustainable development;
- Continuously improve and report performance;
- Enhance product stewardship;
- Extend Responsible Care through the value chain;
- Support national and global Responsible Care governance processes;
- Address stakeholder expectations;
- Provide appropriate resources to implement the program.

In addition to the Charter, the ICCA endorsed a commitment letter for CEOs of ICCA member associations to sign to declare their support for Charter implementation.

Sustainable development as practiced through Responsible Care is increasingly part of the business equation for chemical companies in the global marketplace. Business value is derived in such areas as the following:

- Evolving legal and regulatory practices and standards are becoming more stringent. Responsible Care embodies elements of preventative law that protects companies from legal liability as their performance improves.
- Demonstrating leadership through expanded knowledge for how to better identify and manage chemical-related risks. Such "capacity building" contributes to the development of critical global infrastructure by providing technical information and assistance, management systems, training, and assistance towards institution building in many nations. All of these factors are critical to improving environmental, health and safety performance.
- Improving customer relations as a result of enhanced performance. This practice builds customers' confidence and enables them to communicate progress to their stakeholders.

Responsible Care® is not a substitute for government policymaking or regulation. Rather, it is a necessary and complementary commitment to improving performance that effectively leverages the skills and capabilities of the chemical industry to improved living standards and quality of life. In our rapidly changing world, the value and the need for programs such as Responsible Care will become even more significant as current and future societal needs are addressed through business initiatives and plans.

2.5 CHALLENGES AHEAD

Despite the chemical industry's efforts to improve environmental performance and gain the public's trust, the industry still faces huge hurdles in terms of public acceptance and confidence (see Chapter 3). There are still fundamental issues – such as historical performance, the failure of risk assessment on a broad basis, and the intrinsic nature of materials – that need to be resolved if the industry is to realize the future it envisions.

A genuine commitment to sustainability will mean more than resolving these concerns – it will mean real innovation in the chemical industry. Future generations will continue to need chemicals, and the industrial transformation of chemicals, to meet human needs, will continue to require ingenuity and enterprise. However, the types of chemicals and how they are used must be significantly reconsidered.

This transition will require a new mission for the industry that promotes human health and environmental quality as seriously as the market promotes economic efficiency and product effectiveness. Put conceptually, a sustainable chemicals industry would be one that optimizes value from the use of chemicals, adds no new risks to everyday life, increases natural capital, minimizes the transfer of risks from one generation to another, respects and enhances the natural functioning of the planet's ecosystems, and assures no net loss of valuable resources. Creating such an industry will require government policy and market incentives that promote sustainability. This will require financial investment institutions and international development programs that are committed to developing an industry that is as ecologically sound and socially sensitive as it is economically productive (Geiser, Chapter 7, Section 7.1).

As pressures mount for a world more respectful of resource limits and the material needs of future generations, the chemical industry must find its own path to sustainability.

REFERENCES

(ACC) American Chemistry Council, *The Business of Chemistry: Essential to our Quality of Life and the New Economy.* American Chemistry Council, Arlington, Virginia, 2002.

CCPA, *1999 Emissions Inventory and Five-Year Projections*, Reducing Emissions 8: A Responsible Care Initiative. CCPA, Ottawa; 1999.

(CEFIC) European Chemistry Council, *Facts and Figures*, June 2004. http://www.cefic.org/factsandfigures/. Accessed October 2004.

Earnscliffe Research and Communications, *Results of Key Audience Research*, conducted for the Canadian Chemical Producers' Association. Earnscliffe Research and Communications, Ottawa, Ontario, Canada, 1999.

M. McCoy, M. S. Reisch, A. H. Tullo, P. L. Short, J.-F. Tremblay and W. J. Storck, Facts and figures for the chemical industry. *Chemical & Engineering News*, 82(27), 23–63 (2004).

Innovest, *The Chemical Industry. Uncovering Hidden Value Potential for Strategic Investors.* Innovest Strategic Value Advisors, Inc., April 2002, New York.

(IISD) International Institute for Sustainable Development. http://www.iisd.org/, 2002.

J. Krueger and H. Sein, Governance for Sound Chemicals Management: The Need for a More Comprehensive Global Strategy. *Global Governance*, 8(3), 323–342 (2002).

J. Moffet and F. Bregha, The Responsible Care Program. In: R. Gibson (Ed.), *Voluntary Initiatives* (Broadview Press, Toronto, 1999).

OECD (Organisation for Economic Co-Operation and Development), *OECD Environmental Outlook.* OECD, Paris, 2001.

(S&P) Standard and Poors, http://www.spglobal.com/. Accessed October 2004.

SustainAbility, *External Stakeholder Survey. Final Report for the Global Strategic Review of Responsible Care.* SustainAbility, Washington D.C., February 2004.

(USOMB) US Office of Management and Budget, *Standard Industrial Classification Manual.* JIST Works, Inc., Indianapolis, IN, 1987.

(USOMB) US Office of Management and Budget, *NAICS Desk Reference: The North American Industry Classification System.* JIST Publishing, Inc., Indianapolis, IN, 2000.

T. Yosie, Fifteen Years – Responsible Care. *Environmental Science & Technology*, November 1, 401–406 (2003).

3

VIEWS ON KEY ISSUES FACING THE CHEMICAL INDUSTRY

3.1 OVERVIEW

MARIANNE LINES

BRIDGES to Sustainability

The modern chemical industry has produced many important products that have improved health, the quality of life, and our prospects for more sustainable forms of development. However, the number of chemicals produced in great volumes and the number of products that contain these chemicals have brought serious concerns about the consequences of toxic substances on public health and ecosystems. These concerns, and others (see Box 3.1), boil down to a mistrust of the industry as a whole. Addressing this mistrust remains one of the major challenges facing the chemical industry (Moffet *et al.*, 1999).

BOX 3.1   THE PUBLIC'S KEY CONCERNS – A CASE OF MISTRUST

Accountability and Communication

- Lack of transparency in the industry's communication, for example, poor communication of product risks;
- Lack of toxicity information, public data, and government attention to the large majority of chemicals in commerce, existing chemicals on the market prior to the 1980s;
- Lack of stakeholder involvement in the chemical management process.

Transforming Sustainability Strategy into Action: The Chemical Industry, Edited by B. Beloff, M. Lines, and D. Tanzil

Risk Assessment

- The slow, resource-intensive risk assessment process, which places the burden on governments to demonstrate harm before preventive action can take place.

Risk Management and Chemical Policies

- Lack of regulatory programs that address chemical risks throughout a chemical's lifecycle;
- Continued use of chemicals with inherently dangerous properties such as known or highly suspected carcinogens, mutagens, and reproductive toxicants;
- The lack of funding for research and development of safer and cleaner chemicals, production systems, and products.

Impacts (of products and operational)

- Concern about exposures from unregulated chemicals used in a wide range of consumer products;
- Slow response to the growing information on the impacts of chemicals on ecosystems and health and emerging concerns such as endocrine disruption and the unique vulnerability of children;
- The long-term persistence and accumulation of many chemicals in ecosystems and humans.

Security

- Concern about the vulnerability of security at chemical facilities.

Adapted from *Integrated Chemicals Policy, Seeking New Direction in Chemicals Management* (Lowell Center for Sustainable Production, 2003); *Final Report for the Global Strategic Review of Responsible Care* (SustainAbility, 2004).

The goal of this chapter is to highlight key issues, challenges and opportunities for the chemical industry, as viewed by experts from business, government, and academic and advocacy organizations. This chapter does not provide all the answers, of course, nor is there room to provide an in-depth examination of all possible issues viewed by all key stakeholders. While some of the issues are contentious, they have served to galvanize efforts by the public and forward-looking governments and chemical companies to reduce the impacts of chemicals harmful to ecosystems and human health. Each of the issues highlighted in this chapter is worthy of its own book, but we provide only an overview of the following.

Section 3.2, "The Chemical Industry and the Public: Will the Chemical Experiment Continue?" opens with a summary of public concerns and how the chemical industry is perceived. Without increased scrutiny of chemicals, and their use in commerce and in products, perception of the chemical industry will remain unfavorable. The public sees the next few years as an unprecedented opportunity for the chemical industry to assume a larger burden of assessing and managing chemical risk; to provide environmental and human health toxicity data; and to promote green chemistry, safer substitutes, and innovation in the development of chemicals.

Section 3.3, "Risk Assessment" discusses the traditional scientific approach to the lack of knowledge on chemicals and their potential impacts, which has been to use a technique called quantitative risk assessment to calculate the probability of adverse health effects given particular exposures. Experts examine the science behind the complex process of quantitative risk assessment by which government programs evaluate risks and set regulatory levels or standards. Quantitative risk assessments have been criticized for being time-consuming, costly to complete, narrowing the types of information that go into decision-making, and hiding uncertainties in our knowledge of risks. This section explores recent research initiatives that have led to improved methods for quantitative risk assessment and the interest in reliable large-scale screening and ranking methods as hazard identification tools.

Another view explores the limits of the risk assessment-based approach to decision-making and what a precautionary paradigm might look like. The precautionary principle calls for preventive actions when there is reasonable scientific evidence of harm, although the nature and magnitude of that harm may not be fully understood scientifically. While a highly contentious term, proponents of the precautionary approach see this as a means to make better, more health protective decisions in the face of highly uncertain and complex risks.

Section 3.4, "Limits of Risk Management and the New Chemical Policies" discusses chemical risk management policies, which focus on the intrinsic hazards – the toxicity, chemical stability, and bioavailability – of the industry's chemical products and not on the industry's wastes, pollution, or occupational exposures. Conventional risk management policies are intended to provide agencies with the authority to collect relevant health and safety data on chemical products, require testing where data is missing, and condition and restrict the use of chemical substances so as to reduce "unreasonable risks" to the public and environment. This section evaluates what has worked and not worked in current policies. What are the limits of the current risk management approach? What approaches are governments taking to confront the limits?

In Europe, and internationally, new approaches to chemicals management policies have emerged, which will require basic data on all chemicals in commerce, rapid evaluation of chemical risks, information on risks throughout chemical life-cycles, and substitution of those substances of highest concern. The new European chemicals policy will likely set the standard for chemicals management, affecting all manufacturers, globally.

Section 3.5 considers "Impacts to Human Health and the Environment." While we have known for decades that many industrial chemicals are toxic, mounting evidence indicates that some of these chemicals play a role in the onset of diseases including cancer, developmental and behavioral disorders, respiratory ailments and asthma, neurological disorders, and birth defects, among others. Of particular concern are exposures to the fetus and developing child, where low-level exposures during critical times in development can result in lifelong impairment.

Despite evidence of health effects linked to chemical exposure, a great majority of the chemicals found in our air, water, food, and everyday products lack basic safety data for human and ecosystem health, particularly that of developing organisms. Some of these chemicals are known to persist in the environment for decades, travel great distances, or accumulate in the food chain Even less is known about the potential health effects of exposures to multiple chemicals and other stressors: the reality of our everyday lives (see Tickner, Section 3.3.2).

This section looks at the impact of chemicals that are known to persist in the environment and those that interfere with normal hormone production, damaging animal and human development and reproduction.

Section 3.6 covers "Impacts of, and Issues Associated with, Chemical Production from Manufacture to Final Use and Disposal." Over the entire life of a product produced by the chemical industry (from "cradle to grave") there is a potential for a negative impact on man and the environment. This section will summarize sources of impacts to human health and ecosystems through the chemical lifecycle of manufacture, use, and disposal. It also covers achievements in our efforts to reduce and eliminate the use, production, and release of pollutants to the environment that results from manufacturing processes. Looking at expected trends, this section looks at more holistic approaches to minimize impacts on health and the environment throughout the lifecycle of products.

In Section 3.7, "Closing the Gap on Chemical Plant Security," security issues are discussed. Since September 11, the chemical industry has made tremendous strides in improving the security of its facilities, infrastructure, and IT. While there are no national security regulations in the United States, members of trade groups have completed site vulnerability assessments and are tackling the more difficult steps of implementing security improvements at their plants. Yet, only a fraction of the chemical processing and storage facilities in the United States are members of the trade groups spearheading these efforts. Environmental groups and some security experts are skeptical that enough chemical companies are taking sufficient steps. Surprising breaches in security, brought to light by the media, have served to raise the public's awareness of the vulnerability of security at chemical facilities.

This section covers what the industry is doing to reduce vulnerabilities and the challenges it faces. As the industry shifts its focus from crisis management to critical incident prevention, priorities for the most proactive companies include employee training, threat analyses and vulnerability assessments, optimizing plant emergency operation plans, and integrating them with local emergency services.

Finally, in Section 3.8, "Economic Issues and Competitiveness," we learn how chemical producers are facing a major change in the patterns of trade, globally. In particular, chemical producers in the United States and other OECD countries are facing stiffer competition from newer producers in the Middle East and Asia.

This imbalance of trade in chemicals has come about not only because certain commodity petrochemicals produced in the United States are now less competitive globally, as a result of higher natural gas costs, but also because such Asian countries as China and India have now become substantial, reliable exporters of specialty chemicals and pharma intermediates.

This section concentrates on the effect of changes in manufacturing economics with the attainment of progress toward sustainable development and steps that companies can take to remain competitive include improving energy efficiency and exploring new technologies and alternative feedstock.

3.2 THE CHEMICAL INDUSTRY AND THE PUBLIC: WILL THE CHEMICAL EXPERIMENT CONTINUE?

BEVERLEY THORPE

Clean Production Action

The publication of Rachel Carson's *Silent Spring* in 1962 and her description of pesticide contamination of waterways, land, and wildlife galvanized, for the first time, the American public's concern about the chemical industry. The issue was further fed by the heated response of the chemical industry when Monsanto published and distributed 5000 copies of a brochure parodying *Silent Spring*, which related the devastation and inconvenience of a world where famine, disease, and insects ran amok because chemical pesticides had been banned. Carson's carefully researched work was only vindicated when many eminent scientists rose to her defense, and President John F. Kennedy ordered the President's Science Advisory Committee to examine the issues the book raised (NRDC, 2004). As a result, DDT came under much closer government supervision and was eventually banned.

Today we no longer witness children happily bathing in swimming pools while trucks spray DDT along neighbouring streets or read commercials featuring happy vegetables, animals singing, "DDT is Good for Me-e-e!" (Pennsalt, 1946). Yet the chemical industry continues to suffer from a lack of public confidence.

From 1980 to 1990 in the United States, favorable opinion about the chemical industry fell from 30 percent to 14 percent, while public perceptions of the industry as "unfavorable" grew from 40 percent to 58 percent (CMA, 1993). Polls showed that the public believed the chemical industry had no self-control, did not listen to the public, did not put safety and the environment first, and did not take responsibility for its processes and products (Rees, 1997).

Recent polls in the United Kingdom show that only 22 percent of the UK public has a favorable attitude to the chemical industry. The main reason is concern over environmental pollution. However, issues of social accountability are also growing in importance, specifically perceived secrecy, profit orientation, and bad publicity (Worcester, 1999).

Health problems are increasingly associated with chemicals, particularly since the publication of *Our Stolen Future* by Theo Colborn and her two colleagues.

Colburn presented evidence that connected some chemicals to reproductive and developmental disorders in wildlife and humans (Colborn *et al.*, 1996). Her work, an analysis of over 4000 published papers, argues forcefully that endocrine disrupting chemicals or "gender bending" chemicals may be implicated in reproductive problems such as decreased sperm count, rising infertility and nervous and behavioral problems in children (see Liroff's article in Section 3.5.2). Yet, similar to the response to Rachel Carson's writings, the publication met with stiff chemical industry denunciation. As the authors state, "One of the unfortunate aspects of the public debate about endocrine disruption has been a repeating pattern of distortion of scientific findings by various representatives of chemical interests" (Colborn *et al.*, 2003).

The chemical industry maintains the conclusions are hypothetical, at best, and show inconclusive evidence of harm. Fundamental to this defense is the rejection of the precautionary principle as a scientifically valid methodology and basis for policy. This principle advocates taking precautionary action when chemicals pose possible threats to human health and the environment, rather than waiting for complete scientific proof of cause and effect (see Tickner's article in Section 3.3.2).

However the reality is that we know very little about the chemicals in daily commercial use. Of the 100,000 chemicals in common use, 30,000 are used in volumes of one tonne or more. Of these 95 percent have little or no environmental or human health data simply because prior to 1980 no data were required before marketing in the United States or Europe. Unfortunately, these "grandfathered" chemicals make up the vast majority of chemicals in daily use and our governments continue to rely on a slow process of testing that places the burden of proof on governments to show harm, rather than the chemical industry to prove safety. As a result, the Environmental Protection Agency in the United States has restricted less than ten chemicals in 25 years, even though many chemicals are toxic in animal studies and found in human tissue.

Such was the catalyst for the drafting of a new European Union chemical policy that could fundamentally overhaul the regulation of chemicals in the European Union (see Geiser's article in Section 3.4) and thus have global repercussions. The resulting legislation, REACH, would require the registration, evaluation and authorization of chemicals and would require chemical producers to provide environmental and human health toxicity data. This reverses the onus of proof and effectively ensures that by the year 2012 all chemicals must be registered to remain on the market. Furthermore, chemicals shown to be carcinogens, mutagens, or reproductive toxins will be substituted or require strict authorization for ongoing production and use. The European Commission sees this as a way to stimulate the market for safer substitutes, provide more information for downstream users of chemicals, and protect human health and the environment.

The response of the European chemical industry has been one of outrage, spearheaded by the European Chemical Industry Council (CEFIC). The American Chemistry Council (ACC), fearing that successful reform of European chemicals policy could become a de facto international standard, mounted a vigorous campaign to derail REACH and successfully enlisted the help of various U.S.

government federal departments (DiGangi, 2003). According to senior members of the European Commission, the intensity of lobbying on REACH is unprecedented (Ditz, 2003). The effect of the chemical industry lobby is reflected in the latest version, released in October 2003, which, according to the NGOs and many Parliamentarians now embodies a watered down version of many of the initial ambitions and drafts.

Against this backdrop, new studies examining human levels of toxic chemicals – so-called "body burden" – are continually being published. A 2003 study confirmed the presence of 116 industrial chemicals, most of which are toxic in laboratory animals, in the bodies of average Americans (CDC, 2003). A similar study found even more chemicals in the body of nine volunteers: of the 167 chemicals measured, 76 cause cancer in humans or animals, 94 are toxic to the brain and nervous system, and 79 cause birth defects or abnormal development (EWG, 2002). The United Kingdom's Royal Commission on Environmental Pollution has recommended that where synthetic chemicals are found in elevated concentrations in biological fluids such as breast milk they should be removed from the market immediately (RCEP, 2003).

Over the last few years research has shown that people living in North America have levels of brominated flame retardant chemicals (PBDEs) in their bodies 10–70 times higher than the levels of people in Japan or Europe (Betts, 2001). These halogenated chemicals chemically and toxicologically resemble PCBs and are implicated in liver toxicity, disruption of thyroid hormone levels and developmental neurotoxins, particularly in the young. They are now recognized as global environmental pollutants. Rising public concern in the United States resulted in the United States based bromine industry voluntarily phasing out two brominated chemicals. However, the attempt to control more of these chemicals has met with strong lobby from the Bromine Science and Environment Forum, who maintain that their ubiquitous presence in humans, wildlife, and the environment are harmless and that the benefits of these chemicals in fire prevention are more important. Meanwhile, many downstream users, particularly in the IT sector, have found ways to achieve fire prevention through alternative means either through safer chemical substitutes or product redesign (Thorpe, 2003). However, chemical producers, who could be transitioning to green chemistry alternatives, continue to promote brominated chemical replacements even though halogenated chemicals are often found to be persistent or bioaccumulate in living systems.

Rather than acknowledge the need to fill these large data gaps, bring carcinogenic and other hazardous chemicals under control, stimulate innovation in safer chemicals, and help supply more information to their downstream buyers, the chemical industry argues the cost to supply the data will be too onerous and jobs will be lost. Or at least that is what seems to be the dominant message to outside observers. A study funded by a Federation of German Industries (BDI) predicted huge economic impacts and job losses of over two million workers if the new European chemical policy was passed. However, a panel of economists convened by Germany's Federal Environment Agency (UBA) dismissed the methodology as biased and unrealistic (Ditz, 2003). Similarly the chemical industry counters that production of chemicals post-REACH will be prohibitively expensive, forcing some producers to relocate out of Europe. The European Commission has estimated the testing and administrative costs of registration will be 1.7 to 7 billion Euros spread over more than a decade, which amounts to less than 0.1 percent of the annual sales of chemicals in Europe. This is not counting the economic benefits of reduced health costs, which one study for the United Kingdom calculated could equal to almost 75 billion Euros in savings over the next 17 years (Pearce and Koundouri, 2003).

In response to the defensive positioning of the chemical industry, advocacy groups are shifting their attention to the chemicals in the products we buy. Non-governmental organizations (NGOs) are asking downstream users of chemicals to adopt safer chemicals policies and exert pressure up the supply chain to the chemical suppliers. Campaigns in Europe have measured dust in households and shown correlation of hazardous chemicals in the dust to common household items and furnishings such as carpets, furnishings, flooring, computers, and televisions (GPUK, 2003). Greenpeace's recent analysis of consumer products now reveals priority hazardous chemicals in common brandname articles. Their website, the Chemical House, rates companies according to their chemical policy, thereby offering consumers an informed purchasing choice (GPUK, 2003b).

Similarly, the Friends of the Earth UK's campaign to target safer chemicals use by major retailers has resulted in companies signing the Retailers Pledge to phase out hazardous chemicals within five years (FoE, 2004). As the public loses confidence in the chemical industry it is finding it in retailers and companies who openly profess a commitment to substitute hazardous chemicals for safer alternatives. This trend will continue to grow as groups pressure governments and big purchasers to adopt green procurement guidelines that prohibit carcinogenic, mutagenic or reproductive hazards.

The OsPar Convention for the North East Atlantic has set the generational goal of eliminating hazardous emissions within one generation, or by the year 2020. Such a timeline has also been adopted by the Swedish government in their new chemical policy and this provides a benchmark by which the transition to green chemistry and safer materials can be measured. No other country has detailed such a clear set of criteria and goals to achieve a "nontoxic future."

Based on historical precedent and recent revelations of the chemical industry's suppression of incriminating health data on workers and communities living near chemical production facilities, it is unlikely the public will soon reverse its opinion of the chemical industry's ability to deliver products that are safe (Moyers, 2001).

The next few years offer an unprecedented opportunity for the chemical industry to acknowledge the need for more transparency and accountability for urgently supplying the much needed environmental and health data on chemicals they market. It also provides an opportunity to actively promote green chemistry, safer substitutes and innovation in the development of chemicals that are not persistent and do not bioaccumulate in living systems. The truth will reveal itself by the time the new EU chemical policy is adopted – in whatever form it has been shaped into by industry lobby and public campaigns. That will be the time to conduct a new opinion poll and see if public confidence has indeed risen.

3.3   RISK ASSESSMENT

3.3.1   Chemical Risk Assessment as Used in Setting Regulatory Levels or Standards

BERNARD K GADAGBUI, LYNNE T HABER
and MICHAEL L DOURSON

Toxicology Excellence for Risk Assessment (TERA)

Risk assessment is a process where the magnitude of a specific risk is characterized so that decision-makers can conclude whether the potential hazard is sufficiently great that it needs to be managed or regulated, reduced or removed. The National Research Council (NRC, 1983) of the National Academy of Sciences (NAS) first described the process of human health risk assessment, with an update in 1994 and 1996, as a four-component paradigm (i.e., hazard identification, dose–response assessment, exposure assessment, and risk characterization), with risk communication as a fifth area of study. The first four components are described briefly below.

Hazard identification explores potential concerns about a chemical. It involves an evaluation of the nature, quality, and relevance of scientific data on the specific chemical, the characteristics and relevance of the experimental routes of exposure, and the nature and significance of the observed effects. In this step, scientific studies are reviewed to determine if exposure to an agent could cause increased incidence of adverse health effects (noncancer or cancer effects) in humans, and to identify which effects the chemical can cause.

For noncancer toxicity, the process includes an evaluation of the target organ or "critical" effects (i.e., the first adverse effect or its known precursor that occurs as the dose rate increases). In some cases, one needs to determine whether an effect is adverse or not. The choice of critical effect in the hazard identification process for noncancer toxicity is used as a basis for the dose–response assessment.

For cancer toxicity, hazard identification depends on professional judgment as to the overall weight-of-evidence of carcinogenicity, including epidemiological information, chronic animal bioassays, mechanistic data, mutagenicity tests, other short-term tests, structure–activity relationships, metabolic and pharmacokinetic properties, toxicological effects, and physical and chemical properties. The outcome of this judgment is the placement of the chemical into one of several categories, such as the system developed by the International Agency for Research on Cancer (IARC) in 1978:

Group 1:   Carcinogenic to humans.

Group 2:   Probably carcinogenic to humans (including subgroups 2A for chemicals having limited evidence of carcinogenicity in humans, and Group 2B for chemicals having sufficient evidence of carcinogenicity in laboratory animals, and inadequate evidence in humans).

Group 3:   Cannot be classified as to its carcinogenicity to humans.

U.S. EPA has also published general guidelines in developing and evaluating risk assessments for carcinogens (U.S. EPA, 1986, 1996, 1999, 2003). The most recent guidelines advocate the development of a more comprehensive characterization of carcinogenic hazard in the form of a narrative. Within this context, a cancer hazard characterization should include all information relevant to the weight-of-evidence for carcinogenicity, not just tumor data in humans and animals. This means that mechanistic data can play an integral role in the hazard identification step for carcinogenicity, and may also influence the choice of a dose–response model. Moreover, the hazard characterization can provide specific information about the conditions under which a chemical is likely to be carcinogenic. For example, it may be "likely to be carcinogenic by the route of inhalation but not by ingestion."

Dose–response assessment follows the hazard identification in the risk assessment process. In dose–response assessment, an adverse effect is presumed to either exhibit a threshold or not in the dose–response curve. Depending on the nature of this curve, different approaches are employed to estimate the risk posed by the potential toxic agent.

For example, risk assessment for cancer toxicity generally assumes that no threshold exists below which no adverse effects could be expected. For these agents, the U.S. EPA and others assume that there are no exposures that have "zero risk", implying that even a very low exposure to a carcinogen can increase the risk of cancer. For these agents, the high dose in the experimental animal study is generally extrapolated to the low dose to which humans are generally exposed, or to a dose that is considered to be *de minimus*. Such extrapolation introduces significant uncertainties into the risk assessment process. These uncertainties are recognized in the presentation of the risk. For example, although risk is presented as the 95 percent upper confidence limit, the U.S. EPA and others note that the true risk could be as low as zero.

For noncancer toxicity, it is generally assumed that a threshold exists at or below which no appreciable risk of deleterious effect is expected over a lifetime. Thus, the goal of the risk assessment is to establish a safe dose level for humans based on a no-observed-adverse-effect level (NOAEL) or lowest-observed-adverse-effect level (LOAEL) derived from well-conducted animal toxicity studies with application of uncertainty factors to compensate for (1) differences between experimental animals and humans, (2) differences between average humans and sensitive humans, (3) the lack of a NOAEL, (4) the lack of lifetime studies, and (5) the lack of bioassay that tests a variety of endpoints such as in young experimental animals. This safe dose is used by national and health agencies throughout the world and is variously referred to as reference dose or concentration (RfD or RfC), minimum risk level (MRL), acceptable daily intake (ADI), tolerable daily intake or concentration (TDI or TDC), or tolerable intake (TI) or tolerable concentration (TC) [Barnes and Dourson, 1988; Jarabek, 1994; U.S. EPA, 1994; Dourson, 1994; Pohl and Abadin, 1995; Lu, 1985, 1988; Meek *et al.*, 1994; International Programme on Chemical Safety (IPCS), 1994]. This threshold approach is sometimes used for cancer toxicity when indicated by mode of action data.

In the absence of human data (the most preferred data for risk assessment), the dose–response assessment for either cancer or noncancer toxicity is determined from animal toxicity studies using an animal model that is relevant to humans or using a critical study and species that show an adverse effect at the lowest administered dose. The default assumption is that humans may be as sensitive as the most sensitive experimental species.

The third step in the risk assessment process is exposure assessment. In exposure assessment, the intake of a toxic agent from the environment is quantified using any combination of oral, inhalation, and dermal routes of exposure. This assessment may include a component for each route, such as when an assessor would investigate the potential impact of a point source of pollution. In this case, the magnitude of exposure depends on the amount of chemical used or released, chemical fate and transport, chemical concentration at the point of exposure, the routes and rates of uptake, the duration, the exposure setting (location and number of potential receptors, land use and human activities that could lead to exposure), and characteristics of receptors potentially exposed to the chemical.

In a quantitative risk assessment, these factors are typically combined to estimate a potential human dose rate (and concentrations to which organisms in the environment are exposed). Exposure can be quantified through direct measurement at the point of contact using personal monitoring devices. This method gives the most accurate exposure value for the period of time over which the measurement was taken. Exposure can also be measured indirectly through environmental fate and transport modeling. A number of models are available for use in environmental modeling; many of these are describe in U.S. EPA's exposure assessment guidelines (U.S. EPA, 1992).

An exposure assessment may also be focused on one particular medium and one route of exposure, for example, the oral intake of a disinfectant byproduct from treating water to remove microbes. This type of exposure assessment is often used to determine whether sufficient human exposure to a chemical in a given medium is occurring to warrant regulation. The assessor must estimate the extent to which individuals may be exposed to the byproduct in drinking water, often using default values for these estimates, such as consumption of 2 liters of water daily.

Risk characterization is the most important and final part of a risk assessment. It summarizes and interprets the information from hazard identification, dose–response, and exposure steps, identifies the limitations and uncertainties in risk estimates, and communicates the actual likelihood of risk to exposed populations. The uncertainties identified in each step in the risk assessment process are analyzed and the overall impact on the risk estimate(s) is evaluated quantitatively and/or qualitatively.

Often, protective levels or other criteria for various chemicals in environmental media are developed. This characterization includes an evaluation of the data quality, specific assumptions, and uncertainties associated with each step, and the level of confidence in the resulting criteria. Alternative risk characterizations are also discussed. Specific key qualities, or attributes, of risk characterizations have been identified [American Industrial Health Council (AIHC), 1992; U.S. EPA, 2000b].

These attributes include transparency in decision-making, clarity in communication, consistency, and reasonableness.

The ultimate goal of risk characterization is to provide the risk managers with enough information, presented in a comprehensible fashion, that they understand what is known about the risk from a specific situation. In order for risk assessors to meet this goal, they must understand the needs of the risk managers and should engage them in the process. This engagement often results in an iterative approach, which helps risk assessors meet a level of detail and analysis appropriate for the situation (e.g., initial screening versus national regulation). The iterative approach allows risk managers to better use limited resources and to develop environmental criteria that have a more firm basis in the science.

3.3.1.1 Can Science Effectively Predict Risks Through Quantitative Risk Assessments – A Policy Perspective. Over the years, scientists have gained a great deal of experience, through the conduct of risk assessments, in how to perform each step in the risk assessment more efficiently and accurately. Improvements to risk assessment have been identified, significantly advancing the usefulness of risk assessment.

Our ability to accurately assess risks is affected by the uncertainties inherent in the risk assessment process at each step. Some of the sources of uncertainties in the toxicity assessment include inadequate human or animal data, inappropriate dose – response models, lack of biological basis for the adverse effects, and so on. The impact of these uncertainties is that the risk assessment tends to be conservative. For example, as described above, the U.S. EPA and others generally apply uncertainty factors to adjust the safe dose downwards when data are lacking as a matter of policy, despite the fact that some of these factors might actually adjust the safe dose upwards if sufficient data were available to characterize the uncertainties.

There are many parameters and factors that are components of an exposure assessment. Where there is lack of adequate information on any of these parameters and factors, default assumptions are used. However, some of these parameters and factors (e.g., body weight, exposure frequency, and duration) can be represented by a range of values. If these uncertainties are not reduced, a highly conservative risk may result.

During the risk characterization step, most often risk for a hypothetical sensitive subgroup is reported; historically, no average or central tendency and population risks have been estimated, although recent work also includes such values. Furthermore, risk or hazard from individual chemicals is often added to produce aggregate risk or hazard. These biases provide potential sources of uncertainty during this phase of the risk assessment process. In part because of these biases, new methods for conducting dose – response assessment have been developed, as discussed in the next section.

3.3.1.2 Developing Methods in Risk Assessment and the Impact on the Chemical Industry. In addition to the approaches described above to predict risk to chemicals, a number of recent research initiatives have led to improved methods for

quantitative risk assessment and better incorporation of mechanistic data, both for noncancer and cancer assessments. Structure–Activity Relationships (SAR) can be used to predict chemical toxicity in the absence of adequate data on the chemical of interest. Methods to improve the use of available data include Chemical Specific Adjustment Factors (CSAFs) to replace default uncertainty factors, use of biomarkers to quantify internal dose and variations in host susceptibility, the use of dose–response modeling methods such as Benchmark Dose (BMD) and Categorical Regression, and Physiologically Based Pharmacokinetic (PBPK) and Biologically Based Dose–Response (BBDR) models to incorporate mechanistic data. These approaches are briefly described in the following.[1]

Structure–activity relationship (SAR) analysis is frequently employed as a first step in the analysis of a potentially hazardous toxic agent to predict and characterize its toxicity, particularly in screening assessments when no or very limited information on the chemical's toxicity is available. The SAR is a computer-based modeling method that relates chemical structure to qualitative biological activity. The analysis involves comparing the molecular structure and the physical and chemical properties of the agent with unknown toxicity with those of other, similar chemicals with known toxic or carcinogenic effects. This is based on the assumption that the structure of a chemical determines its physical and chemical properties and reactivities. These properties will determine the biological or toxicological properties when the chemical interacts with a biological system. As a predictive tool, SARs can be used to reduce the need for costly and time-consuming animal and *in vitro* testing to support risk assessment and regulatory action. It is thus a useful tool in screening of chemicals for a wide range of toxicity endpoints.

Chemical Specific Adjustment Factors (CSAFs) provide a consistent approach for incorporating mechanistic data to replace the default uncertainty factors for interspecies extrapolation and intraspecies variability (IPCS, 2001). This framework is based on early work by Renwick (1993) and applied by IPCS (1994), in which the default uncertainty factor of 10 for interspecies differences is divided into two factors of 2.5 and 4.0 for toxicodynamics and toxicokinetics, respectively. The default factor of 10 for intraspecies variability is similarly divided into two equal factors for toxicodynamics and toxicokinetics. Any one or all of these four subfactors can be replaced by chemical-specific data. The CSAFs have been used by the U.S. EPA in deriving an RfD for boron (U.S. EPA, 2004a) and by Health Canada in deriving a TC for 2-butoxyethanol (Health Canada, 2000).

Biomarkers are defined as any cellular or molecular indication of toxic exposure, adverse health outcome or susceptibility (NAS, 1987). From this definition, three distinct biomarkers are evident, that of exposure, effects of exposure, and host susceptibility. As more extensively discussed by Maier *et al.* (2001), this field has progressed to a point where application to risk assessment is possible. For example, biomarkers of effects, exposure, and host susceptibility can give insights into mode of chemical action. These insights are likely to better inform the hazard

---

[1]A useful text to read in addition to this chapter is by Haber *et al.* (2001a), as well as other associated chapters found in the latest edition of Patty's Toxicology.

identification and dose–response assessment parts of risk assessment. Moreover, advances in both genomics and proteomics will likely impact current biomarker research and thus may ultimately change the way risk assessments are currently conducted.

Benchmark Dose (BMD) modeling is an alternative method to the NOAEL/ LOAEL approach (Crump, 1984; Dourson *et al.*, 1985; Barnes *et al.*, 1995; U.S. EPA, 2000a). The method fits flexible mathematical models to the dose–response data and then determines the dose associated with a specified incidence of the adverse effects. Once this dose is estimated, then an RfD is estimated with the use of one or more uncertainty factors or Chemical Specific Adjustment Factors (CSAF) as described above. Advantages over the NOAEL/LOAEL approach include (1) the BMD is not limited to the tested doses; (2) a BMD can be calculated even when the study does not identify a NOAEL; and (3) unlike the NOAEL approach, the BMD approach accounts for the statistical power of the study. Numerous examples of BMD use in the dose–response assessment part of the risk assessment process are available on the U.S. EPA's Integrated Risk Information System (IRIS) (2004b).

Categorical Regression is an analytical method by which a dose–response model may be fit to data where only severity ratings are available (Hertzberg and Miller, 1985; Hertzberg, 1989; Guth *et al.*, 1997; reviewed by Haber *et al.*, 2001b). For example, the only dose–response information available may be that there was mild liver necrosis at $2\,mg/kg/day$, and moderate necrosis at $10\,mg/kg/day$. Thus, an advantage of categorical regression is that it can be used to conduct dose–severity of response analyses in the absence of quantitative data. Categorical regression can also be used for modeling dose–response information from disparate endpoints, by using a common severity metric (in terms of severity categories), although care needs to be taken in how the data are combined (Stern *et al.*, submitted).

Categorical regression can be used to describe the dose–response relationship of a single (e.g., histopathology) endpoint observed in a single study (e.g., Piersma *et al.*, 2000) or as a meta-analytical technique to simultaneously analyze the results from multiple studies (Guth *et al.*, 1997; Dourson *et al.*, 1997). In this latter application, it can be used to develop an overall concentration-duration–response analysis (Guth *et al.*, 1997), providing information of particular interest for evaluation of acute inhalation exposures. Categorical regression has also been used to estimate the risk above the RfD (Dourson *et al.*, 1997; Teuschler *et al.*, 1999), one of the few approaches available for estimating noncancer risk. Such estimation is well founded when the data are in close proximity to the RfD, but greater caution would be needed to estimate risks further from doses at which data exist. The ability to estimate risks above the RfD might be useful, for example, in the case of a chemical with unique usefulness where projected human exposures slightly exceed the RfD, or where it is desirable to set priorities for regulation within a group of chemicals based on the chemical with the greatest likelihood of harm.

Physiologically Based Pharmacokinetic (PBPK) models and Biologically Based Dose–Response (BBDR) models are finding increasing use in risk assessment applications. (See the U.S. EPA's IRIS, 2004b, for several examples.) PBPK models describe the flow and transformation of the chemical by the body (toxicokinetics) using species-relevant organ and tissue volumes, blood flows, and kinetic transformation parameters (reviewed by Clewell, 1995). This allows the estimation of the biologically important dose delivered to the target organ(s). PBPK modeling can be used to calculate tissue dose, to improve interspecies extrapolation, allow route-to-route extrapolation, and evaluate mechanistic questions. While PBPK models only describe a chemical's kinetics, BBDR models also include descriptions of a chemical's interactions with the body (toxicodynamics), including a mathematical description of how the toxic agent interacts with its molecular target and damages the cell. While a number of PBPK models have been developed and used in risk assessments, the number of BBDR models is much lower, due to the very high data demands. Perhaps two of the most famous are the model of formaldehyde (Conolly *et al.*, 2004) and the Moolgavkar–Venzon–Knudson (MVK) model of multistage carcinogenesis (Moolgavkar and Knudson, 1981), a general toxicodynamic model that can be combined with a specific chemical kinetic model for a full BBDR. BBDR models can be used to estimate dose–response at low doses not amenable to experimental evaluation.

Sensitivity analysis is a type of uncertainty analysis that is used to consider the impacts of uncertainty. In such analyses, one input is changed at a time to determine how the results of a model will change over the range of possible values of that single input. Multiple inputs can be varied simultaneously, using a sampling technique called Monte Carlo analysis, to obtain an overall distribution of the result.

3.3.1.3 Use of Screening Methods to Support Sustainability.

While the evaluation of detailed mechanistic data on a chemical is the ideal approach to develop a scientifically rigorous assessment of a chemical, decisions often need to be made based on much less data. For example, a systematic science-based approach for evaluation of a chemical early in its development can help companies evaluate alternatives, identify potential toxicological concerns early in product development, apply science-based strategies for priority setting in their inventory of existing substances, and identify the most sustainable approach. In such cases, often only limited data are available. In these cases, toxicological information is supplemented by information on related chemicals (analogs), as well as professional judgment. Screening and ranking methods are also a major component of large national and international efforts for determining when a more detailed assessment is needed.

In the United States, companies are required to submit premanufacture notice to the U.S. EPA prior to introducing new chemicals to commerce. To encourage the application of pollution prevention principles and the development of inherently low hazard new chemicals, the U.S. EPA has instituted a pollution prevention (P2) framework as part of its Sustainable Futures program (U.S. EPA, 2004c). As part of the program, training and support are provided to companies in the evaluation of new chemicals, with limited available information, using the U.S. EPA's hazard and risk characterization models and protocols. Quantitative Structure Activity Relationship (QSAR) models are used to supplement available data, and used to calculate physical/chemical properties, persistence, bioaccumulation, aquatic toxicity, and potential carcinogenicity. Similarly, when data on the chemical are not available, human health hazards from both noncancer and cancer effects are evaluated based on available data on the chemical and using SARs to extrapolate from analogs, focussing on key reactive structural groups that are likely to determine toxicity.

Broader comparisons of multiple chemicals use chemical hazard and risk ranking methods, of which more than 100 published versions exist, varying in level of sophistication and scope of coverage (reviewed in Pittinger *et al.*, 2003; Swanson and Socha, 1997). Relative hazards could be evaluated based on human toxicity, ecological hazards, physical safety hazards, risk perception issues, regulatory alerts, or a combination of these factors. Interest in reliable screening and ranking methods has increased in recent years, partially due to a variety of new regulatory initiatives. For example, the Canadian Environmental Protection Act (CEPA) of 1999 requires categorization of all of the approximately 23,000 substances on the Domestic Substances List (DSL), a task that will require a variety of simple and complex tools for evaluating toxicity and exposure (Health Canada, 2004). Other programs that are likely to include large-scale screening and hazard identification tools include the European Union's Registration, Evaluation and Authorization of Chemicals (REACH) program, which would require companies to register chemicals in a central database and provide safety information (EU, 2004), and programs evaluating High Production Volume (HPV) chemicals in the United States and Europe.

Another screening approach is to focus on exposure, based on "thresholds of concern." This approach is used to identify exposure levels that, with reasonable certainty based on informed judgment and evaluation of data for existing regulated chemicals, are unlikely to produce adverse health effects under specific conditions of exposure. Building on work by Munro (1990) and Rulis (1986), this concept is used by the FDA to establish "thresholds of regulation" for food additives [U.S. Food and Drug Administration (FDA), 1995]. Similar analyses have been conducted for noncarcinogens by Munro *et al.* (1996).

3.3.2 The Limits of a Risk Assessment-Based Approach to Sustainability in the Chemical Industry and the Need for a New Paradigm Based on Precaution

JOEL TICKNER

Lowell Center for Sustainable Production, University of Massachusetts – Lowell

While we have known for decades that many industrial chemicals are toxic, mounting evidence indicates that some of these chemicals play a role in the onset of diseases including cancer, developmental and behavioral disorders, respiratory ailments and asthma, neurological disorders, and birth defects, among others. Despite evidence of health effects linked to chemical exposure, a great majority of the chemicals found in our air, water, food, and everyday products lack basic safety data for human and ecosystem health, particularly that of developing organisms. Some of these chemicals are known to persist in the environment for decades, travel great distances, or accumulate in the food chain Even less is known about the potential health effects of exposures to multiple chemicals and other stressors: the reality of our everyday lives. While surveys have shown that the public believes that governments would not allow chemicals on the market if they had not been tested and demonstrated safe for use, this is clearly not the case.

The traditional scientific and political response to these data gaps has been to collect more information and use a technique called quantitative risk assessment to calculate the probability of harm given particular exposures, applying numerous assumptions in the process. While this process has been termed the "sound science" approach, it is often far from that. Quantitative risk assessments often narrow the types of information that go into decision-making and hide uncertainties. They are time-consuming and costly to complete and while debates over details of these assessments occur, the default policy option is that no policy action is necessary.

If we are to achieve more sustainable forms of production, we need a new paradigm for environmental science and policy that is based on the best available science (science informed), but also based on professional judgment, inclusion of a wide range of stakeholders, and consideration of the widest range of alternatives to meet particular needs. This approach, called the precautionary principle, is increasingly gaining prominence in international debates over uncertain risks. Put simply, precaution calls for preventive actions when there is reasonable scientific evidence of harm, although the nature and magnitude of that harm may not be fully understood scientifically. While a highly contentious term, we see precaution as simply an avenue to make more health protective decisions in the face of highly uncertain and complex risks.

In this section, we explore the limits of the risk assessment-based approach to decision-making and what a precautionary paradigm might look like.

3.3.2.1 *The Problem of Uncertainty.*

3.3.2.1 *The Problem of Uncertainty.* Our ability to identify adverse human health or environmental effects resulting from chemical substances is limited by our incomplete understanding of science, and therefore makes knowing what to look for and where to look extremely difficult. For example, the U.S. EPA and Environmental Defense Fund issued reports on the lack of basic testing information on toxic substances in 1997 and 1998 (EDF, 1997; U.S. EPA, 1998a). They found that of the 2800 high production volume (HPV) chemicals (over one million pounds in commerce), 93 percent of chemicals included lack some basic chemical screening data, 43 percent have no basic toxicity data, and 51 percent of chemicals on the Toxic Release Inventory lack basic toxicity information. While these reports resulted in the EPA's High Production Volume Challenge (a voluntary effort of the U.S. EPA, the American Chemistry Council and Environmental Defense), which seeks to fill out these gaps, only basic toxicity information for a small number of chemicals will be addressed (for example, data on chemicals used from 10,000 lbs to one million lbs will not be included). The Challenge will not include exposure information or address how data should be used in risk management and reduction. As of fall 2003, one-fifth of the HPV chemicals lacked a commitment from industry for testing (see www.epa.gov/chemrtk).

Our scientific knowledge is especially limited on the effects of pollution on highly variable and complex ecological and human systems. A question for decision-makers is how science can establish an assimilative capacity – a predicable level of harm from which an ecosystem can recover – or a "safe" level of exposure when the exact effect, its magnitude, distribution, and interconnections are unknown (Gee, 1997).

Traditionally, decision-makers have focused on the effects of a single chemical in a single medium, when in reality humans and ecosystems are exposed to a wide variety of physical and chemical stressors, circumstances that environmental science is still struggling to understand. Further, specific populations feel disproportionate impacts of environmental degradation; certain groups may be at higher risk of harm because of genetic disposition, disease, developmental, or social status, and geographic location. For example, children are uniquely susceptible to the effects of toxic substances due to their immature metabolic processes, rapid development, and exposure (Landrigan, 1999). That sensitive subpopulations or high variability in responses to chemical exposures exist within a group are frequently overlooked by decision-makers. Uncertainty is an inevitable condition underscoring all environmental decision-making because humans operate in open, dynamic environments that are difficult to control. For example, variability among individuals generally cannot be reduced. In addition to traditional types of uncertainty, there are also the conditions of indeterminacy (an uncontrollable form of uncertainty due to complex human, technical, and social interactions) and ignorance (the state of not knowing what we do not know; e.g., not knowing what we are uncertain about) (see Wynne, 1993).

Unfortunately, the lack of proof of harm is often misinterpreted as proof of safety. And, because uncertainty complicates decision-making, it is generally played down or ignored. The influence of uncertainty and ignorance, for example, is seen in policy when EPA permits the use and release of chemicals into the environment without having toxicity information. Despite this lack of knowledge, the Toxic Substances Control Act presumes that existing chemicals (those sold before 1980, accounting for more than 99 percent by volume of chemicals on the market today) are safe until proven dangerous. This presumption is not only problematic for health but also a serious limitation to innovation in safer chemistry.

Because uncertainty is underappreciated, early warnings of harm are often overlooked. A review of technology failures saw in 40 percent of the cases, some early sign of harm was overlooked. In 50 percent of the cases, danger signs were known for a similar technology (Lawless, 1977). Many substances once thought benign have been shown to cause severe human or environmental effects. Case studies and common-sense observation often suggest causal links long before they can be proven. For example, concerns about the health hazards of asbestos and benzene were identified as early as 1898 (European Environment Agency, 2000). Lead has been known for centuries to be a neurotoxicant. These cases show how waiting for "convincing" evidence can pose high costs to human and ecological health, and remediation resources.

3.3.2.2 The Problems of Risk Assessment. Over the past 25 years, the United States regulatory and scientific response to environmental damage and uncertainty has focused on the use of quantitative assessment methods. This response was heavily influenced by the U.S. regulatory and political system, as well as the courts, where threats of judicial scrutiny have caused agencies to constantly construct formal, quantitative records (von Moltke, 2000). During the 1970s, risk assessment and cost–benefit analysis were developed to assist decision-making regarding environmental and health risks.

Defined by the National Research Council in 1983, quantitative risk assessment has become a central element of environmental and health decision-making in the United States. Risk assessment is not a scientific discipline, per se. It is a formalized, systematic tool used to integrate and communicate scientific information. The technique of risk assessment has evolved over the years to address different disease endpoints and to incorporate broader notions of exposure and greater analysis of uncertainty. But the general framework for conducting risk assessments remains the same: hazard identification, dose–response assessment, exposure assessment, and risk characterization (National Research Council, 1983). The development of risk assessment has brought substantial advances in scientific understanding of exposure and disease and our ability to predict adverse outcomes from hazardous activities. When much is known about the specific nature of the harm and probabilities are well established, risk assessments provide a science-structured methodology for decision-making. Nearly all decisions involve some weighing of risks, either qualitatively or quantitatively.

The reliance on risk assessment as the sole analytical technique in environmental and health decision-making has significant and widely described disadvantages (see Tickner, 1996; O'Brien, 2000), yet reliance on this technique is increasing. Specific criticisms of the use of quantitative risk assessment as a central, singular tool in regulatory decision-making include the following:

- Risk assessments are generally used to analyze problems rather than to solve or prevent them, and are used to set "safe" levels of exposure that correspond to a predetermined "acceptable" level of risk.
- Risk assessment limits the amount and source of information used in examining environmental and health hazards that may inhibit holistic understanding of complex systems and interactions.
- Risk assessment perpetuates many of the inherent limitations in current scientific methods. For example, research questions may be so narrowly defined that important aspects of the problem are missed.
- Risk assessments limit consideration of uncertainties, multiple exposures, cumulative effects, sensitive populations, or endpoints other than cancer.

- Risk assessments are based on numerous assumptions that are many times based on political or uncertain information. As a result, risk assessments conducted by different groups, even using the same information, can come out with widely differing results.
- Risk assessments can be expensive and time consuming, tying up limited resources. It may be much more cost-effective to prevent exposure in the first place. An example of this is the U.S. Occupational Safety and Health Administration's Methylene Chloride standard, which took seven years to finalize due to debates over a pharmokinetic model and mechanism of action.
- Risk assessment processes often exclude those potentially harmed by environmental degradation as they traditionally do not include public perceptions, priorities, or needs.

Risk assessment poses additional problems when it forms the basis for decision-making because it creates a rigid structure in which narrowly defined evidence of harm is collected, the probability of adverse effects is examined, and a decision is made. Additionally, feedback and follow-up are not guaranteed, alternatives or prevention are not generally considered, and the public is only told about risk once a decision is imminent.

Government agencies have begun to recognize and respond to criticisms about risk assessment and its use, and have noted areas for improvement. Reports by the National Research Council (1994; Stern and Fineberg, 1996) and the Presidential Commission on Risk Assessment and Risk Management (1997) recommended changes to the process of risk assessment that would:

- Better and more comprehensively examine uncertainty and variability;
- Include stakeholders throughout the assessment and management process;
- Consider prevention and options in the risk assessment process;
- Provide more holistic problem definition;
- Consider cumulative/interactive effects and sensitive subpopulations; and
- Better evaluate actions taken.

While these recommendations point to needed change, momentum for their implementation is missing. A trend towards more "reductionist" risk assessments (where action is not taken until the mechanism of action of effects is fully understood) does not bode well for such positive changes. Further, some analysts suggest that all risk decisions fully consider and quantify "risk–risk" trade-offs of regulations – for example, natural versus chemical carcinogens, or the beneficial aspects of pollution (Graham and Wiener, 1995). While it is important to consider potential trade-off risks from decisions and account for them through flexible decision-making structures, such a requirement would create additional burdens that may stall preventive public policies.

3.3.2.3 *Precaution: Response to the Limitations of Current Quantitative Assessment Frameworks, Uncertainty, and Complexity.*

Because risks associated with toxic substances are increasingly complex, and widely distributed, there is a need for a new approach that uses the best available science and informed judgment to protect health and ecosystems and stimulate innovation in safer and cleaner materials. Embodied in the precautionary principle, this approach is rapidly gaining strength in environmental health debates because it encourages policies that protect human health and the environment in the face of uncertain risks. In this broad sense, precaution is not a new concept and is at the heart of centuries of medical and public health theory and practice. The principle originated in response to concerns that science and policy structures were inherently too limited to adequately address evolving risks, and recognized the severe consequences to health and the economy of not taking preventative actions.

As a principle of decision-making, the precautionary principle has its roots in the German word Vorsorgeprinzip. An alternative translation of this word is the "foresight principle" or "forecaring principle," which emphasizes anticipatory action – a proactive idea rather than precaution, which to many sounds reactive and even negative. The Germans saw precaution as a tool to stimulate careful social planning for job creation, innovation, and sustainability. Over the past 20 years the principle has served as a central element in international treaties addressing North Sea pollution, ozone-depleting chemicals, fisheries, climate change, persistent organic pollutants, genetic modification, and sustainable development. The European Union has placed precaution, pollution prevention and the polluter pays principle, as central elements of environmental health policy. While not explicitly mentioned, the precautionary principle is at the core of many environmental and occupational health policies in the United States (Raffensperger and Tickner, 1999).

All definitions of the precautionary principle are similar, essentially stating that when there is uncertainty and credible evidence of a risk, then precautionary actions should be taken. The most widely used definition of the precautionary principle is the 1992 Rio Declaration on Environment and Development, which states

> In order to protect the environment, the precautionary approach shall be widely applied by States according to their capabilities. Where there are threats of serious or irreversible damage, lack of full scientific certainty shall not be used as a reason for postponing cost-effective measures to prevent environmental degradation.

Another widely used definition is the Wingspread Statement on the Precautionary Principle, which states "when an activity raises threats of harm to human health or the environment, precautionary measures should be taken even if some cause and effect relationships are not fully established scientifically." The statement elaborated on earlier definitions by identifying four central components of the Principle: (1) taking preventive action in the face of uncertainty; (2) shifting burdens onto proponents of potentially harmful activities; (3) exploring a wide range of alternatives to possibly harmful actions; and (4) increasing public participation in decision-making (Raffensperger and Tickner, 1999).

3.3.2.4   Components of a Precautionary Approach. Implementing the precautionary principle to protect health and ecosystems from chemical risks demands broad reorganization of environmental science and policy to increase their efficacy in anticipating risks and promoting cost-effective alternatives to risky endeavors. Elements of effective use of the precautionary principle in policy include the following (Tickner, 2002).

1. Shifting the questions asked in environmental and health policy. Instead of scientists and policy-makers asking, "What level of risk is acceptable?" or "How much contamination can a human or ecosystem assimilate?" we must ask, "How much contamination can we avoid while still achieving our goals?" "What are the alternatives or opportunities for prevention?" "Is this activity needed in the first place?" This requires tools to analyze comprehensively not only risks but also feasibility of alternative technologies and products.

This shift reorients regulatory focus from analysis of problems to analysis of solutions and establishment of goals. Goal setting, common in public health, involves backcasting, or establishing long-term goals and shorter term objectives to reduce impacts and exposures of problem substances, and advocates for alternatives. This focus on alternatives and goals allows a product or activity as a whole to be examined as to whether its purpose can be served in a less harmful or more effective way, and may also allow decision-makers to make better use of resources by avoiding costly debates over proof of harm and causality.

2. Shifting presumptions. The precautionary principle favors protecting the environment and public health in the face of uncertain risks, rather than presuming that a substance or economic activities are safe until proven dangerous. This places the responsibility for demonstrating relative safety, analyzing alternatives, and preventing harm on those undertaking potentially harmful activities and allows for legislative disincentives for potentially harmful activities.

3. Transparent and inclusive decision-making processes. Environmental health decisions tend to be primarily policy decisions, informed by science and values. A more participative process for decision-making under the precautionary principle could improve the ability of decision-makers to anticipate and prevent harm to ecosystems and human health. Public participation is vital to development of accountable solutions to environmental health problems because nonexperts see problems, issues, and solutions that experts miss, reflect sensitivity to social and political values and common sense not included in expert models. Broader public participation processes may increase the quality, legitimacy and accountability of complex decisions.

3.3.2.5   Precautionary Actions. There is a misconception that the precautionary principle requires banning an activity. Precautionary actions range from public right-to-know, to phasing out particularly harmful activities. The appropriate precautionary action should come after considering the strength of the evidence, magnitude and potential impact of the risk, uncertainties, opportunities for prevention, and social/economic values. The goal of such actions should be:

- Reducing and eliminating exposures to potentially harmful agents, minimizing trade-offs;
- Redesigning production processes, products, and human activities to minimize risk in the first place;
- Providing information and education to promote empowerment and accountability;
- Establishing a research agenda designed to characterize more comprehensively risks, provide early warnings, and develop alternatives.

Many possible tools for the precautionary approach are outlined in this book, including cleaner production, environmental management systems, and so on.

3.3.2.6 Precaution and Environmental Science. When the precautionary principle is discussed in its relationship to science, it is often portrayed as an antiscience or a risk-management principle that is only used after undergoing conventional scientific processes. As discussed earlier, in practice the limitations of science to characterize complex risks show that precaution is not at odds (Kriebel *et al.*, 2001). Further, precaution is not just about additional safety factors or changing risk assessment default assumptions. Research by U.S. EPA scientists has demonstrated that many of the EPA's Reference Doses – or conservative safe exposures – may correspond to risks of greater than 1 in 1000, meaning that safety factors alone may not protect health (Castorina and Woodruff, 2003).

Environmental science is critical to solving some of our most pressing and complex environmental problems and hence can support precautionary policies. As environmental science faces increasing challenges from more complex risks with greater uncertainty and ignorance, the nexus between science and preventive policy is even more important. In this context, precaution is entirely consistent with good science because it demands more rigorous and transparent science rather than less science. This provides insights into how systems are disrupted by technologies, identifies and assesses opportunities for prevention and restoration, and clarifies gaps in our understanding of risks. A shift to more precautionary policies allows scientists to think differently about the way they research and communicate results (Kriebel *et al.*, 2001). In September 2001, the Lowell Center for Sustainable Production hosted the International Summit on Science and the Precautionary Principle to explore the role of science in the precautionary principle and advocated the following changes to science and policy (Tickner, 2003):

- A more dynamic interface and communication between science and policy;
- A more effective linkage between research on hazards and expanded research on prevention, safer technological options, and restoration;

- Greater use of interdisciplinary approaches to science and policy, including better integration of qualitative and quantitative data;
- Innovative methods for analyzing the cumulative and interactive effects of various hazards, for examining impacts on populations and systems, and the impacts of hazards on vulnerable and disproportionately affected communities;
- Systems for continuous monitoring and surveillance to avoid unintended consequences of actions, and to identify early warnings of risks; and
- Better analysis and communication of potential hazards and uncertainties.

A more precautionary approach should be informed by the most "appropriate science", which can be understood as a framework for choosing methods and tools chosen to fit the nature and complexity of the problem (Kriebel *et al.*, 2003). Critical to this framework are the flexibility to integrate a variety of research methods and data sources into the problem evaluation, and to consult with many constituencies to understand the diversity of views on a problem and seek input on alternative solutions. Appropriate science is solutions-based, focused on broadly understanding risks, but also on finding ways to prevent them in the first place. Under this approach, the limitations of science to fully characterize complex risks are openly acknowledged, making it more difficult to use incomplete knowledge to justify preventive actions.

3.3.2.7 Precautionary Assessment: More Sustainable, Preventive Decisions Under Uncertainty.

Applying precaution to achieve more health and ecosystem-protective decisions under uncertainty requires a set of cautionary considerations that are applied throughout the scientific and policy process, such as analysis of alternative courses of action, expanded scientific tools, incentives for research and innovation, and enhanced public participation. It should encourage decision-making using the broadest possible range of information, stakeholders and scientific and policy tools in identifying and preventing risks. These can ensure a more proactive and positive approach to environmental protection, while improving decision-making and stimulating innovation in safer materials, technologies, scientific approaches, and policies.

One approach to this process is called "precautionary assessment," which focuses on a series of procedural steps to ensure sound health and environmental decision-making, examining all of the evidence on threats as a whole and learning from accumulated experience and understanding (Tickner, 2003). It is a flexible approach, which is critically important in environmental decision-making since each decision is different. Although outcomes will differ with the facts of each case, the approach will be the same. The goal is for governments and entities handling risks to internalize this heuristic approach in their decision-making processes, instituting a precautionary "mindset" with regard to uncertain environmental and health risks. Similar approaches exist and are increasingly used in analyses of complex issues, such as medicine and business decisions.

The steps of a precautionary assessment approach include the following.

1. Determining whether an uncertain risk/problem merits a more thorough review – sometimes a screening process may be useful.

2. Broadly defining problems to capture root sources of risks.

3. Considering and examining all available evidence on exposure, hazard, and risk in an interdisciplinary manner, to take account of variability as well as direct and indirect, cumulative, and interactive effects.

4. Considering simplifying rules of thumb, safety factors, default values, or proxy indicators of exposures and effects when information is lacking.

5. Comprehensively examining and reducing uncertainties and gaps in knowledge.

6. Identifying and examining a wide range of options to reduce risks, as well as their trade-offs, advantages, and disadvantages.

7. Determining an appropriate course of action based on the scientific evidence, examination of alternatives, and public input.

8. Instituting post-implementation follow-up measures to ensure continuous risk reduction and minimize unintended consequences.

Decisions made under a precautionary assessment should not be considered permanent, but part of a continuous process of increasing understanding and reducing overall impacts. Once precautionary actions have been chosen, follow-up and monitoring schemes for the activity should be developed. This type of feedback is critical to understanding the impacts of precautionary actions as well as to stimulate continuous improvement in environmental performance and technological innovation. Follow-up tools include periodic assessment, audit, or prevention planning requirements; regular reporting of environmental impacts (e.g., toxics use reporting); short- and long-term health and exposure monitoring; toxicological testing; and impact statements any time a major change is made to a product, process, or activity.

3.3.2.8 Conclusions. This section has outlined the problem of uncertainty in preventing risks from chemical exposures and the limitations of current scientific and decision tools based on the concept of risk assessment. It has outlined a new paradigm for decision-making for sustainability embodied in the precautionary principle. This approach has several key aspects:

- Expanding scientific tools to define more holistically and characterize risks in an interdisciplinary manner.
- Going beyond border of diagnosis to solutions – focusing on more comprehensive characterization of risks and uncertainties but placing equal emphasis on developing and studying alternatives that can reduce risks.
- Incorporating a diverse range of tools, options, stakeholders, and an ability to build on knowledge – the whole of the evidence – in making the most robust decisions under uncertainty.
- Redirecting research funding to greatly increase budgets for research and development of safer chemicals, processes, and products, and more holistic analyses of risks. For example, the U.S. EPA's budget for development of safer chemistry – green chemistry – amounts to about four million dollars, about the same as the cost of a laboratory cancer study for a single chemical!

The process of precautionary assessment outlined above is not meant to provide rigid rules for invoking the principle, but rather a structure for stimulating consistent and thoughtful application of precautionary thinking. Such a process – one that examines the whole of the evidence from various sources, examines a full range of alternatives, and injects sensible judgment and values – can produce more sound and sustainable decisions. In this sense, precaution is about how do we make more health protective decisions under uncertainty and complexity and how do we achieve the products, services, and technologies we want while minimizing their potential impacts. The ability to adapt current policies and approaches to decision-making that do not readily support precaution will require technical capacity to examine risks and solutions in new ways and capable institutions that comprehensively and fairly apply precautionary procedures. It will also require new mandates and guidelines so that a more precautionary framework and pro-cedures are broadly applied throughout environmental decision-making processes. And, ultimately, it will require important institutional changes in government and industry, and changes in the conduct of environmental science.

3.4 LIMITS OF RISK MANAGEMENT AND THE NEW CHEMICALS POLICIES

KEN GEISER

Lowell Center for Sustainable Production, University of Massachusetts – Lowell

Long before the development of the concepts of sustainable development the environmental impacts of the chemical industry were the subject of environmental concern. The production of caustic soda in England during the mid-19th century caused severe environmental damage from the waste hydrochloric acid released into streams and volatilized into the air. The damage to vegetation and soil was so severe that the British Parliament enacted the Alkali Act in 1863, authorizing special government inspectors to monitor and annually report on the firms behavior (Haber, 1958).

In the United States, early 20th-century concern over the environmental impacts of the chemical industry arose from the emerging fields of public health and sanitary engineering with labor activists as the primary source of attention for the occu-pational health issues. While acute poisoning and observable damage were well recognized, little was known about the more subtle, public health or environmental effects of many of the chemicals manufactured by the rapidly expanding inorganic and petrochemical industries. The professional fields of toxicology and pharma-cology were just emerging during the 1930s, with academic centers opening at the University of Chicago, Johns Hopkins, and Harvard University. In 1933, Dow Chemical Company opened a biochemical research laboratory to study the potential effects of chemicals on humans and other organisms and, two years later, the Du Pont Corporation established Haskell Laboratory, which soon became a world leader in studies of the toxic properties of chemicals. As the scientific information accumulated, the U.S. Public Health Service began issuing recommended environmental exposure standards for hazardous chemicals and the American Conference of Government Industrial Hygienists began publishing "threshold limit values" as recommended guidance for chemicals used in occupational settings (Geiser, 2001).

Environmental concerns were evident as well. During the 1920s the American Water Works Association documented many cases of water pollution from chemical production plants, and in 1935 Franklin Roosevelt's National Resources Committee listed the chemical industry as among the nation's most polluting. In 1945 Purdue University held its first annual conference on industrial chemical wastes and the following year the American Chemical Society sponsored a symposium on "Industrial Wastes: A Chemical Engineering Approach to a National Problem." However, even as concern about the hazards of the chemical industry and its wastes grew, the state and federal governments were slow to enact environmental and public health chemical risk management policies (Colten and Skinner, 1996; Tarr, 1996).

3.4.1 Conventional Chemical Risk Management Policies

The 1970s changed all of this as the emergence of the new environmental movement spurred national governments to enact sweeping new chemical pollution, waste, and product laws. The U.S. Congress enacted the Clean Air Act, the Occupational Safety and Health Act, the Clean Water Act, the Safe Drinking Water Act, the Consumer Product Safety Act, and the Resource Conservation and Recovery Act in a brief six-year period in the early 1970s. Similar laws were passed in Japan and many European countries during this same period. Dealing with chemical production wastes and pollutants and dangerous chemical products, all of these laws had an immediate effect on the chemical industry (see Desai, 2002, for brief case studies from different industrialized countries).

However, none of these statutes were so directly focused on the chemical industry as the 1976 Toxic Substances Control Act (TSCA) in the United States and the Sixth Amendment to the Dangerous Substances Directive enacted by the European Parliament in 1979. These statutes did not focus on the industry's wastes, pollution or occupational exposures as much as on the intrinsic hazards – the toxicity, chemical stability, and bioavailability – of the industry's chemical products as they were used in commerce. These laws were intended to provide government agencies with the authority to collect relevant health and safety data on chemical products, require testing where data were missing, and condition and restrict the use of chemical substances so as to reduce "unreasonable risks" to the public and environment.

These statutes comprised the primary government policies shaping chemical management practices. Both of these laws established a politically defined distinction between those chemicals currently on the market and the chemical industry's proposed new chemicals. Those substances manufactured and used in industrial production prior to 1980 (so-called "existing chemicals") were largely assumed to be safe unless some incident or scientific study proved otherwise. It was only new chemicals coming on to the market after 1980 that were subject to testing and government review. Both statutes provided government agencies with authorities to restrict the use of dangerous substances that posed health and environmental risks and to phase out the worst of them.

Over two decades of experience with these chemical management policies has revealed their weaknesses. While both statutes have had some success in ensuring safety reviews of new chemicals, their impact in terms of generating information on the toxicity of existing chemicals and restricting those considered of highest risk has been limited. While firms were expected to understand the risks associated with their existing chemicals and report on any concerning outcomes, industry has been slow to submit data and quick to cloak it under secretive confidential business information protections. In the United States, legal interpretations of TSCA have hampered risk management initiatives, often requiring long and expensive quantitative risk assessments to justify actions. For many years, little was done by the government to encourage firms to do testing and there was little enforcement of the provisions that required firms to report results from new testing.

Driven by public interest pressure, the U.S. EPA in 1998 inventoried the testing data available on the highest production volume (one million pounds or greater manufactured or imported per year) chemicals in commerce. The agency found that 97 percent of the chemicals considered were missing at least one of six basic toxicity screening tests and some 43 percent had virtually no human health or environmental screening data. Taking advantage of these reports, the U.S. EPA entered into a voluntary "challenge" with the American Chemistry Council and the nonprofit organization, Environmental Defense, to encourage the chemical industry to provide the basic screening data for some 2800 high production volume chemicals by 2005. This program has been modestly successful, with industry consortia "adopting" some 65 percent of the identified chemicals and, in 2002, beginning to submit "robust summaries" of the necessary data. The agency has been unclear about its plans for the remaining 35 percent of unadopted ("orphaned") chemicals, the 700 or more chemicals whose production has crossed the threshold of one million pounds since the original inventory, or the 6000 existing chemicals that are manufactured or used in annual quantities of 10,000 to 1,000,000 pounds (Dennison, 2004).

With so little scientific information on the risks of thousands of common industrial chemicals, many have continued to be used, even in large volumes and under conditions of significant human and environmental exposure, and only a relatively small number have received regulatory controls. Congressional hearings in 1983, 1988, and 1994 have highlighted the inadequacies of the U.S. EPA's existing chemicals program. In 1994 a report from the federal Government Accounting Office found that the U.S. EPA had placed regulatory restrictions on only five chemicals under its authority to prohibit the use of chemicals that "present or will present an unreasonable risk of injury to health or the environment." In overview, one recent administrator of the office overseeing TSCA concluded, "It is fair to state that the results [of TSCA] have come nowhere close to ... the original Congressional intent ... Although Congress has shown little interest in doing so, there are many examples of sections that need to be reformed and strengthened" (Goldman, 2002).

3.4.2 New European Chemicals Policies

After the passage of TSCA, little changed in the basic federal approach to chemical risk management in the United States. This is in marked contrast to other countries that have continued to develop new approaches to chemicals management policies. The Nordic countries, the Netherlands, Germany, and, more recently, the United Kingdom have set the standard for many of these developments with new laws and new programs focused on the chemical industry and its products.

Three basic principles have emerged as common themes in these policies: the "Polluter Pays Principle" clarifies who bears the costs for chemical contamination; the "Substitution Principle" encourages the adoption of the safest chemicals; and the "Precautionary Principle" promotes preventive action even in the face of the uncertainties of risks (see Section 3.3.2 for a more in depth discussion of the Precautionary Principle). Specifically, the new national chemicals policies of Northern European countries have relied on rapid screening tests for determining regulatory actions on chemicals, focused on products and product lifecycles for risk reduction, established lists of undesirable substances, and, in limited cases, employed government authority to phase out the use of the most hazardous substances such as lead, mercury, cadmium, brominated flame retardants and chlorinated paraffins (for a more extensive review, see Tickner and Geiser, 2003, www.chemicalspolicy.org).

Additional government attention in Europe has focused on chemicals in products. The limited facilities and high cost of municipal waste management in Europe have encouraged high rates of recycling for discarded products. With some national recycling rates approaching 50 and 60 percent, end-of-life product return policies are well accepted. First developed for beverage containers and packaging wastes, European-wide directives have been enacted for product recovery for used automobiles and discarded electronic products. Yet, the hazardous chemical constituents of some of these products have inhibited their effective recycling and reuse. Therefore, in order to facilitate the recycling of automobiles and electronic products, the European Union directives passed in 1999 and 2001, respectively, have mandated the phasing out of certain high hazard substances including lead, mercury, cadmium, hexavalent chromium (the electronics product directive also targets brominated flame retardants) in automobiles and electronic products (Fishbein *et al.*, 2000).

However, the boldest new chemicals risk management policy initiative in Europe is potentially the one most likely to seriously challenge the chemical industry. The drive to establish a common market throughout the European Union has forced a major reconsideration of European Union chemicals policies and create a unified, Europe-wide policy. The motivation for this reconsideration has emerged out of a deep sense of frustration with the inadequate information on chemical health effects, the slow pace of exposure-based risk assessments, the growing recognition that commercial products are loaded with hazardous chemicals, and some fairly well publicized lapses in government regulatory oversight.

In 2001, the European Commission issued a "White Paper on Future Chemicals Strategy" outlining a total overhaul of the European chemicals management policy (European Commission, 2001). The White Paper emphasized both increasing the testing of all industrial chemicals and the stricter management of a large number of particularly high hazard chemicals. The specific objectives put forward in the White Paper included the following.

1. Making the chemical industry more responsible for generating knowledge about chemicals, evaluating their risks, and maintaining their safety – establishing a new "duty of care."
2. Extending responsibility for testing and management of substances along the entire product manufacturing chain.
3. Substituting substances of very high concern with chemicals of lesser concern and driving innovation in safer chemicals as substitutes.
4. Minimizing the use of animal testing in determining the potential risks of chemicals.

The central recommendation of the White Paper was the establishment of a new integrated chemicals management scheme called REACH (Registration, Evaluation and Authorization of Chemicals). The REACH proposal would require government registration for all industrial chemicals, testing and evaluation of the largest volume chemicals, and specific government authorization for use of the most dangerous. During 2002 and 2003 the White Paper and the REACH proposal were extensively reviewed and debated and, in May of 2003, draft legislation was released by the European Commission for public comment. After receiving some 6000 comments from a wide array of interest groups, the final language was submitted by the Commission to the European Parliament in October. This final draft would require the following.

1. All chemicals in commerce marketed over ten tons per year would need to be registered with a new central European chemicals management agency, with companies providing basic exposure and risk data.
2. Chemicals produced or imported over 100 tons per year would be required to undergo rigorous new evaluation procedures conducted by the individual European Union member states.
3. Chemicals of greatest concern based on their inherent characteristics (carcinogens, mutagens, reproductive hazards, persistence and bioaccumulative potential) would need to be specifically authorized for continued use by the appropriate agencies of the member states or by the European Union (Tickner and Geiser, 2003).

The REACH proposal marks a broad reorientation of government policy on the use of toxic and hazardous chemicals. The proposal replaces 40 European Union chemicals regulations (including the Dangerous Substances Directive), wipes away the distinction between new and existing chemicals and shifts the burden concerning the testing and safety of chemicals from government to industry. By significantly expanding industry requirements, the European Commission is creating a huge new set of identification, testing, and management responsibilities for the chemical industry. Initially, most of the European chemical and manufacturing industry had been willing to go along with the proposal, recognizing that the current system simply does not work and that the chemical industry has a very poor public image in Europe. Even the German chemical industry (the largest in Europe) begrudgingly supported the general principles of the proposal. As the debates have grown and the American chemical industry registered its concerns, the European chemical industry became more aggressive and some of the original White Paper language has been modified; however, the legislative draft that is now up for consideration by the European Parliament retains a significant redirection of traditional chemicals risk management policy.

3.4.3 New International Chemicals Policies

As these new chemicals management policies have been emerging in Europe, they have been joined by a growing body of international chemicals management policies developed under United Nations agencies. Several of these have emerged in the form of international treaties, or international environmental agreements, drawn up by multinational negotiating conferences convened by the United Nations Environment Program (UNEP). The first such agreements drafted during the 1970s and 1980s focused on wastes and pollution, including treaties protecting regional seas ("Barcelona," "Oslo," and so on), prohibiting ocean disposal of wastes ("London Dumping Convention"), protecting the atmospheric ozone layer ("Vienna Convention"), or regulating international trade in hazardous wastes ("Basel Convention"). However, by the late 1990s, international treaties were also being developed that directly targeted the production and sales of products of the chemical industry.

The Rotterdam Convention on Prior Informed Consent was built out of earlier efforts by UNEP and the international Food and Agriculture Organization (FAO) to encourage voluntary product declarations by exporting companies of toxic substances (particularly pesticides) destined for sales in developing countries. In 1998 a drafting convention meeting in Rotterdam formally signed an agreement requiring that the export of certain listed chemicals could only take place with the "prior informed consent" of the relevant agencies in importing countries. The initial list of chemicals covers 22 pesticides and five industrial chemicals, although the convention permits the addition of more substances over time. This treaty was ratified by the required fifty countries in 2003 and came into force in February of 2004.

In 1996, the International Forum on Chemical Safety concluded that evidence was sufficient to call for international action to ban the use of 12 persistent organic pollutants (POPs). On the basis of this recommendation, UNEP convened a drafting committee in 1997 and, three years later, a final agreement was signed in Stockholm. This Stockholm Convention requires action plans and schedules for the elimination of the production and use of nine pesticides and polychlorinated biphenyls, the severe restriction of the use of DDT, and best control measures for reducing the generation of dioxin and hexaclorobenzene as inadvertent contaminants of certain production and disposal processes. As with the Rotterdam Convention this agreement permits more substances to be included over time. This convention was fully ratified and came into force in May of 2004 (Secretariate of the Stockholm Convention on Persistent Organic Pollutants, 2001).

By providing limited international authority for managing chemical trade and for phasing out the most dangerous substances, the Rotterdam and Stockholm Conventions have laid the basis for more comprehensive policies. In 2002, the administrators of the Basel, Rotterdam, and Stockholm conventions published a proposal for "clustering" their authorities to facilitate a more integrated lifecycle approach to the management of hazardous chemicals. Recognizing this, delegates to the 2002 World Summit on Sustainable Development called for a comprehensive new global strategy for the sound management of chemicals. In November of 2003, an opening conference on a "Strategic Approach to International Chemicals Management (SAICM)" was convened by UNEP in Bangkok. This conference drafted an initial outline for a strategy and set a schedule for drafting a new international agreement by 2005.

It is too early to assess where the emerging international policies on the management of chemicals will be going, but, given the United Nations' strong commitment to the idea of sustainable development, it can be expected that these initiatives will be seeking the development of a sounder and safer chemical industry. The industry will certainly be heavily involved in crafting those policies, but so will many government agencies and nongovernment organizations. There is a long history since the earliest efforts to control pollution and wastes from the chemical industry. Today, the chemical industry and its products have become the central focus of both national and international policy developments. How these policies develop and who bears what costs will substantially determine the prospects for a new, more sustainable chemical industry in the future.

3.5   IMPACTS TO HUMAN HEALTH AND THE ENVIRONMENT

3.5.1   The Challenge of PBTs for the Chemical Industry

JOANNA D UNDERWOOD

INFORM

Over the past decade, growing attention has focused on chemicals that share three characteristics: they are toxic, they persist in the environment, and they bioaccumulate in living organisms. This class of chemicals, called persistent, bioaccumulative toxins (PBTs), represents one of the greatest challenges for chemical companies that manufacture or use them, or that manufacture products that release them to the environment after disposal. In 2000, U.S. EPA made PBTs a top priority because of their ability "to travel long distances, to transfer rather easily among air, water, and land and to linger for generations in people and the environment".[2]

---

[2]U.S. EPA, "Persistent Bioaccumulative and Toxic (PBT) Chemical Program: Multimedia Strategy for Priority Persistent, Bioaccumulative and Toxic (PBT) Chemicals." Available at http://www.epa.gov/pbt/fact.htm (accessed September 2004).

The category of PBTs, as defined by U.S. EPA, includes mercury, lead, dioxins, and several dozen other substances. Mercury and lead have so far been the most widely targeted of these substances because, like other heavy metals (naturally occurring elements contained in rocks, sediments, and soils), they do not degrade at all. Polychlorinated dioxins are also PBTs, and while normally they are not purposefully manufactured, they are often generated as byproducts during the manufacture of chlorinated compounds and during incineration or combustion.[3]

Many PBTs, such as lead, mercury, and polychlorinated biphenyls (PCBs), have been well understood, though poorly controlled, for years. Newer chemicals, such as brominated flame retardants, are now emerging as key targets because of their prevalence in the environment, because they can bioaccumulate, and because there is growing evidence that humans are being exposed. (Environmental Working Group, 2003). Although U.S. EPA has not yet included brominated flame retardants on its official list of PBTs of concern, they are subjects of intense scrutiny because they have been found in human beings and the environment at rapidly increasing levels.[4]

Persistent bioaccumulative toxins pose both immediate and long-term threats to humans. The developing organ systems of fetuses and young children make them more vulnerable to the impacts of PBTs than adults.[5] In addition, indigenous populations and others who rely on locally caught fish and wildlife for food may be more exposed to certain PBTs than the general population.[6]

The specific effects on humans from exposure to PBTs can be subtle or highly visible. A look at four of these substances exemplifies these differences.

- *Mercury.* In human beings, even low concentrations of mercury can be harmful. Each year, nearly 630,000 newborns are estimated to be at risk of developmental problems resulting from their mothers' exposure to mercury.[7] Studies show that exposure to methyl mercury in the womb can cause neurological damage in the fetus and delayed development in young children. In the fetus, in children, and in adults alike, the greatest concern from chronic exposure to methyl mercury is damage to the central nervous system (U.S. EPA, 2001). A national study conducted by the U.S. Department of Health and Human Services found that "approximately 10 percent of women of childbearing age have blood mercury concentrations above the levels U.S. EPA considers safe" (Centers for Disease Control, 2003).

[3]U.S. Agency for Toxic Chemicals and Disease Registry, "Toxicological Profile for Chlorinated Dibenzo-*p*-Dioxins." Available at http://www.atsdr.cdc.gov/toxprofiles/tp104.html (accessed September 2004).
[4]U.S. EPA, "Brominated Flame Retardants to be Voluntarily Phased Out." Available at http://yosemite.-epa.gov/opa/admpress.nsf/0/26f9f23c42cd007d85256dd4005525d2?OpenDocument (accessed September 2004).
[5]U.S. EPA, "Persistent Bioaccumulative and Toxic (PBT) Chemical Program: Frequently Asked Questions." Available at http://www.epa.gov/pbt/faq.htm (accessed September 2004).
[6]U.S. EPA, "Persistent Organic Pollutants: A Global Issue, A Global Response." Available at http://www.epa.gov/international/toxics/pop.htm (accessed September 2004).
[7]Kathryn R. Mahaffey, U.S. EPA, "Methylmercury: Epidemiology Update." Available at http://www.epa.gov/waterscience/fish/forum/2004/presentations/monday/mahaffey.pdf (accessed September 2004).

- *Lead.* The persistence of lead in the environment (such as in homes where lead-containing paint was used before it was banned in 1978) is a key factor in the continuing problem of lead poisoning among children in the United States. Research indicates that children with high levels of lead in their blood suffer from impaired intelligence and are "frequently overactive, aggressive, more distractible, disorganized, and less able to follow directions" (Kurtin *et al.*, 1997). Follow-up studies show that these children have a "lower class standing in the final year of high school, with increased absenteeism, lower vocabulary scores, and impaired motor function" (Kurtin *et al.*, 1997). In adults, exposure to high levels of lead can result in damage to the renal, nervous, digestive, and reproductive systems.[8] U.S. EPA has classified lead as a probable human carcinogen.[9]

- *Dioxins.* Estimates of background levels of exposure to dioxins through food range from 63 to 210 picograms per day for the average person (a picogram is one trillionth of a gram). For many age groups, the average daily intake of dioxins exceeds the level that the World Health Organization recommends as acceptable to protect human health. Low levels of chronic exposure to dioxins have been linked to developmental problems of the brain and thyroid in children exposed in the womb (European Commission, 1999). Background levels of exposure may also be associated with greater susceptibility to infectious diseases in children (Weisglas-Kuperus *et al.*, 2000).

- *PCBs.* Exposure to PCBs can affect memory and learning ability in adults (Schantz *et al.*, 2001), and there is evidence that exposure in the womb can cause lower birth weight, smaller head circumference, poor motor reflexes, a greater inclination to startle, and neurological, behavioral, and developmental problems in children. Eating PCB-contaminated fish has been associated with decreased fertility. Exposure to PCBs can increase the risk of cancer, immune and endocrine system dysfunction, non-Hodgkin's lymphoma, diabetes, and liver disease.[10]

Persistent bioaccumulative toxins have done extensive and long-lasting damage to fish and wildlife in the Great Lakes region, where 43 severely degraded geographic areas (known as "areas of concern") failing to meet criteria set forth by the Great Lakes Water Quality Agreement of 1978 have been identified.[11] For example, despite the fact that PCBs were banned from commerce in the United States more than 25 years ago, they are still causing "reproductive problems, tumors, and

[8]U.S. EPA, "Health Effects of Lead," fact sheet. Available at http://www.epa.gov/dclead/EPA_Lead_Health_Effects_FINAL%20_12.pdf (accessed September 2004).

[9]U.S. EPA, Integrated Risk Information Center, "Lead and compounds (inorganic)," IRIS Summary (CASRN 7439-92-1). Available at http://www.epa.gov/iris/subst/0277.htm (accessed September 2004).

[10]Agency for Toxic Substances and Disease Registry, "Public Health Implications of Exposure to Polychlorinated Biphenyls (PCBs)." Available at http://www.atsdr.cdc.gov/DT/pcb.007.html (accessed September 2004).

[11]Great Lakes Information Network, "Areas of Concern in the Great Lakes Region" (accessed September 2004).

other deformities" in fish and wildlife in most of the Great Lakes, according to the National Biological Service (Minnesota Office of Environmental Assistance, 2002).

Numerous studies have linked a high incidence of fish tumors of the liver and skin to chemical contaminants such as polycyclic aromatic hydrocarbons (PAHs) in sediments of the Buffalo River in New York State.[12] Caspian terns – birds that live year-round in Michigan's Saginaw River and Bay, where there are elevated levels of PCBs and other persistent organochlorine pollutants and heavy metals – have experienced "an unusually high incidence of birth defects and poor reproductive success."[13] The Minnesota Pollution Control Agency found that "lead poisoning accounted for 26 percent of the dead loons sent to research centers for autopsy" (the loon is Minnesota's state bird). Many of these birds had lead fishing tackle in their stomachs (Minnesota Office of Environmental Assistance, 2002). And the list goes on.

Because PBTs do not break down, their levels in our environment are building up. Contamination of water and fish with dangerously high levels of mercury and other PBTs is a challenging national (and international) problem. According to U.S. EPA, fish and wildlife consumption advisories apply to 35 percent of the nation's total lake acreage (not including the Great Lakes) and 24 percent of the nation's total river miles, as well as 71 percent of the nation's coastal waters.[14] The United States has experienced a significant increase in consumption advisories over the past decade, in large part because of ongoing and extensive fish-tissue sampling. In 2003, 48 states – all but Alaska and Wyoming – were under fish advisories for mercury, up from 44 states in 1993, and approximately 12,069,319 lake acres and 473,186 river miles were under advisory.

Nationally, five PBTs – mercury, dioxins, PCBs, chlordane (an insecticide), and DDT – are largely responsible for about 98 percent of all fish consumption advisories, with 76 percent involving mercury.

Recent reports indicate that much more recently manufactured brominated flame retardants appear to be ubiquitous and are increasing in the environment (Sjodin *et al.*, 1999). They have been found around the globe, in the blood of workers at computer assembly and recycling facilities, in breast milk, in whale blubber, and in sewage sludge, sediments, and the air of office buildings (Swedish National Chemical Inspectorate, 1999). According to Environment Canada, "The levels of PBDEs [a form of brominated flame retardant] in North Americans appear to be doubling every two to five years" and "will become a problem" if releases of these substances into the environment are not halted (Betts, 2001).

How are people exposed to PBTs? Most of the public's exposure to chemicals such as mercury and dioxins is from food. However, INFORM research documented in 2000 that the majority of PBTs leaving factories leave in consumer products,

---

[12]U.S. EPA, "Buffalo River Area of Concern." Available at http://www.epa.gov/glnpo/aoc/buffalo.html (accessed September 2004).

[13]U.S. EPA, "Saginaw River/Bay Area of Concern." Available at http://www.epa.gov/glnpo/aoc/sagrivr.html (accessed September 2004).

[14]U.S. EPA, Office of Water, *Update: National Listing of Fish and Wildlife Advisories*, EPA-823-F-04-016, August 2004. Available at http://www.epa.gov/waterscience/fish/advisories/factsheet.pdf (accessed September 2004).

rather than in manufacturing byproducts and industrial waste. These chemicals are therefore likely to be released from trash incinerators, metal smelters, and other solid waste disposal facilities after product disposal. Also, PBTs are also discharged into water via sewage treatment plants when pesticides, products contaminated by mercury or containing mercury as a preservative, and other products are flushed down the drain. People can be exposed to PBTs in the buildings where they live, work, or go to school. The state of Vermont has documented that "health care facilities, educational and research institutions and businesses have experienced significant employee exposures, and incurred significant costs due to accidental mercury releases" (Vermont Senate Natural Resources and Energy Committee, 2001).

Overall, PBTs are most pervasive in consumer goods (U.S. EPA, 1998b). While the use of some PBTs (such as PCBs, DDT, lead in house paint, and mercury in most batteries) is banned or restricted in the United States, many substances with similar characteristics continue to be added to products or generated as byproducts during manufacturing or after disposal. A review (conducted by INFORM) of the most recent data from the expanded right-to-know programs in Massachusetts and New Jersey indicates that the quantity of PBTs going into products is 20 to 25 times greater than the amount generated as industrial waste (data from INFORM, 2000). Hundreds of consumer products contain PBTs; a few examples include electronics with lead solder, nickel–cadmium batteries, fluorescent lamps, thermostats and thermometers, vehicles and appliances with mercury switches, wood preservatives, plastics containing heavy-metal stabilizers, several pesticides, some industrial cleaning supplies, and diesel fuel. (An extensive list of PBT-containing products is available on INFORM's website at www.informinc.org/PBT.htm.)

The bottom line is that, whenever PBT-containing products are made, used, or discarded, the toxins they contain are very likely to be released into the air or water, adversely impacting people and wildlife in the near or long term in subtle or in very visible ways. Hence, the challenge to industry is to seek the most effective near-term steps to ensure that PBT-containing products are retrieved for reuse or recycling and, ultimately, that PBTs as a class are phased out of use in commerce as completely as possible.

3.5.2 Hormone Disrupting Chemicals (Endocrine Disruptors)

RICHARD A LIROFF[15]

Toxics Program, World Wildlife Fund

A remarkable change has occurred in the last fifteen years in scientific knowledge about development, molecular biology, and toxicology. This burgeoning information raises serious questions about the adequacy of existing methods for assessing risks from man-made and naturally occurring chemicals. Hormone disrupting

---

[15] *Richard A Liroff, a political scientist, is Senior Fellow in the Toxics Program at World Wildlife Fund in Washington, DC. The opinions expressed here are his own and not necessarily those of World Wildlife Fund.*

chemicals illustrate this new challenge. ("Hormone disrupting chemicals" frequently are referred to as "endocrine disruptors," substances that affect hormones produced by the endocrine glands. This essay deliberately opts for the broader term, which is more readily understood by nonscientists).

Hormone disruption refers to a chemical's ability to mimic or block the action of the body's own hormones, or its ability to interfere with normal hormone production or breakdown in some way. Chemicals that are able to disrupt sex and thyroid hormones have been particularly under the spotlight, but other hormones, such as those of the adrenal gland, may also be subject to disruption. The overall result appears to be damage to reproductive, immune, and nervous systems, and increases in birth defects, selected cancers, and learning disabilities. For the last three decades, there has been a disconcerting increase in the incidence of such human health disorders as breast cancer, testicular cancer, hypospadias (a birth defect where the urethra does not open at the end of the penis) and learning disabilities.

For some disorders registering increases, such as prostate cancer and autism, at least part of the rise is a consequence of clearly identifiable changes in diagnostic methods and definitions. Nevertheless, scientists suspect that some of these adverse trends in public health may be associated with exposures to chemicals in daily life. Much more is known about effects in wildlife and laboratory animals than about effects on human health, but knowledge about human health effects is growing. The accumulating evidence prompted the International Programme on Chemical Safety, a consortium of the World Health Organization and other United Nations agencies, to convene an expert panel to produce a "state-of-the-science" review of hormone disruptors. In 2002, the panel urged that research on human health effects be given a high priority, because there is strong "biological plausibility," based on evidence from laboratory animals and wildlife, that hormone disrupting chemicals are damaging human development and reproduction (Damstra *et al.*, 2002).

The history of DDT provides useful insight into how scientific understanding of hazards from toxic chemicals has shifted at the close of the 20th century and the dawn of the 21st. DDT has a prominent public profile. It represents the double-edged character of chemical technology; it was welcomed enthusiastically on account of its perceived benefits, but it ultimately was viewed more negatively as its adverse effects surfaced. While mostly used to protect crops from agricultural pests, DDT once was considered a miracle chemical for preventing illness and death from malaria. After its pesticidal properties were recognized in the late 1930s, it was deployed to protect military personnel during World War II. It was then the basis for an ambitious malaria eradication campaign in the 1950s and 1960s. The campaign fell short of its eradication goal in many places, but it nevertheless is credited with saving millions of lives.

Rachel Carson highlighted DDT's adverse environmental effects in her 1962 book, *Silent Spring*. Ten years later, in 1972, U.S. EPA banned DDT for agricultural use in the United States. U.S. EPA's decision was based on DDT's suspected carcinogenicity and evidence that it was thinning eggshells and contributing to population declines in bald eagles and other birds. As of 2004, DDT is used legally in about two dozen countries, where it is sprayed inside homes to control malaria. For such use, DDT historically has been regarded as relatively safe, in the sense that it is not acutely toxic to humans except in very large doses.

In 2002, the U.S. Department of Health and Human Services published an updated toxicological profile of DDT and its breakdown products (ATSDR, 2002). The profile underscores how much has been learned about DDT in the years since U.S. EPA's ban, particularly with respect to DDT's hormone-disrupting characteristics. For example, in two studies published in 1987 and 1995, the length of a mother's ability to breastfeed her children was found to be inversely related to the concentration of p,p-DDE ("para-para-DDE") in her breast milk (ATSDR, 2002). Research published in 2001, based on studies of American women in the 1960s, found increased odds of having pre-term and low birthweight ("small-for-gestational-age") infants among those women having elevated blood concentrations of DDE (ATSDR, 2002). Since it is generally accepted that pre-term birth and low birthweight can contribute to infant mortality, U.S. EPA's decision to ban the use of DDT in 1972 might have increased the survival odds of American infants, even if the agency did not realize it at the time. Conversely, since blood concentrations of DDE have been found to be high in countries that use DDT for malaria control, the toxicological profile reports that "adverse reproductive outcomes may be a concern for women in countries where these chemicals are still used" (ATSDR, 2002).

The remarkable recovery of bald eagle, peregrine falcon, and brown pelican populations in the last 30 years provides strong evidence that U.S. EPA did the right thing in outlawing DDT in 1972. Yet the DDT tale also raises the important public policy question about how much should be known or suspected about a chemical before steps are taken to remove it from the marketplace. While this topic is the subject of more detailed discussion in Section 3.3.2 of this book, it is worth noting that scientists frequently find only associations between exposures and effects. They do not always know exactly how a chemical works, that is, its "mechanism of action." For example, the WHO expert panel determined that "there is strong evidence that eggshell thinning results from exposure to DDE," but added "there is continued uncertainty with respect to the precise mechanism of action of DDE" (Damstra et al., 2002). This begs the question of what might have been the fate of America's bald eagle population if regulation had been contingent upon more precise scientific knowledge about how DDE works.

The DDT story is instructive in other ways as well. For example, scientists historically have stressed that "the dose makes the poison." A corollary of this belief is that while high doses of a chemical may be toxic, lower doses are less likely to be. But scientific evidence gathered in the last 10–20 years indicates that this dogma requires modification – that it is not only the dose that makes the poison, but the timing. For example, in reporting on DDT and its breakdown products, the toxicological profile indicated that adverse developmental effects depended on the dose administered, the timing of exposure during or after gestation (the period in the womb), and the specific chemical administered (ATSDR, 2002). Some of the DDT-related chemicals were associated with female hormonal effects in the reproductive system. One was associated with anti-androgenic (compromising of male hormone) effects.

The importance of timing in making judgments about potential hazards from chemicals was underscored by publication in 1993 of the landmark National Academy of Sciences (NAS) report, *Pesticides in the Diets of Infants and Children* (NAS, 1993). The NAS report found both quantitative and qualitative differences in the toxicity of pesticides between children and adults. The report noted "special windows of vulnerability – brief periods early in development when exposure to a toxicant can permanently alter the structure or function of an organ system" (NAS, 1993).

A principal public policy consequence of the NAS report was the enactment of the Food Quality Protection Act of 1996. This law, amending the Federal Food Drug and Cosmetic Act and the Safe Drinking Water Act, not only called upon U.S. EPA to take special measures to protect children from pesticides, but also required U.S. EPA to establish a screening and testing program for hormone-disrupting chemicals. As of 2004, no such screening and testing requirements had been promulgated. Public and private sector laboratories around the world have been collaborating in designing and validating suitable screens and tests. Progress has been slow in part because so much of what is known and is being learned about hormone disruption is at the frontier of developmental biology and toxicology.

Although a formal screening and testing program is still several years in the future, industry will still face substantial pressure from a concerned public, particularly as biomonitoring programs – measurements of chemicals in human blood, fat, and urine – proliferate. Monitoring contaminants in breast milk provides not only a measure of a newborn infant's exposure, but also an indication of the chemicals that babies have been exposed to in the womb as their brains, testicles, uteruses, and other vital organs have developed. As more and more chemicals are found in mothers' milk, parents will ask with increasing urgency whether the health of their children has been compromised by the chemical industry's products.

3.6 IMPACTS OF, AND ISSUES ASSOCIATED WITH, CHEMICAL PRODUCTION FROM MANUFACTURE TO FINAL USE AND DISPOSAL

RICHARD SIGMAN[16]

Organisation for Economic Co-Operation and Development (OECD)

3.6.1 Use of Natural Resources and Releases to Air and Water and Waste Generation

Over the entire life of a product produced by the chemical industry (from "cradle to grave") there is a *potential* for a negative impact on man and the environment (Figure 3.1). First, as a user of raw materials (e.g., natural gas, coal and coke, minerals, fuel oil, liquefied petroleum gas) as a source for energy and feedstocks, the

---

[16]The opinions expressed in this paper are those of the author and do not necessarily represent the views of the OECD or of the governments of member countries.

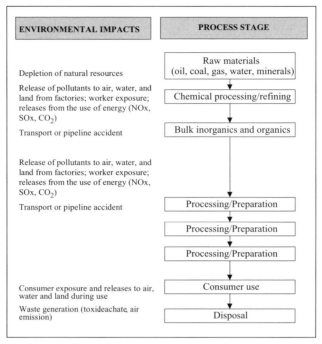

Figure 3.1. Potential impacts from the chemicals industry. (Note: This figure is a modification of figure 13 from OECD, 2001b).

chemical industry can impact on the supply of nonrenewable resources. And, as these materials are in general based on hydrocarbons, their combustion can lead to emissions of carbon dioxide (CO_2) – a greenhouse gas – and volatile organic compounds (VOCs), as well as nitrogen oxides (NOx), which contribute to the formation of tropospheric ozone or "smog." Processing the raw materials and feedstocks can result in the release of hazardous pollutants to the environment (e.g., benzene emitted from a factory) as can their actual use, either by other industries or consumers (e.g., benzene in gasoline emitted during fuelling of automobiles). Finally, hazardous waste can be generated by the chemical industry as a byproduct of manufacturing and from products that work their way through the supply chain and are eventually disposed of after final use.

The chemical industry is a major energy user (7 percent of world energy use in 1998) (IEA, 2000), and yet it contributes only 4 percent of overall emissions of CO_2 from fossil fuel combustion (IEA, 1999). However, when compared to other industries (e.g., pulp and paper contributes just 1 percent), the chemical industry in OECD[17] countries is a major industrial emitter of CO_2. Over the last 15 years,

----

[17]The Organisation for Economic Co-Operation and Development (OECD) is an intergovernmental organisation made up of 30 member countries from North America, Europe and the Pacific, as well as the European Commission.

the chemical industry has nonetheless made important energy efficiency gains, resulting in a stabilization of CO_2 emissions at a time when production has been increasing. However, according to the OECD Reference Scenario,[18] global chemical industry emissions are projected to increase in the future, primarily because of growing chemical production in non-OECD countries, which use less energy-efficient technology and are more reliant on coal as a fuel. However, if greater energy efficiency gains are achieved in the chemical industry, CO_2 may increase at slower rates or continue to stabilize in OECD countries.

Another major raw material used by the chemical industry is water. Compared to all other manufacturing industries, the chemical industry in OECD countries is the largest consumer of water; however, agriculture is a much larger user of water than all manufacturing industries put together (OECD, 2001a).

Overall, the chemical industry in OECD countries has made significant progress in reducing releases of pollutants to the environment that result from manufacturing processes. Although there are no consolidated data on emissions of known *hazardous* substances across OECD countries, it is probable that, overall, such releases from the chemical industry in these countries are declining. Over the last two decades, the industry may have greatly reduced its releases of hazardous substances per unit of output, but, compared to other industrial sectors (e.g., electronics, automobile, textiles), it still ranks high today in the intensity of the toxic chemicals and bioaccumulative metals it releases to air and land in terms of weight of emissions per production output (Hettige *et al.*, 1994). The situation on releases in non-OECD countries is unclear since no past trends data are available.

Few global data are available on the total contribution by the chemical industry in OECD countries to the release of substances that promote the formation of tropospheric ozone (VOCs, NOx) and acid rain (SOx) and the generation of hazardous waste. However, reported data suggest that emissions from the chemical industry are generally decreasing due to technological changes that are influencing energy use and the operation of chemical plants (OECD, 2001b). Since the adoption of the Montreal Protocol in 1987, tremendous progress has been made in phasing out the production and consumption of chemicals that deplete the stratospheric ozone layer (e.g., CFCs).

The chemical industry contributes to the generation of total industrial or manufacturing waste in several ways. First, hazardous substances generated during manufacturing may be disposed of on land or incinerated, or treated by physical/chemical means. Materials can also be recovered from this waste and they can be used as a source of energy. Hazardous chemicals produced by the chemical industry and incorporated into products that work their way through the supply chain will eventually be disposed of after final use. In addition, the chemical industry produces

---

[18]The Reference Scenario was developed for the OECD *Environmental Outlook* report using the OECD JOBS model and the PoleStar Framework of the Stockholm Environment Institute – Boston. For more information on the assumptions used in the Reference Scenario and the specifications of the modelling exercise, please see Annex 2 of the OECD *Environmental Outlook* (2001a).

nonhazardous waste. Unfortunately, only limited information is available that distinguishes total hazardous waste generated by industrial sector.

The final use of chemical products can result in the release of hazardous substances affecting man and the environment. Chemicals can be released from products used in the indoor environment, released to the atmosphere when a product is used outside, and released to groundwater if the chemicals leach out of a product in a landfill. Data are extremely limited on how many chemical products there are on the market today, on their chemical content, and on whether – and how – they may be releasing hazardous substances to the environment. Recent studies have shown that, in some cases, the majority of emissions released during the entire lifecycle occur during the service life of the product, rather than during production. A study conducted by TNO estimated that over 75 percent of total emissions to the environment from phthalates in plastics, brominated flame retardants in plastics, and zinc from tires occurred during the service life (TNO, 2001).

3.6.2 Environment, Health and Safety Issues, and Policies

The OECD *Environmental Outlook for the Chemicals Industry* (2001b), which examined the past and expected economic trends for the chemical industry and its environmental impacts, concluded that while good progress had been made in developing methodologies for (1) generating and collecting data, (2) conducting risk assessments, and (3) making risk management decisions, these methodologies have mainly been developed for and applied to reducing releases of pollutants to the environment that result from manufacturing processes. The *Outlook* concluded that "Overall, the chemical industry in OECD countries has made significant progress in reducing releases of pollutants to the environment that result from manufacturing processes." However, when it comes to the thousands of chemicals that are sold or used in products today "limited information exists on the volumes released to the environment, the targets of exposure and the toxic properties." As a general rule, the amount of information on the releases that occur during each of these stages of the chemical product lifecycle varies widely. In general, the best information is available for the production stage of a chemical, with less information being available as the chemical works its way through commerce (e.g., as it is transformed into preparations and consumer products).

Historically, most of the management approaches used for controlling emissions during production have dealt with "end-of-pipe" solutions. Recently, governments and industry have been considering more holistic approaches to minimize impacts on health and the environment throughout the lifecycle of a product by designing more environmentally benign chemicals and adopting integrated product policies, including extended producer responsibility.

Considering these past and expected trends, OECD has launched a new initiative, called "Chemical Product Policy" (CPP), to support Member countries and others manage lifecycle environmental risks from chemicals. The first phase of this initiative involved identifying activities that could support each phase in the chemical management process (i.e., data generation/collection, risk assessment, and risk management) for each stage in the chemical product lifecycle (i.e., from production of a chemical through to the use of a final product that contains that chemical – (e.g., an article) – recovery and recycling of the chemical where possible, and eventual disposal of that product).

To learn more about how chemicals in products work their way through commerce, and to what extent chemicals are released to the environment along the way, during the second phase of the project on CPP, OECD will conduct case studies to track, along the supply chain, specific chemicals/products from manufacture to final consumer. This will examine, among other things, the various uses and exposures along the supply chain for a chemical/product. The CPP project is also developing methodologies for *estimating* such releases by using emission scenario documents (ESDs).

This initiative by OECD and its member countries will bring a new perspective to chemical management that will identify new approaches and build on existing experience to manage chemical risks in a more comprehensive and cost-effective manner.

3.7 CLOSING THE GAP ON CHEMICAL PLANT SECURITY[19]
(The Industry is Building a Security-Conscious Culture, but Some Facilities are Behind the Curve)

AGNES M SHANLEY

Pharmaceutical Manufacturing Magazine

> You can't put a dome over a chemical plant, but you can take reasonable steps to keep it safe
>
> Security consultant Richard Sem

Since September 11, and the discovery of Al Qaeda's experiments with chemical weapons, the chemical industry has made tremendous strides in improving the security of its facilities, infrastructure, and IT. Chemical production plants pose a potentially huge security risk. According to the U.S. EPA, acts of terror at any one of 123 plants nationwide could threaten over 1 million people; at each of 700 facilities, a terror strike would risk 100,000 human lives; and at each of 3000 facilities, at least 10,000 lives could be affected.

At the point this was written in 2004, there are no national security regulations (Box 3.2), although a change in Administrations could alter this picture. Nevertheless, members of the American Chemistry Council and the Synthetic Organic Chemicals Manufacturing Association have completed site vulnerability assessments, and are implementing security improvements at their plants. This year, the industry has worked to synchronize efforts in facility and IT security, and to take

---

[19]A previous version of this article appeared in *Chemical Processing Magazine*, February 2004.

action to assure that process-control and IT departments are on the same page when it comes to security (Box 3.3).

Currently, however, only a fraction of the chemical processing and storage facilities in the United States are members of the trade groups spearheading these efforts. "At this point, it's safe to say that most chemical plants haven't even done site vulnerability assessments. They're behind the curve," says Richard Sem, a 34-year security veteran and former senior Pinkerton executive who now runs his own consulting firm in Plainfield, IL. Atlanta-based consultant Sal De Pasquale, who helped develop one of the most widely used site vulnerability assessment methods, puts it bluntly: "At most chemical plants handling hazardous materials today, existing security simply could not stand up to a guy with a six shooter and a bomb."

For some companies, the problem may not be assessment, but the next, and most difficult step: implementation. Some plants are dragging their feet, due to budgetary and political issues, Sem says. An average sized chemical plant can easily spend from hundreds of thousands to millions of dollars for site security – then, some plants have fence lines over a mile long, he adds. "Right now, many companies know they should do more but can't see the forest for the trees." Generally, the most critical business processes should determine how to prioritize the intensity and focus of security efforts," says Troy Smith, senior vice president and practice leader for IT Security with the risk-assessment specialists, Marsh Inc. (New York City).

Environmental groups are skeptical that enough chemical companies are taking sufficient steps. Greenpeace first raised this issue in 1992, pointing out surprising gaps in security when its members gained access to sensitive process-control equipment and restricted areas at a number of facilities. Problems persisted at some facilities. "At one large chlorine storage facility in Jersey City, the only thing the company did was to take its sign down and put up a 'no trespassing' sign," says Rick Hind, spokesman for Greenpeace USA.

Journalists have also "visited" facilities after hours, raising public awareness. As late as November 2003, a report by the television news magazine, *60 Minutes*, found significant security loopholes at chemical plants near New York, Los Angeles, Chicago, and Houston. At one facility outside of Pittsburgh, PA, owned by the Neville Chemical Co., reporters found an open gate near an anhydrous ammonia facility and no additional protection for a boron trifluoride storage unit. More disturbing, perhaps, was the fact that, until someone at the plant called the police, who arrested the reporters on trespassing charges, some employees initially acknowledged their presence without taking any steps to remove them from the premises.

Neville was in a situation that many companies may find themselves in today. It had completed a site vulnerability assessment that found weaknesses in perimeter security, with unlocked gates in need of repair along the rail line that runs through the plant site.

The company had already taken concrete steps to improve security and had an employee training program in place, says Jack Ferguson, vice president of manufacturing and plant manager. After September 11, Neville had decreased the maximum

---

BOX 3.2   SECURITY LEGISLATION LAGS

Efforts to develop comprehensive national security regulations for chemical facilities remain on a back burner. The first comprehensive chemical plant security legislation (S 1602) proposed by Senator Jon Corzone was blocked after it had been unanimously approved in the Senate; industry currently supports a bill first proposed by Senator James Inhofe (R-Okla.), chairman of the Senate's Environmental and Public Works Committee.

A major issue with Corzine's bill was its requirement that manufacturers substitute "inherently safer" products for chlorine, ammonia, and hydrogen fluoride. Industry argues that the concept of safer product substitution has already been built into key safety and environmental regulations, including the Occupational Safety and Health Administrations Process Safety Management standard. Industry's stance has provoked criticism from environmental groups. "There seems to be an ideological issue here," says Jeremiah Baumann, policy director at the U.S. Public Interest Group. "Industry wants to draw the line between regulating releases from its plants and regulating the types of chemicals it uses, even though such regulations on the state level have resulted in dramatic reductions of waste, and saved the companies money."

BOX 3.3   CYBERSECURITY: GETTING IT AND PROCESS CONTROL ON THE SAME PAGE

As they work to improve their physical plant security and to harmonize product stewardship programs to reflect ACC's Responsible Care code, the trade groups representing chemical plants are also addressing cybersecurity. Coordinating these efforts is the chemical sector cybersecurity program (CSCP), which works with 10 trade groups representing more than 2000 chemical companies.

A key effort for the group, which is working closely with the chemical industry data exchange (CIDX) on this project, is getting process control and IT to work more closely together to solve IT security problems. It has not been easy, since, at most chemical companies, process control and IT are separate worlds. "In many companies, process control groups don't all report through a single coordinated executive. There's a lot of autonomy within plant sites," says CSCP's manager Christine Adams.

"Each group has different vendor sets and different requirements at the plant level and you need to integrate them," says Jeff Frayser, director of cybersecurity at CIDX. "You don't want an operator locked out at 3 a.m. during a crisis because his IT password has expired."

A process control team within the program and CIDX has been working with ISA to develop guidance and practices to supplement SP-99, a cybersecurity document published in 2003. CIDX is already working to integrate physical and cybersecurity assessments that companies can deploy to complete their assessments on site. The new project will add guidance for process control, which, Frayser says, had been somewhat overlooked in the past.

CSCP tackled a number of projects in 2004, including:

- Supporting guidance documents based on ISO 17799, which were first released in 2003.
- Working with Sandia National Labs, Albuquerque, NM and AIChE's Center for Chemical Process Safety (CCPS) to develop a combined methodology for physical and cybersecurity methodology.
- Studying ways to develop Information Sharing and Analysis Center capabilities, through the ACC's Chemtrec program, to enhance the repository of validated information on cybersecurity threats.

Vulnerabilities Remain

Networking continues to be a sore spot at many chemical companies, Adams says. It is essential to ensure that software is updated with the latest patches. Another weakness is remote access, she says. Some companies may be operating or maintaining systems remotely, but need to have the right firewall and protection points in place.

Cybervandalism is another area of growing concern. "Companies may not even realize that attacks are being launched from their own servers," says Troy Smith, senior vice president and IT practice leader at Marsh Consultants (New York City).

quantity of boron trifluoride on site by 50 percent, and the average quantity on site by 65 percent. It had also moved on improving perimeter security, negotiated with the railroad, developed a plan, and put out design bids for the gate repair work. "New gates were not the answer," Ferguson says, "we needed new fences encompassing a rail spur."

Once Neville found that railway track repairs were also needed, the company applied for funding from the Federal Rail Administration and decided to postpone the repair work, as chief operating officer Thomas McNight explained in a letter to *60 Minutes'* parent network, CBS. When Neville learned that funding would not be granted, the company decided to go ahead with the repairs anyway when the "visitors" arrived. Neville installed the gates in November, Ferguson says, and has also completed plans for work required to meet Coast Guard regulations for waterfront facility security. If the staff had not recognized the *60 Minutes* crew, he says, they would have reacted right away.

Ferguson would not comment on economic pressures. "Security measures have to fit plant location, raw materials and products. There can't be a 'one size fits all' approach," he says.

3.7.1 Doing the Right Thing

The industry has taken a lot of heat since the story aired, observers say. "These companies are under a lot of pressure right now and are trying to do the right thing," says Gary Salmans, senior vice president and leader of Marsh's critical incident prevention practice. "Companies that we work with are extremely proactive."

However, others believe industry can learn some valuable lessons from the report. For one thing, more companies need to move from "response" to "deterrence" mode, Sem says. The first deterrent, and the most powerful and inexpensive solution, is employee awareness training, yet it is often overlooked.

Other obvious steps are ensuring that gates are locked and well lit, and that outsiders are not permitted on site, or into sensitive areas. "I often test security awareness by trying to gain access to control rooms," Sem says. "In some cases, until staff have been adequately sensitized and trained, I have been allowed to walk in without question."

Lighting should not be neglected, either. It is usually fine near operating areas, Sem says, but can be quite poor along fence lines. Technology can also help. Sem recommends closed-circuit TV monitors with good motion-detection capabilities. "CCTV makes intruders worry they're being watched and greatly expands the range of security officers. Instead of having 10 monitor a site, you only need one or two."

3.7.2 The Human Factor

For many companies, the failure to consider "the human factor" within their facilities is one of the biggest obstacles to improving security. "All the hardware in the world can't address this," says Marsh's Salmans. "You can buy the most sophisticated control system in the world, but if someone props the plant door open – [a surprisingly common ventilation practice at some facilities] – where is the benefit?" he asks.

"The threats are internal, not external, and can play havoc with systems, like when a disgruntled employee changes a process recipe," says Marsh's Smith.

Security requires looking more closely into hiring and training practices. "Companies are getting much better on this front, particularly in contractor hiring, which had been a real Achilles heel two years ago," Sem says.

However, there have been cases where companies have hired unpredictable workers without a criminal record. In early 2004, in Texas, for example, a contract security guard at a BASF ammonia facility apparently shot himself, then blamed it on a foreign intruder, only to be arrested on other unrelated charges two weeks later. His story had been covered widely by the media, raising public concerns about plant safety.

Around the same time, Pilot Chemical Co. discovered that a senior Houston plant foreman and exemplary employee of 20 years was a fugitive from California with a burglary record. Most companies do background checks, and more than half do criminal record checks on potential hires, but some may only do them for their specific region, Sem adds.

Overall, many chemical plants are making great progress on the long, hard road to improved facility security. Proactive companies and members of ACC are "ahead of the curve," says Marsh's Salmans, and the industry is shifting its focus from crisis management to critical incident prevention (see Section 6.6).

In addition to employee training, threat analyses and vulnerability assessments, priorities now, Salmans says, are optimizing plant emergency operation plans, and integrating them with local emergency services. "Instead of depending solely on the government, each plant, post-9/11, needs to be independent to some degree," he says. "That need wasn't there in the past" (Box 3.4).

Midland, Michigan, and Wilmington, Delaware, home to Dow and Dupont, have become models of preparedness. Both companies work closely with local emergency responders and law enforcement, and have staged several major Weapons of Mass Destruction (WMD) drills. Dow is also focusing on bioterror readiness.

3.7.3 Implementation is a Weak Point

Most chemical companies could improve the way they are handling security implementation, Sems says, since it is easiest for a security program to fall through the cracks at this stage. Sem says, "The low hanging fruit, like perimeter fences, access control and lighting, can be addressed easily enough, but longer-term fixes can hit budgetary walls as companies move from crisis to crisis." Implementation needs a budget and management support.

Third parties can be extremely helpful in vetting site assessments and implementation plans. Currently, as site assessments have been set up within the Responsible Care Code, local law enforcement and fire chiefs can look at facility plans. They may be very familiar with the emergency response aspects of a site security plan, but they may not always be the best choice when determining other security strategies or assessing investments.

Management buy in is another critical factor for assuring a successful security plan. In particular, companies need to ensure that security is an integral part of their management of change program, Sems says.

3.7.4 Need to Look Offshore

National security regulations, if they pass, will underscore the need for chemical companies to do more. Current draft regulations could require temporary shutdown of plants that do not meet security standards, and penalties of up to $300,000 per day of noncompliance. In addition, U.S. companies with facilities abroad will have to address vulnerabilities at these plants, which, at this point, may be even more attractive terror targets than domestic plants, Sem says.

BOX 3.4 SMALL COMMUNITIES "ON THEIR OWN"

Chemical companies need to realize the new post-9/11 realities for the small communities where many of their facilities are based. These issues were summarized in a report issued in March 2003 by the Joint Committee on Accreditation of Healthcare Organizations (JCAHO) called "Healthcare at the Crossroads: Strategies for Creating and Sustaining Community-Wide Emergency Preparedness Systems." The report found smaller communities especially vulnerable to biological, nuclear and chemical terror attacks, pointed to typical gaps and flaws in most emergency response systems, and suggested responses.

A key finding was the need for clear tie-in between police and fire fighters, local chemical manufacturing plants and medical facilities, and a clear grasp of "surge capacity," the additional resources that would be needed in the event of a terror emergency. The study suggested developing a strong community organization made up of representatives from local government, emergency management, industry, police fire utilities, and emergency medical services. It also suggested that teams adopt fluid incident management approaches that allow for simultaneous involvement by the different groups involved.

The best part about the site vulnerability analyses developed for chemical plants, Sem says, is that, while addressing terror risks, they also force companies to look at issues that are far more likely to occur: workplace violence and theft of trade secrets. "Companies shouldn't get so hung up on terror that they ignore internal risks."

Given the wide range of threats today, it would be impossible for companies to prevent all possible scenarios. "You can't put a dome over a chemical plant, but you can take reasonable steps to keep it safe," Sem says.

3.8 ECONOMIC ISSUES AND COMPETITIVENESS

PETER H SPITZ
Chemical Advisory Partners

A combination of factors is bringing about a major change in world patterns of chemical trade, with negative consequences for the traditional chemical producers, who are facing stiff competition from newer producers in the Middle East and Asia. This comes at a time when there is a growing world sentiment to advance Sustainable Development (SD) issues. While some advances towards SD will, in fact, help traditional producers become somewhat more competitive, their drive to achieve the lowest possible manufacturing costs in general runs counter to an SD-favorable move to renewable materials as feedstocks, even if technologies for such processes become available. Historically, the often multistage fermentation-type technologies employed to make chemicals from biomass are substantially more expensive than processes using hydrocarbon feedstocks.

Chemical producers in the United States and, to a greater or lesser extent, in the other OECD countries are facing a rapidly changing dynamic with respect to their traditional favorable balance of trade with Asia, as will be explained in more detail below. This situation will in some respects make it less attractive for these producers to implement certain aspects of sustainable development, although other aspects will continue to make economic sense. The fast rising production of petrochemicals from inexpensive natural gas in the Middle East and elsewhere can be considered as favorable to SD in one respect in that natural gas is considered a more plentiful worldwide resource than crude oil, the currently more widely used petrochemical feedstock. Conversion of so-called "stranded gas" in remote locations is another case where natural gas is replacing crude oil in the manufacture of products such as diesel fuel and kerosene, among others. On the other hand, while some firms in the United States and elsewhere are working on technologies to produce chemicals from renewable, natural (i.e., biomass) feedstocks in an effort to reduce the use of (scarcer) hydrocarbon feedstocks, their efforts, even if successful, will be more than counterbalanced by the increasing amount of petrochemicals (still) made from cheap natural gas in the Middle East and parts of Asia, which will satisfy much of the normal global growth expected for these materials. (Figure 3.2).

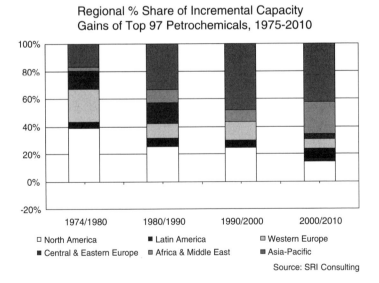

Figure 3.2. The growing importance of Asia and Latin America in petrochemicals. (*Source*: Kevin Swift, American Chemistry Council, presentation to CDMA meeting 28 April 2003.)

This section chapter will mainly concentrate only on the effect of changes in manufacturing economics with the attainment of progress (or lack thereof) toward sustainable development.

The unfavorable balance of trade in chemicals, notably as experienced in the United States (Figure 3.3) has come about not only because certain *commodity* petrochemicals produced there are now less competitive globally as a result of higher natural gas costs, but also because such Asian countries as China and India have now become substantial, reliable exporters of *specialty chemicals* and *pharma intermediates*. The latter development probably does not have important SD consequences, but is a sobering fact for North American, European, and Japanese chemical firms. The more serious problem for U.S. Gulf Coast producers is the fact that exports of certain natural gas derivatives are now in serious jeopardy due to their relatively high manufacturing cost and can no longer be used to keep plants at high operating rates, as was the case in the past. U.S. companies have historically exported large amounts of ethylene derivatives to Asia and elsewhere. These exports have now declined substantially and, in fact, there are growing imports of these materials, both as chemicals and polymers as well as in the form of fabricated products (e.g., plastic shopping and garbage bags). Ethylene production costs in the Middle East are far lower than those in other important producing areas, and this advantage is transformed into low-cost polyethylene (Figure 3.4). Comparing Asian versus U.S. prices for polyethylene resin shows a typical 10–15 cents per pound disadvantage for U.S. producers for certain commodity grades (Copley *et al.*, 2002).

There is almost no production of ammonia and methanol in the United States any more, with such countries as Trinidad, Chile, and Saudi Arabia having become key suppliers. With natural gas expected, in the future, to stay in the range of $4–5 per million BTU versus 50–75 cents per million BTU in countries with plentiful gas and less internal use, ammonia and methanol will largely be produced abroad, having become global commodities with substantial quantities imported into the United

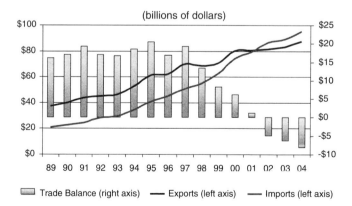

Figure 3.3. U.S. balance of trade in chemicals (1989–2004). (*Source*: Kevin Swift, American Chemistry Council, presentation to CDMA meeting 28 April 2003.)

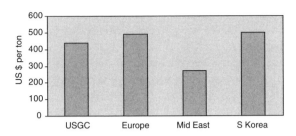

Figure 3.4. The Middle East has a major comparative advantage in the production of ethylene and polyethylene. (*Source*: Philpot, 2003.)

States. Canada, which for a long time enjoyed very low natural gas prices, is now such a large gas supplier to U.S. utilities, industrial users, and residential/commercial users, that Canadian gas prices now are not much lower than those below the border. Thus, Canada is no longer competitive with the above countries as a producer of ammonia and methanol.

The situation is different with respect to chemicals made from propylene and from benzene-toluene-xylene (BTX) aromatics. The crackers making ethylene in the Middle East and Asia from natural gas liquids produce much less propylene than ethylene, with additional propylene to some extent made via a relatively costly process involving dehydrogenation of propane. These two facts make propylene relatively less plentiful and more expensive in Asia, a favorable situation for U.S. Gulf Coast producers of propylene derivatives. The BTX derivatives are largely based on crude oil-derived naphtha, which is subject to global pricing. Thus, U.S. and European producers of styrene and other such aromatics derivatives are on a more level worldwide playing field.

Steps that U.S. and other OECD producers can take to deal with the competitiveness issue related to natural gas pricing are not strongly relevant to SD, although improving energy efficiency and closing uncompetitive, smaller plants are steps in the right direction. Clearly, improving the energy efficiency of plants that are based on now more costly natural gas becomes even more important in terms of reducing production costs, while at the same time implementing an important principle of SD. And certain uncompetitive plants subject to closure often have lower energy efficiencies, meaning a more wasteful use of raw materials and energy sources than more modern plants.

Improving plantwide energy efficiency can take many forms. As an example, BASF's *Verbund* structure has resulted in schemes to link units that produce energy (e.g., in the form of byproduct high-pressure steam) with units that consume steam, resulting in maximum energy efficiency (Spitz, 2003).

The chemical industry is a very large consumer of energy and much is needed to be done to reduce consumption (see Section 3.6 for additional energy reduction statistics). The largest consumers of energy in the chemical industry are represented by six "chemical chains," namely agricultural fertilizers, ethylene, benzene/toluene/xylene, chlor-alkali, propylene, and butadiene. Together, they consume 1646 trillion

BTU per year, representing 54.3 percent of total energy use by the process industry (Johnson, 2000). The US DOE Office of Industrial Technologies (OIT) has given 50–50 matching grants totalling $170 million per year to companies that conduct pilot projects to reduce energy consumption. A report "Energy & Environmental Profile" of the U.S. chemical industry was produced by OIT to guide decision-makers as to which pilot projects should be funded.

The use of coal as a petrochemical feedstock needs to be examined in the context of a world where producers consider steps toward SD. The advantage of using coal resides in the fact that these resources are vast and less subject to depletion in the long run than hydrocarbons (i.e., crude oil and natural gas). Two problems, however, arise. First, while researchers have developed technologies to make "petrochem-icals" from coal, the economics are relatively unfavorable, except where coal is extremely cheap and subsidies are available (mainly South Africa). Crude oil prices need to rise to $50–75 per barrel before coal-based ethylene derivatives, for example, make economic sense. Sustained prices in the $50–75/barrel range and predictions of continuing supply–demand pressures yielding such high prices will be required for firms to consider such plants. Another problem is that coal lique-faction or gasification, both proposed for chemicals production, tends to be more highly polluting than processes based on hydrocarbons and give rise to high amounts of greenhouse gases, contributing to global warming. Many countries are committed to reducing emission of greenhouse gases. For these two reasons, it is doubtful that coal-based processes will be adopted for some time to come. China, which has very large coal reserves and very few hydrocarbons, is slated to become the second largest global consumer of crude oil for its vast industrialization program.

It is important to reflect that a number of chemicals now produced from hydro-carbon feedstocks were, in fact, at one time made from methanol and ethanol derived from such feedstocks as corn and other grains, as well as wood. Others were derived from coal liquids, as a byproduct of coke ovens employed for iron and steel production. Coal was also used to make acetylene using a high voltage arc. The advent of "petrochemicals," that is, chemicals made from hydrocarbons, rapidly phased out these earlier production methods, which were less efficient, less economical, or both. This has left fermentation technology as essentially the only commercial route to a few commercially produced chemicals, with citric acid, high fructose corn syrup, and corn-based ethanol as the main examples of natu-ral materials-derived chemicals. But this situation will start to change to some extent as chemical firms develop new technologies to expand the use of natural materials. A notable example is the joint venture between Dow Chemical and Cargill to make a biopolymer based on corn-based starch converted to polylactic acid. A $300 million plant was announced for this venture (Westervelt, 2000). The product, expected to cost around 50 cents per pound, will seek markets in packaging and fiber appli-cations. Cargill is also a large producer of soy-based polyols for producing poly-urethane foams (McCoy, 2003).

Dupont has also been active in developing chemical intermediates based on natural materials. The most prominent example is glucose-derived 1,3-propanediol, a raw material for polymethylene terephthalate, a polyester with promising applications as a carpet fiber material, which is under development by Shell Chemical (Spitz, 2003: 79). Dupont's sustainable development goals for 2010 include holding energy use to 1990 levels, sourcing ten percent of its energy from renewable sources and generating 25 percent of total revenues from nondepletable sources (Holliday, 2001).

A number of other companies, including Dow, BP, and Shell have taken steps in promoting SD initiatives within their firms. This has included developing business cases for the integration of SD into business strategy (Challener, 2001). See Section 8 for the business case presented by several chemical companies.

The use of "greener" syntheses has also received considerable attention. A good example of this is the work to eliminate the use of organic solvents as used in current reactions. Examples of potential replacement solvents under study include supercritical carbon dioxide and so-called ionic liquids (nitrogen- or phosphorus-containing cations coupled with inorganic anions). Processes that might be so transformed include those for terephthalic acid, adipic acid, and others (Ritter, 2001). Elimination or reduction of use of solvents is an important goal in SD. 3M claims to have saved $827 million in finding friendlier alternatives to solvents (Thomas, 2001).

Another issue relative to use (or reuse) of natural materials is the question of whether some natural or, more broadly speaking, traditional material applications, now replaced by plastics, will again make their appearance as the price of hydrocarbon-derived polymers rises. Paper, cardboard, wood, glass, metal, and natural rubber continue to compete with plastics in some applications and their importance could again grow. Even at this time, paper cups compete with foamed polystyrene, paper bags with polyethylene shopping bags, and so on. It is not difficult to visualize the resurgence of traditional materials as crude oil and natural gas prices keep escalating. This would be a real win for SD, although some studies have shown that petrochemical polymer production is more energy-efficient than that used for natural materials (Spitz, 2003: 17). In some cases, hybrid (i.e., part natural/part synthetic polymer) applications may develop.

In summary:

- Chemical producers are facing a major change in the patterns of trade, globally.
- Chemical producers in the United States and other OECD countries are facing stiffer competition from newer producers in the Middle East and Asia.
- The imbalance of trade in chemicals has come about not only because certain *commodity* petrochemicals produced in the United States are now less competitive globally, as a result of higher natural gas costs, but also because such Asian countries as China and India have now become substantial, reliable exporters of *specialty chemicals* and *pharma intermediates*.
- Steps that companies can take to remain competitive include improving energy efficiency and exploring new technologies and alternative feedstocks.
- As crude oil and natural gas prices continue to escalate, it is not difficult to visualize the resurgence in use of more traditional materials.

- While some companies are taking steps in this direction, as shown by the few examples in this brief analysis, the move from hydrocarbon to renewable feedstocks is a long way off.

REFERENCES

AIHC, "Improving Risk Characterization. Summary of Workshop held September 26 and 27, 1991," American Industrial Health Council, 1992.

ATSDR (Agency for Toxic Substances and Disease Registry), *Toxicological Profile for DDT/DDD/DDE* (Update), Agency for Toxic Substances and Disease Registry, Department of Health and Human Sciences, Atlanta, Georgia, 2002, pp. 205–207.

D. G. Barnes and M. L. Dourson, "Reference Dose (Rfd): Description and Use in Health Risk Assessments," *Regulatory Toxicology and Pharmacology*, 8, 471–486 (1988).

D. G. Barnes, G. P. Daston, J. S. Evans, A. M. Jarabek, R. J. Kavlock, C. A. Kimmel, C. Park and H. L. Spitzer, "Benchmark Dose Workshop: Criteria for Use of a Benchmark Dose to Estimate a Reference Dose," *Regulatory Toxicology and Pharmacology*, 21(2), 296–306 (1995).

K. Betts, "Rapidly Rising PBDE Levels in North America," December 7, 2001, *Environmental Science and Technology Online (2001)*. Available at http://pubs.acs.org/subscribe/journals/esthag-w/2001/dec/science/kb_pbde.html (accessed September 2004).

R. L. Carson, *Silent Spring*, 40th anniversary edition (October 22, 2002), Houghton Mifflin, New York, 400pp.

R. Castorina and T. Woodruff, "Assessment of Potential Risk Levels Associated with U.S. Environmental Protection Agency Reference Values," *Environmental Health Perspectives*, 111(10), 1318–1325 (2003).

CDC (Center for Disease Control), *Second National Report on Human Exposure to Environmental Chemicals*. Department of Health and Human Services, Atlanta, Georgia, January 2003. Available at http://www.cdc.gov/exposurereport.

C. Challener, *Sustainable Development at a Crossroads*, Chemical Market Reporter, July 16, 2001, pp. 3–4.

H. J. Clewell, "The Application of Physiologically Based Pharmacokinetic Modeling in Human Health Risk Assessment of Hazardous Substances," *Toxicological Letters*, 79(1–3), 207–217 (1995).

CMA (Chemical Manufacturers Association), *Improving Responsible Care Implementation, Enhancing Performance and Credibility*, Washington, DC, 1993.

T. Colborn, D. Dumanoski and P. J. Myers, *Our Stolen Future*, Dutton Publishing, New York, March 1996.

T. Colborn, D. Dumanoski and P. J. Myers, *Our Stolen Future – Myths versus Reality*. Available at http://www.ourstolenfuture.org/Myths/myths.htm (accessed November 2003).

C. Colten and P. Skinner, *The Road to Love Canal: Managing Industrial Waste before EPA*, University of Texas Press, Austin, TX, 1996.

R. B. Conolly, J. S. Kimbell, D. Janszen, P. M. Schlosser, D. Kalisak, J. Preston and F. J. Miller, "Human Respiratory Tract Cancer Risks of Inhaled Formaldehyde: Dose–Response Predictions Derived from Biologically-Motivated Computational Modeling of a Combined Rodent and Human Dataset," *Toxicological Sciences*, 82, 279–296.

G. L. Copley and H. I. Ahmed, *U.S. Chemicals. The End of the Good Old Days*, Bernstein Equity Research Analysis presented at the International Petrochemical Conference of the National Petroleum Refiners Association, San Antonio, March 25, 2002.

K. S. Crump, "A New Method for Determining Allowable Daily Intakes," *Fundamental and Applied Toxicology*, 4, 854–871 (1984).

T. Damstra *et al.*, Eds., *Global Assessment of the State-of-the-Science of Endocrine Disruptors*, WHO Publication No. WHO.PCS/EDC/02.2, World Health Organization, Geneva, 2002.

R. Dennison, *Orphan Chemicals in the HPV Challenge: A Status Report*, Environmental Defense, Washington, DC, June 2004.

U. Desai, Ed., *Environmental Politics and Policy in Industrialized Countries*, MIT Press, Cambridge, MA, 2002.

J. DiGangi, *US Intervention in EU Chemical Policy*, Environmental Health Fund, Jamaica Plain, MA, September 2003.

D. Ditz, *EU Chemicals Policy Offers a Model for Managing Risk.* World Wildlife Fund, Washington, DC, April 4, 2003.

M. L. Dourson, "Methods for Establishing Oral Reference Doses (RfDs)," in W. Mertz, C. O. Abernathy and S. S. Olin, Eds., *Risk Assessment of Essential Elements*, ILSI Press, Washington, DC, 1994, pp. 51–61.

M. L. Dourson, L. K. Teuschler, P. R. Durkin and W. M. Stiteler, "Categorical Regression of Toxicity Data: A Case Study using Aldicarb," *Regulatory Toxicology and Pharmacology*, 25, 121–129 (1997).

M. L. Dourson, R. C. Hertzberg, R. Hartung and K. Blackburn, "Novel Approaches for the Estimation of Acceptable Daily Intake," *Toxicology and Industrial Health*, 1(4), 23–41 (1985).

EDF (Environmental Defense Fund), *Toxic Ignorance: The Continuing Absence of Basic Health Testing for Top-Selling Chemicals in the United States*, Environmental Defense Fund, Washington, DC, 1997.

Environmental Working Group, "Mother's Milk: Record Levels of Toxic Fire Retardants Found in American Mothers' Breast Milk," September 2003. Available at http://www.ewg.org/archives/reports/mothersmilk/es.php (accessed September 2004).

EU (European Union), "The New EU Chemicals Legislation – Reach," 2004. Available at http://europa.eu.int/comm/enterprise/reach/overview.htm (accessed October 2004).

European Commission, *Compilation of EU Dioxin Exposure and Health Data*, produced for European Commission DG Environment and UK Department of the Environment Transport and the Regions (DETR), October 1999. Available at http://europa.eu.int/comm/environment/dioxin/summary.pdf (accessed September 2004).

European Commission, *White Paper on a Strategy for a Future Chemicals Policy*, European Commission, Brussels, April 2001.

European Environment Agency, *The Precautionary Principle 1897–2000: Late Lessons From Early Warnings*, European Environment Agency, Copenhagen, December 2000.

EWG (Environmental Working Group), http://www.ewg.org/reports/bodyburden/, 2002. (Accessed September 20, 2004).

B. K. Fishbein, J. R. Ehrenfeld and J. E. Young, *Extended Producer Responsibility: A Materials Policy for the 21st Century*, INFORM, New York, 2000.

FoE (Friends of the Earth UK), http://www.foe.co.uk/campaigns/safer_chemicals/ (Accessed September 20, 2004).

D. Gee, "Approaches to Scientific Uncertainty," in A. Fletcher and A. McMichael, Eds., *Health at the Crossroads: Transport Policy and Urban Health*, John Wiley & Sons, London, 1997, pp. 27–50.

K. Geiser, *Materials Matter: Toward a Sustainable Materials Policy*, MIT Press, Cambridge, MA, 2001.

L. Goldman, "Preventing Pollution? U.S. Toxic Chemicals and Pesticides Policies and Sustainable Development," *Environmental Law Reporter*, 32, 11018–11041 (2002).

GPUK (Greenpeace UK), "Hazardous Chemicals in House Dust as an Indicator of Chemical Exposure in the Home," in D. Santillo *et al.*, Eds., *Consuming Chemicals*, Department of Biological Sciences, University of Exeter, Exeter, UK, 2003a.

GPUK (Greenpeace UK), www.greenpeace.org.uk/, 2003b. (Accessed September 20, 2004.)

J. Graham and J. Wiener, Eds., *Risk vs. Risk: Tradeoffs in Protecting Health and the Environment*, Harvard University Press, Cambridge, MA, 1995.

D. J. Guth, R. J. Carroll, D. G. Simpson and H. Zhou, "Categorical Regression Analysis of Acute Exposure to Tetrachloroethylene," *Risk Analysis*, 17, 321 (1997).

L. F. Haber, *The Chemical Industry in the Nineteenth Century*, Oxford University Press, New York, 1958.

L. T. Haber, J. S. Dollarhide, A. Maier and M. L. Dourson, "Noncancer Risk Assessment: Principles and Practice in Environmental and Occupational Settings," *Patty's Toxicology*, 5(1), 169–232 (2001a).

L. Haber, J. A. Strickland and D. J. Guth, "Categorical Regression Analysis of Toxicity Data," *Comments on Toxicology*, 7(5–6), 437–452 (2001b).

Health Canada, "Priority Substances List Assessment Report: 2-Butoxyethanol. Draft for Public Comments," Ministry of Public Works and Government Services, Ottawa, August 2000. Available at http://www.hc-sc.gc.ca/hecs-sesc/exsd/psl2.htm (accessed October 2004).

Health Canada, "Categorization of Substances on the Domestic Substances List," Available at http://www.hc-sc.gc.ca/hecs-sesc/exsd/categorization_dsl.htm (accessed October 2004).

R. C. Hertzberg and M. Miller, "A Statistical Model for Species Extrapolation Using Categorical Response Data," *Toxicology and Industrial Health*, 1, 43–57 (1985).

R. C. Hertzberg, "Fitting a Model to Categorical Response Data with Application to Species Extrapolation of Toxicity," *Health Physics*, 57, 405–409 (1989).

H. Hettige, D. Shaman, D. Wheller and D. Witzel, *The World Bank Industrial Pollution Projection System*, World Bank Policy Research Working Paper #1431, 1994.

C. O. Holliday, *The Challenge of Sustainable Growth*, Chemical Business, May 2001, pp. 19–23.

IEA (International Energy Agency), *CO$_2$ Emissions from Fuel Combustion 1971–1997*, IEA, Paris, 1999.

IEA (International Energy Agency), *Energy Balances of OECD Countries and Energy Balances of Non-OECD Countries, 1971–1998*, on-line service and IEA secretariat estimates, IEA, Paris, 2000.

IPCS (International Programme on Chemical Safety), "Environmental Health Criteria No. 170: Assessing Human Health Risks of Chemicals: Derivation of Guidance Values for Health-Based Exposure Limits," World Health Organization, Geneva, 1994.

IPCS (International Programme on Chemical Safety), "Guidance Document for the Use of Data in Development of Chemical-Specific Adjustment Factors (CSAF) for Interspecies Differences and Human Variability in Dose/Concentration Response Assessment," World Health Organization, Geneva, 2001. Available at http://www.ipcsharmonize.org (accessed October, 2004).

A. M. Jarabek, "Inhalation Rfc Methodology: Dosimetric Adjustments and Dose–Response Estimation of Noncancer Toxicity in the Upper Respiratory Tract," *Inhalation Toxicology*, 6(Suppl), 301–325 (1994).

J. Johnson, *Getting a Grip on Wasted Energy*, Chemical and Engineering News, pp. 31–33 (July 24, 2000).

D. Kriebel, J. Tickner et al., "The Precautionary Principle in Environmental Science," *Environmental Health Perspectives*, 109, 871–876 (2001).

D. Kriebel, J. Tickner and C. Crumbley, "Appropriate Science: Evaluating Environmental Risks for a Sustainable World," presented at Education for Sustainable Development, Committee on Industrial Theory and Assessment, University of Massachusetts Lowell, Lowell, MA, October 23–24, 2003.

D. Kurtin et al., "Demographic Risk Factors Associated with Elevated Lead Levels in Texas Children Covered by Medicaid," *Environmental Health Perspectives*, 105(January): 66–68 (1997). Available at http://www.eph.niehs.nih.gov/members/1997/105-1/kurtin.html.

P. Landrigan, "Risk Assessment for Children and Other Sensitive Populations," in J. Bailer and J. Bailar, Eds., *Uncertainty in the Risk Assessment of Environmental and Occupational Hazards*, Annals of the New York Academy of Sciences, New York, 895 (1999).

E. Lawless, *Technology and Social Shock*, Rutgers University Press, New Brunswick, NJ, 1977.

Lowell Center for Sustainable Production, *Integrated Chemicals Policy, Seeking New Direction in Chemicals Management*, Lowell Center for Sustainable Production, University of Massachusetts Lowell, Lowell, MA, 16 pp.

F. Lu, "Safety Assessments of Chemicals with Thresholded Effects," *Regulatory Toxicology and Pharmacology*, 3, 121–132 (1985).

F. Lu, "Acceptable Daily Intake: Inception, Evolution, and Application," *Regulatory Toxicology and Pharmacology*, 8, 45–60 (1988).

A. Maier, D. G. Debord and R. E. Savage, "Incorporating Biomarkers into 21st Century Risk Assessments," *Comments on Toxicology*, 7(5–6), 519–540 (2001).

M. McCoy, *Starting a Revolution*, Chemical and Engineering News, pp. 17–18 (December 15, 2003).

M. E. Meek, R. Newhook, R. G. Liteplo and V. C. Armstrong, "Approach to Assessment of Risk to Human Health for Priority Substances Under the Canadian Environmental

Protection Act," *Environmental Carcinogenesis and Ecotoxicology Reviews*, C12(2), 105–134 (1994).

Minnesota Office of Environmental Assistance, "Let's Get the Lead Out! Non-lead Alternatives for Fishing Tackle," May 2002. Available at http://www.moea.state.mn.us/reduce/sinkers.cfm (accessed September 2004).

J. Moffet, F. Bregha and M. Middelkoop, "Responsible Care: A Case Study of a Voluntary Environmental Initiative," in R. Gibson, ed., *Voluntary Initiatives*, Ch. 6, Broadview Press, Toronto, Canada, 1999.

S. H. Moolgavkar and A. G. Knudson, "Mutation and Cancer: A Model for Human Carcinogenesis," *Journal of National Cancer Institute*, 66, 1037–1052 (1981).

B. Moyers, "The Chemical Papers: Secrets of the Chemical Industry Exposed," in *Trade Secrets*, www.pbs.org/tradesecrets/ (accessed September 20, 2004).

I. C. Munro, "Safety Assessment Procedures for Indirect Food Additives: An Overview," *Regulatory Toxicology and Pharmacology*, 12(1), 2–12 (1990).

I. C. Munro, R. A. Ford, E. Kennepohl and J. G. Sprenger, "Correlation of Structural Class with No-Observed-Effect Levels: A Proposal for Establishing a Threshold of Concern," *Food Chemistry and Toxicology*, 34(9), 829–867 (1996).

NAS (National Academy of Sciences), "Committee on Biological Markers of the National Research Council," *Environmental Health Perspectives*, 74(3) (1987).

NAS (National Academy Press), *Pesticides in the Diets of Infants and Children*, Board on Agriculture and Board on Environmental Studies and Toxicology, National Academy Press, Washington, DC, 1993.

National Research Council, *Risk Assessment in the Federal Government: Managing the Process*, National Academy Press, Washington, DC, 1983.

National Research Council, *Science and Judgement in Risk Assessment*, National Academy Press, Washington, DC, 1994.

NRC (National Research Council), *Risk Assessment in the Federal Government: Managing the Process*, National Academy Press, Washington, DC, 1983.

NRDC (Natural Resources Defense Council), *The Story of Silent Spring*, 2004. Available at http://www.nrdc.org/health/pesticides/hcarson.asp (accessed September 20, 2004).

M. O'Brien, *Making Better Environmental Decisions: An Alternative to Risk Assessment*, MIT Press, Cambridge, MA, 2000.

OECD (Organisation for Economic Co-Operation and Development), *OECD Environmental Outlook*, OECD, Paris, 2001a.

OECD (Organisation for Economic Co-Operation and Development), *OECD Environmental Outlook for the Chemicals Industry*, OECD, Paris, 2001b.

D. Pearce and P. Koundouri, *The Social Cost of Chemicals. The Cost and Benefits of Future Chemicals Policy in the European Union*, World Wildlife Fund – UK, Godalming, Surrey, UK, 2003.

Pennsalt, Advertisement in Life Magazine (Pennsalt later became Elf Atochem), 1946.

J. Philpot, *An Overview of the Asian Petrochemical Business*, ARTC Petrochemical Conference, Singapore, 20–21 March 2003.

A. H. Piersma, A. Verhoef, J. D. te Biesebeek, M. N. Pieters and W. Slob, "Developmental Toxicity of Butyl Benzyl Phthalate in the Rat Using a Multiple Dose Study Design," *Reproductive Toxicology*, 14, 417–425 (2000).

C. A. Pittinger, T. H. Brennan, D. Badger, P. J. Hakkinen and M. C. Fehrenbacher, "Aligning Chemical Assessment Tools Across the Hazard-Risk Continuum," *Risk Analysis*, 23(3), 529–535 (2003).

H. R. Pohl and H. G. Abadin, "Utilizing Uncertainty Factors in Minimal Risk Levels Derivation," *Regulatory Toxicology and Pharmacology*, 22, 180–188 (1995).

Presidential/Congressional Commission on Risk Assessment and Risk Management. *Framework for Environmental Health Risk Management, Final Report.* The White House, Washington, DC, 1997.

C. Raffensperger and J. Tickner, Eds., *Protecting Public Health and the Environment: Implementing the Precautionary Principle*, Island Press, Washington, DC, 1999.

RCEP (Royal Commission on Environmental Pollution), *Chemicals in Products – Safeguarding the Environment and Human Health*, RCEP, UK, June 2003.

J. Rees, "The Development of Industry Self-regulation in the Chemical Industry," *Law and Policy*, 19(4) (1997), quoted in A. King and M. Lennox, "Industry Self-regulation Without Sanctions: the Chemical Industry's Responsible Care Program," *Academy of Management Journal*, 43(4): 698–716 (August 2000).

A. G. Renwick, "Data Derived Safety Factors for the Evaluation of Food Additives and Environmental Contaminants," *Food Additives and Contaminants*, 10(3), 275–305 (1993).

S. K. Ritter, *GreenChemistry*, Chemical and Engineering News, pp. 27–34 (July 16, 2001).

A. M. Rulis, "*De Minimis* and the Threshold of Regulation," in C. W. Felix, Ed., *Food Protection Technology*, Lewis Publishers, Inc., Chelsea, MI, 1986, pp. 29–37.

S. Schantz *et al.*, "Impairments of Memory and Learning in Older Adults Exposed to Polychlorinated Biphenyls via Consumption of Great Lakes Fish," *Environmental Health Perspectives*, 109, 605–611 (2001).

Secretariate of the Stockholm Convention on Persistent Organic Pollutants, *Stockholm Convention on Persistent Organic Pollutants, Text and Annexes*, GE.01-02667, United Nations Environment Program, Geneva, 2001.

A. Sjodin *et al.*, Flame retardant exposure: Polybrominated diphenyl ethers in blood from swedish workers, *Environmental Health Perspectives*, 107(8), 643–648 (1999). Available at http://www.ehp.niehs.nih.gov/docs/1999/107p643-648sjodin/abstract.html (accessed September 2004).

P. H. Spitz, *The Chemical Industry at the Millennium*, Chemical Heritage Press, Philadelphia, pp. 334, 2003.

B. R. Stern, M. Solioz, D. Krewski, P. Aggett, T.-C. Aw, S. Baker, K. Crump, M. Dourson, L. Haber, R. Hertzberg, C. Keen, B. Meek, L. Rudenko, R. Schoeny, W. Slob and T. Starr, "Copper and Human Health: Biochemistry, Genetics, and Strategies for Modeling Dose–Response Relationships," *Journal of Toxicology and Environmental Health. Part B. Critical Reviews.* Submitted.

P. Stern and H. Fineberg, Eds., *Understanding Risk: Informing Decisions in a Democratic Society*, National Academy Press, Washington, DC, 1996.

SustainAbility, *External Stakeholder Survey. Final Report for the Global Strategic Review of Responsible Care*, SustainAbility, Washington, DC, 2004.

M. B. Swanson and A. C. Socha, "Chemical Ranking and Scoring: Guidelines for Relative Assessments of Chemicals," SETAC Press, Pensacola, FL, 1997.

Swedish National Chemicals Inspectorate, "Phase-out of PBDEs and PBBs," March 15, 1999. Available at http://www.kemi.se/aktuellt/pressmedd/1999/flam_e.pdf (accessed September 2004).

J. Tarr, *The Search for the Ultimate Sink: Urban Pollution in Historical Perspective*, University of Akron Press, Akron, OH, 1996.

L. K. Teuschler, M. L. Dourson, W. M. Stiteler, P. McClure and H. Tully, "Health Risk Above the Reference Dose for Multiple Chemicals," *Regulatory Toxicology and Pharmacology*, 30, S19–S26 (1999).

G. H. Thomas, *Green – The Color of Nature, Success and Innovation*, Chemistry Business, p. 17, October 2001.

B. Thorpe, *Safer Chemicals within Reach. Using the Substitution Principle to Drive Green Chemistry*, Greenpeace Environmental Trust, UK, 2003.

J. Tickner, "Hazardous Exports: The Export of Risk Assessment to Central and Eastern Europe," *New Solutions*, Summer, 3–13 (1996).

J. Tickner, "Precaution and Preventive Public Health Policy," *Public Health Reports*, Nov/Dec, 117, 493–497 (2002).

J. Tickner and K. Geiser, *New Directions in European Chemicals Policies: Drives, Scope and Status*, Lowell Center for Sustainable Production, University of Massachusetts Lowell, Lowell, Massachusetts, 2003. Available at www.chemicalspolicy.org.

J. Tickner, Ed., *Precaution, Environmental Science, and Preventive Public Policy*, Island Press, Washington, DC, 2003.

TNO, *Emissions of Hazardous Substances from Finished Products*, TNO, The Netherlands, 2001.

U.S. EPA (U.S. Environmental Protection Agency), "Guidelines for Carcinogen Risk Assessment," *Federal Register*, 51(185), 33992–34003 (1986).

U.S. EPA (U.S. Environmental Protection Agency), "Guidelines for Exposure Assessment," *Federal Register*, 57, 22888–22938 (1992).

U.S. EPA (U.S. Environmental Protection Agency), "Methods for Derivation of Inhalation Reference Concentrations and Application of Inhalation Dosimetry," Office of Health and Environmental Assessment, Washington, D.C., EPA 600-8-90-066F, 1994.

U.S. EPA (U.S. Environmental Protection Agency), "Proposed Guidelines for Carcinogen Risk Assessment," Office of Research and Development, Washington, D.C., EPA 600-P-92-003C, 1996.

U.S. EPA (U.S. Environmental Protection Agency), *What Do We Really Know about the Safety of High Production Volume (HPV) Chemicals*, United States Environmental Protection Agency, Office of Pollution Prevention and Toxics, Washington, DC, 1998a.

U.S. EPA, "New Jersey Material Accounting Data"; "Massachusetts Toxics Use Reductions Act Data," 1998b.

U.S. EPA (U.S. Environmental Protection Agency), "Guidelines for Carcinogen Risk Assessment (SAB Review Copy, July 1999)," Washington, D.C. Available at http://www.epa.gov/ncea/raf/crasab.htm (accessed October 2004).

U.S. EPA (U.S. Environmental Protection Agency), "Benchmark Dose Technical Guidance Document (External Review Draft)," Risk Assessment Forum, Washington, D.C., EPA/

630/R-00/001, 2000a. Available at http://cfpub2.epa.gov/ncea/cfm/recordisplay. cfm?deid = 20871 (accessed October 2004).

U.S. EPA (U.S. Environmental Protection Agency), "Science Policy Council Handbook," Office of Research and Development, Washington, D.C., EPA 100-B-00-002, 2000b.

U.S. EPA, "Mercury Update: Impact on Fish Advisories," June 2001. Available at http:// www.epa.gov/ost/fishadvice/mercupd.pdf (accessed September 2004).

U.S. EPA (U.S. Environmental Protection Agency), "Draft Final Guidelines for Carcinogen Risk Assessment External Review Draft," Office of Research and Development. Washington, D.C., 2003. Available at http://cfpub2.epa.gov/ncea/cfm (accessed October 2004).

U.S. EPA (U.S. Environmental Protection Agency), "Boron and Compounds," 2004a. Available at http://www.epa.gov/iris/subst/0410.htm (accessed October 2004).

U.S. EPA (U.S. Environmental Protection Agency), "Integrated Risk Information System (IRIS)," 2004b. Available at http://www.epa.gov/iris/subst/0410.htm (accessed October, 2004).

U.S. EPA (U.S. Environmental Protection Agency), "Pollution Prevention (P2) Framework," 2004c. Available at http://www.epa.gov/oppt/p2framework/ (accessed October 2004).

U.S. FDA (U.S. Food and Drug Administration), "Food Additives; Threshold of Regulation for Substances Used in Food-Contact Articles. Final Rule," *Federal Register*, 60, 36581– 36596 (1995).

Vermont Senate Natural Resources and Energy Committee, proposed mercury legislation (Senate Bill 91) as passed by the Vermont Senate Natural Resources and Energy Committee on March 21, 2001. Available at http://www.mercvt.org/s91final.htm (accessed September 2004).

K. von Moltke, *The Precautionary Principle, Risk Assessment, and the World Trade Organization*, International Institute for Sustainable Development, Winnipeg, 2000.

N. Weisglas-Kuperus *et al.*, "Immunologic Effects of Background Exposure to Polychlorinated Biphenyls and Dioxins in Dutch Preschool Children," *Environmental Health Perspectives*, 108, 1203–1207 (2000).

R. Westervelt, *Bet $300 Million on Corn-based Polymers*, Chemical Week, p. 9, January 19, 2000.

R. Worcester, *British Science and the Chemical Industry*, Chemical Industries Association, London, UK, November 1999.

B. Wynne, "Uncertainty and environmental learning," in T. Jackson, Ed., *Clean Production Strategies*, Lewis Publishers, Boca Raton, 1993, pp. 63–84.

4

PLANNING FOR SUSTAINABILITY

4.1 PLANNING OVERVIEW

BETH BELOFF

BRIDGES to Sustainability

Sustainable development is a complex concept that incorporates and integrates the concepts of environmental stewardship, economic growth, and social progress. These are considered interrelated factors around which one optimizes to contribute to the growth and even to the survivability of the business enterprise. Because sustainable development is, by its nature, interdisciplinary and cross-functional, frameworks have emerged to help companies align their stakeholders around common concepts and integrate these broad concepts into action, enterprisewide.

4.1.1 Planning Framework

The intent of all planning frameworks is to transform complex systems into something that is simplified so that it can become comprehensible to many and therefore becomes something that can be managed. In an effort to contribute to the many simplifications, we offer the BRIDGES Sustainability Framework™ to define and describe the elements of sustainability. The Framework (Fig. 4.1) offers a perspective that incorporates the three dimensions of sustainability – environmental stewardship, economic growth, and social development. Environmental stewardship can be conceptually broken down into resource use and pollutant/waste streams, which are inputs and outputs of industrial processes. Economic growth can be organized conceptually into that which is internal to the business and that which is external to the broader community. Likewise, social development can also be broken

Transforming Sustainability Strategy into Action: The Chemical Industry, Edited by B. Beloff, M. Lines, and D. Tanzil

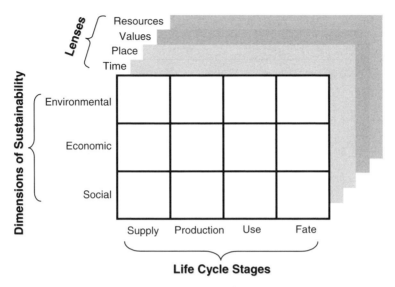

Figure 4.1. BRIDGES Sustainability Framework (Beloff and Beaver, 2000).

down into workforce considerations and external stakeholder/community considerations. Table 4.1 offers some suggestions as to how to further consider the dimensions of sustainability as aspects that need to be included in sustainability planning. The table also expands from the three traditional dimensions of sustainability to include general/institutional considerations, such as sustainability management. Determining what is of importance to the particular organization is also the necessary first step in selecting indicators of sustainability and related metrics that can be used to track for enhanced management decision-making and communications purposes, as described further in Section 6.1.1 on Indicators and Metrics.

The Framework suggests that impacts from each of the dimensions of sustainability be considered along the lifecycle stages and/or supply chain steps. This is due to the fact that impacts from decisions made in business enterprises often extend beyond their "fencelines" and affect the larger system in which they are embedded. Companies can shift their impacts to their suppliers or their customers by the decisions that they make regarding procurement of raw materials, systems selected for transportation and distribution, design of products that push impacts into product use, or end-of-life disposition. Not only can impacts shift, but also there is the issue of who bears the cost of sustainability-related investments, who derives the benefits, and when in time do those costs and related benefits occur. In order to consider the sustainability of the system in which a company is embedded, it is essential to look at the broad set of impacts – environmental, economic and social – resulting from management decisions along the value chain in which that business is embedded.

In the Framework, the lenses represent variables that affect the way in which the sustainability issues are further characterized and the boundary conditions that are set. Depending on the enterprise, there may be more lenses than shown in

TABLE 4.1. Identifying the Important Aspects of Sustainability in a Company

Environmental Stewardship	
Resources	• Material intensity
	• Energy intensity
	• Water usage
	• Land use
Pollutants & Waste	• Wasted resources
	• Releases to air, water, and land
	• Material recycling
	• Impacts on human health
	• Impacts on ecosystem and biodiversity
Economic Development	
Internal	• Eco-efficiency
	• Internal costs to the company
	• Revenue opportunities
	• Access to capital and insurance
	• Shareholder value
External	• Costs of externalities
	• Benefits to local community
	• Benefits to society
Social Progress	
Workplace	• Workplace conditions
	• Employee health, safety, and well-being
	• Security
	• Human capital development (education & training)
	• Aligning company values with those of employees
Community	• Social impacts of operations
	• Stakeholder engagements
	• Quality of life in community
	• Human rights
General/Institutional	
Sustainability	• Commitment to triple-bottom-line
Management	• Product stewardship and supply-chain leadership programs
	• Accountability and transparency
	• Product and service development
	• Employees' impact on environment

Source: ©BRIDGES to Sustainability, 2003.

Figure 4.1, but these are the most common ones in considering sustainability issues, and they include:

- *A time dimension.* Is the time horizon short- or long-term? Will the costs be borne today but the benefits occur in the long term, or vice versa? Are the desired improvements planned to be incremental/continuous improvement or revolutionary/discontinuous over time? How do we account for future generations?

- *A place dimension.* At what scale are the considerations being addressed: local, national, global? Where in the world are we considering sustainability issues and what is the socioeconomic, political, and physical context? What part of the whole system are we considering: the process, facility, corporation, value chain? How do the elements of the system roll up?
- *Values.* Location in the world influences the attitudes and values of the locale. What is important to the local community? What are the key sustainability concerns? This social context shapes and weights the importance of certain issues in a particular place for a particular group of stakeholders?
- *A resource context.* What is the ecosystem context of the place? What is the nature of the resource integrity: scarcity, overabundance, or potential to be disrupted now or in the future?

The value of this Framework is to identify the sustainability considerations that are important to the company and its stakeholders in order to plan for sustainability comprehensively.

4.1.2   Corporate Sustainability Learning Curve

To business, sustainability thinking is really about creating new knowledge, connecting the dots differently, mining through the plethora of information and data points that all firms have available, but looking at them differently with respect to management decision-making. This almost always requires connecting different functional areas, working between business silos. A company moving along the sustainability learning curve must extend the boundaries of consideration from traditional EHS *aspects* to those that include workplace and labor practices, often found in Human Relations, and community concerns, often handled through Community Relations or other related functional groups. Similarly, the *impacts* of the aspects extend from more traditional considerations of risks and their reduction to the area of opportunity enhancement and increased related benefits. Impacts are also extended out along the *value chain*, from consideration of those impacts generated at the facility level to impacts produced by suppliers upstream and in use by customers downstream. The *time* dimension is also extended beyond traditional boundaries of consideration when moving up the sustainability learning curve. Rather than just focusing on the short term of quarterly earnings, companies must anticipate the longer term view. Where does the company want to be positioned in the future; what will be the future demand of the company as a socially responsible actor; how can future generations be protected by current management decisions; and so on?

The progression of corporate practices – from *compliance* with government policies to *risk management*, to *ensuring sustainability* by building integrated sustainability approaches and systems, to *creating new opportunities* through technical, business process, and social innovations, and finally to *leadership*, leading by example, mentoring other players in the supply chain, extending corporate social responsibility – are all logical building blocks for moving toward corporate sustainability. They represent organizational steps commonly found in practice, from inward, exclusive and reactively focused, to outward, inclusive and more proactive. Such a journey is necessary to produce the organizational and system changes required to promote greater sustainability. Understanding a company's history in moving along these dimensions can contribute to a comprehensible planning effort in moving a company toward sustainability. Figure 4.2 captures the elements of the corporate sustainability learning curve.

The Planning for Sustainability section of this book offers a range of planning frameworks, from an elaboration on the elements to consider in planning for sustainability and steps to be taken, to broader conceptual frameworks regarding the systems in which business operates, what contributes to their unsustainability, and how to make the systems as well as the companies operating within them more sustainable.

It should be noted that these frameworks are offered for consideration of how an enterprise may address the issue of planning and implementation of sustainability. Any of these approaches may be used or, alternatively, a company may decide to utilize an existing business management framework and "fit" sustainability planning and implementation into what is already in place. Regardless of the approach a company takes, this chapter attempts to capture the essence of these alternative "frameworks" as a way of demonstrating the breadth of ways to move sustainable development into management practice.

Figure 4.2. Corporate sustainability learning curve. (Adapted from a conceptual model developed for The Stanley Works by BRIDGES to Sustainability and Convergence Consulting, 2004.)

Section 4.2 is titled "GEMI's Approach to Sustainable Development Planning." The Global Environmental Management Initiative (GEMI), a nonprofit association of leading companies dedicated to fostering environmental, health and safety excellence, developed a strategic planning approach supported by a software tool called *SD Planner*™ to support companies embarking on the SD journey. The approach enables companies to better understand the context of SD, create internal awareness of the importance of addressing SD, assess SD status, provide a basis for developing a SD strategy, highlight opportunities to create business value, and provide a roadmap for taking action.

Section 4.3 is titled "Environmental Management Systems (EMS) Frameworks for Sustainability." At the very basic level, a management system provides a structure that allows an organization to systematically manage a particular issue by:

- Defining, assessing and understanding the issue within its boundaries;
- Planning and implementing controls and improvements; and
- Monitoring and evaluating performance of associated efforts.

These functions of a management system are reflected in specific processes, procedures, and tools necessary to manage relevant issues. These can easily be adapted to include issues that are of importance to sustainability practice. In this section, an overview of adapting management systems to sustainability will be presented, but specific management systems and approaches that have been adapted for these purposes are also highlighted. They include RCMS, EMAS and ISO 14001, OHSAS 18001, and Six Sigma management initiatives.

Section 4.4 is titled "The Natural Step Framework: Backcasting from Principles of Sustainability." In the late 1980s, a group of Swedish scientists set out to develop an approach to sustainability that would not be based on an exhaustive understanding of all variables, but rather on the discovery and disciplined use of *principles* by which sustainability-oriented decisions are determined upfront and "upstream," or at the original point of decision of whether or not to use a particular input – not after the damage has already occurred. This approach, called The Natural Step Framework for strategic sustainable development, has been developed and published in numerous scientifically peer-reviewed papers. It has also been applied within businesses to enable decision-makers to understand sustainability issues and begin planning based on these new insights. This piece describes this planning and decision-making framework and how it is developed and applied within a business context.

Section 4.5 outlines "Natural Capitalism for the Chemical Industry." Despite the successes of the Industrial Revolution and the growth of the chemical industry, the conventional economic theories that were the foundation for growth and prosperity had a serious flaw. Natural capital, the asset that was the cornerstone of the industrial machine, was left off the balance sheet and has resulted in significant environmental consequences. Most financial and business systems account and manage for physical and financial capital – the assets that have represented wealth. But there have been other forms of capital that have been ignored in the decision-making and managerial algorithms as well: natural, human, and social capital. The sustainability framework, Natural Capitalism, addresses this omission.

In Section 4.6, "Sustainable Value in the Chemical Industry" is discussed. This article describes the financial and societal challenges the chemical industry faces and offers a practical approach to building enduring value through simultaneously creating shareholder and stakeholder value. Stakeholder value – based on the economic, environmental, and social impacts a company has on its diverse constituents – is a rapidly growing source of business advantage, fuelled by rising societal expectations and the swelling ranks of social change agents newly empowered by information and communication technologies. Taking advantage of this source, however, requires a change in the mindset of leadership and a disciplined approach to planning for and integrating stakeholder value throughout the business.

Section 4.7 details "CSR/SRI Reporting Complexity and the Future 500 CAP Gap Audit^{TM}: An Opportunity for Improved Strategic Business Planning and Stakeholder Alignment." Today, leadership companies are expected to respond to a proliferation of corporate social responsibility (CSR) and socially responsible investing (SRI) standards. Leading companies report not just their return to shareholders but also their "return to stakeholders." But with more than a dozen major standards organizations – each representing a variety of stakeholder perspectives – and even more questionnaires for reporting on CSR/SRI accountability and sustainability practices, how should a company approach this emerging field? And, even more critical for the business bottom line, what strategic opportunities are indicated? The Future 500, a nonprofit environmental and stakeholder engagement organization, offers an approach and tool, the CAP Gap Audit^{TM}, to help cut through the confusing array of guidances regarding stakeholder expectations.

4.2 GEMI'S APPROACH TO SUSTAINABLE DEVELOPMENT PLANNING

ELIZABETH C GIRARDI SCHOEN
Pfizer, Inc.

STEPHEN POLTORZYCKI
The Boston Environmental Group

The Global Environmental Management Initiative (GEMI), a nonprofit association of leading companies dedicated to fostering environmental, health and safety excellence, developed a strategic planning approach supported by a software tool called *SD Planner*^{TM} to support companies embarking on the SD journey. The approach enables companies to better understand the context of SD, create internal awareness of the importance of addressing SD, assess SD status, provide a basis for developing an SD strategy, highlight opportunities to create business value, and provide a roadmap for taking action. The overall objective of the approach is to help companies integrate SD into their business practices and realize business value from SD. The approach is designed to be highly flexible and customizable to each company's unique situation.

There are a number of specific capabilities that *SD Planner*™ provides to help companies start on their journey toward SD.

- *Establish generic elements of sustainable business practices.* The tool synthesizes a broad range of generally accepted SD concepts into seven major elements. This provides a comprehensive basis for companies to select the desired scope and focus of their own SD efforts.
- *Enable company assessment of current status.* The tool provides a straightforward, structured approach toward assessment of a company's current status relative to broad industry norms. For each element, a company can assess its position relative to a five-stage model of evolution. This will help companies to set context and formulate their SD goals.
- *Enable formulation of SD goals and gap analysis.* The tool provides a flexible means for establishing company-specific goals with regard to any of the major SD elements. It also enables the identification of gaps between the company's current and desired position, which provides the basis for developing action plans.
- *Clarify the potential business value of SD.* Depending on the chosen emphasis, the tool will help companies to understand the potential value associated with particular SD initiatives and thence formulate a business case for action. These insights are enriched by examples drawn from the experiences of GEMI member companies.
- *Provide guidance and support development of an action plan.* For each selected element of SD, the tool assists companies in developing an agenda for action, by recommending appropriate actions that can help to achieve the company's goals. Again, these recommendations draw upon the experiences of GEMI member companies.

The intent of *SD Planner*™ is to provide these capabilities in a manner that is generic, yet highly customizable to the needs of an individual company. In addition, *SD Planner*™ strives for clarity and simplicity, so that it is intuitive and easy to use with minimal training. The graphical user interface has been designed accordingly.

4.2.1 Planning Framework

The GEMI approach synthesizes a broad range of generally accepted SD concepts into seven major *elements* that cover the full range of social, environmental, and economic aspects of SD (Table 4.2). There are significant interrelationships between these *elements* and they should be addressed in an integrated manner. However, it is important to note that it was not GEMI's intent to prescribe a Code of Conduct, or to suggest that companies should be actively engaged in all of the SD *elements*.

The approach also defines five *stages* of evolution in corporate SD practices that companies may take in moving towards their SD goals, from initial preparation to acting as an SD champion within industry (Table 4.3).

TABLE 4.2. Elements of Sustainable Development

Category	Element	Definition
Social	1. Employee well-being	Protecting and preserving the fundamental rights of employees, promoting positive employee treatment, and contributing to employee quality of life.
	2. Quality of life	Working with public and private institutions to improve educational, cultural, and socioeconomic well-being in the communities in which the company operates and in society at large.
	3. Business ethics	Supporting the protection of human rights with the company's sphere of influence, and promoting honesty, integrity and fairness in all aspects of doing business.
Economic	4. Shareholder value creation	Creating value for the company's shareholders. Includes securing a competitive return on investment, protecting the company's assets, and enhancing the company's reputation and brand image through integration of sustainable development thinking into business practices.
	5. Economic development	Building capacity for economic development in the communities, regions, and countries in which the company operates or would like to operate.
Environmental	6. Environmental impact minimization	Minimizing and striving to eliminate the adverse environmental impacts associated with operations, products, and services.
	7. Natural resource protection	Promoting the sustainable use of renewable natural resources and conservation and sustainable use of nonrenewable natural resources.

I. Prepare: Minimize SD efforts while assessing the business drivers.

II. Commit: Commit to addressing SD and choose a strategic direction.

III. Implement: Launch programs consistent with the SD strategy.

IV. Integrate: Make SD part of everyday business processes.

V. Champion: Encourage others within the industry to pursue SD.

Although these *stages* represent increasing sophistication of the business processes and practices in place to address SD, companies need not necessarily strive for the "highest" *stage*. Rather, companies need to determine which *stage* will deliver the most business value at a given time.

4.2.2 Who Should Use *SD Planner*™?

SD Planner™ is designed for use by a variety of different groups within a company. It is optimized for use by a business unit, but it can also be used by a corporate group on behalf of the company as a whole. Typically, a functional group will assume a

TABLE 4.3. Stages of Sustainable Development

	I. Prepare	II. Commit	III. Implement	IV. Integrate	V. Champion
Thrust	Minimize SD efforts while assessing opportunities	Choose a strategic direction for SD actions	Launch programs consistent with strategy	Make SD part of daily life in the organization	Act as SD champion within industry
Objective	Comply, avoid surprises, and be ready for next steps	Improve stakeholder relationships and explore the SD territory	Reduce adverse impacts and realize business value	Enhance quality of life, and create a sustainable enterprise	Influence industry to act more sustainably
Management processes	Track issues and respond to external questions or challenges	Develop a strategy, plan actions, and pilot initiatives intended to realize business value	Incorporate SD issues as an extension of EH&S management	Innovate, learn, and integrate SD into business processes	Engage in external partnerships, external advocacy, and public policy development
Organizational scope	Engage company functional specialists	Align key business leaders	Integrate into internal company operations	Align distribution value chain (suppliers, contractors, customers)	Collaborate with other organizations to foster industrial ecology
Organizational alignment	Engage company functional specialists	Endorse commitment to SD at senior management and key middle management levels	Consider routinely SD as part of key business decisions	Effectively integrate SD thinking into all aspects of the business	Extend SD thinking to industry and other partnerships
Stakeholder engagement	Monitor company stakeholder concerns and respond to pressures	Establish dialogue with key stakeholders	Emphasize transparency in dialog with stakeholders	Integrate stakeholder engagement into all business processes	Form partnerships with stakeholders

lead role for applying the tool, and a specific individual or team will be responsible for actually entering information and generating results. However, because of the broad scope of the tool, it is likely that this team will need to interface with representatives from several functional areas, potentially including strategic planning, marketing, communications, human resources, operations, distribution, research and development, finance, and environmental, health and safety. The team will also need to define the organizational scope of *SD Planner*'s application (e.g., wholly owned subsidiaries, partially owned subsidiaries, joint ventures, contractors, investments).

4.2.2.1 Overall architecture and functionality. The SD planning approach involves the following process steps (Fig. 4.3):

1. *Screening* – helps to identify the company context, key issues, and initial focus. This information helps the user customize the approach to the user's unique situation, particularly with regard to potential goals and actions.
2. *Assessment* – the heart of the approach, this consists of three steps that can be performed iteratively:
   (a) *Self-evaluation* – to characterize the company's SD status (along the five *stages* of SD) for specified *elements*.

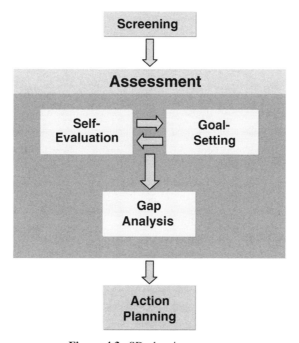

Figure 4.3. SD planning process.

(b) *Goal-setting* – establishment of goals for specified SD *elements*.

(c) *Gap analysis* – comparison of self-evaluation status against goals for specified SD *elements*, thus identifying gaps.

3. *Action planning* – identification of action plans to address the gaps identified above.

Screening. The Screening window (Fig. 4.4) requests several items of information regarding characteristics of the user's company or business unit. Issues that the user ranks as high in importance on this screen, such as "stakeholder scrutiny," are used later to suggest goals or action priorities. For most questions, the user simply selects a button from the choices provided.

Self-Evaluation. The Self-Evaluation window (Fig. 4.5) is used to assess the current status of the user's company or business unit relative to the elements defined in *SD Planner* ™. This information is used subsequently to determine whether the organization is meeting its stated goals and to recommend potential actions.

In this window, the user can view any of the seven *elements* in the form of a matrix. The columns correspond to business practices in the five progressive *stages* of SD. The rows, or *subelements*, correspond to the different *issues* that are included in each element.

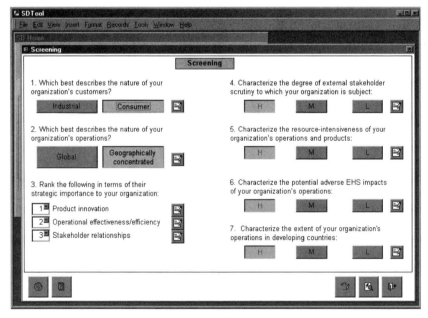

Figure 4.4. *SD Planner* ™ Screening window.

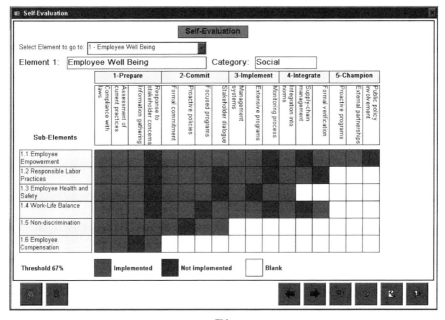

Figure 4.5. *SD Planner*[TM] Self-Evaluation window.

This window offers the user various options to bring up additional information, including:

- Definitions of terms.
- Further guidance regarding the interpretation of specific cells.
- Actions that you can consider implementing to address specific cells.
- Case studies of company experiences that exemplify the relevant practices.
- Comments that you can edit and retain in order to annotate specific cells.

Goal-Setting. The Goal-Setting window (Fig. 4.6) is used to establish specific performance goals relative to the elements defined in *SD Planner*[TM]. This information can subsequently be compared to the self-evaluation results in order to determine whether the organization is meeting its stated goals and recommend appropriate actions.

Gap Analysis. The Gap Analysis function is the heart of *SD Planner*[TM]. It is used to compare goals against self-evaluation results. Whenever the organization's current practices fall short of the goals, a *gap* is said to exist. The tool allows the user to view gaps in several different ways, one of which (by Element) is depicted in Figure 4.7.

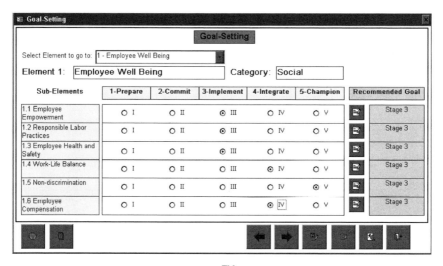

Figure 4.6. *SD Planner* ™ Goal-Setting window.

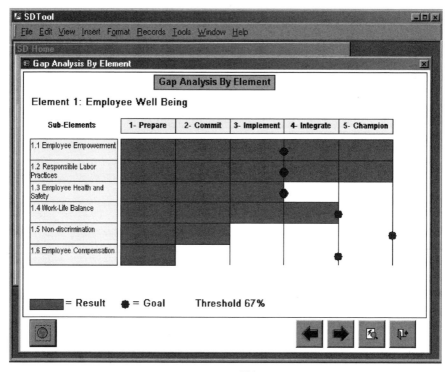

Figure 4.7. *SD Planner* ™ Gap Analysis.

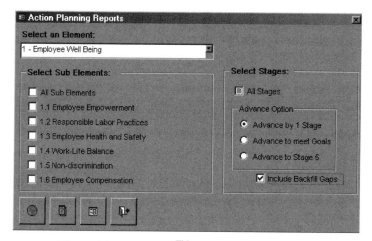

Figure 4.8. *SD Planner* ™ Action Planning window.

Action Planning. The Action Planning window (Fig. 4.8) enables the user to generate a customized list of proposed actions based on the identified gaps. It is expected that the user will take this list and use it to generate ideas for an action plan that is appropriate for the user's organization. After the user selects the elements of interest, *SD Planner* ™ can generate a set of proposed actions for every stage where a gap exists between goals and current practices. An example of an Action Planning report is illustrated in Figure 4.9.

4.2.3 Company Experience with *SD Planner* ™

Companies have adapted *SD Planner* ™ to suit their unique situations. Some examples of how companies have used the tool include the following.

- One company formed a cross-functional team to use the tool to help evaluate corporate SD status. Some gaps were identified (e.g., supply chain management and some stakeholder communication topics) that helped to develop corresponding programs. In addition, one of the key benefits of the approach was that it enabled learning and productive discussion within the cross-functional team that enhanced everyone's understanding of the SD issue and its importance for the company.

- A Canadian energy company formed an *SD Planner* ™ team consisting of Corporate SD staff, the vice-president of Communications and Public Affairs, the director of Employee Benefits, the manager of Corporate Community Investment, and the Supply Chain Manager, Major Projects. This team used the tool to identify SD gaps and to prioritize those gaps. One of the unforeseen and positive outcomes of using the tool was the dialog it fostered between people in different parts of the organization, thereby building sustainability-related knowledge capacity among the participants. The team also found the

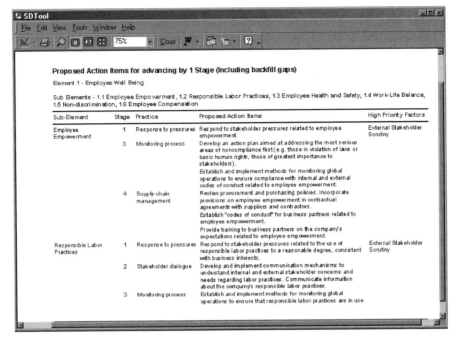

Figure 4.9. An Action Planning report generated by *SD Planner* ™.

tool useful in creating *SD Planner* ™ process at the corporate level, and some of the team discussions associated with using the tool revealed differences in performance and approach between business units. This led the team to delve further into the reasons for the differences and consider whether a consistent approach across the company was needed.

- Another company used the tool to help senior management think more globally about SD, to identify gaps in its product lifecycle management process, and to help its public affairs function to deal more effectively with stakeholders.

- Another company incorporated the tool into its training program for new environmental, health, and safety managers. Participants in the training use the tool to develop a comprehensive approach to addressing sustainable development for a hypothetical acquired business unit.

- A company that was formed as the result of a merger between two large industry rivals used the tool to develop SD metrics for the newly merged company.

- A business unit of another company used the tool to conduct a competitive assessment of the business unit versus its peers.

4.2.4 Conclusion

Companies around the globe are just beginning their journeys toward sustainability. The GEMI approach embodied in the *SD Planner* ™ tool was created to help companies address environmental, economic, and social issues in a way that creates business value. By considering the many possible pathways towards applying SD principles, companies can discover the pathway that best suits their unique characteristics and business conditions.

4.3 ENVIRONMENTAL MANAGEMENT SYSTEMS (EMS) FRAMEWORKS FOR SUSTAINABILITY

4.3.1 Expansion of Environmental Management Systems to Address Sustainability

ART GILLEN

First Environment, Inc.

At the very basic level, a management system provides a structure that allows an organization to systematically manage a particular issue by:

- Defining, assessing and understanding the issue within its boundaries;
- Planning and implementing controls and improvements; and
- Monitoring and evaluating performance of associated efforts.

These functions of a management system are reflected by specific processes, procedures, and tools necessary to manage the relevant issues.

4.3.1.1 Broadening the Scope of the EMS. Each particular management system design identifies key system elements as well as provides a focus important to addressing the management issue. Expanding management systems to address sustainability, in general, involves expanding the focus of the system, using a balanced approach, to include environmental, social and economic impacts and influence of the organization.

A key consideration regarding this expansion of environmental management systems is to recognize that the basic arrangement and the operation of the system does not considerably change. The scope of the management system is simply expanded to include issues, topics, ideas, and concepts representing social and economic concerns. For example, in addition to employee health and safety, the system could consider employee education programs. In addition to coordination of solid waste disposal services, an expanded management system could address community health care services. Or instead of driving hazardous waste reduction, a sustainability-expanded environmental management system could increase employee diversity.

Environmental management systems have an advantage over many management systems with respect to expansion toward sustainability. Environmental management systems already address one of the components of sustainability. This aspect makes them a perfect starting point for further advancements.

4.3.1.2   Plan, Do, Check, Adjust. Most management systems have the common foundation of the plan, do, check, and adjust (PDCA) cycle, which drives the organization's continual improvement of its performance regarding the issue. This common foundation provides a basic process for advancing an organization with respect to the focus management issue. After these boundaries of the system are expanded, the stages and steps of the PDCA cycle can also be enlarged to address the broader focus required for a corporate sustainability program.

In the first step of the planning stage, the expansion of the management system's scope can be reflected through the identification of sustainability in the policy statement and specific commitments to social, environmental, and economics issues that comprise it. This change generally identifies management recognition and support of sustainability that can improve the program's success.

The planning stage of a management system also generally includes an assessment of activities and impacts of the organization. Although environmental management systems focus only on impacts to the environment, the assessment can be expanded to consider both social and economic impacts. These issues can include but are not limited to employee diversity, working conditions, wages and benefits, education and advancement opportunities, supply chain, philanthropy, and community development.

The planning stage also includes the identification of objectives and targets for the organization. These objectives and targets often include taking action on significant issues identified during the impact assessment. The introduction of social or economic objectives and targets provides the first steps for an organization to move beyond considering sustainability and working on the path toward it.

The doing stage of the PDCA cycle typically involves development of programs to achieve the objectives and targets identified during the planning stage. Many organizations develop sustainability programs independent of a management system. Developing sustainability processes procedures and programs within a management system, in addition to ensuring these efforts are aligned with articulated goals of the organization, also ensures that these activities remain under robust document control and record keeping common to all management systems.

The doing stage of the cycle also addresses roles and responsibilities within the system. Expansion of the management system to address sustainability may require the assignment of role and responsibilities to individuals whose activities were not primarily associated with the original system activities. These may include representatives from human resources, accounting, investor relations, or even marketing departments.

The check stage of the cycle involves monitoring and internal and external assessment of performance. Organizations can use this step to establish metrics that address social and economic issues, in addition to the environmental measures that they already use.

The check stage also involves assessment of the effectiveness of the efforts. This assessment can take many forms, from internal auditing to independent verification. These checking activities are important to sustainability programs in providing confirmation and improved credibility to efforts in social activities that often are difficult to quantify.

Perhaps most important to the effective function of a management system, the check stage involves a process for recognizing and formally identifying problematic issues or potential opportunities, as well as taking actions to develop solutions or obtain additional benefits. A robust environmental corrective action process should require little modification to be applied to social and economic concepts.

Finally, the adjust stage of the PDCA cycle returns an organization's activities, impacts, performance, and progress to review by organization management. Within a sustainability system, this stage ensures that environmental, social, and economic issues remain under ongoing consideration in the development of strategy, resource allocations, and major operational decisions.

4.3.1.3 Specific Strengths for Sustainability of Existing EMS Designs.

Beyond the applicability and flexibility of a PDCA management system structure for corporate sustainability efforts, specific types of system designs also possess components that could be modified or enhanced to facilitate an organization's pursuit of sustainability. The subsequent contributions of this section provide overviews of the following existing management designs, whose focus is protection of human health and safety and the environment:

- *The American Chemistry Council's Responsible Care Management System*: This section introduces a management system that was customized to address the needs of the chemical industry.
- *EMAS versus ISO 14001*: This section addresses the additional aspects of the European Union's EMAS system, which builds upon the classic environmental management system structure of ISO 14001.
- *Occupational Health and Safety Assessment Series*: This section provides an overview of a management system design with a specific focus on human health and safety as compared to the broader scope of the environment.
- *Six Sigma*: This section covers using the quality management system and tools that assist organizations in implementing a process improvement methodology.

The strengths of each of these are as follows.

The ACC's Responsible Care Management Systems as well as EMAS systems include requirements for the public reporting of performance information. RCMS reportable metrics already include some economic measures for some organizations. Public reporting of performance information can increases transparency and facilitates stakeholder involvement. Organizations that use RCMS or an EMAS system can seek expansion to address sustainability by publicly reporting social and additional economic metrics and information.

RCMS includes requirements for community and stakeholder communications and outreach. While the basis of these communications and outreach under RCMS are typically environmental issues, an organization seeking expansion to sustainability could enlarge this program to address other issues of social and economic priority.

RCMS also requires third-party certification of any organization's efforts, which increases stakeholder confidence in reported information. This certification process can also be expanded to encompass a company's performance in social and economic areas.

EMAS requires more extensive involvement of employees in system operation and includes provisions regarding both suppliers and contractors. These aspects demonstrate important steps regarding stakeholder outreach. To expand the system to further support corporate sustainability, organizations with EMAS systems could include other stakeholders such as relevant government representatives, members of the communities in which they operate, their customers, in the operation of the management systems. This involvement would extend beyond external communication common to management systems and include participation in system operation such as the assessment of impacts, the setting of objectives and targets, or the development of sustainability programs.

Strengths of OHSAS 18001 include the system's component of proactive monitoring and measurement. This requirement could be more effective in addressing some social and economic issues that exist beyond the direct boundaries and control of an organization.

Six Sigma focuses on the integration of the human and process aspects of business improvement. By utilizing structured methods that link analytical tools into an overall program framework, there is now a standardized method for resolving work problems.

4.3.1.4   Summary. As presented in the following sections on various management systems, along with the basic PDCA structure, each of these management systems have strengths that can be leveraged to address the broader scope of sustainability's social, environmental and economic dimensions. In general, management systems provide a useful structure in which to develop sustainability programs. The conversion of an environmental management system to address sustainability involves expanding the focus of the system rather that significant modification. The stages and steps of the Plan, Do, Check, Adjust cycle can be expanded to effectively include social, environmental, and economic issues, which companies seek to address. Environmental management system designs, such as RCMS, EMAS, and OHSAS 18001, possess specialized requirements and components that can be further leveraged to more effectively pursue corporate sustainability.

4.3.2   The American Chemistry Council's Responsible Care® Management System

ART GILLEN

First Environment, Inc.

The chemical industry's major trade association, the American Chemistry Council (ACC), has been implementing a program called Responsible Care® for over 15 years. This program is the industry's comprehensive initiative to address environmental, health, and safety (EHS), security, outreach, process safety and "cradle to grave" product management. For more information about Responsible Care®, visit the ACC's web site at www.americanchemistry.com.

4.3.2.1   What is RCMS? RCMS is defined as a company's overall system to manage its Responsible Care® program. It replaces previous a Responsible Care® program that was based on meeting the 106 management practices of the initial six Codes. The Guiding Principles and management practices, including the seventh Code, Security are incorporated in the RCMS. As many ACC members plan to pursue ISO 14001 certification to meet customer or internal requirements or to further differentiate themselves from other stakeholders, a hybrid option of the RCMS, RC14001 was developed. As a result, there are two separate tracks for meeting the RCMS requirement, conformance with the RC14001 Technical Specification or the RCMS Technical Specification.

4.3.2.2   What is RC14001 Technical Standard? The RC14001 standard is based entirely on the ISO 14001 technical specification, including all 17 elements under the sections of Policy, Planning, Implementation and Operation, Checking and Corrective Action, and Management Review. It contains the whole of ISO 14001 with "boxes" under each element. These boxes describe additional Responsible Care® requirements. The technical specification is thus organized in a manner that allows for a company to receive an ISO 14001 only or an ISO 14001 and an RC14001 certificate.

4.3.2.3   What Are the Specific Additional Requirements in RC14001 above ISO 14001? These "boxes" of additional requirements differentiate RC14001 from ISO 14001. While there are a number of specifics, the boxed requirements fall into a few categories. They are:

- Health, safety and security;
- Community and stakeholder communications and outreach;
- Product stewardship;
- Transportation; and
- Commercial partner interactions.

4.3.2.4   RCMS Technical Specification. The RCMS standard is organized in the same manner, and contains mostly the same requirements. There are 30 elements divided into five sections: Policy and Leadership; Planning; Implementation, Operation and Accountability; Performance Measurement, Corrective and Preventive Action; and Management Review and Reporting.

4.3.2.5   So Why Two Specifications? The ACC Membership is very diverse in terms of the nature of their operations. They also have very different business needs. As a result, the association did not want to prescribe one specific method of certifying a company's Responsible Care® management system (i.e., ISO). The difference in the content of the standards, while minimal, includes terminology in the RCMS standard that more directly relates to the industry. For example, instead of using aspects and impacts and significance from ISO 14001, RCMS uses hazard, risk and prioritized. The most significant difference between the two specifications is the way each system is audited/certified.

Both RCMS technical specifications follow the Plan–Do–Check–Act framework after top management develops the organization's Responsible Care® Policy.

- *Plan* – Identify and evaluate environmental, health, safety, security, distribution, product, and process aspects and impacts, or hazards and risks. Establish goals, objectives, and targets and programs to meet them.
- *Do* – Establish roles, responsibilities, and accountabilities to implement the programs. Identify and provide necessary training, communication, documentation, and operational control.
- *Check* – Monitor and measure performance, investigate, correct and prevent nonconformances, including auditing the management system, and identify and maintain records.
- *Act* – Conduct a top management review of the management system to evaluate effectiveness and driving toward continual improvement.

4.3.2.6 ACC Certification Timeline. Regardless of which track is taken, ACC member companies are required to certify their Responsible Care® management systems. The Certification timeline is shown in Table 4.4.

4.3.3 EMAS versus ISO 14001

RAINER OCHSENKUEHN

First Environment, Inc.

Environmental management systems (EMS), and also ISO 14001, are becoming household names in many organizations. But what about EMAS? Almost unknown in the United States, EMAS is the acronym for the European Eco-Management and Audit Scheme, a voluntary system based on European Union (EU) regulations and

TABLE 4.4. Certification timeline

Company size	Minimum required	12/2005	12/2007	1/2008– 12/2010
1–25 sites	33% + HQ (max of 4)	HQ audited (minimum)	All required site audits	Next cycle
26–40 sites	6 sites + HQ	HQ audited (minimum)	All required site audits	Next cycle
41+ sites	8 sites + HQ	HQ audited (minimum)	All required site audits	Next cycle

harmonized principles. The two systems are complementary, but EMAS is more rigorous in some areas. EMAS continues to be seen as more prestigious than ISO 14001 in many EU Member States. It is open to any organization in the public or private sector that is committed to improving its environmental performance. It is open to Member States of the European Union and the European Economic Area (Norway, Iceland, and Liechtenstein). An increasing number of candidate countries are also implementing the scheme in preparation for their accession to the EU. Originally issued as European Council Regulation #1836 in 1993, which allowed voluntary participation by industrial company sectors in the EMAS, this Regulation was replaced by (EC) N°761/2001 OJ L 114 in 2001, which now allows participation in the scheme by *all* organizations.

New European Union environmental policies encourage the wider use of the EMAS logo (seen in Fig. 4.10) as a consumer-oriented tool to promote environmental improvements. Many organizations welcome such opportunities to make their environmental achievements more visible to stakeholders. Even the European Commission has committed itself to implement EMAS in its own services and buildings. As of July 2004, Germany had the most EMAS certifications, with over 2000 registered sites and additionally almost 1700 registered organizations (which included multiple sites). This is partly based on government/regulatory incentives, that is, grants for EMAS certifications, the increased environmental performance focus of EMAS, and also the historical emphasis of other countries within the European Union, such as the United Kingdom, to favor ISO 14001.

The objective of EMAS, similar to ISO 14001, is to ensure continuous improvements in environmental performance by getting companies and organizations to commit themselves to monitoring and improving their own environmental impacts (Fig. 4.11). The general aim of EMAS is to ensure that the European Community develops policies and implements actions to promote sustainable development and environmental issues. EMAS also covers the distribution of relevant information to the public. Independent auditors regularly check that participating companies and organizations fulfill their commitments, enabling them to keep their EMAS registration. ISO 14001 is a global standard created for corporate environmental management systems by the International Organization for Standardization (ISO). The main difference between the systems is that EMAS requires public reporting, while ISO 14001 does not. Furthermore, EMAS is considered a "stricter" regulation than ISO 14001, because the EMAS organization commits itself to the constant

Figure 4.10. EMAS logo.

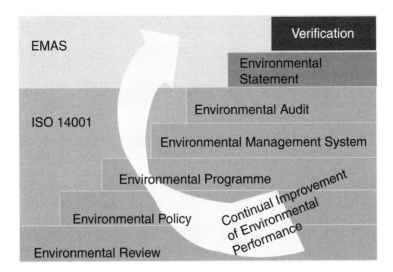

Figure 4.11. Objectives of EMAS. (*Source*: European Commission Environment Directorate-General.)

improvement of the level of its environmental performance. ISO 14001 "only" requires the continual improvement of the management system, with the implied improvement of the environmental performance. EMAS requires organizations also to undertake an initial environmental review and to actively involve employees in implementing EMAS. Further differences are listed in Table 4.5.

If an organization is already ISO 14001 certified, the recent revisions have made it easier to register for EMAS. Minor modifications will need to be made to the core ISO 14001 elements as well as some additional steps specific to EMAS.

The additional steps for EMAS registration include the following:

1. *Initial Environmental Review* – EMAS requires that an initial environmental review be performed to identify an organization's environmental aspects. If the organization already has an Environmental Management System (EMS) that is ISO 14001 certified, it does not need to conduct a formal environmental review when implementing EMAS (so long as specific environmental aspects in Annex VI are fully considered in the certified EMS).

2. *Environmental Statement* – The organization will need to prepare an environmental statement, based on the outcome of the EMS. At this point, the organization will need to check that the environmental statement fulfils the requirements of Annex III and examine all the data generated by the environmental management system to ensure it is represented in a fair and balanced way in the environmental statement.

TABLE 4.5. Comparison of EMAS and ISO 14001

	EMAS	ISO 14001
Preliminary environmental review	Verified initial review	No review
External communication and verification[a]	Environmental policy, objectives, environmental management system and details of organization's performance made public	Environmental policy made public
Audits	Specified frequency and methodology of audits of the environmental management system and of environmental performance	Audits of the environmental management system (frequency or methodology not specified)
Contractors and suppliers	Required influence over contractors and suppliers	Relevant procedures are communicated to contractors and suppliers
Commitments and requirements	Employee involvement, continuous improvement of environmental performance and compliance with environmental legislation	Commitment of continual improvement of the environmental management system rather than a demonstration of continual improvement of environmental performance

[a]With the proposed revision of ISO 14001 in late 2004, some changes in Clause 4.4.3, Communication, specifically address the European Requirements (EMAR/EMAS), related to the organization's external communication of its environmental aspects (environmental performance). This is strictly voluntary globally, although within the European Union it is required.

3. *Verifying the Environmental Statement and environmental performance* – For the organization to attain EMAS registration, the Environmental Statement must be independently validated (usually through the registrar). This process will check that the statement meets the requirements of Annex III and is publicly available.

Modifications to ISO 14001 to meet EMAS requirements include:

1. *Environmental Policy* – ISO 14001 includes a commitment, but not a provision, to comply with relevant environmental legislation. The organization must strengthen its statement of commitment included in its environmental policy to make provision for regulatory compliance. If more than one site is registered under EMAS then continual improvement must be demonstrated on a site-by-site basis.

2. *Planning* – EMAS has very specific requirements on the type of environmental aspects that may need to be addressed within the environmental
   management system, while ISO 14001 is less prescriptive in this area. The
   organization should ensure that in identifying its environmental aspects in
   the planning stage of ISO 14001, it has addressed the items listed in Annex
   VI that are applicable.

The organization should also ensure that all the elements of the initial environmental
review, detailed in Annex VII, have been considered and incorporated where necessary in the ISO 14001 process. It is possible that the areas and the scope covered by
ISO 14001 and EMAS may be different. The organization should take steps to
ensure that the scope to be covered by the EMAS registration is covered by the
ISO 14001 certificate.

3. *Implementation* – One of the additional requirements of EMAS is the open
   dialog with interested parties (employees, local authorities, suppliers, and
   so on). This includes the active participation of employees in the environmental improvement program.

The organization should also take steps to ensure that any suppliers and contractors
used also comply with the organization's environmental policy.

4. *Checking and corrective action* – Since the frequency of the audit cycle is not
   specified in ISO 14001, it is necessary for the organization to check that the
   frequency of the audit cycle is in compliance with Annex II of the EMAS
   Regulation and takes place at intervals of no longer than three years.
   In addition to the EMS being audited, the organization's environmental
   performance must also be addressed annually to demonstrate continual
   improvement.

5. *Certification of ISO 14001* – In order to comply with the requirements of
   EMAS, the ISO 14001 certificate must be issued under one of the accreditation procedures recognized by the European Commission.

4.3.4 Occupational Health and Safety Assessment Series (OHSAS 18001:1999)

ART GILLEN

First Environment, Inc.

Parallel to the development of ISO 14001:1996 – Environmental Management
Systems – several organizations developed guides, draft specifications, and requirements for occupational health and safety management systems (OHSMSs). Most of
this development has been done by management system registrars with the expectation that an OHSMS be accepted and issued by a national or internationally accredited standards body, that is, International Organization for Standardization (ISO),
British Standards, and so on.

The most recognized version of an OHSMS is British Standards Institution (BSI) Occupational Health and Safety Series OHSAS 18001 – Occupational Health and Safety Management Systems – Specification. This specification, prepared to assist organizations in managing their health and safety programs and to improve performance, was released in the spring of 1999. OHSAS is not an ISO specification, nor has it been accepted by any other accredited standards body. Organizations can, however, obtain third-party certification for their OHSMS.

According to the BSI, the OHSAS 18001 specification was developed in response to urgent customer demand for a recognizable occupational health and safety management system standard against which their management system can be assessed and certified. OHSAS 18001 has been developed to be compatible with ISO 14001. They both contain 17 elements and 95 percent of the differences between the two are replacing "environment" and "environmental" from ISO 14001 with occupational health and safety (OH&S) for OHSAS 18001. Other differences are on emphasis and format. OHSAS 18001 stresses communication to employees and interested parties, top management's commitment and role, workplace safety and ergonomics within operational controls. It also distinguishes between reactive and proactive measurement and monitoring and adds accidents and incidents to nonconformance and preventative and corrective actions.

In addition, there are minor phrasing or syntax changes to OHSAS 18001 from the ISO 14001 specification. These include the addition of risk to many sections, and presenting the sections in outline rather than paragraph form. OHSAS 18001 also takes advantage of its later publication date to add a few words here and there for clarification.

The major change is in Section 4.3.1 – Environmental Aspects under ISO 14001 and Planning for hazard identification, risk assessment and risk control under OHSAS 18001. OHSAS is much more detailed and prescriptive in how hazards are identified, and how risks are identified, assessed and controlled compared to how aspects and impacts are managed under ISO 14001. The environmental benefits derived from implementation of ISO14001 would similarly be expected to be realized by including occupational health and safety into the applicability of an environmental management system. For the record, many organizations have chosen to add occupational health and safety into the scope of their ISO 14001 EMS.

For that reason, many manufacturing and service organizations are resisting the standard's accreditation of OHSAS 18001. In many circles this standard is viewed as an additional accreditation expense with no value added. The resistance to developing ISO 18001 within many nations further demonstrates this point. On the other hand, the recent proposed changes to ISO 14001 are expected to call for similar changes to OHSAS 18001. Still others are stating that OHSAS 18001 is too new to be revised and are requesting status quo for the time being.

Comparisons between OHSAS 18001 and the new Responsible Care® Management System (RCMS) Technical Specifications are expected. While the RC14001 specifications do add health and safety to ISO 14001, there are other topics included in both RC14001 and the non-ISO RCMS specification.

4.3.5   Using Six Sigma Management Initiatives

ROBERT B POJASEK

Pojasek and Associates

Sustainability programs are usually focused on "projects" and working in a more general fashion to improve predesignated sustainability indicators. It is much too easy to isolate the sustainability program in its own "silo" with this approach. Much of the rest of the organization may already be using quality management techniques and tools to organize its improvements that are designed to keep the company competitive in the marketplace. This section specifically addresses how six sigma techniques may help improve the sustainability effort while helping to integrate it into the core business.

Three popular management initiatives are "lean," "six sigma," and lean six sigma. Lean is process improvement methodology that emphasizes gains in quality by eliminating waste and reducing delays and total costs. It fosters an organizational culture in which all workers continually improve production processes and their own skill levels. Six sigma focuses on eliminating wasteful mistakes and rework using a measurement approach, statistical analysis methods, and alignment to organizational priorities to increase customer satisfaction and enhance the bottom line. The designation, six sigma, denotes that the operation is capable of having only 3.4 defects per million operations. Using sigma as a common metric across processes permits the comparison of relative quality across similar and dissimilar products, services, and processes. This ability to measure in a consistent manner is largely absent from sustainable development initiatives. Lean six sigma represents a recent morphing of these two process improvement programs into a single effort.

Neither lean nor six sigma are standardized in practice. The terms have been used to describe many variants of these programs. No matter what form of lean six sigma program is in place, there are four specific programmatic aspects that are important to a sustainability program:

1. *Integration of the human and process aspects of business improvement.* This includes operating with a sense of urgency and correcting problems focusing on customer and other interested party concerns. All work is conducted in project teams seeking bottom-line results and emphasizing continuous improvement and innovation. Lean six sigma creates a constancy of purpose in an organization by adding a new dimension to business process measurement – variation as an indicator of process improvement. Everyone is involved in the program.

2. *Concentration on determining bottom-line results using structured methods that link analytical tools into an overall program framework.* Management understands that it is their responsibility to foster and encourage improvement efforts. This is done by making the improvement of the company's business processes and products/services a part of every employee's job, and providing appropriate training at all levels of the organization.

3. *A standardized method for resolving work problems.* Workers use quality management tools on projects that are approved by management. Projects

are chosen to improve the business processes (i.e., those processes that provide competitive advantage for the enterprise).

4. *An integrated, phased approach to applying standard quality management tools for problem solving and decision-making.* The quality tools are used in a prescribed order to optimize the use of these tools on each lean six sigma project. By using the same tools in all projects, it is much easier to transfer information and share best practices.

These four programmatic aspects would help organize sustainability projects into a coherent program, as opposed to conducting separate projects. In other words, recognition and use of these aspects would help operationalize the sustainability program.

The use of quality as the means of *communicating* the program between workers, management, and outside parties is very important since this has been the language of business for many years. It is very important that the sustainability program learns to translate many of its objectives and targets into the well-known language.

Lean six sigma also promotes a culture that motivates employee teams to work on common problems in an organization to achieve higher levels of performance effectiveness and productivity at a lower cost. Once an organization has operated a lean six sigma program for a number of years, the concepts of management by fact, root cause analysis, and the definition of problems according to their source of variation all become part of the organization's business language and form a common bond between employees at all levels.

DuPont became an early adopter of the six sigma philosophy. Many other companies in the chemical industry have followed. The following groups are also actively promoting lean six sigma: the National Institute for Science and Technology (NIST), the American Society for Quality, and a number of trade associations. It is very likely that the use of these business improvement methods will continue to expand in the future.

All sustainability projects could be easily adapted to be consistent with the lean six sigma framework. Integration of these management initiatives into the sustainability program (or vice versa) will require changes in how sustainability programs are operated.

1. Sustainability needs to develop a process focus. All sustainability issues must be assigned to a process, that is, back to the source. The process needs to be mapped so everyone will understand it and see the connection. Hierarchical process maps (Pojasek, 2003a) are particularly helpful since they encourage systems thinking – an important concept for sustainability programs that is largely absent in lean six sigma. All resources, activities, and information for the selected work step can be viewed as the employee team initiates its work on the project. It is easy to identify everyone with a connection to this issue at its source.

2. The sustainability project teams need to convince the lean six sigma project directors (often referred to as value stream managers in lean and "black

belts" in six sigma) for permission to begin work on these projects within the system. These program "gatekeepers" are often uninformed on environmental and social responsibility issues. They will need to be convinced that there is a strong relationship to a process and will look for ways to change the process while meeting strict financial return criteria. These program managers are often judged largely on their financial returns. It is important that the sustainability program people become educated in their company's lean six sigma programs. They need to be trained to a level referred to as "green belt" in six sigma. It certainly would not hurt to have at least one sustainability black belt to lead the integration effort.

3. All sustainability projects should be conducted using quality management tools to help the program become more mainstream in the business organization. Six sigma uses a five-step problem-solving process: define, measure, analyze, improve, and control (i.e., DMAIC). The use of this process helps establish the business context for a six sigma change management process. Each of the DMAIC steps uses quality management tools. A Systems Approach to process improvement has been proposed (Pojasek, 2002) that has selected from the body of quality management tools those tools that are the most visible and the most interactive in their use. Because it takes considerable skill to use quality management tools effectively, it is very important that a lean six sigma program seek to standardize its use of tools while limiting them to a small number of tools. The systems approach helps accomplish this task. These tools have been used in ISO 14001 programs (Pojasek, 2002) that also use a quality approach to project development.

4. Many lean six sigma programs have formal supply chain programs and "design for six sigma" (DFSS) components. The objective of the DFSS is to ensure that the processes, products, or services consistently meet customer needs and to anticipate the changing requirements of the future market. This aspect of six sigma is mirrored in lean by the development of "future state" value stream process maps. Sustainable development programs can use these aspects of lean six sigma to help keep their programs focused on continuous improvement and innovation.

If your company does not have a lean or six sigma management initiative, you could also use one of the other management systems described in this chapter for process improvement to apply such a management structure to the sustainable development program and provide it with a business language. This can be done within the ISO 14001 program or independently using one of the other systems approaches. When your company does decide to implement lean six sigma, the sustainability program will already have a consistent approach that can be easily accommodated into these popular process improvement programs.

Should you already have a lean six sigma program in your company, it is important that you understand this program and learn how to use its operationalization methods. A more detailed examination of the use of these programs for guiding sustainability and other process improvement projects has been prepared (Pojasek,

2003b). A number of books that describe these programs have been provided in the "suggested reading section" of this chapter that can start you on the path to either integrating lean six sigma into your sustainability program or integrating your sustainability program into the lean six sigma effort. The triple bottom line focus of sustainable development is a natural fit no matter which way you go.

4.4 THE NATURAL STEP FRAMEWORK: BACKCASTING FROM PRINCIPLES OF SUSTAINABILITY

KARL-HENRIK ROBÈRT

Blekinge Technical University and The Natural Step, Stockholm

SISSEL WAAGE

The Natural Step, USA

DICKSEN TANZIL

BRIDGES to Sustainability

Most people are familiar with CFCs (chlorofluorocarbons) and how they eventually became doomed as an input into modern industrial products. Ironically, these compounds were initially introduced as environmentally perfect alternatives due to their nontoxic and nonbioaccumulative nature. This is one of many examples of decisions – in this case about "safe" materials – that have been made on large scale, only to be followed by a late awakening and significant costs to society and individual organizations. Some of the more recent examples now looming on the horizon may be even worse due to their direct impacts on humans – antibiotic-resistant strains of microbes from antibiotics in biota, hampered kidney function from cadmium in foods, and endocrine disruption from plastic additives, to mention just a few.

On the principle level, society is repeating the same kind of mistake over and over again. The industrial history of such events tells us a few things that should be kept in mind for future planning. It has shown that impacts generally occur through very complex interactions in the biosphere, which generally cannot be determined beforehand. At best, a certain impact can – after it has occurred – be clearly linked to a certain activity or process. However, these findings are usually associated with delays between the time of initial use of the compound and discovery of impacts.

Given this situation, it seems advisable to develop another approach to planning and decision-making that could more adequately take into account these realities. Such an alternative approach would not be based on an exhaustive understanding of all variables. Rather, an alternative method could be based on the discovery, and disciplined use of, *principles* by which sustainability-oriented decisions – such as the selection and management of materials – are determined upfront and "upstream," or at the original point of decision of whether or not to use a particular input instead of after damage has already occurred.

In the late 1980s, a group of Swedish scientists set out to develop such an approach. This approach, called The Natural Step Framework for strategic sustainable development, has been developed and published in numerous scientifically peer-reviewed papers (e.g., Robèrt, 1994, 2000; Holmberg and Robèrt, 2000; Robèrt *et al.*, 2002). It has also been applied within businesses to enable decision-makers to understand sustainability issues and begin planning based on these new insights (Bradbury and Clair, 1999; Nattrass and Altomare, 1999, 2002). This chapter describes this planning and decision-making framework and how it is developed and applied.

4.4.1   "Backcasting" from Principles of Success: Introduction to The Natural Step Framework and Approach

The Natural Step Framework is built on "backcasting" from "basic principles of success." Backcasting refers to a planning procedure by which a successful outcome of the planning is imagined in the future, followed by the question: "what was it that we did today, that allowed us to get there?" The term "basic principles of success" denotes principles that

- are general enough to cover the successful outcome (i.e., social and ecological sustainability) and to be independent of scale or field of activity;
- are concrete enough to guide problem analysis and creative solutions;
- are not overlapping, so that comprehension is supported, and so that metrics for the monitoring of progress can later be developed.

The methodology has been elaborated from "backcasting from scenarios" (Robinson, 1990), a planning approach that begins with envisioning a picture of success. This scenario approach is similar to working on a jigsaw puzzle while being guided by a specific image of the finished puzzle.

Backcasting from scenarios, however, has some disadvantages when applied to sustainable development. First, it is difficult to develop detailed scenario descriptions of a successful *sustainable* outcome upon which many people will agree. Secondly, there is often a resistance to making very detailed plans in the light of the ongoing technical development that may subsequently change the conditions for the planning. Finally, how do we know if a detailed description of a sustainable enterprise or society really *is* sustainable? Decision-makers appear to face great uncertainty.

These shortcomings, however, can be addressed by an adapted approach of 'backcasting from basic principles of success.' This approach resembles a game of chess rather than a jigsaw puzzle. It is the principles of success – such as the principle of checkmate – that guide the game. This is a dynamic way of planning, whereby each move takes the current situation of the game into account while at the same time optimizing the possibility of winning, which can come about in many different ways. One of several advantages of this method is that it is easier to agree on basic principles for success – as well as some concrete steps that can serve as flexible stepping stones in that direction – than to agree on detailed descriptions of a desirable final outcome. Finally, the two methods can be combined – scenarios can be scrutinized by basic principles of sustainability.

4.4.2 Understanding Complex Systems and Thinking Upstream: Rationale for the Principles of The Natural Step Framework

Within a framework for planning in complex systems, it is essential to keep five hierarchial levels of decision-making and not confuse them with each other (Robèrt, 2000) (see Fig. 4.12):

Level 1: System – articulation of how the system is constituted.

Level 2: Success – setting vision and identifying desired outcomes in the system.

Level 3: Strategies – to achieve vision and move purposely toward success.

Level 4: Actions – concrete measures that will lead to the desired outcomes.

Level 5: Toolbox – set of tools to assess, manage, and monitor the actions.

For example, the use of renewable energy belongs to the Action category (level 4), not the Success category (level 2). Changing to renewable energy is not a principle, but something that we do. While it is an important step forward, it is important to

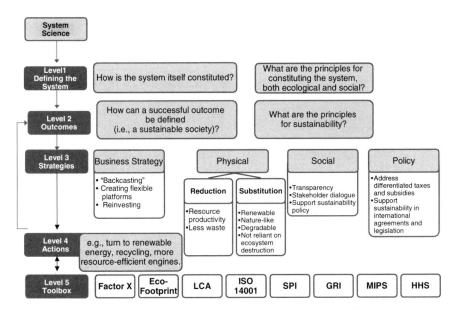

Figure 4.12. Overview of the "Strategic Sustainable Development" decision-making approach (adapted from Robèrt *et al.*, 2002).

note that it may lead to significant ecological alterations, such as flooding areas in generating hydropower, or deforesting lands through extensive use of biomass fuel sources. Therefore, this action is not in itself a principle for sustainability. Rather, renewable energy should be introduced in a way that will comply with principles for sustainability.

This example highlights that at the heart of planning is the Success level, which should inform strategies, actions, and the design of our tools. This level is understood through "backcasting from the principles of success," or imagining that the conditions for success are complied with and then asking: "What shall we do now to optimize our chances to reach this successful outcome?"

As detailed in Figure 4.12, in the case of sustainability, to arrive at a principle definition of Success (level 2), we must know *enough* about the System (level 1), which is both the biosphere and the human societies as well as the interactions and flows of materials between the two. Since the concept of sustainability (level 2) becomes relevant only as we understand the *non*sustainability inherent in the current activities of society, it is logical to design principles for sustainability as restrictions that determine what human activities must *not* do in order to avoid destroying the interrelated ecological and social systems. CFCs, for example, were thought of as "harmless" in the relatively recent past. It is incumbent upon us to ask what compounds are perceived of as harmless today, but may be understood differently tomorrow. In what ways – expressed as principles – could we destroy the system's (of both the biosphere and society) ability to sustain us?

The negative impacts related to nonsustainability that we encounter today can – on the basic principle level – be divided into three separate mechanisms by which humans can destroy the biosphere and its ability to sustain society:

1. A systematic increase in concentration of matter that is net-introduced into the biosphere from outside sources.
2. A systematic increase in concentration of matter that is produced within the biosphere.
3. A systematic degradation by physical means.

Sustainability of society also depends on addressing human needs through the maintenance and robust functioning of social systems, including both formal institutions as well as the informal structuring of civic society at large. These issues relate not only to sustaining society, but also to complying with the three ecological constraints. Therefore, these interrelated social and ecological dynamics necessitate a fourth basic condition that takes social aspects of sustainability into account. By adding "not" before each of these mechanisms, it is possible to identify a set of basic principles for defining sustainability in the system (i.e., the biosphere and society). These four principles are articulated as system conditions in Box 4.1. Taken together, the first three system conditions define an ecological framework for any sustainable society and the fourth principle is the basic social condition.

BOX 4.1   THE NATURAL STEP'S FOUR SYSTEM CONDITIONS

In the sustainable society, nature is not subject to systematically increasing...

1. concentrations of substances extracted from the Earth's crust;
2. concentrations of substances produced by society;
3. degradation by physical means;

and, in that society...

4. the ability of humans to meet their needs worldwide is not systematically undermined.

Correcting errors at this basic systems level (level 1), or "upstream" in cause–effect chains, is the only way to both come to grips with current problems and *avoid new problems*. Understanding this level – the "basic rules of the game" from a biophysical and societal perspective – makes it possible to ask the right questions and to structure all the details in a way that makes sense for decision-making. If CFCs had been scrutinized up front against these basic principles, it is obvious that they would not have passed questions related to the second principle: CFCs are made to increase in concentration in the biosphere as long as they are used on a large scale without rigorous control to hinder them from leaking out into the biosphere. Any compound that is "left over" and increases systematically as waste in the system is inherently unsustainable. Sooner or later it will pass its ecotoxic threshold and since complexity makes it impossible to foresee such thresholds, we need to solve the problem up front and on the principle level – looking through the front screen and not chasing after reality in the rear view mirror after problems and impacts have already occurred. Thousands of compounds are now increasing systematically in the system and are consequently not managed in a sustainable way.

4.4.3   Applying The Natural Step Framework: A Strategic Sustainable Development Decision-Making Process

The Natural Step has placed these principles into a framework for sustainable development, which is used in conjunction with "backcasting" techniques. Actions are launched, step by step, in a strategic way to serve as viable "stepping stones" towards compliance with the sustainability principles. The framework is systematized as an approach to facilitate the brainstorming sessions and team-planning that is presented in the "A,B,C,D" methodology.

The process begins with *phase A*, in which the framework (with the system conditions shown in Box 4.1) is explained to offer a shared mental model for "community building" among the participants of a planning process. This step enables the group to play the "chess game" of sustainable development by the same rules.

The second phase, *phase B*, includes conducting an assessment of where an organization is *today* in terms of sustainability. This task is done by developing a set of diagrams, lists, and analyses of all current flows and practices that are critical from a sustainability perspective. At the same time, the group considers all of the assets that are in place to deal with the problems.

Thirdly, *phase C*, is the creation of solutions and visions of *tomorrow*, including applying the constraints of the system conditions to decision-making processes in order to spark creativity and both list and scrutinize suggested solutions. This process is undertaken in a type of brainstorming session where top management should be actively engaged. At this point, no significant constraints are applied, other than what is theoretically feasible.

Finally, the fourth phase, *phase D*, is making priorities from the list developed in the previous step and launching concrete programs for change. This step provides the framework with its strategic component. Suggestions from the list generated in the C phase are prioritized to be launched relatively early on – to serve as stepping stones for further improvements – based on responses to the following three questions.

1. *Does this measure move in the right direction as regards all system conditions?* Sometimes a measure represents a trade-off, moving in the right direction as regards one of the system conditions, but at the expense of some of the others. Asking this question enables decision-makers to see the full picture and find complementary measures that may be needed to take all system conditions into account. Consequently, this question often leads to increasing the lists of opportunities and potential solutions.

2. *Does this measure provide a stepping stone for future improvements?* It is important that investments can be further elaborated or completed in line with the system conditions, so that they do not lead into dead ends. For example, it would be unwise to invest heavily in a technology that will cause fewer impacts in nature in the short term, but would be difficult to later adapt, or move away from, to achieve ultimate and complete compliance with the system conditions.

3. *Is this measure likely to produce good return on investment soon enough to fertilize the further process for future improvements?*

It is the combination of "yes" to all three questions that provides the strategic element of the framework. Each suggested investment is scrutinized as regards its potential to reduce impacts, develop further towards sustainability, and allow for funding of ongoing work, particularly within a business context. Each individual organization must draw its own conclusions from these basic principles as regards problems, solutions, goals, and objectives. However, the framework provides a systematic way of guiding this intellectual process, through an "A,B,C,D" process.

4.4.4 Integrating Sustainability Thinking into Action: The Dynamics of Dematerializations and Substitutions

To consider how the principles of sustainability may be put into action, let us consider the future sustainable management of materials. Complying with the system conditions requires combined dematerializations and substitutions (transmaterializations) (Robèrt *et al.*, 2002). These actions mean that when society is managing materials in a sustainable way, all compounds will have ceased to systematically increase in the biosphere, not only the ones that are currently causing identified impacts. This include matters that are net-introduced from the Earth crust (system condition 1), such as metals and minerals, as well as substances produced by society (system condition 2), such as chemical products and unintentionally produced emissions and releases from the entire product lifecycle.

Furthermore, managing materials sustainably also requires that the biosphere not be degraded by physical means (system condition 3). Thus, renewable materials will not be overharvested and/or purchased from poorly managed ecosystems or from companies that are failing to restore ecosystem. In addition, infrastructure for transport will not be growing systematically and erode the landmass of productive ecosystems.

Finally, to comply with system condition 4, materials will not wasted and/or made unaccessible by other means to people in less affluent areas of the society. Nor does the extracting, manufacturing, transporting, warehousing, distributing, and marketing of any items contribute to social behavior or abuses that undermine people's capacity to address their human needs and to live a fulfilling life.

In turn, from an industrial perspective, this future will require the following:

- *Dematerializations*, by means of higher resource productivity and less waste. Such dematerializations, such as recycling or improvements of design, will allow for higher material performance per unit of service. These actions will avoid accumulation of waste (system condition 1 and system condition 2) and reduce the physical pressure on productive ecosystems (system condition 3). In addition, these actions will increase resource productivity and reduce waste, which will feed into the possibility of sufficient resources for people on the global scale (system condition 4).
- *Substitutions* will also be needed, as many of the currently used materials and management routines are so problematic from a sustainability perspective that they will be too expensive to safeguard within the constraints provided by the system conditions. Consequently, dematerializations will not be enough to reach sustainability. Examples of such needed substitutions include:
  - heavy metals that are normally very scarce in ecosystems (e.g., cadmium; Azar *et al.,* 1995) (system condition 1);
  - chemicals that are relatively persistent and foreign to nature (e.g., bromine organic antiflammables) (system condition 2); and
  - materials that are extracted in ways that do not restore natural systems (e.g., strip-mining and timber from poorly managed ecosystems) (system condition 3).

Such flows should not only be dematerialized – which is necessary during a transition period – but in the end should be phased out and substituted with other materials and practices.

New materials should be selected in a way that maximizes the benefits for a global society and presents opportunities for future generations that will be easier to adapt within the constraints of the system conditions. This means that the flows of certain other materials may not be dematerialized, but will be *increased* in relation to current uses in order to arrive at a sustainable society. Other materials may be scarce and foreign to nature, and yet their respective flows may be essential for sustainability and, consequently, need to be *increased* in a sustainable way (i.e., safeguarded by extraordinary societal means and "closed-loop" processes). Examples could be scarce metals in thoroughly recycled photovoltaic cells.

The practices of dematerialization and substitution are not only important independently, but are also interrelated in a dynamic way that should be utilized for planning. For example, the less degradable a material is, the more it must be safeguarded and/or dematerialized within the "techno-sphere," or industrial systems, particularly if it is relatively scarce in natural systems. For scarce metals the assimilation is slow and occurs as sedimentation and biomineralization. For chemicals that are relatively persistent and foreign to nature, assimilation occurs also as degradation with relatively long half-lifetimes.

Finally, it is essential to note that there are economic relationships between dematerialization and substitution. For example, when very profound dematerializations are not sufficient for sustainability goals – perhaps because materials are so relatively nondegradable and/or impact levels in natural systems are already trespassed (e.g., CFCs or PCBs) – then substitutions may be relatively expensive early on. This cost ratio is a function of the early production volumes of the substitutes, which are likely to be relatively small in the initial stages of a transition. Furthermore, these changes will often require investments in new infrastructure. One example is the development of new coolants in refrigerators, in the shift away from CFCs, and requiring new refrigerators to fit those new coolants. Making the *substitutions* affordable and the implementation of new technologies is often made possible through various kinds of dematerializations, such as higher resource productivity and less waste within the new and more expensive production lines and products (Holmberg and Robèrt, 2000; Robèrt *et al.*, 2002). In short, dematerializations support substitutions, and substitutions will prompt dematerializations.

4.4.5   Industry Example: PVC Production at Hydro Polymers

PVC has been the target for serious attacks from NGOs during at least two decades. It is often perceived as inherently nonsustainable due mainly to pollution throughout the whole lifecycle of the material, but also as regards its contribution to the greenhouse effect. Top management of Hydro Polymers, one of the leading manufacturers of PVC in Europe, decided to take on the intellectual challenge of scrutinizing PVC from a sustainability perspective. They are now pioneering a major sustainable development programme for the plastics industry using The Natural Step Framework.

The B and C analyses in the A,B,C,D methodology displayed a very wide window of opportunities. On the one hand, PVC has currently a number of positive qualities from a sustainability perspective, the most important of which are its long lifetime, lightness, weather resistance, low flammability, and the fact that it requires little maintenance and is easy to mold and color. On the other hand, those aspects are built on the chlorine and the additives that make PVC currently the subject of heated debate. Backcasting from compliance with the system conditions made Hydro Polymers endorse the following challenges:

- The industry should commit itself long term to becoming carbon neutral (system condition 1 – plastics are currently produced with petroleum and natural gas as raw materials which amount to around 3 percent of the total use of these fossil raw materials).
- The industry should commit itself long term to a closed-loop system of PVC waste management (system conditions 1 to 4 – in relation to a back-casting perspective, today's use of PVC in society is highly wasteful, with around 50 percent ending up on land deposits).
- The industry should commit to ensuring that releases (emissions) of metals and persistent organic compounds from the whole lifecycle do not result in systematic increases in concentration in nature (system condition 2).
- The industry should review the use of stabilizers and additives consistent with attaining full sustainability, and especially commit to phasing out, long term, substances that can accumulate in nature, or where there is reasonable doubt regarding toxic effects (system condition 1, as such compounds include heavy metals as stabilizers, and system condition 2, as a number of organic additives are persistent in the environment and foreign to nature).
- The industry should commit to the raising of awareness about sustainable development across the industry, and the inclusion of all participants in its achievement.

The D part of the analysis, that is, some early flexible stepping stones, can be exemplified by a number of "low-hanging fruit" that are already picked and related to the long-term goal:

1. Education of personnel throughout the Hydro factories in Europe, making A,B,C,D analyses part of the education and training, as well as a source of new ideas to top management.
2. Dematerializations of flows, for instance making exothermic processes endothermic and utilizing scrap as raw materials.
3. Hydro is actively working to develop a new paste process involving a particle distribution that means less plasticizer needs to be added, and in which new types of plasticizers can be used (flexible platform). Hydro is also involved in international research on new PVC production methods and the development of new PVC materials.

Over the past few years, Hydro Polymers and the PVC processing industry have, in cooperation with stabilizer manufacturers, developed new, modern calcium/zinc stabilizers. A few product areas, such as cables, have already changed from using lead stabilizers to using calcium/zinc ones. In the long run, as revealed by a backcasting analysis, zinc is also a doubtful stabilizer because it is relatively sparse in ecosystems in comparison to the societal flows of the metal. This is a particular issue since zinc is purposely used by society in a dissipative way to protect iron from oxidizing, that is, 'not contributing to materials from the Earth's crust increasing in the Biosphere' has made Hydro consider the calcium/zinc stabilizers as a platform for other solutions.

4.4.6   Discussion and Conclusion

The Natural Step Framework and approach are elaborated from the constraints determined by basic principles of sustainability. Backcasting from basic principles of sustainability is a framework that covers relevant aspects of how to plan ahead in a complex system, such as for societies within the biosphere. The approach brings a sustainability perspective to analyses of current practices and materials, suggested solutions and visions, and the strategic evaluation of various solutions and paths to arrive at sustainability. And it brings this new perspective in with opportunities for improved economic outcomes.

A framework for sustainable development is neither an alternative to scientific studies and facts, nor specific concepts and tools to deal with such facts and inform actions. All these elements are essential. Rather, a framework stitches it all together, creates comprehension, and provides direction to the planning. Without a full systems-based approach and framework, it is difficult to:

- ensure that all aspects of sustainable use of materials are considered from a full systems perspective;
- enable decision-makers to assess current data and information on sustainability in a structure that is relevant for strategic decisions;
- discover areas where more information is necessary – or unnecessary – for making relevant decisions;
- focus problem-solving upstream at the source of problems, in order to design problems out of the system;
- evaluate alternative materials solutions and visions from a strategic point of view, so that blind alleys can be avoided;
- deal with trade-offs in a strategic way;
- build creative assessment and problem-solving communities through shared mental models;
- involve all aspects of business in a cohesive manner, including leadership, management, programs of activities, product-development, choice of materials, indicators, and so on.

Sometimes there are many possible choices that fit the presented framework and can serve as a strategic stepping stone towards consonance with the system conditions. How can a decision-maker determine priorities among various options? Is it do-able to come up with checklists or manuals to support decisions beyond the overall framework with its guidelines for dematerializations and substitutions throughout the lifecycle of materials? Given that complete compliance with the system conditions is the ultimate goal, on what grounds can trade-offs during the transition be managed? How are uncertainties as regards compliance with the principles addressed?

From other complex systems, such as chess, a couple of essential conclusions can be drawn in this respect. First, once basic rules are clear, the individual's potential to deal with trade-offs and to optimize chances in multidimensional and complex situations is very large. Secondly, one of the most essential elements to utilize this potential of the individual, and to become professional, is learning and getting more and more experienced. And, finally, beyond a certain level of specificity, checklists confuse more than they help.

Therefore, it is unlikely that very detailed checklists or manuals can replace any of the time-consuming training it takes to be a professional planner in a complex system. The reason is that when decision-makers choose between various strategic options for sustainable development, there are so many categories of criteria that are simultaneously in play and that present themselves as gray areas, which results in each situation having a tendency to be unique. Or in other words – attempts to come up with very detailed hands-on manuals that are layered on top of a framework of basic principles and their respective guidelines have a tendency to result in so many feedback loops and footnotes and exceptions to the rule that they risk confusing more than they help.

The conclusion is clear. It is not possible to create up-front comprehensible and easy-to-handle and very detailed checklists or manuals for the detailed management of complex systems. Problems are generally multidimensional, and each dimension presents itself as gray areas. Instead, the overall recommendation would be to make principles for success very clear up front, as well as create smart overall strategies and guidelines to approach those principles (i.e., to apply a framework for decisions as a shared mental model among team members), and then to get on with the learning and playing the game. This process allows for deep experience in seeing the large picture of the goal and selecting stepping stones in that direction.

Finally, as the process unfolds – and the marginal costs in relation to utility and profit decrease, as more and more "low hanging fruit is picked" – it is likely that a need for more sophisticated tools will evolve, including, ISO 14001, lifecycle assessments (LCA), tools for product development, purchase manuals, and so on. To ensure, however, that all efforts are continuing to move in the same direction, all of these tools should be informed by the same framework as is informing the business program – backcasting from basic principles of success.

4.5 NATURAL CAPITALISM FOR THE CHEMICAL INDUSTRY

CATHERINE GREENER

Rocky Mountain Institute

4.5.1 What is Natural Capitalism?

Natural and capital are two words that are not usually found together. Capital is typically defined as the money or financial assets and machinery that are needed to produce goods and services needed to create wealth. Capitalism is the organization of society supporting capital as the central driver for material acquisition necessary to produce large-scale wealth and economic growth with no end. It is an economic system in which the means of production and distribution of goods are privately or corporately owned and distribution is proportionate to the accumulation and reinvestment of profits gained in a free market. With profits defined from the distribution of goods, economists consider manufactured capital – money, factories, and so on – the principal factor, and perceived natural capital as a marginal contributor. The exclusion of natural capital from balance sheets was an understandable omission. There was so much of it, it did not seem worth counting. Nature seemed free, or a gift to be used for the creation of wealth. The success of the industrial revolution reinforced and popularized the existing model of capitalism and economics, but nature suffered.

The Industrial Revolution enabled people to be vastly more productive at a time when previously the per capita output was limited by time, technology, process, but not limited by the vastly abundant natural world. Beginning with the Industrial Revolution the raw materials used by the chemical industry existed in the form "natural capital": coal, crude oil, natural gas, sulfur and other materials that resulted from millions of years of natural processes. The Industrial Revolution turned those raw materials – in massive quantities – into refined fuels, new materials, medicines, explosives for war, mining, and construction; plastics for aircrafts and automobiles, buttons, and toys; medicines to cure the wounded or the ailing; textile dyes for fabrics, inexpensive clothing; and fertilizers to increase food production for nations.

Despite the successes of the Industrial Revolution and the growth of the chemical industry, the conventional economic theories that were the foundation for growth and prosperity had a serious flaw. Natural capital, the asset that was the cornerstone of the industrial machine, was left off the balance sheet and has resulted in significant environmental consequences. Most financial and business systems account and manage for physical and financial capital – the assets that have represented wealth. But there have been other forms of capital that have been ignored in the decision-making and managerial algorithms as well: natural, human, and social capital. The sustainability framework, *Natural Capitalism* addresses this omission.

Companies are beginning to evaluate and account for nature and the true environmental costs and risks associated with chemical refining and manufacturing, the impact to natural capitalism. *Natural capital* refers to all of the resources used and consumed by humans (water, trees, fish, soil, air, animals, and so on) and extends to the ecosystems that support these resources: wetlands, rivers, coral reefs, prairies, forests, jungles, oceans, mountain ranges, deserts – the entire Earth. *Human capital* refers to the skills, innovation, and unbounded human potential that we are capable of realizing given the appropriate system and culture. *Social capital* refers shared culture, norms, stories, and experiences that support the generation of wealth and well-being. These services are of immense economic value, and many are priceless, as there is no known substitute. These companies recognize that we are entering into a period of a new industrial revolution, a sustainable industrial evolution. This sequel to our industrial success is a response to the changing pattern of scarcity and recognition of the lasting ecological degradation, extraction, and liquidation. Natural capital will be the limiting factor to future economic growth. A hundred years ago, who would have envisioned a world constrained by the quality of water needed for industrial processes?

Companies that adopt these principles will do very well, while those that do not won't be a problem, since ultimately they won't be around.

– Edgar Woolard, former Chair of DuPont

4.5.2 Four Principles of Natural Capitalism

Four principles construct a planning and design framework in an effort to value natural capital and consider true environmental impact.[1] Each of these four principles is powerful enough alone to impact an organization's progression toward sustainability, but the greatest impact is when these principles are adopted as a complete framework. Additionally, the principles will be presented linearly, but are best used as a circuitous framework, similar to the quality improvement framework of the Deming Cycle: Plan, Do, Check, Act.[2]

Principle 1 (P1): Radically Increase the Productivity of Resource Use – creating a lot more with far fewer resources.

Principle 2 (P2): Biomimicry – what would Nature do here?

Principle 3 (P3): Migrating Business Models from Products to Services – give the customer what they want, when they want it, not the product.

---

[1]The phrase, natural capitalism alone may appear paradoxical, bringing to mind the atrocities foisted on the natural world by capitalists in the name of profits. In 1997, an article by Paul Hawken appeared in the March issue of *Mother Jones* titled "Natural Capitalism: We can create new jobs, restore our environment, and promote social stability." Some say that it was the first time that the word *capitalism* appeared in the magazine in a positive light. The collaboration and thinking continued and, in 1999, with Amory Lovins and Hunter Lovins as co-authors with Paul Hawkin, the book *Natural Capitalism* was published, and a framework towards approaching a new model for capitalism, in a sustainable world emerged.

[2]The Deming cycle is a set of activities that fall into the categories of Plan, Do, Check, Act that are designed to drive continuous improvement and innovation. The continuous improvement activities were first developed by Walter Shewhart, but are commonly called the Deming cycle, or the PDCA cycle. The Deming cycle was initially implemented in manufacturing environments, but has broad applicability.

Principle 4 (P4): Re-invest in Natural and Human Capital – good capitalists always invest in what will provide the best return over the long run.

The first principle, Radical Resource Productivity, provides a problem-solving lens and offers a basic challenge: how can we do much more with much less? Through fundamental changes in production design and technology, natural resources can be stretched five, ten, even hundreds times further than before. The resulting savings in operational costs, capital, and time quickly pay for themselves, and in many cases, the changes can reduce initial capital investments. The second principle, Biomimicry, is a shift to biologically inspired production with closed loops, no waste, and no toxicity. Natural Capitalism seeks not merely to reduce waste but also to eliminate the concept altogether. Closed-loop production systems, modeled on nature's designs, return every output harmlessly to the ecosystem or create valuable inputs for other manufacturing processes. Industrial processes that emulate nature's benign chemistry reduce dependence on nonrenewable inputs, eliminate waste and toxicity, and often allow more efficient production. The third principle, Product to Service, is the shifting of the business model away from the making and selling of "things" to providing the service that the "thing" delivers to the customer. The business model of traditional manufacturing rests on the sporadic sale of goods. The Natural Capitalism model delivers value as a continuous flow of services – leasing an illumination service, for example, rather than selling light bulbs. This shift rewards both provider and consumer for delivering the desired service in ever cheaper, more efficient, and more durable ways. The final principle, Reinvest in Natural and Human Capital, allows for the restoration and investment of future resources. Any good capitalist reinvests in productive capital. Businesses are finding an exciting range of new cost-effective ways to restore and expand the natural capital directly required for operations and indirectly required to sustain the supply system and customer base.

4.5.3   The Natural Capitalism Framework

4.5.3.1   Principle 1: Radically Increase the Productivity of Resource Use. The simplest definition of P1 is doing *significantly more with significantly less*. P1 can be applied to feedstocks, processing, or to the resource and utility streams that support manufacturing. This principle is the most widely adopted principle of the framework. It is where most organizations transitioning from unsustainable practices to sustainable futures find the most traction. Waste is also very expensive. Implementation of P1 has also gained some nicknames such as "free money," or "hidden assets." Money saved from resource productivity could be used for other sustainability initiatives, training programs, productivity improvements, or other long-term capital projects. In many process industries, the utility costs are not assigned directly to the production activity they support (through accounting) and therefore they are treated as fixed costs. They are assumed to be uncontrollable, and are reviewed periodically (not episodically) and they are not correlated with production activities. By implementing P1, companies not only demonstrate a commitment to sustainability, but also demonstrate how expensive waste really is, make operational improvements, and begin to transform how their organization designs, engineers, manages, and produces.

So where does one begin? Inefficiency often starts "in the beginning," that is in how plants are designed and constructed. The resource efficiency designed in will impact the plant's ability to implement P1 and sustainability. The chemical and plant engineers working for the large engineering firms such as Stone and Webster, Bechtel, and Foster Wheeler have the ability to control the plant's energy efficiency, as well as resource efficiency. Plants are often designed and engineered to meet specifications for production specification and initial acquisition price and not operating costs. Additionally, support systems are regularly over sized to ensure that production specifications are met. Most plants are not designed and engineered to optimize both throughput and resource efficiency. Additionally, when a company has been successful with one layout, the design is then copied from facility to facility, as more plants are needed. Most plants are designed in a sequence that lays out the large equipment, then the utilities are designed around the equipment, yielding designs that include many bends in the pipes and big pumping systems that need to (over) compensate. Additionally, engineers are often rewarded for their ability to reduce the initial design costs of the plant; they are not paid to optimize the entire system of the plant, or to reduce the operating costs.

Engineers and designers using P1, or also commonly called eco-efficiency, use a set of algorithms to find and eliminate waste. Some of the rules are:

- Review of system boundaries and what are the interactions of the flows of the inputs, outputs and resources.
- Begin designing as a whole system, but beginning downstream and moving backwards upstream, reviewing demand before supply, application before equipment, and thinking passively before actively.
- Conserve mechanical energy where possible. Review flows, friction sources, as well as pump and motor requirements and placing.
- Conserve thermal energy where possible. Review heating and cooling requirements where the waste from one process can become an input for another.

One of the best examples of whole-system engineering for resource efficiency cited is the Interface Shanghai carpet plant. There was, in the initial design, a fluid loop that called for a 95-horsepower pump. Jan Schilham, an Interface engineer, made design changes that reduced the requirement to only 7 horsepower, reduced the capital cost of the system, and involved no new technology than that already specified. Instead of using thin, low-cost piping, and angling the piping around the equipment, Jan switched to fat, straight, short pipes. The result? While there was probably an increased price per foot for the larger diameter pipes, the reduction in friction in the system significantly reduced the pumping requirements so that the process could work reliably with smaller motors, smaller pumps, less control equipment, and fewer welds (and less welder time!). Since the pipe runs were straight and covered smaller distances, and did not have to negotiate large pieces of equipment such as tanks and boilers, the pipes used less overall material and were easier to insulate (Lovins *et al.*, 1999).

If you do not have the luxury of having a brand new facility, a significant retrofit, or new capital project, there are often many opportunities to radically improve the resource use within an existing facility. Pumps, motors, steam, and compressed air are regularly the top four energy and resource "pigs" in a production facility. Companies are now "replacing for the environment," as well as other factors including reliability, cost, and ease of maintenance. Instead of replacing equipment as it fails with like equipment, they are replacing it with the most energy and resource efficient models available.

Many pumps installed in industrial facilities are oversized for their actual loads. If possible, reduce the pump size during replacement to increase operating efficiency and reduce maintenance costs. If purchased from a wide variety of vendors, the highest efficiency pumps in each size and class are not necessarily more expensive than lower efficiency models.

Many of the new facility design algorithms can be applied to existing operations, but there are others as well that incorporate the wisdom and experience of the people running and maintaining the plants. Some of these rules marry nicely with other operationational improvement programs such as TPM (total productive maintenance) and six sigma (quality) and have been applied to boiler improvement projects, steam system improvements, and improvements of wastewater treatment processes and facilities. Examples of some of the algorithms are:

- Examine where there are losses, both with materials and energy. A loss equates to an opportunity?
- Examine what breaks, often and predictably.
- Understand what goes up the stack and what goes down the drain. There are significant opportunities in waste streams.
- Review data outputs from control systems. Do you understand the variation occurring in the utilities as well as production information?
- Replace inefficient equipment with efficient equipment during routine maintenance versus as a project.
- Understand where all the bottlenecks are occurring in the facility, and not just in the production system, but also the utility system as well.

Companies such as Royal Dutch Shell, Dupont, Dow, BASF, and many others have been leading the processing industry in significantly reducing the energy and resource use without any negative impacts to production, and in many cases have seen an improvement in yield. With innovation, creativity, and improvements, chemical engineers and professionals have reduced the U.S. chemical industry's energy use since 1970 almost by half (Hawken *et al.*, 1999). And some would speculate that is only the beginning.

4.5.3.2 Principle 2: Biomimicry, Shift to Biologically Inspired Production Models and Products.

Biomimicry, the second principle of Natural Capitalism, is based upon the work of Janine Benus and other scientists who look to nature first to solve complex and system problems. Nature provides numerous examples of fascinating materials and processes that the processing and chemical industry are still studying. For example, the material that spiders use to make their webs is often studied for strength, elasticity, and durability. The substance that the spider uses to make the web is made from pure protein and water in a gland below the spider's abdomen. Compared to steel, ounce for ounce, the spider's silk is five times stronger, and compared to plastics, it is able to absorb several times the impact without breaking. The small spider is able to do this at ambient temperature, under normal pressure, without the use of toxic chemicals, and without producing toxic wastes (Geiser, 2001). If you decided to make nylon string to make your own web, you would need chemicals such as sebacoyl chloride, methylene chloride, sodium hydroxide, 1,6-hexanediamine, and ethanol (see http://www.kolias.com/science/nylon.htm). Lotus flowers are self-cleaning. Mussel adhesive works under water and will adhere to anything – without a primer. And the inner shell of abalone has twice the hardness of many ceramics.

Nature continues to serve as the learning laboratory for scientists and engineers regarding products and processes, but nature manufactures, filters, lubricates, cleans, and performs many other services without creating terminal waste. Nature lives where it works. How can we learn more from nature and apply the lessons to how we manufacture, filter, and lubricate, and do it simply, with less cost, and without the burdens of toxicity?

In Nature, every output from one process becomes an input for another. Our ingenuity and technology are helping us mimic nature, and to ask the simple, yet vital question, "What would nature do here?" to address some of our more persistent problems, such as toxic clean up with bioremediation and membrane technology for water filtration.

In addition to turning to nature to solve discrete engineering problems such as acceleration and pumping, more and more engineers and planners are turning to natural models to solve problems regarding waste management and waste elimination. We humans are the only creatures that create terminal waste and primary products that serve no purpose (other than to fill disposal sites). Nature neatly takes care of its messes. There is no output of a process or system that is not an input to another. It is one big closed loop. Waste is not treated as an externality, but is treated as an asset.

As more and more companies implement the tenets of biomimicry, examples of innovative waste sharing and product take-back programs bare the actions of the first adaptors. Considering waste as a resource opens possibilities for entrepreneurial opportunities and creative solutions. Excess steam can be used for district heating, waste citric acid can be used for wastewater treatment buffers, and plastics that decompose in the compost pile can be used for organic matter. There are many other examples in which nature-based technology and cooperation have reduced industrial waste streams. By repeatedly asking, "Who wants what you don't want?" solutions for grouping reactions and processes together based on resource use (back to principle 1) and exothermic and endothermic reaction come to light. Who has the heat? What needs coolness? What is already flowing downhill? The questions are endless, as are the solutions they foster.

As biomimicry is introduced into chemical engineering, process engineering, operations, and design, one can imagine a set of design parameters that can borrow from Nature's design specifications (Benyus, 1997):

- Nature runs only on sunlight.
- Nature uses only the energy it needs.
- Nature fits its form to function.
- Nature recycles everything.
- Nature rewards cooperation.
- Nature banks on diversity.
- Nature demands local expertise.
- Nature curbs excesses from within.
- Nature taps the power of limits.

Nature can guide strategy and business decisions as well. By adopting P2, a petrochemical plant manager set a goal of a facility-wide output of zero waste by 2006. This decision impacted how he purchased feedstocks, what type of catalysts he bought, and what is done with contaminated soil and oily wastes. Through careful decisions, the plant now achieves many of the natural design principles of biomimicry.

4.5.3.3 Principle 3: Product to Service, Move to a Solutions-Based Business Model.
Traditional capitalism generates profit from the flow of goods through distribution. The third principle of Natural Capitalism amends the traditional model and emphasizes a side of capitalism that generates wealth from the services the goods provide. P3 involves a strategic shift of a company's business model away from the making and selling of things to providing the service, the value to the customer, that the thing delivers. Pursuing a shift from goods to services has implications throughout the supply chain. The shift rewards provider and customer for delivering the desired service in ever cheaper, more efficient and durable ways. The provider is responsible for the lifecycle of the goods, and therefore is more interested in designing and deploying it for extended lifecycles. Extended durability and reliability become crucial.

Consider a scenario where it was no longer possible to obtain the materials or feedstocks needed. There are several likely outcomes. First, you would cease operating. Secondly, you would innovatively find alternative sources of the previous input, or find ways to reclaim the raw material from existing goods. And thirdly, you would creatively find ways to deliver the same service to your customers, with alternative or existing products. P3 assists companies think through alternative scenarios, which are becoming more likely as we progress through the 21st century. Raw materials such as copper are becoming scarce. Water rights and use are becoming contested and producer responsibility is becoming increasingly popular. As companies react to these and other external forces and move to make their products and processes sustainable, they will ultimately consider their product portfolio and the long-term goals they are attempting to achieve. Indeed, this means many will rethink the value of the product that they are offering to the market, and what purpose it serves. In many instances, the value of the goods themselves will be secondary to the service that good provides, and the market will be commensurately adjusted.

As a company adapts to a service-based business model, the focus becomes solving customer's problems without selling products. Chemicals provide many services including cleaning, coating, and lubrication. What would the business model look like if only the service was provided? Studies have examined this question and have concluded that the savings could be significant, up to 15 percent for the user of business model (Perthen-Palmisano and Jakl, 2004). Customers do not pay for the chemicals or the equipment, but pay for the performance of the machines.

Dow Chemical's subsidiary SAFECHEM's core business is the supply, use, and reclamation of chlorinated solvents produced by Dow Chemical. The business is organized around the principle of "the cradle to the next cradle" (Scholl, undated). The customer pays a combined price for the fresh solvent, recollection, and recycling. What the customer is really paying for is the degreasing service that the solvent is providing and insurance that the solvent is being handled according to the guidelines of Responsible Care. Companies are also leasing many of the support activities. The cleaning, testing, and in some cases the hauling, of the bulk materials are commonly leased services. At the core of P3 is the extension of the overall assets and materials that have created economic activity. But the economic activity also was built on the model of take-make-waste. By moving to service-based economic activity, the turnover of capital goods decreases without sacrificing either financial performance or disposing of assets before the end of their useful life. Additionally, the owners of the physical asset have an incentive to move beyond take-make-waste, to one of a closed loop (P2) and radical resource productivity (P1). The fewer assets and equipment used and disposed of, the less resources used, and the fewer emissions, effluents and negative environmental impacts occur. It can be a triple win, that is, a win for suppliers, a win for customers, and a win for Nature.

4.5.3.4 *Principle 4: Reinvest in Natural and Human Capital.* Good capitalists reinvest in the capital that has created their wealth. Implementing the first three principles will improve operating costs, reduce process waste, likely improve throughput and operational effectiveness, create new products, and open up new markets. But there is one long-term activity remaining before an organization can declare that it is operating at a sustainable level: reinvesting in natural and human capital. Reinvesting in natural and human capital involves the restoration and renewal of the natural and cultural systems that created the environment for the organization's initial success. It may not show a return on quarterly statements, but it will ensure sustainability decades from now.

BASF, a known leader in the chemical, plastics, and paint industries, has been involved in several successful natural restoration projects, including the successes at Fighting Island on the Detroit River. John B. Ford of the Michigan Alkali Company originally purchased Fighting Island in 1918. The Michigan Alkali Company became Wyandotte Chemicals, which was later purchased by BASF, giving BASF ownership of this three-square mile island. Fighting Island had been used from 1918 through the early 1980s as a waste disposal site. The island has special ecological significance due to its location in the lower Detroit River, which is a major migration corridor for many types of fish, butterflies, shorebirds, and waterfowl. The island stored alkaline byproducts primarily from the manufacture of soda ash and other lime-based products. The byproducts are disposed of in three settling beds on the southern three-quarters of the island. The settling beds were active from 1924 to 1982. The beds now hold approximately 20 million m^3 of material, mostly calcium chloride, sodium chloride, coke ashes, unreacted limestone, and limestone impurities such as silica, alumina, and metallic oxides (Greater Detroit American Heritage River Initiative). For over twenty years, BASF has significantly invested in the restoration and renewal of Fighting Island. These once barren settling beds, where nothing could live, are being rehabilitated and restored. A variety of prairie plants, wetlands, and wooded habitats once again exist on the southern end of Fighting Island. Pheasant, turkeys, deer, foxes, rabbits, owls, and now bald eagles are once again flourishing on this island. The metrics for this project has not only been in financial capital terms, but expressed in natural capital by:

- Decreased runoff of alkaline waters to the Detroit River;
- Decreased incidents of dust rising from the settling;
- Increased habitat for resident and migratory birds;
- Enhanced habitat for fish; and
- Increased biodiversity on the island.

The examples of P4 are as diverse as the companies that are implementing them. Companies that are large greenhouse gas emitters are often planting trees and other plants as carbon offsets. Argonne National Laboratory near Chicago has migrated from a typical corporate Kentucky bluegrass blanket to one of natural prairie grasses and local plants, saving over $50,000 per year, mostly from the cost of mowing and fertilizers. There are wetland restoration projects around the world that are not only restoring the land, but are acting as natural wastewater treatment facilities. For example, the company Elf Atochem in Tacoma, Washington, was able to identify over $100,000 in savings by reducing waste disposal costs just by soliciting wisdom and knowledge from the organization's human capital, their workforce (EnviroSense, October 31, 1995). As companies invest in the training that allows the workforce to address former unsustainable practices and inefficiencies, companies not only recoup loses, but allow for organizational barriers to dissolve and allow engineers, designers, managers, and other stakeholder to work as a system and innovate.

4.5.4 Implementing Sustainability through Natural Capitalism

Natural Capitalism, as with any framework, allows an individual to reframe their thinking to look at the previous problems in a different way, and move the actions of an individual, department, or organization in a new and more sustainable direction. However, then the question is asked, "What do I do, Monday morning?" Migrating from our current models of capitalism to natural capitalism will involve transitions – in products, processes, and business models. The power of creation and innovation is at the center of the new mental models and the solutions that will guide us sustainably and consciously through this transition.

P1 addresses this question of "What can I do now?" with the urgency that most companies need and yields operational efficiency and opportunities for design for the environment. P1 can liberate hidden cash stashes that had been hidden in the operations and fund new capital projects, workshops, or longer term projects that will positively impact environmental and financial performance. The first step is to understand the stocks and flows of the system you are trying to understand and improve. Most operations have process flows of the operations and utility diagrams supporting operations. What most operations do not have are simple and comprehensive diagrams looking at what is flowing in, flowing out, what is value, and what is waste, including frequency and volumes, as well as identifying exothermic and endothermic reactions. This is not an easy task, but one that will assist in all future operationational environmental improvements. By creating this mass balance of the overall system, there is a picture of the impact of the process, not only for throughput, but also for the planet.

P2 will allow operators, designers, engineers, facility managers, and even business managers to find creative and innovative solutions by not only thinking differently, but also possibly looking to alternative sources for inspiration and drawing on the science biology and biologists. Who knows what will be invented when the engineers and biologists routinely work together to solve product and process problems?

P3 moves a company in a direction of seeking new business opportunities that generate wealth for all. No longer are customer supplier relations a win–lose proposition, but they can be structured for true partnership. There will be direct gains from the reduction of operating waste and the extension of the useful life of equipment, but there will be indirect benefits from deeper supplier–customer relationships and the elimination of adversarial procurement practices.

And finally, P4 supports true capitalism and investment. Investment for a long-term planning horizon will guarantee results for generations of future capitalists.

The four principles, taken together as a framework, support the implementation of a sustainability post-industrial age. Chemical engineering, as a profession, is a product of the industrial age, but has the discipline and the skill sets to lead the sustainable post-industrial age. The four principles of Natural Capitalism, along with the other tools available, can, over time, continue to support the increase in the standard of living that the chemical industry has provided without the strain on air and watersheds and the impact to human health resulting from accidental exposure.

The chemical industry can be one of the leaders of the sustainable post-industrial age if the following challenge issued from The Institution of Chemical Engineers, IChemE (Melbourne Communique, 2001) is met.

> To use our skills to improve the quality of life: foster employment, advance economic and social development, and protect the environment. This challenge encompasses the essence of sustainable development. We will work to make the world a better place for future generations.

4.6 SUSTAINABLE VALUE IN THE CHEMICAL INDUSTRY

DAVE SHERMAN
Sustainable Value Partners

This article describes the financial and societal challenges the chemical industry faces and offers a practical approach to building lasting value through simultaneously creating shareholder and stakeholder value. Stakeholder value – based on the economic, environmental, and social impacts a company has on its constituents – is a rapidly growing source of business advantage, fuelled by rising societal expectations and the growing ranks of social change agents newly empowered by information and communication technologies. Taking advantage of this source, however, requires a change in the mindset of leadership and a disciplined approach to integrating stakeholder value into the business.

4.6.1 The Chemical Industry's Challenge

The chemical industry has a long history of product and process innovation and has provided enormous benefits to society. Since the late 19th century the industry has continually developed new products and processes, improved its functionality and cost-effectiveness, and displaced many traditional materials. Its products are now integrated into many economic sectors and it has a history of collaborating with its customers to develop innovative solutions. As a result there has been a dramatic increase in the use of chemicals and plastics and these have provided tremendous cost savings as well as improved functionality. (For details, see "Industrial Biotechnology and the Chemical Industry's Sustainability Challenge," presented by Dave Sherman at the World Congress on Industrial Biotechnology and Bioprocessing, Orlando, FL; available at www.SustainableValuePartners.com.)

Despite this history or innovation, today the industry faces significant challenges. The industry is mature, many of its products face commoditization pressure, and its financial returns are not competitive in attracting investment capital in today's increasingly global capital markets. The industry's products and processes also have resulted in unintended consequences, many of which were not known until years after the damage was done. Consequently, the industry has contributed to the spread of toxic chemicals, the creation of large numbers of hazardous waste sites, climate change, and risks to public health.

Increasingly, excellent companies are expected to perform socially as well as economically. The CEOs of leading companies understand this as their responses to a recent World Economic Forum CEO survey on Global Corporate Citizenship indicate:

> In this survey CEOs saw stakeholder pressures as a key driver of social performance and cited employees, government, customers, communities and NGO's as being most the most important stakeholders in shaping their action.
>
> Societal expectations for the chemical industry will continue to increase driven by a number of forces such as the following:

- Education, knowledge, information, freedom of choice
- Liberalization, globalization, privatization, and growing scope of corporate activity
- Technology and communication
- Increased role played by civil society
- Increased environmental footprint as emerging economies develop

> (From "Responding to the Leadership Challenge: Findings of a CEO Survey on Global Corporate Citizenship," conducted by the World Economic Forum in partnership with the Prince of Wales Business Leadership Forum, January 2002.)

Global growth will further compound the societal challenge. Figure 4.13, shows chemical industry sales and population by country/region of the world. The arrows show the incremental sales that developing countries/regions would have if they were to reach one-half of the developed world's chemical intensity.

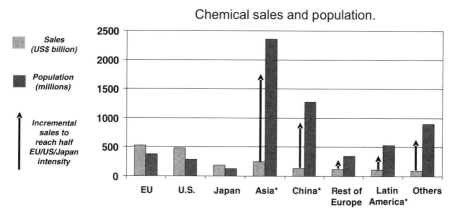

Notes: Sales: Based on 2002 data. *CEFIC Estimates; Asia excludes China and Japan.
Population: Based on 2002 data; Population Reference Bureau.

Figure 4.13. Chemical sales and population. (*Source*: Sustainable Value Partners, Inc., 2004.)

Asset-based strategies will not solve the chemical industry's challenges. Traditionally, companies across a range of industries measured their success by the returns on their physical and financial assets. Strategies were built by finding favorable market opportunities, constructing plant and equipment, and producing, selling, and realizing a strong return on assets employed. However, research sponsored by the Council for Chemical Research shows that today physical assets in the chemical industry only return their cost of capital – they do not create economic value. This same research shows that all of the industry returns above the cost of capital are attributable to intangible assets such as R&D. In fact industry returns for R&D are 17 percent, well in excess of the cost of capital.

Cross-industry studies provide similar results. For example, a recent analysis presented in the CFO magazine indicates that changes in Economic Value Added (the measure that has been associated with shareholder value) explain only 35 percent of changes in market value. Another study at New York University's Stern School of Business indicates that Intangibles drive up to 75 percent of market value.

The research is clear – asset-based strategies will not solve the chemical industry's financial challenge. Moreover, such strategies will only exacerbate the industry's societal challenges. Clearly new thinking is required.

Industry players need to capture new strategic opportunities compatible with declining traditional sources of advantage and rising societal expectations. We propose that such opportunities can be identified and captured by reframing stakeholder value in terms of competitive advantage. By systematically integrating stakeholder considerations into business strategy and operations to reduce costs, differentiate products and services, develop new markets that serve unmet societal needs, and influence industry "rules of the game," chemical companies can create new sources of sustainable advantage. Success in capturing these opportunities requires a new leadership vision and the courage to understand and engage a diverse set of stakeholders.

4.6.2   Who are Stakeholders?

Anyone who risks something of value (such as capital, health, welfare, or happiness) in interacting with a company can be said to have a stake in it (Kochan and Rubenstein, 2000). A stakeholder can be considered to be a person or group who can help or hurt the company. Stakeholders who wield sufficient power to materially affect business performance either favorably or unfavorably are important to the company's future; they are *key* stakeholders (Laszlo *et al.*, 2004). In Figure 4.14, we distinguish between economic and societal stakeholders. Economic stakeholders participate directly in the value chain by contributing to the company's product or operations. Societal stakeholders are external to the value chain, but they can nevertheless exert significant positive or negative influence on a company's value, as is often the case with governments and NGOs.

The power of economic stakeholders is well documented in the case of customers, employees, and suppliers; they are often important partners for companies increasingly dependent on collaboration and networking to create and sustain competitive advantage (Heskett *et al.*, 2003). Societal stakeholders have increasingly impacted

Figure 4.14. Stakeholders. (*Source*: Sustainable Value Partners, Inc., 2002.)

the business equation. Local communities and NGOs have only recently become broadly recognized as important to corporate success on issues such as governance, environmental protection, human health, and quality of life (Waddock, 2002).

With advances in information and communications technology, the cycle time for identifying and internalizing negative stakeholder impacts has been shortened. A company's impact on a stakeholder group in any part of the world can be instantly communicated across the globe, with direct consequences for its ability to conduct business.

Globalization and instantaneous communication of information, exploited by a cadre of highly committed and effective NGOs,[3] have contributed to the heightened risk of negative publicity arising from controversial events.

The rise of complex and amorphous NGO alliances serves as an emerging indicator of a global business environment that is on the cusp of fundamental change. Increasingly, previously marginal actors create opportunities to realize their collective potential as networked actors in ways unimagined in the past.

Rising societal expectations for corporate responsibility have created a new class of socially-responsible investors, customers, and employees and, more generally, a broader public awareness of human health and ecological risks.

4.6.3   The Need for a New Approach

This new environment requires a new approach (Laszlo, 2003); one that forces managers to think "outside-in" about how their companies can create and sustain competitive advantage. Outside-in thinking, which sees the world from the perspective of stakeholders, is a powerful new lens through which managers can discover new business opportunities and risks. Leaders who engage stakeholders and proactively

---

[3]*The San Francisco Chronicle* (January 21, 2003) reported that the number of international NGOs quadrupled in the last decade to over 50,000 organizations.

address stakeholder issues can better anticipate changes in the business environment, discover new sources of value, and avoid being surprised by emerging societal expectations that can put shareholder value at risk

Business leaders are familiar with managing financial value, whether in terms of economic value added (EVA) or other measures driving stock price performance. They are less knowledgeable about measuring and managing stakeholder value. Because a company's impacts on stakeholders are often unintentional, it faces hidden risks and opportunities that managers can no longer afford to ignore. To succeed in a stakeholder-driven business environment, business leaders must think and operate in new ways, shaping strategies and actions with full awareness of their impacts and the implications of these on key stakeholders.

Figure 4.15 describes company performance along two axes: shareholder value *and* stakeholder value. Managing in two dimensions represents a fundamental shift in how managers must think about business performance. Companies that deliver value to shareholders while destroying value for other stakeholders (or exploiting externalities) have a fundamentally flawed business model. Those that create value for stakeholders are cultivating sources of extra value that can fuel competitive advantage for years to come. Sustainable value occurs only when a company creates value that is positive for its shareholders and its stakeholders.

Starting in the upper left of Figure 4.15 and moving counterclockwise, consider the following four cases of value creation:

1. *Upper left quadrant*: When value is transferred from stakeholders to shareholders, the stakeholders represent a risk to the future of the business. Leaded paint and asbestos are two well-known examples; however, a much broader range of products and services face this situation. For example, chlori-

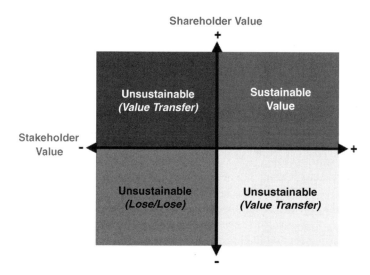

Figure 4.15. The sustainable value framework. (*Source*: Sustainable Value Partners, Inc., 2002.)

nated plastics contain chemicals that can present a variety a human health hazards even at low exposure level and are found in everything from children's toys, to sneakers, to plastic food wrap.

2. *Bottom left quadrant*: When value is destroyed for both shareholders and stakeholders, this represents a "lose/lose" situation of little interest to either. Monsanto and its European competitor Aventis lost large sums of money by underestimating stakeholder rejection of their GMO crop products. Before Aventis sold its CropSciences division to Bayer in 2001, it is believed to have lost $1 billion in buyback programs and other costs associated with its genetically modified corn StarLink. StarLink was approved only for use in animal feed but was found by NGOs to have contaminated a number of human food products.

3. *Bottom right quadrant*: When value is transferred from shareholders to stakeholders, the company incurs a fiduciary liability to its shareholders. Actions intended to create stakeholder value that destroys shareholder value put into question the company's ability to create societal value over time. Avoiding offshore sourcing to protect American jobs is an example of actions that could create value for some stakeholders while destroying value for shareholders. Bringing back offshore jobs and union strikes to "Keep Jobs in America" may create job security for American workers in the short term, but can hurt companies whose operating cost structures become uncompetitive.

4. *Upper right quadrant*: When value is created for stakeholders as well as shareholders, stakeholders can represent a source of hidden value. For example, Shaw Industries, the world's largest carpet manufacturer with over $4.6 billion in annual sales, found a way to create a new carpet backing that offers benefits to both shareholders and stakeholders. Rising concerns among stakeholders about the environmental and health risks associated with traditional PVC backing led Shaw to search for an alternative. Its solution was EcoWorx backing, in which a thermoplastic polyolefin reinforced by fiberglass provides the same functionality as PVC. The new backing is half the weight of the old, resulting in savings on shipping costs. Shaw has made a commitment to pick up any EcoWorx product at the end of its life, at no charge to the customer, and recycle it into more EcoWorx, enabling the company to use these materials in a perpetual loop. Receiving a call when the customer's product reaches end of life also presents the company with a selling opportunity for new products. Within 36 months of launch date, EcoWorx production exceeded 50 percent of Shaw's total tile backing production and the company publicly committed to ending all PVC backing by the end of 2004 (Shaw Industries, 2004).

Companies can apply the sustainable value framework to their existing portfolio of products and services to identify risks and opportunities. Managers assess the overall value created for a business or product in both shareholder and stakeholder terms. For example, an industrial paints producer identifies solvent-based industrial paints as positive for shareholders but negative to stakeholders due to the presence of harmful volatile organic compounds (VOCs). By switching to water-based paints that are classified as non-VOC, it has the opportunity to create value for shareholders and stakeholders. By profitably recycling its water-based paints, it has the opportunity to move its product portfolio even further up into the upper right quadrant of Figure 4.15.

4.6.4 A Disciplined Process

The opportunity for industry today is to understand its impact on stakeholders, anticipate changing societal expectations, and use its capacity for innovation to create additional business value from superior social and environmental performance. Capitalizing on this opportunity requires that companies apply the same systematic discipline in managing social and environmental performance as they do in managing other aspects of business performance.

4.6.4.1 Why Stakeholder Value is Poorly Managed Today. Stakeholder value is often poorly managed. Several factors contribute to this, including:

- An incomplete awareness about the company's impacts on stakeholders and how these impacts might in turn affect future business value;
- Fragmented responsibility and knowledge of social and environmental issues;
- Lack of ownership within the core management team;
- Lack of practical tools to measure and manage the business implications of social and environmental performance;
- Line managers who are focused on traditional drivers of shareholder value and view stakeholder-related issues as a distraction from their business objectives.

The most critical barrier to managing stakeholder value is the dominant mental model. A new mindset is needed to capture the systemic interrelationships between a company and its societal context. With this mindset it is not enough to just compete with industry rivals, but it is also important to understand and manage the changing expectations of an ever growing and diverse set of stakeholders. It no longer makes sense to see nature and society as external and peripheral to the business.

The process described below requires leadership that is willing and able to alter the dominant mental models of their organizations. While it is not necessary that every employee buy into a stakeholder view, it is important that the CEO and key senior executives actively promote it.

4.6.5 Three Key Phases

A disciplined process to create sustainable value can be built in three phases

1. *Diagnosis*

2. *Value creation*
3. *Value capture*

4.6.5.1 The Diagnosis Phase. The diagnosis phase expands the organization's view of value to include stakeholder-related risks and opportunities. This requires a process of identifying and segmenting stakeholders, deciding which ones are important, and gaining a clear understanding of the issues that matter the most to each of them. The organization must develop a clear picture of where it is creating and destroying value for them. The company must also understand how value flows *from* stakeholders to the company and where and how stakeholders impact the organization both positively and negatively? The current picture of value flows should be augmented by exploring how it might change in the future. Such assessments result in foresight and "pressure" to create solutions to improve both financial and societal performance. Through performing such assessments, chemical companies can begin to realize the full scope of their sustainability challenges and create the will to address them.

When diagnosing stakeholder value it is important to understand that perceptions can be as important as scientific facts. For example, polyvinyl chloride (PVC) producers defend PVC on the basis of scientific arguments such as energy efficiency, low biomass accumulation, and product safety in normal use. Customers such as Nike, Sony, and Shaw Industries that have committed to eliminating PVC in their products as a precaution for their customers due to perceived health and environmental risks, are unlikely to change their perspective based on additional scientific facts provided by the chemical industry. As in the PVC case, manufacturers are vulnerable to value loss as a result of their customers' perceptions of risks.

4.6.5.2 The Value Creation Phase. The pressure to create solutions should be directed at both strategies and operations. Figure 4.16 lays out the range of operational and strategic risks and opportunities that the chemical industry faces today. The left side shows the operational risks and opportunities, the right side shows the strategic risks and opportunities, and the boxes at the bottom show the benefits in terms of brand value and reputation, license to operate, attracting and retaining talent, cost savings, and revenue enhancements.

The operational opportunities include incremental improvements in environmental, health and safety, and social performance. The two major strategic risks are persistent toxins such as dioxin and mercury that accumulate in the food chain and are now found in mammals and fish in even the most remote parts of the world, and global climate change. Green chemistry and biotechnology provide the potential for overcoming the two major strategic risks through radical innovation for new products and markets.

In creating solutions, managers need to consider potential value from multiple levels of strategic focus. These are shown in Table 4.6. Often companies focus on the bottom two levels concerned primarily with eco-efficiencies from reducing energy or waste; avoidance of fines, penalties, and litigation due to regulatory noncompliance; and reducing risks related to license-to-operate.

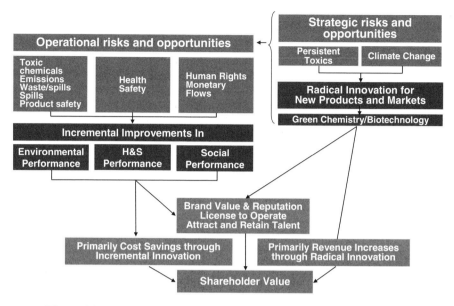

Figure 4.16. Operational and strategic. (*Source*: Adapted from IMD, 2003.)

The top four levels in Table 4.6 represent opportunities that can be significantly larger than those represented by the bottom two. These opportunities can drive innovation and top-line growth based on solutions that integrate financial and societal performance.

BASF's Vebund network of integrated industrial sites is an example of creating value at the uppermost level. It changes the "rules of the game" by co-locating with suppliers. DuPont's integrated agriculture plan for Columbian farmers reaches a new previously underserved market. The replacement of metal with high-performance

TABLE 4.6. Levels of Strategic Focus

Focus	Sources of Value
Business context	Changing the "rules of the game" so that sustainable strategies are both feasible and competitive
Reputation/brand	Gaining stakeholder preference and recognition as well as employee motivation
Market	Addressing new markets driven by customer and societal needs
Product	Using stakeholder pressure as a driver of product innovation. Creating product differentiation based on technical and environmental/social features
Process	Reducing energy, waste, other process costs; improving quality
Risk	Avoiding fines and penalties, protecting license to operate

Source: Sustainable Value Partners, Inc. (2002).

TABLE 4.7. Key Questions for each Phase

Diagnosis	Value Creation	Value Capture
• Who are your stakeholders? • What are their interests? • Where are you creating value of destroying value for them? • What potential future developments might change this stakeholder value picture? • What are the business risks and opportunities associated with this picture? • Which risks and opportunities warrant action?	• What actions will simultaneously create shareholder and stakeholder value? • At what level of strategic focus will they create value: risk, process, product, market, brand, business context? • What financial value will result: profitability, capital utilization, lower cost of capital, growth, intangibles, market confidence? • What are the critical success factors for the actions?	• What existing programs or systems could be adapted to inclue the stakeholder dimension? • What stakeholder alignment and support is required? • How will you engage line managers? • What financial and human resources are required? • How will you track progress, measure results, and share learning?

Source: 2004 Sustainable Value Partners, Inc. (2004).

plastics in vehicles provides environmental and economic benefits from product use, while 3M's pollution prevention pays is a well-known example of shareholder value created at the process level. New processes using biotechnology can provide dramatic benefits. For example the cost of L-lysine was reduced from $1.17 to 44 cents per pound through the use of biotechnology.

4.6.5.3 The Value Capture Phase. In the value capture phase, attention is focused on the conditions for successful implementation. A key consideration is how to use actions to change the dominant mindset and embed the stakeholder value perspective into the organization's management processes and operating model. In many cases this can be accomplished by expanding the frame of existing programs such as Six Sigma to include the full stakeholder perspective. The ability to measure in a credible way the impact of actions on stakeholder value is also critical.

Table 4.7 summarizes key questions you can use to guide your efforts in each phase.

4.6.6 Conclusion

The chemical industry faces significant financial and societal challenges. Like the U.S. auto industry of the early 1980s that believed higher quality meant higher costs, new thinking is required. The Japanese demonstrated to the auto industry that it was possible to achieve higher quality and lower costs simultaneously.

Today companies across a range of industries are finding that they can achieve high quality, fast speed to market, high customer service, and low cost, all at the same time. The leaders of tomorrow will demonstrate the same thing about stakeholder and shareholder value. They will find ways to create business value while delivering value to stakeholders. Integrating the full range of stakeholders into strategic and operational decision-making will become best practice.

Tomorrow's leaders can create competitive advantage today by understanding their key stakeholders' interests, anticipating societal expectations and using the insight, skills, and relationships developed through an outside-in process to design new products and services, shape new markets, develop new business models, and ultimately reshape the business context itself to one that supports the creation of truly sustainable value. The new leadership vision and a disciplined approach to creating stakeholder value are key success factors in tomorrow's marketplace.

4.7 CSR/SRI REPORTING COMPLEXITY AND THE FUTURE 500 CAP GAP AUDIT™: AN OPPORTUNITY FOR IMPROVED STRATEGIC BUSINESS PLANNING AND STAKEHOLDER ALIGNMENT

CATE GABLE
Future500

4.7.1 A Business Case Introduction

The business environment of this new century is characterized by an emerging paradigm with several salient features:

- Increased scrutiny of any industry with a heavy environmental impact – especially the chemical industry.
- Building momentum for inclusion of "cost and benefit externalities" in business accounting, or a requirement to consider the "triple bottom line."
- Leverage-companies (i.e., purchasers or retailers) demanding upstream and downstream supply chain collaboration on greening initiatives.
- Identification and strategic management of key stakeholders.
- Stakeholder engagement as part of a competitive business strategy.
- Proliferation of Corporate Social Responsibility and Socially Responsible Investing (CSR/SRI) standards and the compliance with key standards as a requirement of doing business.

An anecdote involving stakeholder interaction and supply chain collaboration will serve to illustrate several of these points and begin the dialog about how these reporting requirements can lead to improved business planning. Polyvinyl chloride, also called PVC or vinyl, is one of the most common synthetic materials. Over 14 billion pounds of PVC are produced every year in North America. Approximately 75 percent of all PVC manufactured is used in construction materials. Dioxin, one of the most potent carcinogens known, ethylene dichloride, and vinyl chloride are unavoidably created in the manufacturing of PVC (see PVC fact sheet at http://www. healthybuilding.net/pvc/). PVC materials used in buildings create the phenomenon of "off-gassing," which is the main contributor to indoor air pollution.

Additionally, since most medical paraphernalia – swabs, syringes, gloves, IV tubes, bed pans, and so on – is made of some variety of PVC, hospital waste incineration delivers other silent dangers into the air. Dioxin, hydrogen chloride, and hydrochloric acid are released and created in the process of combustion. (The incineration and collapse of the World Trade Center, for instance, sent a deadly dioxin plume over all of lower Manhattan.)

Over the past decade in the CSR arena, a combination of environmental, healthcare advocacy and governmental groups like Health Care Without Harm, Healthy Building Network, and U.S. EPA have begun to raise issues relating to hospital materials used both for construction and day-to-day operations.

PVC has floated to the top of the list as one of the materials with the highest negative environmental impact. As a result of these efforts, reducing the use of PVC, particularly in healthcare facilities both in construction and operating supplies, is one of the goals of a variety of emerging healthcare and green building policy documents (see USGBC, LEED: http://www.usgbc.org/leed/leed_main.asp; Green Guideline for Healthcare: www.gghc.org; Sustainable Hospitals Project: http://www.sustainablehospitals.org).

4.7.1.1 A Healthcare Provider Gains Strategic Marketplace Advantage. Kaiser Permanente, headquartered in Oakland, CA, a private not-for-profit healthcare provider serving 8 million members in 11 states and a leader in the healthcare delivery field, follows CSR recommendations wherever possible. Kaiser began efforts to replace PVC in carpets, vinyl composition tiles, and particleboard several years ago.

Carole Antle, co-chair for the LEED^{TM} healthcare committee and Director for Capital Projects at Kaiser, says, "We wanted a cradle-to-cradle assessment of the most polluting, most commonly used hospital products." After a comprehensive discovery process, Kaiser identified selected manufacturers to work with on new product and materials R&D in order to replace PVC and related products (Gable, 2004).

Some of these material challenges have provided opportunities to source cheaper *and* greener products. Kaiser found a particleboard without formaldehyde that was below their previous price point. Nontoxic flooring, although more expensive up front, has proven to be less expensive over the life of the product, as both the labor needed for clean-up and cleaning supplies have provided cost benefits.

Other collaborations like this one have produced a Dow product called Woodstalk, a formaldehyde-free polyurethane resin particleboard from a harvested wheat straw fiber, a renewable resource. Dow acknowledges the importance of stakeholder collaboration in their business strategy. "Green building practices such as those defined by the LEED^{TM} rating system have helped to establish the direction for product research" (see, Dow, *The Synergy of Nature and Science*, at http://www.buildings.com/Articles/detail.asp?articleid=1579). Supply chain collaboration between high-volume healthcare providers, or purchasing coops, and savvy manufacturers is a powerful force for change. This collaboration provides early movers – both manufacturers and product purchasers – with strategic marketplace advantage. As Antle puts it, "It helps us when activist stakeholders get out in front of the issues."

Environmental risk factors identified by activists or policy organizations can provide an early-warning system to both producers and buyers. New policies often establish standards that catalyze innovation, reform the marketplace, create new product categories and propel strategic planning refinements. Consumers benefit from safer buildings and products; manufacturers, suppliers, and service-delivery providers benefit from refinements in strategic business direction.

In the balance of our chapter, we will consider how a company can best understand the CSR/SRI world in order to take advantage of these business opportunities and, at the same time, deflect potential risk. Does the proliferation of CSR and SRI standards offer an opportunity for business planning refinement or simply a disruption to the status quo?

4.7.2 The CSR/SRI Standards, the Future 500 CAP Gap Audit^{TM}, and Business Planning

Today, leadership companies are expected to respond to a proliferation of corporate social responsibility (CSR) and socially responsible investing (SRI) standards. Leading companies report not just their return to shareholders but also their "return to stakeholders." But with more than a dozen major standards organizations – each representing a variety of stakeholder perspectives – and even more questionnaires for reporting on CSR/SRI accountability and sustainability practices (see the Standards Directory; Appendix Item A), how should a company approach this emerging field? And, even more critical for the business bottom line, what strategic opportunities are indicated?

The evolving arena of CSR/SRI standards provides a growing challenge to leadership companies. There is a plethora of information to manage – much of it confusing, disorganized, and duplicative.

Among the many questions that companies need to address in planning for CSR/SRI and sustainability are the following.

1. What are the key business objectives or benefits to be achieved with CSR/SRI programs? Which standards set(s) should be selected?
2. Is there rigorous documentation on performance against the many CSR/SRI standards used today?
3. Does compliance embed the processes of citizenship into the culture of the business?

4. How is CSR/SRI compliance implemented on a daily basis and by whom?
5. How does a company assure consistency; are all departments and functional areas of the company on the same page?
6. What are the key issues to which a company must respond in its industry, and where does the company stand relative to those issues?
7. What are key stakeholders, by stakeholder category, most concerned about?
8. How should corporations track issues and solve dilemmas in risk areas at the same time they increase their exposure by reporting publicly?
9. How do companies manage the "control" issue when often reputation – so critical to brand – is in the hands of external stakeholders?

The Future 500, a nonprofit environmental and stakeholder engagement consulting firm, provides tools and services that enable best practices in stakeholder engagement at the world's leading companies. One of the Future 500 tools, the Corporate Accountability Practice (CAP) Gap Audit^{TM}, delivers an answer to the dilemma of CSR/SRI "survey fatigue," and supports the planning effort to manage the myriad of issues most important to key stakeholders.

The CAP Gap Audit^{TM} is a "meta-tool" that uses a powerful software platform to consolidate the criteria of 17 leading standards into one 195-point survey (Future 500 website, www.future500.org). The leading CSR/SRI standards systems included in the CAP Gap Audit^{TM} are the following:

- New York Stock Exchange (NYSE), Section 303A, corporate governance standards
- Goldman Sachs Best Practices recommendations
- Malcolm Baldrige National Quality Award
- Global Reporting Initiative (GRI)
- Social Accountability 8000 (SA 8000)
- Boston College Center for Corporate Citizenship (BCCC)
- International Chamber of Commerce (ICC)
- BCSD and Corporate Governance Principles
- Dow Jones Sustainability Index
- FTSE4 Good Index Series
- Global Sullivan Principles
- Domini Social Investments
- Calvert Group
- UN Global Compact
- Coalition of Environmentally Responsible Economies (CERES)
- Caux Roundtable
- Smart Growth Network
- Other referred indexes such as, AA1000, BITC, ICCR, OECD, Innovest

The methodology for the compilation of the 17 standards into The CAP Gap Audit™ incorporates a process that allows business planning by stakeholder grouping or business benefit categories. After the selection of the top standards was made, each complete standard spread was painstakingly read and analyzed in order to group like questions. The language of each question grouping was "homogenized" so that one carefully phrased question could stand in for those that were similar. In some cases, questions needed to be re-framed to accommodate a simple YES/NO answer; in other cases, multiple entries could be grouped together in a multipart question.

To illustrate, here are two representative questions from the CAP Gap Audit™ and the standards sources that these were taken from:

- Fulfill company and legal requirements by conducting regular environmental audits and compliance assessments (including facilities and suppliers), and insure transparency by providing information to the Board and other stakeholders such as outside authorities, employees, and the public? (Standards sources for this particular question: ICC-BCSD, Calvert, Innovest, DJSI.)
- Have initiatives to increase energy efficiency and use of renewable energy, such as in operations, goods and services sold, facilities design and development of alternative energy sources? (Sources: ICC-BCSD, GRI, Domini, Innovest, DJSI, CERES.)

Additionally, the CAP Gap Audit™ correlates answers on the set of business policy questions to a set of planning criteria, such as the following business value and benefit categories.

Internal and Operational Benefits: Operational/Cost Efficiencies

1. Decrease costs
2. Increase employee loyalty
3. Improve employee training and skills
4. Improve employee skills, training, teamwork
5. Increase employee productivity
6. Promote continuous improvement
7. Promote breakthrough innovation
8. Increase regulatory compliance
9. Reduce risks and legal liabilities

External and Marketplace Benefits

1. Attract new customers
2. Expand market share
3. Access new markets
4. Improve supply chain management

5. Enhance company reputation/brand
6. Protect against negative consumer action
7. Strengthen community relations
8. Increase customer goodwill/loyalty
9. Enhance overall stakeholder relations
10. Attract investors

Audit questions are "tagged" with these underlying benefit and value criteria and can be aggregated into groupings for planning purposes depending upon which aspects of managing the business an executive team wishes to focus on.

If improving human resource management is a key factor for success in a company's business model, all questions and policies that relate to this area (tags 2, 3, 4, and 5 above) can be sorted from the whole and grouped together for analysis. Both strengths and opportunities in this query area can be studied and utilized in order to set action plans into motion.

Alternatively, a process of double sorting can be accomplished whereby first human resources questions are sorted for, then a market-based sort within the first selected group can be applied. The result would be all policies or procedures that relate to human resources *and* new market sourcing or expansion.

The key to further organizing the questions for stakeholder analysis is a structure that accommodates the placement of each question into one of five quadrants organized by theme or stakeholder grouping – corporate governance, workplace, community, marketplace, and environment (Fig. 4.17). This organization framework allows for a numeric grade by themed area and allows a company to begin an analysis of its stakeholder strengths and opportunities in a very targeted way.

In the Future 500 stakeholder engagement schema, the *corporate* or shareholder stakeholder terrain is at the center of the sphere. The heart of any corporation is its mission and values, its policies and procedures for doing business, and its value proposition – that is, what product or service is being creating. A company's cultural values and ways of doing business are created in this terrain. A company's culture emanates from this center and little can be done to transform activities in the outer layers if there is no alignment with what exists at the heart of the stakeholder framework. Primary stakeholders in this terrain include the executive management team, the board of directors, and all shareholders.

The *workplace* is the stakeholder terrain of employees and their families, all policies concerning them, and benefits that accrue to them. This category includes employees' feelings about their company and their commitment to their place of employment. A company's cultural values and attitudes and all skills, experience, and training needed for an employee to do his or her job are passed on and exist in this terrain.

Community represents the actual physical location of the place of employment, the footprint of the physical plant and its impact on the community around it. Stakeholder perspectives in this category include local politics, ordinances, public affairs, and the relations with a company's physical neighbors and local government officials.

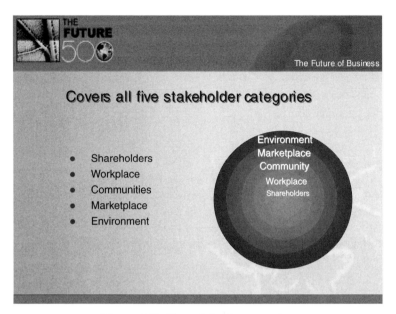

Figure 4.17. The stakeholder categories.

Marketplace is the arena of a company's product or service; its supply chain; production processes; distribution; competitors; alliances and partnerships, local, national and international; and all industry requirements and regulations. The quality of the product is part of this circle, and the attitude or beliefs consumers have about this product. Also included are the inherent dangers for the whole "product category." Shifting values in the arena of food products, for example, would be included here – like the growing awareness of parents about the dangers of sugar consumption for children; the caution about beef purchases; the growing demand for organic foods; or the European aversion to GMOs.

The *environment* comprises the final terrain in the stakeholder model, yet provides the underlying support for and encompasses all other terrains. The environmental terrain therefore influences and affects all others. It includes all natural macro and micro systems and the quality of those systems: soil fertility and micro organisms; pollination; water quality and availability; air quality; availability of natural resources – minerals, timber, flora and fauna, fisheries. In short, the environmental terrain includes any natural aspect of our planet that supports life or that is a requirement for life. Stakeholders in this terrain are comprised of all the advocacy groups that represent or speak for the natural processes of the earth, its systems, materials, habitats, and creatures.

Currently in development is a follow-on tool that assists a corporation in a more complete stakeholder mapping process that assigns stakeholder partners to roles within each stakeholder category, identifies top stakeholders in each area, and collates those to key success factors for a particular business and industry. The result is a powerful mapping process that outlines the next steps for a comprehensive stakeholder engagement program.

Because the audit is a software-based tool, information that is gathered digitally can be sorted in a variety of different ways to indicate areas of action for supporting selected business results. Thus, the strength of creating this software "meta-tool" as a higher order analysis device that aggregates data across many CSR/SRI standards means that the range of analysis approaches available to a corporation is increased. Business planning can be accomplished using the standard business criteria listed above, or by the five stakeholder categories, or by each individual contributing CSR/SRI standard.

The CAP Gap Audit™ assessment tool grades a corporation's CSR/SRI performance and generates "readiness" indicators for the top-ranking standards. Risk areas are flagged, performance is documented, and results are delivered in a concise high-impact executive report or on-site presentation. Figure 4.18 illustrates one of the reporting features – Readiness Indicators – for the major accountability standards that the CAP Gap Audit reprises.

The analysis features of the CAP Gap Audit™ allow participating companies to understand their current business opportunities and risks and use the compiled data to direct their efforts to bottom-line business planning.

An additional benefit is one reporting framework that can relieve "survey fatigue," which is often experienced at leading companies that want to report to a

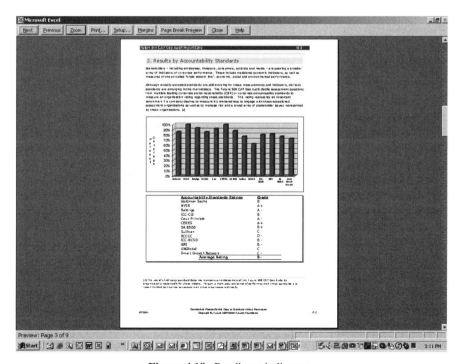

Figure 4.18. Readiness indicators.

variety of standards organizations. The CAP Gap Audit™ can provide one comprehensive executive report that includes:

- Benchmarks of a company's performance against the 17 standards at once;
- Repository of CSR/SRI information from which data can be sorted for a variety of purposes;
- Information needed to create an annual report to meet world class standards;
- Flagged areas of performance and risk – information that will assist in setting priorities and aligning cross-departmental activities;
- Customized responses to standards groups and NGO questionnaires;
- Benchmark measurements against 20 business benefits;
- "To-do" lists by business areas to create linkage and assist in transformation across business units;
- Graphically presented and easy-to-understand data.

4.7.3 Conclusion

The CAP Gap Audit™ is the product of a five-year Future 500 project supported by member companies and organizations such as Bank of America, Mitsubishi Electric, Nike, Det Norske Veritas, ERM, Manning Selvage & Lee, Pitney Bowes, WSP, Coca-Cola, and the Coors Foundation. Leaders from these participating organizations collaborated in the creation of the audit tool either by providing supporting grants for its development or becoming beta-users for its execution and refinement.

Additionally, a group of CAP alliance partners have been convened to provide follow-up consulting or mitigation measures that assist companies in implementing policy and operational changes in key areas targeted by the CAP Gap Audit™ or improved business planning processes.

The audit is generally conducted on site; key executive team members are brought together to discuss aspects of their stakeholder strategy. In the experience of the Future 500 consultants, this is often the first time a cross-disciplinary group of managers has assembled to share critical path information on environmental health and safety issues. This information gathering session is often the beginning of a process of cross-functional collaboration and stakeholder planning within the corporation.

Fundamentally, this audit tool is designed to improve accountability and business planning processes, which, in turn, boost corporate adaptability and market responsiveness to enhance the economic, social, and environmental "triple bottom line" of participating corporate partners. As the field of CSR/SRI continues to develop and mature, we anticipate that tools like the CAP Gap Audit™ will be refined to assist industry leaders in benchmarking best-in-class standards for particular industries.

The CAP Gap Audit™ is one of a new generation of CSR tools that has been developed specifically to accelerate the process of stakeholder engagement, which "is increasingly a part of mainstream business practice" (Net Impact for the Forum for Corporate Conscience, 2004). The CAP Gap Audit™ methodology improves a company's strategic decision-making processes by illuminating stakeholder partner categories and highlighting areas of potential concern, as well as those areas where a company is top-of-class.

In the current business environment

Leadership companies no longer see stakeholder engagement as an optional means to 'touch base' with a variety of interest groups; but rather as a critical part of their business strategy. In many cases, companies are using stakeholder engagement to enter challenging markets, resolve or head off confrontations with stakeholder activists, and improve or preserve their reputation in communities and the marketplace.
— Net Impact for the Forum for Corporate Conscience, 2004

In summary, understanding and responding to the CSR/SRI world of standards is no longer considered a "cost of doing business" by industry leaders but rather as an approach to strategic planning that is critical to business survival in the current climate. The Fortune 500 CAP Gap Audit^{TM} is an important tool to consider in the creation of strategic business policies, practices, and stakeholder alliances.

REFERENCES

C. Azar, J. Holmberg and K. Lindgren, "Socio-Ecological Indicators for Sustainability," *Ecological Economics*, 18, 89–112 (1995).

B. Beloff and E. Beaver, "Sustainability Indicators and Metrics of Industrial Performance," paper SPE 60982, *Proceedings of Society of Petroleum Engineers International Conference on Health, Safety, and Environment in Oil and Gas Exploration and Production*, Stavanger, Norway, 26–28 June, 2000.

J. Benyus, *Biomimicry, Innovation Inspired by Nature*, Quill Publishing, 1997, p. 7.

H. Bradbury and J. Clair, "Promoting Sustainable Organizations with Sweden's Natural Step," *Academy of Management Executive*, 18(4), 63–74 (1999).

G. Davis, *Corporate Reputation and Competitiveness*, Routledge, London, 2003.

C. Gable, "Do No Harm: Greening the Healthcare Industry," *Sustainable Industries Journal*, August 2004.

K. Geiser, *Materials Matter, Toward a Sustainable Materials Policy*, MIT Press, Cambridge, MA, 2001.

P. Hawken, A. Lovins and L. H. Lovins, *Natural Capitalism: Creating the Next Industrial Revolution*, Little, Brown and Company, New York, 1999.

J. Heskett *et al.*, *The Value Profit Chain*, Free Press, New York, 2003.

J. Holmberg and K.-H. Robèrt, "Backcasting – A Framework for Strategic Planning," *International Journal for Sustainable Development and World Ecology*, 7(4), 291–308 (2000).

IMD (International Institute for Management Development), "CSM/WWF Research Project: The Business Case of Sustainability," IMD Working Paper Series, IMD, Lausanne, Switzerland, 2003.

T. Kochan and S. Rubenstein, "Toward a Stakeholder Theory of the Firm: The Saturn Partnership," *Organizational Science*, 11(4), 373 (2000).

C. Laszlo, *The Sustainable Company*, Island Press, Washington, DC, 2003.

C. Laszlo, D. Sherman and J. Whalen, "Expanding the Value Horizon: Stakeholders as a Source of Competitive Advantage," Sustainable Value Partners, Vienna, VA, 2004. Available at www.sustainablevaluepartners.com/svp_stakeholder_value.pdf.

A. B. Lovins, L. H. Lovins and P. Hawken, "A Road Map for Natural Capitalism," *Harvard Business Review*, 77(3), 145–148 (1999).

B. Nattrass and M. Altomare, *The Natural Step for Business: Wealth, Ecology, and the Evolutionary Corporation*, New Society Press, Gabriola Island, British Columbia, 1999.

B. Nattrass and M. Altomare, *Dancing with the Tiger*, New Society Press, Gabriola Island, British Columbia, 2002.

Net Impact for the Forum for Corporate Conscience, "Communications with Internal and External Stakeholders," Forum in Action Workshop, May 19–20, 2004.

B. Perthen-Palmisano and T. Jakl, "Chemical Leasing – The Australian Approach," Workshop, Sustainable Chemistry, 27–29 January 2004, Dessau, Austrian Federal Ministry of Agriculture, Forestry, Environment and Water Management, Austria.

R. B. Pojasek, "Creating a Value-Added, Performance-Driven Environmental Management System," *Environmental Quality Management*, 12(2), 81–88 (2002).

R. B. Pojasek, "Selecting Your Own Approach to P2," *Environmental Quality Management*, 12(4), 85–94 (2003a).

R. B. Pojasek, "Lean, Six Sigma, and the Systems Approach: Management Initiatives for Process Improvement" *Environmental Quality Management*, 13(2), 85–95 (2003b).

K.-H. Robèrt, "ICA/Electrolux: A Case Report from 1992," Presented at the 40th CIES Annual Executive Congress, June 5–7, 1997, Boston, Massachusetts.

K.-H. Robèrt, "Tools and Concepts for Sustainable Development, How Do They Relate to a Framework for Sustainable Development, and to Each Other?" *The Journal of Cleaner Production*, 8(3), 243–254 (2000).

K.-H. Robèrt, *Den Naturliga Utmaningen* ("The Natural Challenge" in Swedish, translated to Japanese), Ekerlids Publisher, Stockholm, Sweden, 1994.

K.-H. Robèrt, B. Schmidt-Bleek, J. Aloisi de Larderel, G. Basile, L. Jansen, R. Kuehr, P. Price Thomas, M. Suzuki, P. Hawken and M. Wackernagel, "Strategic Sustainable Development – Selection, Design and Synergies of Applied Tools," *The Journal of Cleaner Production*, 10(3), 197–214 (2002).

J. B. Robinson, "Futures Under Glass – A Recipe for People Who Hate to Predict," *Futures*, 22(8), 820–842 (1990).

S. Satarug, J. R. Baker, P. E. B. Reilly, M. R. Moore and D. J. Williams, "Cadmium Levels in the Lung, Liver, Kidney Cortex, and Urine Samples from Australians without Occupational Exposure to Metals," *Archives of Environmental Health*, 57(1), 69–77 (2002).

C. Schmidt, "Antibiotic Resistance in Livestock: More at Stake than Steak," *Environmental Health Perspectives*, 110(7), 396–402 (2002).

G. Scholl, "Eco-efficient Service Innovations – The Example of Safechem," *Innovationsprofile*, IÖW, Berlin (undated). Available at www.ioew.de/dienstleistung/publikationen/safechem.pdf.

Shaw Industries, Inc., "Shaw Ceases Production of Polyvinyl Chloride: Replacement Carpet Tile Backing Recognized by the EPA," press release on June 14, 2004.

A. Sjodin, D. Patterson and A. Bergman, "Brominated Flame Retardants in Serum from U.S. Blood Donors." *Environmental Science & Technology*, 35(19), 3830–3833 (2001).

B. Vastag, "'Cipromania' and 'Superclean' Homes are Now Increasing Antibiotic Resistance," *The Journal of the American Medical Association*, 288(8), 947–948 (2002).

S. Waddock, "Responsibility: The New Business Imperative," *Academy of Management Executive*, May 2002.

SUGGESTED READING

G. Eckes, *Six Sigma for Everyone,* John Wiley & Sons, Hoboken, New Jersey, 2002.

G. H. Watson, *Six Sigma for Business Leaders: Agenda to Implementation*, Goal QPC, Salem, New Hampshire, 2004.

5

DESIGNING FOR SUSTAINABILITY

5.1 DESIGNING FOR SUSTAINABILITY: OVERVIEW

DICKSEN TANZIL AND EARL R BEAVER

BRIDGES to Sustainability

Design is a critical element in the implementation of sustainable development. Meeting the needs of the growing global community while minimizing negative impacts to the environment and to societal well-being requires that we develop alternative patterns of resource utilization, production, and consumption. Obviously, innovations in multiple fields, including policy, business management, infrastructure planning, and social sciences, are necessary for such transformation. Re-designing the products, services, and manufacturing processes that serve people's needs, however, is arguably one of the best places to start.

To companies that have embraced the goal of sustainable development, incorporating sustainability considerations into design also makes practical sense. As often pointed out in standard design textbooks, impacts of a product or a process are largely "locked in" during early design stages as decisions regarding product specifications, materials, technology, and so on, are made. While corrective or remedial actions may be taken after the product or manufacturing process is commercialized, options available at these later stages tend to be limited, less effective, and more costly. Thus, with greater emphasis on reputation and risk management experienced in industry today, designing for sustainability is becoming increasingly important.

This section provides an introductory overview of some of the concepts and ideas that underlie the development of sustainable design approaches. This overview is followed by a discussion of "cradle-to-cradle" design, a lifecycle-based approach to material assessment and product design, and a brief discourse on the emerging principles of sustainable engineering. These correspond to some of the emerging thoughts and approaches that may serve to guide more sustainable design in the chemical industry.

5.1.1 Designing for Sustainability: What It Means

Designing for sustainability represents the broadening of the design objectives to include aspects not customarily considered by the chemists, material scientists, and engineers involved in the design process. Traditionally, design objectives are largely limited to economic ones. Products are designed primarily to provide the greatest market value, while processes are designed to minimize costs. Environmental considerations are conventionally applied as design constraints and accounted for only at the end of the design process. For manufacturing processes, this typically results in end-of-pipe treatment – an important element of environmental costs that can be reduced through more efficient use of material and energy resources. Furthermore, uncertainties related to long-term availability of nonrenewable resources, shipping of hazardous materials, and health and safety effects of chemicals in the environment, to name a few, are driving the incorporation of sustainability considerations into the design of chemical products and processes.

Incorporating sustainability into design requires a more systemic view (see Section 5.3.2 for detailed descriptions on systems thinking) beyond the boundaries of a chemical facility. This often includes the consideration of a product's lifecycle as well as the needs of a broader set of stakeholders, such as:

- consumers' need for products and services that improve their quality of life;
- shareholders' need for economic gains to be generated from the products and services; and
- society's need for products and services to be sourced, manufactured, delivered, used, and disposed of with no negative effects on their health, safety, and environment.

The linkage between consumers' and shareholders' needs is obvious to any company that remains in business. Economic gains can be realized only when the products are marketable and match people's needs (including the needs of the industrial customers). However, societal issues, including the societal impacts of a company's environmental, health, and safety (EHS) performance, are becoming increasingly important due to the demonstrable impacts of EHS liabilities, public relation, and corporate reputation to the bottom line. Even when considering customers' needs, viewing the market through the lens of sustainable development may result in the identification of new opportunities, such as applying the company's expertise and products to satisfy the needs of people in the less-developed economies. In short, a designer may view designing for sustainability as design with the multiple objectives of maximizing benefits and minimizing risks to all stakeholders.

Related to the inherent ambiguity in the concept of sustainable development, "designing for sustainability" may be understood differently by different people.

In this chapter, "designing for sustainability" is used as a broad term that refers to the integration of all of the sustainability considerations into the design of industrial products and processes. As such, the term encompasses a broad range of environmentally conscious chemical process design approaches that have been developed in the literature as well as emerging approaches that push the boundaries of design considerations in the chemical industry. They include approaches that originated in the field of industrial ecology, such as lifecycle design and design-for-environment, which are based on a systems view and tend to focus on design to improve environmental performance of a product system along the product's lifecycle (see Table 5.1 for descriptions of terminologies). Other approaches that focus specifically on aspects of chemical product and process design, including green chemistry, green engineering, and inherently safer chemical processes, are also considered here as

TABLE 5.1. Several Terminologies Related to Designing for Sustainability

Term	Descriptions	References
Industrial ecology	A systems-oriented subject that seeks to optimize resources, energy, and capital through the study of industrial and economic systems and their interactions with the natural ecosystems.	Graedel and Allenby, 2003
Design for Environment (DfE)	A product design approach originated in industrial ecology and a concurrent design approach called "Design for X" (DfX, where X is any desirable product characteristics such as safety, manufacturability, recyclability, etc.); typically focuses on reduction of environmental impacts and resource consumption throughout the product lifecycle.	Graedel and Allenby, 1996; Fiksel, 1996
Lifecycle design	Similar to DfE, emphasizing the integration of lifecycle and environmental impact consideration at each stage of product development cycle.	Keoleian and Menerey, 1993
Green chemistry	Design of chemical products and processes that reduce or eliminate the use and generation of hazardous substances.[a]	Anastas and Warner, 1998
Green engineering	Design, commercialization, and use of processes and products that are feasible and economical while minimizing generation of pollution at the source and risk to human health and the environment.[b]	Allen and Shonnard, 2002
Inherently safer chemical processes	Eliminating or significantly reducing process hazards through material substitution, alternative reaction routes, process intensification/simplification, etc.	Hendershot, 2004

[a]http://www.epa.gov/greenchemistry/whats_gc.html, accessed October 23, 2004
[b]http://www.epa.gov/oppt/greenengineering, accessed October 23, 2004

part of designing for sustainability. Viewed more broadly, designing for sustainability reflects a crucial paradigm shift for the 21st century: the transition from environmental management to systems design – coming up with solutions that integrate environmental, social, and economic factors and radically reduce resource use, while increasing health, equity, and quality of life for all stakeholders.

The different elements of designing for sustainability are currently practiced in industry to varying degrees. In order to gain relevance to the industry, the design approaches must be effective in reducing costs and risk exposure and result in reasonable short-term as well as long-term financial benefits. Thus, green chemistry and engineering and inherently safer chemistry appear to have gained greater acceptance as they directly address environmental, health, and safety issues specific to the chemical industry. However, industrial ecology and lifecycle design approaches are only gradually becoming more relevant as the industry begins to discover opportunities for new products and services found through working with the value chain, that is, with suppliers and industrial customers, in reducing impacts of their processes and products. Development of high-performance, light-weight polymers for transportation represents an oft-cited example. Obviously, numerous policy and societal aspects are also critical in driving these changes.

5.1.2 Design and Lifecycle Assessment

Systems thinking and, in particular, lifecycle concepts (Box 5.1: *How Sustainability Uses Lifecycle Concepts*) underlie many of the different approaches for more sustainable designs. These approaches are becoming increasingly relevant to the industry as companies strive to work with their suppliers and customers in reducing impacts. Arguably the most widely recognized tool for incorporating systems thinking into design is lifecycle assessment (LCA). The ISO 14040 series of standards from the International Organization for Standardization defines LCA as "the consideration of inputs, outputs, and potential environmental impacts of a product system throughout its life-cycle." While skewed mainly towards the environmental aspects of sustainability, LCA forms a basis of a broader analysis of sustainability impacts.

Lifecycle assessment was born out of the realization that there are significant environmental impacts throughout the lifecycle of an industrial product. Making decisions without considering the entire lifecycle often lead to suboptimization; that is, one may simply shift the adverse environmental impacts outside the boundaries of one's process or facility. Outsourcing toxic production to the suppliers, for instance, is as poor a decision as improving process safety while making the product less safe. In both cases, adverse impacts are reduced within the facility boundaries. However, as one takes a systemwide view over the entire lifecycle, it becomes clear that the total impacts are not reduced.

In its formal configuration, LCA is almost always used for a comparative study – evaluating multiple designs or decision alternatives. The comparison is performed on the basis of a "functional unit," which may be a physical unit (e.g., a pound of a chemical product or a piece of gadget) or a service unit. The latter is usually more representative of the value delivered by the product system. For example, in performing LCA for textile dye alternatives, BASF (Saling *et al.*, 2002) uses a unit "m^2 of jeans dyed" (a service unit) as the basis of the comparison, instead of a unit kg of dye. In this case, the lifecycle performance of a technology or a product is judged by the service that it provides, instead of product weight.

Typically, all stages in the lifecycle of a product system are included in LCA. As illustrated in Figure 5.1, this consists of extraction (including processing of the raw materials), production, use, and end-of-life. Transportation, which occurs between the stages and within each stage, is also generally considered. Such assessment is often termed "cradle-to-grave," that is, encompassing the entire product lifecycle from raw material extraction (cradle) to end-of-life (grave). Options related to end-of-life, such as reuse and recycling, are inherently part of the assessment.

For products that go to a vast number of outlet uses, including most commodity chemicals such as acids, bases, and engineering polymer resins, a "cradle-to-grave" analysis is often impractical. Lifecycle assessment remains relevant, however, especially in the form of "cradle-to-gate" analysis. Instead of including all the typical lifecycle stages, the analysis may stop at the manufacturer's exit gate, or to the point that the product is shipped to customers. Such analysis is particularly important in evaluating alternative processes that uses different starting materials, or the same set of materials but at significantly different amounts or compositions. In addition, lifecycle thinking may be employed qualitatively to identify various uses and end-of-life issues that the manufacturers may face in the future.

Methodology for performing lifecycle assessment has been discussed in many books dedicated to the subject (e.g., Goedkoop, 1994; Curran, 1996) and will not be elaborated in great depth here. In essence, LCA begins with the definition of goal and scope, and is followed by the development of "lifecycle inventory" (LCI), where information on the inflows and outflows of materials and energy for

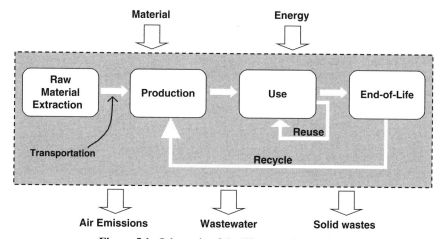

Figure 5.1. Schematic of the lifecycle of a product.

each relevant lifecycle stage are collected and tabulated. Energy is considered in terms of the consumption of primary energy.[1] Thus, secondary forms of energy, such as electricity and steam, are considered not in terms of their inherent energy content. Instead, it is accounted in terms of the energy contained in fuels and other primary sources consumed in generating the secondary energy sources.

The lifecycle inventory is then analyzed for impacts to the environment through a methodology called "lifecycle impact assessment" (LCIA). Each flow in the inventory is analyzed for its contribution to multiple impact categories, such as human and ecosystem toxicity, global warming potential, air acidification, water eutrophication, resource depletion, and so on. When desired, one may use a set of weighting factors to aggregate the various impact categories into a single environmental performance index (sometime called the "environmental footprint"). The various metrics and indicators used in LCIA will be discussed in greater details in Chapter 6 (Section 6.1).

Finally, as the last step in LCA methodology, the findings obtained from LCIA are analyzed for their significance, statistical uncertainty, and so on. Primary concerns may be identified and recommendations made based on the analysis.

Lifecycle assessment provides a formal, systematic methodology for assessing and quantifying impacts along the lifecycle of a product system. The process, however, can easily become time-consuming and costly. Consequently, the full LCA is generally performed only for cases where there are significant differences in the lifecycle (e.g., differences in the primary raw materials or in how the products are used and disposed). Furthermore, the full LCA is usually completed only during the later stages of design when much of the information required is more readily available and there is greater chance of commercialization to justify the costs.

Nevertheless, the lifecycle thinking formalized in LCA lies behind many design strategies and guidelines as well as screening methodologies used in the early stages of product and process design. These LCA-inspired screening methodologies are sometime called "streamlined LCA" (Graedel, 1998) and typically rely on the use of a two-dimensional matrix, with lifecycle stages on one dimension and impacts on the other. The streamlined LCA also tends to be more qualitative, with a relative score assigned for each cell in the impact versus lifecycle stage matrix. The use of a more qualitative matrix system also allows considerations that are not commonly included in the formal LCA, such as noise and other nuisance factors, public perception, market position, and so on.

5.1.3 Sustainable Design Strategies and Implementations

Motivated by systems thinking and the lifecycle concepts, strategies for sustainable design typically focus on the entire product system, including design of the product

---

[1]The Intergovernmental Panel on Climate Change (IPCC) defined "primary energy" as "energy embodied in natural resources (e.g., coal, crude oil, sunlight, uranium) that has not undergone any anthropogenic conversion or transformation." See http://glossary.eea.eu.int/EEAGlossary/P/primary_energy, accessed February 9, 2004.

itself, the manufacturing process, and the facility/industrial processes where the process is sited.

5.1.3.1 Product Design. Product design has not traditionally been the focus of the chemical industry. As mentioned earlier, some chemical products, especially commodity chemicals, are used for the production of myriad other products, sometime outside the control of the chemical producers. However, even in the manufacture of commodity chemicals, there is a value in employing sustainability and systems thinking in formulating the product. Xylene, for example, has been one of the top 25 chemicals by volume for decades. It may be produced as pure isomers or a mixture of all its three isomers, and each has its distinct uses (*o*-xylene for phthalic anhydride, *m*-xylene for isophthalic acid, and *p*-xylene for terephthalic acid). Mixed xylenes and relatively impure isomers are significantly less expensive than high-purity individual isomers. A xylene producer can separate isomers via selective crystallization, but a user can also process mixed xylenes and separate their desired product from a mixture resulting from their reaction step. Thus, at least theoretically, the commodity chemical supplier can optimize the overall energy use, material use, emissions, and costs by working together with the user. In practice, this concept often requires a close alliance and effective communication between the supplier and user as well as relatively stable market demands for the products.

Product design in the chemical industry today, however, is much more than determining the required purity of the shipped products. Stephanopoulos (2003) identified an ongoing shift in the chemical industry: from a process-centered industry towards a more product-centered industry. Increasingly, chemical companies are collaborating with their downstream industrial customers to provide compounds and materials that make their customers' products better. Performance polymer manufacturers, for example, now routinely work with automakers to provide lighter and stronger materials that improve the automobiles' fuel efficiency and safety. With such trends, the chemical industry's chemists, material scientists, and engineers are increasingly involved in many aspects of product design.

Material safety is certainly one important dimension of product design. The chemical industry has made considerable progress in recent years in formulating more benign alternatives that meet or exceed the functionalities of those that they replaced. For example, Engelhard Corporation (Iselin, NJ) is phasing out its production of 1.2 million pounds of heavy metal pigments and replacing them with its newly developed, more benign alternatives containing calcium, strontium, and sometime barium. The new pigment formulations also provide additional benefits, such as improved heat stability and improved color strength. While products are rarely commercialized based solely on superior environmental performance, the additional advantages in the product's performance greatly facilitate the commercialization. This new set of formulations has won the company the 2004 EPA Presidential Green Chemistry Challenge award for designing safer chemicals. (For more on the Presidential Challenge award winners, visit www.epa.gov/greenchemistry/presgcc.html.)

Designing more sustainable product also requires focus on the product's perform-ance, specifically: designing the product to deliver maximum function during its use and maximum value at end-of-life. The designer must ask a number of important questions, such as:

- During life, can the product be easily maintained, repaired, and remanufac-tured to extend its service life?
- At the end-of-life, can the product perform a different service?
- Can the product be dissembled into parts useful for another purpose?
- Can components be recycled?
- Can portions be used as fuel without great cost or pollution?
- Does the nature of the product solve problems for the manufacturer while causing greater impacts upstream or downstream?

By extending the service life, impacts per unit of service can be greatly reduced. Energy consumption in delivering the desired function, material consumed during operation, ability to perform more than one function, and amount of undesirable environmental or ecological consequence along the lifecycle stages are also attri-butes of importance in sustainable product design.

At end-of-life, value recovery is of the utmost importance. The highest value is obtained when the product is used for yet another service with little or no additional expenditure of resources. For example, a Pentium I personal computer considered outdated in the United States may be shipped to a less developed economy to pro-vide simple services such as e-mail access. The worst end use would be casting the device into a landfill. Figure 5.2 shows the end-of-life values recovered by different

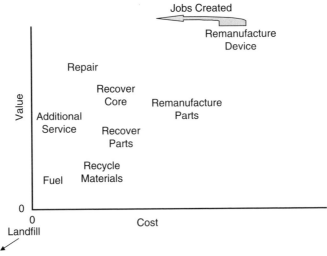

Figure 5.2. Product end-of-life value versus cost for multiple end-of-life options.

end-of-life options and the costs they incur. Remanufacturing a device retains most of the original product value. However, it is a costly process, requiring three to five times the number of labor hours than original manufacture. Nevertheless, the labor requirement provides valuable job opportunities and substitutes for material, water, and energy – resources that are becoming increasingly scarce. In addition, pollutants are reduced over the total lifecycle.

Chemical product designers play an important role in providing better end-of-life options. For example, Shaw Industries (Dalton, GA), as another recent EPA Presidential Green Chemistry Challenge award recipient, introduced new carpet tile backings that make the carpet tiles recyclable. Conventionally, carpet tile backings are made with PVC or polyurethane (PU). As a chlorine-based product, PVC is associated with various health concerns, especially upstream in the product's lifecycle. Polyurethane, on the other hand, is very difficult to recycle due to its thermoset cross-linking. In contrast, the new product, made with a polyolefin mixture from Dow Chemical, provides superior properties and is easy to recycle. In fact, the company claims that the cost of collection, transportation, and the remanufacture process is less than the cost of making the products from virgin raw materials.

Such strategy in product design is often termed "cradle-to-cradle," that is, recycling the material back to the cradle where it can gain a new "life." This concept is discussed in greater depth by Lauren Heine is Section 5.2.

5.1.3.2 Process Design. Practical implementation of sustainable development also requires the incorporation of sustainability concepts into the design of chemical processes. Principles and strategic guidelines have been formulated for "greener" or more sustainable design of chemical processes. One of the most widely recognized sets is that of Green Chemistry (Anastas and Warner, 1998), listed in Table 5.2. In addition to its requirement for safer chemical products, the principles highlight the need for processes that are resource-efficient and inherently safe.

TABLE 5.2. Twelve Principles of Green Chemistry

1. Prevent waste
2. Maximize the integration of all process materials into the finished product
3. Use and generate substances with little or no toxicity
4. Design chemical products with less toxicity while preserving the desired functions
5. Minimize auxiliary substances (e.g., solvents, separation agents)
6. Minimize energy inputs
7. Prefer renewable feedstocks over nonrenewable ones
8. Avoid unnecessary derivations and minimize synthesis steps
9. Prefer selective catalytic reagents over stoichiometric reagents
10. Design products for post-use decomposition and no persistence in the environment
11. Use in-process monitoring and control to prevent formation of hazardous substances
12. Use inherently safer chemistry that minimizes the potential for accidents

Source: Anastas and Warner (1998).

Hierarchical design methodology informs much of the efforts in sustainable design of chemical processes. Proposed by Douglas (1988), the methodology recognizes a hierarchy of decisions in the design of chemical processes. One begins with a simple, conceptual design. Successive layers of detail are then added to the process flowsheet with each design iteration. The hierarchical method has been adapted for the purposes of waste minimization and pollution prevention by various authors (e.g., Alva-Argáez et al., 2001; Chen et al., 2003).

Chemistry generally constitutes the first step in hierarchical process design. This relates to process synthesis, specifically the selection of reaction routes and separation agents (Chen et al., 2003). Anastas and Allen (2002) proposed a set of strategies based on the Green Chemistry principles discussed above, which include:

- Identifying synthesis pathways with superior environmental performance;
- Selecting alternative feedstocks that are benign and leads to higher selectivity and yield; and
- Selecting solvents that are less hazardous, safe, with less environmental impacts.

The strategies also seek to come up with chemistries that are inherently safe. The considerations of alternative feedstocks and catalysts as well as new technologies such as biotechnology are necessary in identifying the superior synthesis route. Safety and environmental performance of the reaction byproducts are also important.

The EPA Presidential Green Chemistry Challenge also provides some illustrations on how Green Chemistry contributes to sustainability of chemical processes. The 2004 awards recognized, among others, the development of an alternative synthesis pathway for Taxol®, the anticancer drug. Previous technologies require paclitaxel, the active ingredient in Taxol®, to be isolated from the barks of the Pacific yew trees – plants that take 200 years to mature and are part of sensitive ecosystems – or produced semisynthetically from the twigs and leaves of European yew. Scientists and engineers at Bristol-Myers Squibb succeeded in developing an alternative pathway through the latest plant cell fermentation technology. This biotechnology alternative significantly simplifies the synthesis route, with no chemical transformation required beyond fermentation, and thus reducing the use of solvents and energy. Not only does the alternative improve environmental performance, it is also good business, ensuring yearlong harvests and a sustainable supply of paclitaxel.

Once the chemistry is known, one must then select the technologies and equipments for the various unit operations that make up the chemical process. Guidelines and heuristics have been developed to improve the environmental performance from this design step. Many have been summarized by Allen and Rosselot (1997) and more recently by Shonnard (2002). Choices of catalyst technology (e.g. homogeneous versus heterogeneous catalysts) and reactor technology (e.g., fixed bed versus fluid bed) can have a tremendous effect on efficiency and environmental performance. In designing the reactor, one may further optimize the reaction conditions, mixing conditions, and reactant concentration to maximize conversion, yield, and selectivity toward the intended product. The choice of separation technology is also important, as separations typically constitute the most energy- and resource-intensive unit operations in a chemical process. While distillation remains the most widely used separation technology in the chemical industry today based on volume of materials separated, numerous alternative separation technologies have evolved over the past century (Fig. 5.3). New separation technologies, such as dilute solution separation and simulated moving bed chromatography, tend to be more efficient, benign, but costly. However, these technologies are increasingly used when very high value-added materials are involved, such as in the separation of individual enantiomers of pharmaceuticals.

Another important step in improving environmental performance in chemical process design is process integration, which is "a holistic approach to process design and operation that emphasizes the unity of the process" (Dunn and El-Halwagi, 2003). Design of "heat exchanger networks" (HENs) is perhaps the most familiar example of process integration. To optimize energy use, heating and cooling requirements of the process are analyzed systematically using a method called "pinch analysis" (e.g., Shenoy, 1995). The term "pinch" refers to a key system temperature constraint that thermodynamically limits heat recovery and thermal energy efficiency. Pinch analysis enables the matching of cold streams (i.e., streams that need heating) with hot streams (i.e., streams that need cooling) in the most effective way, hence minimizing the need for additional process heating and cooling. Similar approaches have been developed to optimize other aspects of a chemical process, such as the "mass exchange networks" (MENs) that optimize

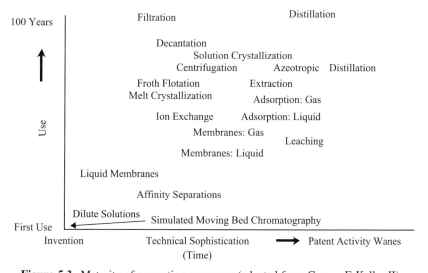

Figure 5.3. Maturity of separation processes (adapted from George E Keller II).

the design of end-of-pipe separation and recycle structure and the water conservation networks that optimize water recycle and reuse (Dunn and El-Halwagi, 2003).

Lifecycle perspective is also increasingly used in chemical process design, especially in evaluating processes involving different feedstocks or auxiliary materials, or significantly different quantities of the same materials. Even when one focuses only within the boundaries of the chemical process, various impact assessment categories in LCIA are frequently used to assess its environmental performance. Computational tools such as the Waste Reduction algorithm (WAR) (Young and Cabezas, 1999) and Environmental Fate and Risk Assessment Tool (EFRAT) (Shonnard and Hiew, 2000) automate the impact assessment process. Midpoint analysis (described in Section 5.3.2) is used in these automated tools, where impacts are categorized in terms of human and ecosystem toxicity as well as pollutant impacts such as global warming potential, acidification, and water eutrophication. Integration of these tools into chemical process simulators such as Chem-Cad (ChemStations, Inc., Houston, TX), Aspen Plus and HYSIS (both of Aspen Tech, Cambridge, MA) significantly simplifies the consideration of environmental impacts in chemical process design.

Figure 5.4 illustrates the use of some of the above considerations in the production of maleic anhydride from n-hexane. The process comprises a reactor and a separation system that includes an organic solvent absorber and a series of distillation columns (Lagace, 1995; Chong, 1995). Energy use shown for the chemical processes in Figure 5.4 includes the feedstock energy consumed by the process (i.e., the difference between the heating values of the product and the feedstock, ΔH_c). Using pinch analysis, an optimum heat exchange network (HEN) can be constructed, which significantly reduces the energy consumption by the chemical process. Energy consumption is further reduced when a different reactor technology is employed (fluid bed instead of fixed bed). This, obviously, represents a fundamental redesign of the process. Despite the lower energy consumption, the fluid-bed

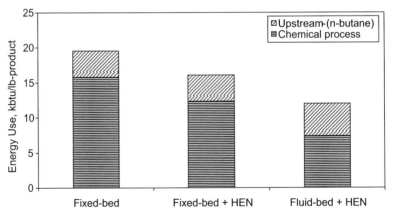

Figure 5.4. Comparison of energy use among alternatives for the production of maleic anhydride from n-butane. (*Source:* BRIDGES, 2001.)

technology consumes more of the n-butane feedstock compared to the fixed-bed technology. This calls for the integration of lifecycle thinking: the consideration of energy use required to provide the n-butane. As shown in Figure 5.4, the fluid-bed process indeed requires more energy in the upstream processes to produce n-butane. However, the lifecycle consideration also shows that the fluid-bed technology still consumes less energy overall. Other impacts can also be calculated for the three alternatives (BRIDGES, 2001).

5.1.3.3 Facility Design. Industries' quest for greater sustainability has fueled the growth of industrial ecology, which builds on the notion that nature "can serve as a useful metaphor for industrial systems, which can be used to help industry become more efficient and more sustainable" (Allen and Butner, 2002). In nature, there exists an optimal network of nutrient exchanges as the waste from one organism becomes food to others. With nature as a source for inspiration, a similar network of material and energy flows may be created where the waste from one process serves as feedstock or energy source for other processes, hence increasing resource efficiency and reducing releases to the environment.

An industrial network in the small Danish town of Kalundborg is an example of the efficient integration of material and energy flows. For more than 30 years, this multi-industry "eco-park" has been swapping material and energy between processes (Garner and Keoleian, 1995; Kaiser, 1999; Allen and Butner, 2002). For instance, steam and waste heat from power generation are transferred for industrial and municipal heating such that up to 90 percent of the heat from the coal-burning power plant can be utilized (Allen and Butner, 2002), as opposed to the industry average of approximately 34 percent efficiency. Further, gypsum from the power plant is utilized by a drywall factory, a biotech's fermentation waste is shipped to farmers for fertilizing fields, and cooling water from a refinery is used as boiler water by the power plant (Kaiser, 1999).

These examples may not be directly applicable in the U.S. settings. Legal issues related to the transfer of liability often hamper such efforts. However, it illustrates that, ideally, a well-designed industrial network would allow not only very high material and energy efficiency, but also minimal environmental costs as only limited waste is generated. This concept of industrial ecology also underlies the "cradle-to-cradle" design strategy discussed in Section 5.2.

5.1.3.4 Further Considerations. Principles and strategies that have been developed to date for more sustainable design have largely focused on reducing the environmental impacts of the industrial products and processes. However, design of truly sustainable products and processes also requires the consideration of societal impacts beyond the impacts created through resource use and environmental degradation. This has been recognized by many engineers and scientists, including those that gathered at the 2003 Green Engineering conference (Nguyen and Abraham, 2003). Described by Dr. Martin Abraham in Section 5.3, the conference participants produced a preliminary set of principles of green engineering that recognizes the needs to "develop and apply engineering solutions, while being cognizant of local geography, aspirations, and cultures" and to "actively engage communities and stakeholders in development of engineering solutions." Such considerations will ensure the design of chemical products and processes to optimally meet society's needs.

BOX 5.1   HOW SUSTAINABILITY USES LIFECYCLE CONCEPTS

BRUCE W VIGON

Although there is controversy about the actual definition of sustainability, it has been argued that by its very nature the concept requires a lifecycle perspective. The basic premise is as follows. If sustainability ultimately is a characteristic of society as a whole, then a systems approach has to be taken regarding the products, processes, and services that support that society and its functioning. In other words, an organization, a product or a process cannot be locally "sustainable" in isolation from its broader surroundings. Any method for assessing the contributions of various products and processes to sustainability needs to be able to take into account the full lifecycle from raw materials extraction through production, use, and, eventually, final disposition. Otherwise, trade-offs cannot be judged among different lifecycle stages, different time periods over which consequences may occur, or different locations where costs or benefits may accrue. It is hardly sensible to improve in one aspect if this improvement is offset in other parts of the lifecycle or other geographic areas or if problems are shifted from the present to the future (Klöpffer, 2003). These disparities created by a perspective limited to the firm's own operations are clearly inconsistent with the popular definition of sustainable development as that type and rate of development that satisfies the needs of the present without compromising the ability of future generations to meet their needs (WCED, 1987).

Lifecycle concepts span a range of formalization and quantification. At one end of the range is lifecycle thinking and at the other very formal systems modeling. Lifecycle thinking creates awareness of the interconnected nature of the constituent stages and operations inherent in products and processes and supports decision-makers in broadening their perspectives. While necessary for sustainability assessment, lifecycle thinking is not sufficient. To actually operationalize lifecycle concepts for sustainability, quantification is necessary. It is generally recognized that sustainability has three dimensions – environmental, social, and economic – although the relative importance of these is much debated. Lifecycle concepts, and to some extent methods and tools, exist for measurement and quantification along each of these dimensions.

Lifecycle concepts also operate at varying application levels – at a management system level, a program level, or an analytical technique level. Life cycle management (LCM), as the name implies, operates at the highest of these levels and produces the broadest overlay with sustainability. It evaluates, according to various environmental, market, financial, health and safety-related metrics, the impacts of a product or service throughout its lifecycle (Hunkler *et al.*, 2001). Further, it systematizes the management of products as required in programs such as Integrated Product Policy (IPP) and Environmentally Preferred Products (EPP) procurement (Executive Order 13101, 1998). The following sections of this chapter will add some details on the nature, strengths, and limitations of these programs and tools.

In addition to playing a role in assessing generally whether developments are more or less sustainable or not, lifecycle concepts and methods provide a basis for supplier and customer value definition in the product chain, and allow an organization to relate its goals and activities to national or global needs and interests. By using a lifecycle-based approach, an organization is more aware of the consequences of its decisions up and down the value chain and can take credit for creating value outside the limited purview of its own facility or company boundaries. Discussion below will show the intersection of lifecycle concepts and approaches with chemical industry programs and actors, including Responsible Care® and ISO 14000 Environmental Management Systems.

Benefits of Engaging Value Chain Partners through Lifecycle Practices

Many companies, both within the chemical industry and in other sectors, have recognized benefits from involving their suppliers and customers in various aspects of their business. The area of supply chain management (SCM) has become a critical element in the overall business strategy of improved productivity, reduced costs, and better control of the quality and potential risks associated with raw materials and intermediates. Proactive management of supplier environmental performance, as practiced by Hewlett Packard, can lead to product and process simplification, improved resource efficiency, product quality enhancement, reduced liability, and customer perception of the company as an industry leader.

Within the chemical industry, BASF and Bayer, among others, have instituted aggressive programs to extend their activities up the lifecycle chain. Product safety and environmental evaluation begin at the buying stage, when purchase contracts are negotiated for raw materials for the company's worldwide production facilities. Hundreds of staff members comprising the global purchasing team are not only responsible for buying raw materials and negotiating contract terms, they also assess product and supplier risk using company-developed, lifecycle tools. Assessing environmental and safety standards of suppliers means that a BASF employee from Raw Materials Purchasing visits suppliers and carries out an environmental and safety assessment to ascertain whether they operate effluent treatment plants to minimize pollution and use safety standards that comply with Responsible Care® (BASF website, http://www.basf.de/en/corporate/sustainability). The aim is not to disqualify suppliers, although in certain instances that may occur. Rather it is to consider suppliers as production partners and use BASF expertise to help suppliers deal with environmental and safety issues and work with them to reduce any potential risks. Although not fully quantified, such partnerships can mean real savings in the form of stabilizing supply, ensuring that environmental issues cannot force shutdown of a critical raw material production process, and reducing the liability of the purchaser due to use or disposal issues (BASF, 2002). Suppliers themselves gain by obtaining access to a larger base of expertise on hazards and risks and can reduce costs by reengineering their business processes and operations.

Bayer recently instituted lifecycle cost reduction for use by procurement personnel with suppliers (Purchasing Magazine Online, http://www.manufacturing.net/pur/index.asp?layout = article&stt–000&articleid = CA310878). The database and software allow purchasing staff and the business units they are supporting to identify the most promising cost-reduction opportunities globally and to track the financial consequences of the resulting agreement during the lifecycle of the contract. By doing so, visibility is gained into which contracts with which suppliers are producing the greatest cost reduction benefits.

Extending this type of involvement to the social dimension of SCM is a bit more challenging. Nevertheless, some companies have begun to take the employment and community relations aspects of their suppliers into account. As part of its commitment to the principles of sustainable development, BASF has now supplemented its environmental criteria with minimum social standards. Their purchasing activities comply with the UN's Global Compact initiative – insisting that suppliers reject child labor and do not use forced or bonded laborers. Conditions for the purchase of technical equipment and goods specify that suppliers must comply with the International Labor Organization's (ILO) employment standards.

Other examples of customers using lifecycle approaches to involve and energize their suppliers can be found. The U.S. automotive manufacturers, both individually and collectively through the USAMP initiative, have undertaken to prepare lifecycle inventories for an entire generic midsize passenger vehicle and various components parts of other vehicles. In doing so they needed to engage their supply chain in providing data for the analyses. By itself, this would have been considered a burden and a cost for their suppliers. However as a participant in this exercise, Visteon recognized that they could adopt lifecycle practices in the design and marketing of their own products and leverage the awareness and willingness of the automakers toward lifecycle considerations in product development (http://www.visteon.com/). One tangible result of this, among several, is a new, long-life air filter that Visteon designed around lifecycle considerations and using lifecycle tools to estimate the environmental consequences.

5.2 CRADLE-TO-CRADLE MATERIAL ASSESSMENT AND PRODUCT DESIGN

LAUREN HEINE
GreenBlue

The current prosperity of the Western world is largely a product of the Industrial Revolution. While the industrialization of the past two centuries produced enormous benefits, it has also left us with a legacy of unintended, negative effects. Among its many impacts, the conventional "cradle-to-grave" model of the Industrial Revolution – with its "take-make-waste" pattern – produces vast quantities of waste and broadly exposes people and ecosystems to toxic materials. In the past few decades we have begun to address the deficiencies of this unsustainable model. A recent report by the National Academy of Sciences calls for creating and maintaining material flows accounts for developing sound public policy (Board on Earth Sciences and Resources, 2004). But we cannot wait until these resources are in place before taking action.

One practical approach to what has come to be called "sustainability," is a new paradigm for positive industrial activity known as "cradle-to-cradle" design. As we consider millions of years of nature's biological and ecological activity, we see astounding and prolific creativity. Cradle-to-cradle proposes a model for human industry based on principles gleaned from these natural systems. The key "design principles" cradle-to-cradle adopts from nature's example include the following:

- *Use current solar income.* With very few exceptions, life on Earth is ultimately fueled by energy from the Sun. We are only beginning to learn to harness current solar energy, directly and indirectly, to human purposes.
- *Celebrate diversity.* Natural systems thrive on richness and diversity. Likewise, industry should promote development of diverse products that are fitting for different cultures and ecosystems.
- *Waste equals food.* There is no "waste" in nature. The product of one organism is food or structure for another. Human systems can also be designed to circulate materials productively, eliminating the concept of waste.

By redesigning industry based on nature's principles, economic activity can reinforce, rather than compromise, social and environmental prosperity.

5.2.1 The Cradle-to-Cradle Model

At the heart of cradle-to-cradle practice is the design of safe and effective materials maintained in productive, cyclical flows. The key principles of cradle-to-cradle were first systematically outlined as the Intelligent Product System (IPS), developed and articulated in 1992 by Michael Braungart and colleagues (1992) (see also www.epea.com). The IPS provides a framework for cradle-to-cradle product conception and material flow management. Just as in natural systems one organism's "waste" becomes nutrients for another, IPS applies effective nutrient cycles to the design of human industry. It recognizes two metabolisms within which materials are conceived as nutrients, circulating safely and productively – the biological metabolism and the technical metabolism.

The *biological metabolism* is the system of natural processes that support life. These processes are cyclical, ultimately fueled by the energy of the sun, and include the biodegradation (and possibly other forms of degradation) of organic materials and their incorporation into organisms. Materials that contribute to the productivity of the metabolism are *biological nutrients* that are rapidly renewable, biodegradable, and ecologically safe. Products of industry made from biological nutrients can be integrated into the biological metabolism via organic processing techniques such as composting or anaerobic digestion, leading to soil amendments and potential energy production.

Industry can be modeled on natural processes to create *technical metabolisms* that productively cycle industrial materials. These materials, valuable for their performance qualities and typically nonrenewable, are *technical nutrients*, designed to circulate safely through product lifecycles of manufacture, use, recycling, and remanufacture. Some companies engage in leasing programs whereby valuable components made from technical nutrients can be recovered and reused in new products. Technical nutrients can also be designed to be shared over an industry or multiple industries, depending on the application and the value of the material. Products made from technical nutrients should be recoverable at their highest value with minimal expenditure of energy and cost.

5.2.2   The Context for Cradle-to-Cradle

In its positive, design-based approach to sustainability, cradle-to-cradle connects principles from a number of disciplines. It embraces Green Chemistry in calling for the selection and design of new molecules and materials that are inherently benign with respect to human and ecological health. As with Industrial Ecology, cradle-to-cradle conceives materials and energy flows as "nutrients" processed within "metabolisms" (Ayres, 1994). It incorporates Design for the Environment (DfE), directing design to eliminate human and ecological hazards at all phases of the product lifecycle, and facilitating the efficient recycling of products and materials. It also involves the engineering disciplines – green engineering and/or engineering for sustainability – by challenging designers and engineers to integrate the use of benign materials into products and subsequently into material flow metabolisms, moving away from cradle-to-grave thinking and towards continual recycling and/or use for energy and value recovery in biological systems. In its broad vision, cradle-to-cradle is consistent with other programs for sustainability such as the System Conditions of The Natural Step (see Chapter 3).

Cradle-to-cradle design differs from other initiatives and approaches in some significant ways. First of all, it is both poetic and practical. While that sounds profoundly nontechnical, it is profoundly effective in the business world. Its optimistic approach to sustainability does not argue that we should do less, make less, and use less. It is not primarily concerned with efficiency, but with effectiveness. It sets forth a positive vision of healthy, sustaining industry toward which businesses can progress by redesigning products and processes. And it offers practical methods for creating products made of nutrients to flow safely in metabolisms. Within this model, efficiency serves to maximize positive effects rather than minimize damage. Cradle-to-cradle moves the discussion away from comparative risk assessment, lifecycle assessment or even whether or not to be precautionary – and towards a positive design vision that leads to sustainable material flows. This is a fundamental distinction and shift. Risk assessment, lifecycle and impact assessment, and the precautionary principle are all important tools, but they have different intents and purposes than cradle-to-cradle design. Cradle-to-cradle asks different questions of chemists, engineers, and other designers. Rather than assessing and comparing products for the lowest risk or least environmental impact, cradle-to-cradle asks how best to design a product with particular functional attributes so that it can productively and safely circulate within material flow metabolisms. The cradle-to-cradle approach assesses hazard, exposure, and to some extent energy, but within a framework of sustainable material flows.

5.2.3 The Cradle-to-Cradle Approach to Material Assessment and Product Design

The practice of the cradle-to-cradle approach involves both analytical and design-centered activities working in parallel to define and create healthy materials and productive flows. These two types of activities can be broadly defined as (1) gathering knowledge for discernment, and (2) designing for cradle-to-cradle metabolisms.

5.2.3.1 Gathering Knowledge for Discernment. A surprising number of product manufacturers have little knowledge about the chemical composition of the materials they purchase and use. Understanding the composition and potential human and environmental health effects of materials is critical to creating sustainable industry.

Companies can begin to address this need by conducting a detailed inventory of the entire palette of materials and chemical ingredients used in a given product, and the substances they may give off in the course of manufacture and use. With this information, they can determine the human and environmental health characteristics of substances, and their fitness for use as biological or technical nutrients.

Once a material's chemical ingredients have been identified, each ingredient is profiled based on the attributes listed in Table 5.3. The attribute set is used to provide a profile of a chemical's potential human and environmental health effects and the profile can be modified as new tests, data, and discoveries become available. Some of the data relating to the attributes of chemical ingredients are available from numerous "lists," that is, federal and state regulatory lists, international classifications, lists of carcinogens (National Toxicology Program Report on Carcinogens, NTP; International Agency for Research on Cancer Carcinogens, IARC; U.S. EPA Integrated Risk Information System, IRIS; American Conference

TABLE 5.3. Chemical Assessment Attributes

Priority human health criteria (known or suspected): • carcinogenicity, • endocrine disruption, • mutagenicity (accidental and/or engineered), • reproductive and developmental toxicity (teratogenicity). Additional human health criteria: • acute toxicity, • chronic toxicity, • irritation of skin/mucous membranes, • sensitization, • other (e.g., skin penetration potential, flammability, etc.).	Ecological health criteria: • bioaccumulation, • climatic relevance, • content of halogenated organic compounds, • fish toxicity, • algae toxicity, • daphnia toxicity, • heavy metal content, • persistence/biodegradation, • other (e.g., mobilization of metals, regulatory issues, toxicity to soil organisms, etc.). Natural systems equilibrium criteria: • global warming potential, • ozone depletion potential.

of Governmental Industrial Hygienists, ACGIH; etc.), teratogens (National Toxicology Program, NTP), ozone-depleting substances (Section 602(b) of the Clean Air Act), and so on. Other data must be gathered from various standard test methods (aquatic toxicity [OPPTS Harmonized Guidelines; 850.1075 and 850.1400], biodegradability [OPPTS Harmonized Guidelines, 835.3500 and 835.3400], and so on) and others from credible scientific literature (fate in the natural environment, known biodegradation breakdown products, etc.). Some of the data are publicly available, while others must be obtained from raw material manufacturers at the request of the product manufacturers. Some are based on expert assessment, models, and evaluation of analogs.

For ease in communicating levels of potential health effects, color coding is used to designate whether or not the data exceed criteria for each attribute. For list-based data, a chemical either is or is not on a particular list, for example, the National Toxicology Program Report on Carcinogens. For attributes such as aquatic toxicity, data are evaluated and cut-off criteria limits are set. Red would indicate toxicity to fish based on LC_{50} <10 mg/L. Yellow would indicate moderate aquatic toxicity to fish with LC_{50} between 10 and 100 mg/L. Green would indicate that based on test results, the chemical is considered relatively nontoxic to aquatic life at LC_{50} > 100 mg/L. A color coding of orange is used to indicate the absence of data.

The first step most companies and industries take on the path to cradle-to-cradle is to identify and move away from substances that are widely recognized as harmful. A number of companies have developed Restricted Substances Lists based both on regulatory requirements and design ideals to support their designers and purchasers in this commitment. This approach has resounding impacts up and down the supply chain.

In the cradle-to-cradle assessment process, criteria are prioritized. The "Priority Human Criteria" (known or suspected) include:

- carcinogenicity;
- endocrine disruption;
- mutagenicity (accidental and/or engineered);
- reproductive and developmental toxicity (teratogenicity).

If an ingredient is found to violate the Priority Criteria, then every effort is made to avoid its use. Criteria are also prioritized based on the type of product and exposure routes. For example, cleaning products are generally released to the air and/or sewerage system. In addition, there is opportunity for inhalation, and skin and eye exposure. Therefore, criteria such as biodegradability, aquatic toxicity, skin and eye irritation, chronic toxicity (asthma), and sensitization would be priority criteria in designing cleaning products.

At a second level of commitment, companies can select ingredients by comparing existing substances based on their human and environmental health and safety (EH&S) profiles. In order to do so, they will need full product ingredient disclosure for ingredient assessment, as outlined above. Challenges in comparing ingredients include trade-offs and lack of data. Sometimes ingredients will appear to be relatively equivalent with respect to EH&S criteria. Companies must decide what criteria are most important and relevant based on the product's intended use, fate, and exposure routes and determine which attributes best reflect the company's highest aspirations. Which attributes are more important to the company and its stakeholders? What is the best decision that can be made now, knowing that more information in the future will continue to inform decisions? This requires decision-making under uncertainty, but still pushes forward the intention to design using materials with the most favorable human and environmental health profiles. For example, the Port of Portland worked with their Janitorial Service provider to convert at least 65 percent inventory to environmentally preferable cleaning products (EPP). They chose products that were either already eco-labeled or distinguished via partnerships such as U.S. EPA's Design for the Environment Formulator Initiative, products screened via State, City or Federal environmentally preferable purchasing programs, or products for which full ingredient disclosure was obtained and ingredients were screened for carcinogens, endocrine disruptors, hazardous air pollutants, ozone-depleting chemicals, persistent, bioaccumulative, and toxic substances (PBTs), teratogens, and alkylphenol ethoxylates. The screening was also informed by the Indiana Relative Chemical Hazard Score (http://www.ecn.purdue.edu/CMTI/IRCHS/) and the Janitorial Products Pollution Prevention Project (http://www.wrppn.org/Janitorial/jp4.cfm).

In the end, the goal is to specify and use only the safest, healthiest chemicals and materials. This can involve the development of lists of preferable chemicals (what we call "P-lists") for use in specific applications. The development of P-lists is useful and practical within a company, as well as within and between industries, as many of the same raw materials are used by different manufacturers for different applications. For example, dyes and surfactants are used so broadly that the identification of positively assessed ingredients in these categories could support the use of more benign ingredients in multiple sectors. There is a need to share data and information on ingredients based on application and performance to support product design, especially because the chemical assessment work is resource intensive.

Where materials do not currently exist, there is a clear need for new green chemistries to fill the void. Criteria for designating green or positive chemicals must be defined along with associated predictive human and ecotoxicology tests and physical chemistry assessment in order to support sustainable product design. And of course, one cannot assume that a new chemical is benign when no negative data are available simply because it has not been thoroughly tested.

Figure 5.5 illustrates levels at which data are used and weighted to support organizations in optimizing chemicals and materials in their products and processes.

Green chemistry faces many challenges in the next decade as data related to physical-chemical properties, human and ecological toxicology are collected internationally, and new tools emerge for predicting toxicology. Tools used for drug design and new chemical review are being adapted for identifying potential human and environmental health concerns for chemical ingredients, prior to their use in formulated products. In addition, green chemistry must address the development of materials with fundamentally new properties, such as those derived from nanotechnology and processes using genetically modified organisms.

5.2.3.2   Designing for Cradle-to-Cradle Metabolisms.   Along with analyzing and selecting chemicals and materials for favorable human and ecological health profiles, the practice of cradle-to-cradle design sets out to design products from beginning to end to circulate productively and safely in biological or technical metabolisms. The design process includes the treatment of value recovery potential and energy considerations. Design for cradle-to-cradle metabolisms begins with the

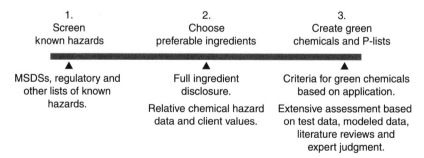

Figure 5.5. Applications and information needs for assessing human and environmental health.

identification of which metabolism, biological or technical, is appropriate for the product under design. Evaluation of the value recovery potential of a material is based on the following considerations:

- Is it technically feasible to recycle the material? What is the nature of the recycling process (chemical, mechanical)?
- Is the material biodegradable and potentially beneficial when discharged to environmental media or to an organic recovery system such as anaerobic digestion or composting?
- Does a recycling or organic recovery infrastructure exist for the material?
- What is the resulting quality of the recycled material or biodegraded nutrient?
- How can the materials and products be designed to enhance post-use value recovery?
- What is the value of the material for use in energy recovery?
- Has the material been designed for safe incineration in currently existing energy recovery systems?

For example, because it is difficult at this time to conceive of recovering cleaning products after use, it is sensible to design cleaning products as biological nutrients. Therefore, the cleaning product formulation must be amenable to wastewater treatment via the criteria of biodegradability and low aquatic toxicity. The packaging of these products, however, can be designed for a technical metabolism.

Asking the question, "How can this material or product be conceived as a nutrient?" drives different avenues of thinking. For example, eutrophication by excess phosphorus in manures is considered a pollution problem, while phosphorus is mined elsewhere for use as fertilizer. Thinking in terms of nutrients would suggest recovering P from manures for use as fertilizer, whether from anaerobically digested manures or as precipitated struvite. Likewise, nitrogen is a pollutant when emitted from cement kilns and coal-fired plants at the same time it is needed for growing food. Thinking in terms of nutrients would suggest that NOx could be recovered and transformed (it can be) for use as a nutrient.

The challenge of nutrient-metabolism thinking is building infrastructures for metabolizing materials to capture their highest value. While a material may be fully recoverable or recyclable, its value will not be realized if the infrastructure does not exist for recovery and recycling. On the large scale we envision for a truly sustainable industry, this cost can rarely be borne by individual companies, but groups of companies can organize, agree on common goals, and pool their resources to promote common infrastructures. There are already examples of such groups. Starting in summer 2003, GreenBlue organized a working group of companies in the packaging value chain – from consumer brands like Nike and Esteé Lauder to commodity materials companies like MeadWestvaco and Dow – to explore cradle-to-cradle concepts and opportunities. In 2004 they formed the Sustainable Packaging Coalition to collectively outline and pursue strategies for creating cradle-to-cradle packaging products and systems (www.sustainablepackaging.org).

As industry moves toward renewable energy sources and materials, materials and products can be evaluated for their effective use of energy. This process includes the following considerations.

- What energy sources are used in its creation, distribution, use, and value recovery processes?
- How much energy is required for creation of the virgin material?
- How much energy is required for recovery by recycling versus other forms of value recovery? (For example, Dewulf and van Langenhove, 2002, evaluate metabolic efficiency of different solid waste treatment options including recycling, incineration with energy recovery, and landfilling with biogas/energy recovery using exergy analysis.)

Figure 5.6 illustrates a value recovery hierarchy that can be used to inform decisions about material cycling strategies.

5.2.4 What the Chemical Industry Can Do

The chemical industry has a unique role to play in the creation of cradle-to-cradle products and systems. To begin with, the chemical industry must understand the design ideals of and adopt a framework for designing sustainable technology. Cradle-to-cradle can offer a unifying vision, framework, and approach. Next, it is important to understand what the chemical industry's customers want, and why. Are clients specifying environmentally preferable materials? What is encouraging this? Some of the drivers for environmentally preferable products based on sustainable materials include:

- U.S. Green Building Council – Leadership in Energy and Environmental Design (LEED) standards for:
  - New construction,
  - Existing buildings, and
  - Commercial interiors.
- State and federal government (Executive Order 13101), environmentally preferable purchasing programs.
- Private industry with sustainability initiatives

An example of a company moving toward the use of environmentally preferable raw materials is SC Johnson, which has created a GreenList™ similar to the P-list approach described above, and is inviting raw material manufacturers to supply new green chemistries   (http://www.greenbiz.com/news/news_third.cfm?NewsID = 26889). Criteria for green or positive ingredients are based on application and performance. The goal of the GreenList is to help SC Johnson continue its progress away from "restricted use materials," toward ingredients that are "best in class" based on performance, human and environmental health concerns.

The chemical industry would offer a great service to clients such as SC Johnson by providing materials that meet positive design criteria and providing appropriate data to support the characterization and prioritization of material attributes. It would also help formulators and product manufacturers assure their customers that the products they are making meet environmentally preferable criteria identified by external markets.

Finally, according to Roger McFadden, vice president of Technical Services at Coastwide Laboratories.

Chemical manufacturers should recognize the opportunity that the sustainabililty movement is creating for innovation. The green movement is exciting in part because it is driving real product innovation and development of new raw materials for formulation after a long period of incremental change.

—Personal communication, 2004, Roger McFadden, vice president Technical Services, Coastwide Laboratories.

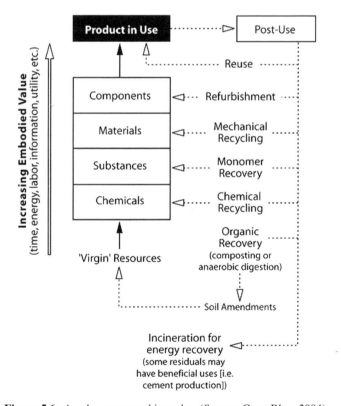

Figure 5.6. A value recovery hierarchy. (*Source*: GreenBlue, 2004).

While some may argue that there is no good time for innovation, others see the potential benefits and understand that sustainable product design is a driver for innovation and will profit those who engage in it. The cradle-to-cradle approach can be an especially potent catalyst for innovation because it offers both a practical structure for application, and a compelling and hopeful vision.

Acknowledgment

The author gratefully acknowledges Phil Storey for his valuable contributions to both content and editing.

5.3 PRINCIPLES OF SUSTAINABLE ENGINEERING

MARTIN A ABRAHAM

University of Toledo

The ECI-sponsored conference "Green Engineering: Defining the Principles", San-Destin, was an attempt to bring together a group of engineers and scientists from a diverse set of backgrounds to develop a set of principles that underpin the concepts of green engineering. Shortly before the activity of the conference, Anastas and Zimmerman (2003) published an article in which they defined 12 principles of green engineering. This article became a basis for discussion at the conference, along with other statements of green chemistry, green engineering, and sustainability. Ultimately, the attendees used the existing statements to build up their own set of principles that can be broadly applied throughout industry. Although the original intent was to develop principles of green engineering, the tone of the resulting statements were that of sustainable engineering, and the defining principles are rewritten here in that context.

What distinguishes sustainable engineering from green engineering? In my recent article (Abraham, 2004), I provide a brief discussion of the different contexts in which these elements can be considered. In particular, I state that "while green chemistry addresses issues of natural capital, and green engineering addresses both natural capital and economic viability, sustainability also addresses the human condition and implores the individual to improve the quality of life for all habitants." Thus, we see that sustainable engineering seeks solutions that are beyond those of green engineering, by looking outside the scope of the process or product being developed and considering the system as a part of the global ecosystem, which includes all of humanity.

Within this section, we review the Principles of Sustainable Engineering that were developed at the SanDestin conference, and provide a brief discussion of each of the principles. First, the principles are stated in their entirety. Then, an elaboration is provided to put these principles into an appropriate engineering context.

5.3.1 The Principles

Sustainable engineering transforms existing engineering disciplines and practices to those that promote sustainability. Sustainable engineering incorporates development and implementation of technologically and economically viable products, processes, and systems that promote human welfare while protecting human health and elevating the protection of the biosphere as a criterion in engineering solutions.

To fully implement sustainable engineering solutions, engineers use the following principles:

1. Engineer processes and products holistically, use systems analysis, and integrate environmental impact assessment tools.
2. Conserve and improve natural ecosystems while protecting human health and well-being.
3. Use lifecycle thinking in all engineering activities.
4. Ensure that all material and energy inputs and outputs are as inherently safe and benign as possible.
5. Minimize depletion of natural resources.
6. Strive to prevent waste.
7. Develop and apply engineering solutions, while being cognizant of local geography, aspirations, and cultures.
8. Create engineering solutions beyond current or dominant technologies; improve, innovate and invent (technologies) to achieve sustainability.
9. Actively engage communities and stakeholders in development of engineering solutions.

There is a duty to inform society of the practice of sustainable engineering.

5.3.2 Discussion of the Principles

5.3.2.1 Systems Thinking. The optimal design of a product or process will be one in which the design considers the unit as a system, which in turn is part of a larger system. What is a system? A system is a collection of objects that receive inputs from the external environment and/or from other objects of the system, and transforms the inputs to outputs (Olson, 2005). The system consists of the object being analyzed, plus any inputs and outputs; thus, it is imperative to identify system boundaries in order to define the system. A system can be large or small, and it can be part of a larger system. Systems analysis describes the potential interactions that an engineer can evaluate in designing a process or product to minimize its impact on the environment. In addition, a systems analysis allows the engineer to gain a complete understanding of the interactions of the process or product with the environment. The first step in a systems analysis is to define the goals and objectives (Olson, 2005), always keeping in mind that the product must perform the function for which it is designed. The systems engineer completes an iterative process, beginning with what is already known, creating a diagram for study, communicating and verifying the information, expanding the information based on targeted objectives, solving the problem, and then determining if the solution is adequate. Systems analysis is the only method that allows the engineer to holistically design a product to meet a set of design criteria that can include both function and sustainability.

5.3.2.2   Systems and Ecosystems. Regardless of the process or product that is being designed, it is important to consider that the unit will be part of the larger ecosystem. Thus, sustainable engineering challenges the designer to include the interactions of the unit with the ecosystem as a key design parameter. What is an ecosystem? The Natural Resources Board of Canada defines an ecosystem as "a dynamic set of living organisms (plants, animals and microorganisms) all interacting among themselves and with the environment in which they live (soil, climate, water and light)" (http://www.cfl.scf.rncan.gc.ca/ecosys/def_eco_e.htm). Thus we see that all living creatures are part of the ecosystem, as are all parts of the Earth. The ecosystem includes elements of the carbon cycle, the water cycle, the natural progression of plant and animal life in a region, and numerous other activities that can be defined through the science of ecology. Humans, and human activity, are also part of the ecosystem. What distinguishes an ecosystem from a natural ecosystem? The European Environment Agency defines a natural ecosystem as "an ecosystem where human impact has been of no greater influence than that of any other native species, and has not affected the ecosystem's structure since the industrial revolution" (http://glossary.eea.eu.int/EEAGlossary/N/natural_ecosystem). Thus, the goal of the engineer who designs for sustainability should be to maintain the influence of the system being designed to that which is no greater than any other component of the ecosystem. This challenge requires that we define an engineered system within the ecosystem, and design our engineered system to have minimal flows of energy and materials across its boundaries.

5.3.2.3   Life Cycle Thinking. The impacts that a product or process might have on the surrounding ecosystem vary throughout the life of that particular product. Life-cycle analysis requires an evaluation of the interactions during the manufacture of the material, the use of the unit, and eventually the disposal of that object. It includes all of the energy and material interactions between the engineered system and the ecosystem. As a natural consequence, it is essential to define the boundaries of the engineered system. Although it is often possible to identify the interactions between the engineered system and the ecosystem, aggregation of the impacts throughout the lifecycle are often difficult, and require complex analysis tools. One approach is the concept of the ecological footprint (Wackernagel and Rees, 1993), which attempts to describe the ecological impacts of a community based on the land area that would be required to maintain the current lifestyle within that community. This type of metric provides an overall measure of the consumption of a particular community, and presents it in a fashion that is easily translated by the general public. Another technique, based on exergy analysis, uses a thermodynamic measure to analyze and improve the efficiency of chemical and thermal processes, and aims to couple interactions between the system and ecosystem through rigorous calculations (Hau and Bakshi, 2004). The U.S. EPA, through its Waste Reduction Algorithm (WAR), evaluates chemical interactions and defines a term known as potential environmental impact to evaluate the interactions between the system and the ecosystem (Young *et al.*, 2000) and including the impacts of energy utilization by the system. Regardless of the method used by the designing engineer, the goal of reducing the impact of the engineered system throughout its entire lifecycle, not just during the manufacture of the product, is an important component of sustainable engineering.

5.3.2.4 Use Safe and Benign Materials. Since the 1970s, the U.S. government has developed regulations and guidelines to limit the use of specific materials that have been demonstrated to be harmful to the public, or to society at large. The Montreal Protocol, which curtailed the use of fluorocarbons for refrigeration, is an example wherein groups of governments have identified a hazardous material and restricted its use. While restricting the use of known hazardous materials helps to avoid ecological damage, a proactive approach that seeks to use inherently benign materials ensures long-term environmental propriety. The design of chemicals that are inherently benign is a central concept of green chemistry (Anastas and Warner, 1998), which proposes that chemicals should be designed using the principles of health and environmental sciences so that they contain minimal toxicity for both humans and the ecosystem. Green chemistry seeks to minimize the risk associated with the use of chemicals by eliminating the hazard, reducing the need for extensive risk management or health and safety precautions. The sustainable engineer should seek to utilize chemicals that are as harmless as possible, and consider the potential impact of every material used throughout the lifecycle of the product or process. A chemical that may not be hazardous during the manufacturing stage but is hazardous to the environment during the disposal stage should clearly be avoided.

5.3.2.5 Minimize Depletion of Natural Resources. Although there may be disagreement on the quantity of fossil reserves, there is general agreement that the reserves are limited. Thus, in order to promote sustainable designs, one must focus on the use of nonfossil reserves. Renewable resources should be promoted and used wherever technically and economically possible to replace products and processes that are derived from fossil resources. The use of renewable resources, however, should not occur without a concern for the level of reserves of these materials. In order to be truly sustainable, energy and material inputs must be derived from renewable resources at a rate that does not exceed the regenerative capacity of the ecosystem (Heusseman, 2004). If one consumes renewable resources at a rate that exceeds the regeneration rate, then renewable reserves will decline, and society will once again be faced with the need to identify new stocks of raw materials. For electricity generation, renewable sources can take the form of solar energy production in photovoltaic systems, wind energy, or the conversion of biomass. If one requires liquid fuels, then biomass is the most viable renewable resource, although improvements in crop and conversion technology are required in order for the production of fuels from biomass to become economically viable (Towler *et al.*, 2004). Likewise, biomass is the most viable renewable material for the production of chemicals and plastics. However, manufactured goods that are based on metal components cannot be derived from biomass-derived resources without substantial reengineering, and thus must derive from increased recycling and reuse of already consumed materials. Design techniques that promote materials reuse and recycle, such as design for remanufacturing, in which the form of the product is retained and the product is reused for the same purpose or for a secondary purpose (Bras and McIntosh, 1999), seek to minimize the consumption of new natural resources and maximize the utilization of previously mined materials.

5.3.2.6 Strive to Prevent Waste. A waste is an unusable or unwanted substance or material, or something, such as steam, that escapes without being used. Regardless of the definition, waste is something that is lost without recovering an appropriate value. Whether from an economic or an environmental perspective, waste is a lost opportunity to convert a raw material into a profitable product. Waste prevention returns to one of the original environmental paradigms known as the three Rs: reduce, reuse, recycle (see http://www.epa.gov/epaoswer/osw/index.htm for more information). It is codified in the Pollution Prevention Act of 1990, which states that source reduction is the ultimate goal of environmental protection. Waste reduction in consumer goods can be achieved by decreasing the amount of material that is consumed in packaging; in manufactured goods we can redesign the manufacturing process to use fewer raw materials. Reuse involves using a product many times; for example, use ceramic coffee mugs instead of Styrofoam cups, or converting an abandoned office building into apartments or new retail space. Recycling turns materials that would otherwise be waste into valuable resources, such as the conversion of waste cooking oil into fresh biodiesel fuel. Although waste cannot be completely eliminated, it can be substantially reduced through the development of new technologies that minimize the consumption of unnecessary inputs (either material or energy) and maximize utilization of all raw materials to achieve a desired function (Zimmerman and Anastas, 2005).

5.3.2.7 Employ Good Engineering in the Context of Societal Desires. Sustainable development requires the assimilation of sound scientific knowledge applied with the acceptance and support of the Earth's population. The technical community serves as a catalyst to build the social capital needed to change the social environment from one of waste production and disposal, to that of recognizing the need to reuse and recycle products (Neace, 2003). As individuals, we support regulations and manufacturing techniques that lead to improved environmental performance; however, as a society there is a tendency to adopt the attitude that resolution to a problem will be achieved through the actions of other people. For example, decreasing supplies of fossil energy and increasing evidence of the impact of energy consumption on global climate change has raised public awareness of the need for increased energy conservation. However, individuals still seek to drive large vehicles with poor fuel economy, arguing that their individual choice will have no substantial impact on these global challenges, but failing to recognize that societal behavior is the aggregation of individual choices. On the other hand, social capital is high in formalized groups in which people have the confidence to invest in collective activities, knowing that others will do so too (Pretty, 2003). In the context of sustainable engineering, social capital is needed to induce societal changes. Social capital can be developed when the engineer can pose technically feasible alternatives in the context of community goals, thereby inducing societal change with minimal political and social upheaval. Engagement with the local community is a central element of sustainable engineering. As stated by Engineers without Borders, "We also believe that the non-engineering components of local needs are almost always more complicated than the engineering aspects, and we seek to instill this reality within the engineering students that are an integral part of the entire process" (for more about Engineers without Borders, see http://www.ewb-usa.org/index.htm).

5.3.2.8 *Look for New and Innovative Alternatives.* Sustainable development will require the reform of existing institutions and policies, combined with the introduction of new, clean technologies. However, new technologies are often complex and require long lead times for commercialization. As a result, successful implementation will depend on the creativity and ingenuity of the scientific and engineering community (Coles and Peters, 2003). Opportunities for the development of sustainable technologies are only limited by the imagination of the designers. New developments in biotechnology are providing opportunities to grow crops in marginal regions, and to convert ever larger portions of these crops into useful materials. Nanotechnology is creating new electronic devices that are smaller and more energy efficient. History provides excellent examples of how new engineered products have created completely new paradigms that can enhance sustainability. For example, twenty years ago the cellular telephone was first being developed; today, this technology can provide digital communications from anywhere on the globe using commercial satellite systems. From the standpoint of sustainable technologies, the application of photovoltaic systems in remote regions is bringing electricity to remote communities that are far removed from grid access. The Coalition to Access Technology and Networking (CATNet) recycles older computer technologies to bring Internet access to urban and disadvantaged communities at public access points, making connectivity possible to all individuals, regardless of economic and regional conditions (see http://uac.utoledo.edu/metronet/catnet/Default.htm). Several of these examples illustrate that we can deliver services without the need to consume new products. This concept is embedded as a central element in the design of product service systems, in which a company works with a client to identify the service that is required, as opposed to the product that will perform the service. Thus, instead of purchasing carpeting from a vendor for an equipment show, it is now possible to rent carpet tiles that can be returned to the vendor at the conclusion of the show (Manzini and Vezzoli, 2003).

5.3.2.9 Engage Communities and Stakeholders. In order to design a product for use in a community, it is critical to understand the needs of the community. Engaging stakeholders is an effective way to clarify and prioritize a community's needs, which promotes sustainability by explicitly including the goal of human welfare in design (Heine and Willard, 2005). The community stakeholders help to identify their acceptance of a particular technology and their willingness to operate within a constrained ecosystem. Stakeholders will identify concerns and issues in a project that can be clarified before planning decisions are made. In this way, proponents can address stakeholder concerns and goals, minimizing costs and maximizing social benefits, and producing a better overall proposal (Bender and Simonovic, 1995). Stakeholder involvement is an important management tool in developing an engineered product that can find a market niche with the public at large, but is also an essential business component during interactions with supply chain partners. Consider, for example, the product service systems concept, described above. In a product-based market, each stakeholder optimizes the manufacture of their own product. Unfortunately, this scenario often fails to optimize the delivery of the service, resulting in waste and excess costs. However, by engaging all of the stakeholders throughout the lifecycle of the delivered service, the entire process can become optimized, creating the most efficient mechanism for delivering the desired outcome. Thus, we see that stakeholder involvement in the design stage can create a more sustainable engineering solution.

5.3.3 Concluding Comments

According to the Engineering Council of South Africa, engineering design is the creative, iterative, and often open-ended process of conceiving and developing components, systems, and processes. A designer works under constraints, taking into account economic, health and safety, social and environmental factors, codes of practice, and applicable laws (http://www.ee.wits.ac.za/~ecsa/gen/g-04.htm# Engineering_Design). The Principles of Sustainable Engineering provide a paradigm in which engineers can design products and services to meet societal needs with minimal impact on the global ecosystem. The principles cannot be taken as independent elements, but rather should be considered as a philosophy for the development of a sustainable society. The principles are not prescriptive. They do not provide engineers with a definitive methodology for deriving a sustainable design. Rather, they provide engineers with overarching concepts that can be used along with traditional design principles to develop new products and services to be applied for the growth and development of human society, while simultaneously minimizing the impact of these designs on the global ecosystem.

REFERENCES

M. A. Abraham, "Sustainable Engineering: An Initiative for Chemical Engineers," *Environmental Progress*, 23(4), 261–263 (2004).

D. T. Allen and D. R. Shonnard, Eds., *Green Engineering: Environmentally Conscious Design of Chemical Processes*, Prentice Hall, Upper Saddle River, NJ, 2002.

D. T. Allen and K. S. Rosselot, *Pollution Prevention for Chemical Processes*, Wiley-Interscience, New York, 1997.

D. T. Allen and R. S. Butner, "Industrial Ecology: A Chemical Engineering Challenge," *Chemical Engineering Progress*, 98(11), 40 (2002).

A. Alva-Argáez, A. C. Kokossis and R. Smith, "Process Integration, Synthesis, and Analysis for Clean Processes," in S. K. Sikdar and M. El-Halwagi, Eds., *Process Design Tools for the Environment*, Taylor & Francis, New York, 2001.

P. T. Anastas and D. Allen, "Green Chemistry," in D. T. Allen and D. R. Shonnard, Eds, *Green Engineering: Environmentally Conscious Design of Chemical Processes*, Prentice Hall, Upper Saddle River, NJ, 2002.

P. T. Anastas and J. C. Warner, *Green Chemistry: Theory and Practice*, Oxford University Press, New York, 1998.

P. T. Anastas and J. B. Zimmerman, "Design Through the 12 Principles of Green Engineering," *Environmental Science & Technology*, 37(5), 94A–101A (2003).

R. U. Ayres, *Industrial Metabolism: Theory and Policy. The Greening of Industrial Ecosystems*, National Academy Press, Washington, DC, 1994. (Reprinted from *Industrial Metabolism*, R. Ayers and U. Simonis, Eds., United Nations University Press, Tokyo, Japan, 1993.)

J. C. Bare, G. A. Norris, D. W. Pennington and T. McKone, "TRACI: The Tool for the Reduction and Assessment of Other Environmental Impacts," *Journal of Industrial Ecology*, 6(3–4), 49–78 (2003). Software available at http://epa.gov/ORD/NRMRL/std/sab/iam_traci.htm#download.

BASF, *Environmental Health and Safety Report*, BASF, 2002.

M. J. Bender and S. P. Simonovic, *IAHS Publication (International Association of Hydrological Sciences)*, 231, 159–168 (1995).

Board on Earth Sciences and Resources, *Materials Count: The Case for Material Flows Analysis*, Board on Earth Sciences and Resources, National Academies Press, Washington, DC, 2004. Available at www.nap.edu.

B. Bras and M. W. McIntosh, *Robotics and Computer Integrated Manufacturing*, 15, 167–178 (1999).

M. Braungart and J. Engelfried, "An 'Intelligent Product System' to Replace 'Waste Management'," *Fresenius Environmental Bulletin*, 1, 613–619 (1992).

BRIDGES to Sustainability, "A Pilot Study of Energy Performance Levels for the U.S. Chemical Industry," report to the U.S. Department of Energy, June 2001. Available at http://www.bridgestos.org/Publications.htm.

H. Chen, T. N. Rogers, B. A. Barna and D. R. Shonnard, "Automating Hierarchical Environmentally-Conscious Design using Integrated Software: VOC Recovery Case Study," *Environmental Progress*, 22(3), 147–160 (2003).

V. Chong, "Update on 'Maleic Anhydride from *n*-butane via Fluid Bed Reactor with an Organic Solvent Recovery'," PEP Review 94-2-3, Process Economics Program, SRI International, Menlo Park, CA, 1995.

A.-M. Coles and S. Peters, *International Journal of Environmental Technology and Management*, 3(3–4), 278–289 (2003).

M. A. Curran, *Environmental Life-Cycle Assessment*, McGraw-Hill, New York, 1996.

J. P. Dewulf and H. R. van Langenhove, *Environmental Science and Technology*, 36(5), 1130–1135 (2002).

J. M. Douglas, *Conceptual Design of Chemical Processes*, McGraw-Hill, New York, 1988.

R. F. Dunn and M. M. El-Halwagi, "Process Integration Technology Review: Background and Applications in the Chemical Process Industry," *Journal of Chemical Technology and Biotechnology*, 78, 1011–1021 (2003).

Executive Order 13101, "Greening the Government Through Waste Prevention, Recycling, and Federal Acquisition," September 1998.

J. R. Fiksel, *Design for Environment: Creating Eco-Efficient Products and Processes*, McGraw-Hill, New York, 1996.

A. Garner and G. A. Keoleian, *Industrial Ecology: An Introduction*, National Pollution Prevention Center for Higher Education, University of Michigan, Ann Arbor, Michigan, 1995.

M. Goedkoop, *Life Cycle Analysis for Industrial Designers*, European Design Centre, Eindhoven, The Netherlands, 1994.

T. E. Graedel and B. R. Allenby, *Design for Environment*, Prentice Hall, Englewood Cliffs, NJ, 1996.

T. E. Graedel and B. R. Allenby, *Industrial Ecology*, Prentice Hall, Englewood Cliffs, NJ, 2003.

T. E. Graedel, *Streamlined Life-Cycle Assessment*, Prentice Hall, Englewood Cliffs, NJ, 1998.

J. L. Hau and B. R. Bakshi, *Environmental Science and Technology*, 38(13), 3768–3777 (2004).

L. G. Heine and M. L. Willard, in M. Abraham, Ed., *Sustainability Science and Engineering: Defining Principles*, Elsevier, New York, 2005.

D. C. Hendershot, "A New Spin on Safety," *Chemical Processing*, 67(5), 16–23 (2004).

Heusseman, Env. Prog., December 2004.

D. Hunkler *et al.*, *Life Cycle Management – Definitions, Case Studies, and Corporate Applications*, Society of Environmental Toxicology and Chemistry, Pensacola, FL, 2001.

J. Kaiser, "In This Danish Industrial Park, Nothing Goes to Waste," *Science*, 285(5428), 686 (1999).

G. A. Keoleian and D. Menerey, "Life Cycle Design Guidance Manual: Environmental Requirements and the Product System," EPA 600/R-92/226, Risk Reduction Engineering Laboratory, Office of Research and Development, U.S. Environmental Protection Agency, Cincinnati, OH, 1993.

W. Klöpffer, "Life-Cycle Based Methods for Sustainable Product Development," *International Journal of Life Cycle Assessment*, 8(3), 157–159 (2003).

L. S. Lagace, "Maleic Anhydride from *n*-butane via Fixed Bed Reactor; Organic Solvent Recovery," PEP Review 93-2-3, Process Economics Program, SRI International, Menlo Park, CA, 1995.

E. Manzini and C. Vezzoli, *Journal of Cleaner Production*, 11(8), 851–857 (2003).

M. B. Neace, *Advances in Ecological Sciences*, 18, 603–611 (2003).

N. Nguyen and M. A. Abraham, " 'Green Engineering: Defining the Principles' – Results from the Sandestin Conference," *Environmental Progress*, 22(4), 233–236 (2003).

W. Olson, "Systems," in M. A. Abraham, Ed., *Sustainability Science and Engineering: Defining Principles*, Elsevier, New York, 2005.

J. Pretty, "Social Capital and the Collective Management of Resources," *Science*, 302(5652), 1912–1914 (2003).

P. Saling, A. Kicherer, B. Dittrich-Krämer, R. Wittlinger, W. Zombik, I. Schmidt, W. Schrott, and S. Schmidt, "Eco-Efficiency Analysis by BASF: The Method," *International Journal of Life-Cycle Assessment*, 7(4), 203–218 (2002).

U. V. Shenoy, *Heat Exchanger Network Synthesis*, Gulf Publishing Co., Houston, 1995.

D. R. Shonnard and D. S. Hiew, "Comparative Environmental Assessments of VOC Recovery and Recycle Design Alternatives for a Gaseous Waste Stream," *Environmental Science and Technology*, 34(24), 5222–5228 (2000).

D. R. Shonnard, "Unit Operations and Pollution Prevention," in D. T. Allen and D. R. Shonnard, Eds., *Green Engineering: Environmentally Conscious Design of Chemical Processes*, Prentice Hall, Upper Saddle River, NJ, 2002.

G. Stephanopoulos, "Invention and Innovation in a Product-Centered Chemical Industry: General Trends and a Case Study," 55th Institute Lecture presented at the AIChE Annual Meeting, San Francisco, November 16–21, 2003.

Towler *et al.*, *Env. Prog.*, December 2004.

M. Wackernagel and W. Rees, *How Big is Our Ecological Footprint; A Handbook for Estimating a Community's Appropriated Carrying Capacity*, University of British Columbia, Vancouver, 1993.

WCED – UN World Commission on Environment and Development, *Our Common Future* (also known as the Brundtland Commission Report), WCED, New York, 1987.

D. M. Young and H. Cabezas, "Designing Sustainable Processes with Simulation: The Waste Reduction (WAR) Algorithm," *Computers & Chemical Engineering*, 23(10), 1477–1491 (1999).

D. Young, R. Scharp and H. Cabezas, *Waste Management*, 20(8), 605–615 (2000).

J. B. Zimmerman and P. T. Anastas, in M. Abraham, Ed., *Sustainability Science and Engineering: Defining Principles*, Elsevier, New York, 2005.

6

IMPLEMENTING SUSTAINABLE DEVELOPMENT: DECISION-SUPPORT APPROACHES AND TOOLS

6.1 ASSESSING IMPACTS: INDICATORS AND METRICS

6.1.1 Overview

DICKSEN TANZIL and BETH R BELOFF

BRIDGES to Sustainability

As sustainable development is put into operation in our society and industry, there emerges the need to understand the pertinent key indicators and how they can be measured to determine if progress is made. The ability to measure sustainability becomes even more crucial – as the business adage of "only what gets measured gets managed" suggests – as we make decisions and navigate towards the sustainability development goals.

Efforts to develop sustainability indicators and metrics have been made at various scales, ranging from global down to local community, business unit, and technology levels. In general, indicators and metrics are designed to capture the ideas inherent in sustainability and transform them into a manageable set of quantitative measures and indices that are useful for communication and decision-making. In this section, we provide an overview of sustainability indicators and metrics, especially as they relate to the chemical industry. This section will be followed by two specific case studies on the uses of sustainability metrics and indicators in product/process development in chemical companies.

Regarding nomenclature, the terms "indicators" and "metrics" are both commonly used, often interchangeably, in referring to measurement of sustainability.

Transforming Sustainability Strategy into Action: The Chemical Industry, Edited by B. Beloff, M. Lines, and D. Tanzil

Copyright © 2005 John Wiley & Sons, Inc.

"Indicators," however, are typically used more broadly, encompassing both quantitative measurements and narrative descriptions of issues of importance as well as, in certain cases, the key aspects that needs to be managed. The term "metrics," on the other hand, is used almost solely in referring to quantitative or semi-quantitative measurements or indices.

6.1.1.1 Framing Sustainability Indicators and Metrics.
Since the "Earth Summit" in 1992, there has been a global consensus that sustainable development encompasses at least economic growth, social progress, and stewardship of the environment. As these so-called three pillars or dimensions of sustainable development have long been separately managed, the significance of the concept lays primarily in the integration of the various concerns.

Various metrics within each of the three dimensions of sustainability have long been well developed and used. Gross domestic product (GDP) per capita, literacy and poverty rates, and ambient concentration of urban air pollutants are some examples of economic, social, and environmental metrics employed at the national level. In businesses, financial metrics such as return on investment as well as certain metrics that reflect employee well-being and environmental performance, such as health and safety incident rates and regulated toxic releases, are also conventionally used. Sustainable development, however, requires further cross-functional integration of these metrics as well as the inclusion of additional metrics that facilitate more systemic and comprehensive multidisciplinary communication and thinking.

Integrating Economic, Social, and Environmental Dimensions. Figure 6.1 illustrates how the three dimensions of sustainable development interlink. At the intersections, one finds:

- *Socio-economic* considerations, such as jobs creation and other impacts (positive and negative) of the relationship between the economy and societal well-being;
- *Socio-environmental* considerations, including effects of natural resource degradation and environmental releases on livelihood, health, and safety of people today and generations to come; and
- *Eco-efficiency* considerations, that is, the generation of greater (economic) value using fewer natural resources and with less environmental impact.

While each of the above considerations nominally focuses on only two of the dimensions of sustainable development, it is in fact closely connected to all three. Environmental degradation or improvement, for example, is one important factor related to how economic activities affect societal well-being (a socio-economic consideration). Similarly, socio-environmental impacts have their related cost and economic implications. Further, judicious consideration of eco-efficiency may result in various societal advantages, including (a) the benefits derived from the products

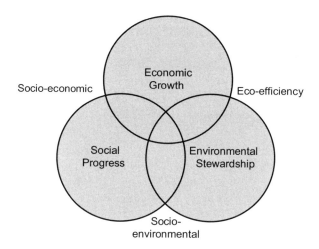

Figure 6.1. Socio-economic, socio-environmental, and eco-efficiency considerations at the intersections of the triple dimensions of sustainability.

and services in satisfying human needs and improving people's livelihoods; (b) the reduction of public health and safety concerns associated with environmental impacts; and (c) the more equitable use of natural resources enabled by their more efficient uses.

Global and National Metrics. To effectively support decision-making, indicators and metrics must be linked to key goals of an organization and important aspects that need to be managed. Global agreements, such as Agenda 21 and the more recent Millennium Development Goals, both developed under the umbrella of the United Nations, are eminent foundations for identifying the important aspects of sustainability. Agenda 21, adopted by over 178 nations at the 1992 Earth Summit in Rio de Janeiro, calls for actions to address disparities in social and economic development among and within nations, especially in terms of poverty, hunger, adequate housing, and public health, as well as education and institutional capacity. In addition, it recognizes the need for social and economic development to be accompanied by the conservation and management of natural resources in terms of protecting the atmosphere, forests, fragile ecosystems, oceans, and freshwater resources, managing land use and releases of wastes and toxic chemicals, and conserving biodiversity. Furthermore, additional issues such as energy, material use, transportation, and global climate change have gained attention in the dozen years since. Many of these aspects have been translated into indicators and metrics, such as those compiled by the United Nations Commission for Sustainable Development (UNCSD, 2001). National governments may adopt these indicators and metrics and adapt them to their needs in order to track, manage, and report on their countries' progress in sustainable development.

Some useful proxy indices of sustainability have also been proposed in the literature. One example is the "ecological footprint accounts" developed by Dr. Mathis Wackernagel and co-workers at the Ecological Footprint Network, which aggregate the ecological impacts of human economic activities in terms of "biologically productive land and water required to produce the resources consumed and to assimilate the waste generated by humanity" (Wackernagel *et al.*, 2002). Recent data on the ecological footprint account for countries with the world's largest economies, and populations are shown in Table 6.1. Biologically productive land and water areas used and impacted by agriculture and herding, forestry, fishing, infrastructure, and the burning of fossil fuels are included in the index. While far from providing the entire picture of sustainability, the calculations effectively highlight how many countries, both developed and developing, have overshot their ecological capacities. The issue of equity in the use of natural resources is also evident. Developed countries such as the United States, for example, use more than four times the world's average in global acres per person. Details on the calculation of ecological footprint accounts are available elsewhere (Loh and Wackernagel, 2004).

Corporate-Level Metrics. Indicators and metrics of sustainability have also been identified and developed at the community, corporate, business unit, or even process or technology levels. Information regarding sustainability indicators at the community level may be found, for instance, at the International Sustainability Indicators Network's website (www.sustainabilityindicators.org). The remainder of this

TABLE 6.1. Ecological Footprint and Biocapacity per Person by Country[a]

	Population (2002) (million)	Ecological footprint (1999) (global acres/pers)	Biological capacity (2002) (global acres/pers)	Ecological deficit (if negative) (global acres/pers)
World	6,210.1	5.6	4.5	−1.1
Brazil	174.5	6	15	9
China	1,284.2	4	2.6	−1.3
France	59.3	13	7	−6
Germany	82.2	12	5	−7
India	1,053.4	1.9	1.6	−0.3
Indonesia	217.3	3	4.4	1.4
Japan	127.2	11	2	−9
United Kingdom	60.2	14	4	−10
United States of America	288.3	24	15	−9

[a]Provided by Dr. Mathis Wackernagel, Ecological Footprint Network (www.footprintnetwork.org). Numbers may not always add up due to rounding. Note that Ecological footprint results are based on 1999 data, since consumption data reporting has a longer time lag. However, Ecological footprint numbers only change slowly over time.

section will focus on sustainability indicators and metrics used by companies in tracking and managing their performance and in assessing different technologies.

A list of important aspects of sustainability shown in Chapter 4 (Table 4.1) offers a starting point for companies to identify important aspects that need to be considered in developing sustainability indicators and metrics. By necessity, this list will differ somewhat among companies and organizations. How aspects of sustainability are prioritized in terms of importance is a function of many factors, including differences in types of business, operational conditions, history, community issues, and other aspects specific to a company or an organization. Generally, resource use and waste, monetary costs and benefits internal and external to the company, and well-being of people in the workplace and in the community affected by the company's operations are the environmental, economic, and social aspects relevant to most companies. In addition, the company's strategy in managing sustainability is of foremost importance in ensuring the company's satisfactory performance in all aspects of sustainability.

Leading versus Lagging. Effective management of sustainability in an organization requires the use of both lagging and leading indicators and metrics. A lagging metric is reported after an impact has occurred, and reflects past (including recent) outcomes in relation to the organization's performance goals. Leading metrics and indicators, on the other hand, assess activities that occur prior to the impact and affect the future performance of the organization, usually identified through a cause-and-effect analysis.

Operational metrics that quantify a company's environmental, health, and safety (EHS) and sustainability performance are usually considered lagging. In contrast, management metrics, which include the characterization of a company's EHS and sustainability management system, are commonly regarded as leading metrics. Examples of management metrics include the number of audits performed and the incorporation of global EHS and sustainability management standards. However, as pointed out by Epstein and Wisner (2001), leading and lagging are relative terms in a continuum of cause and effect relationships. For instance, incorporation of an environmental management system is a leading indicator of environmental performance, which in turn can be leading indicators of a company's reputation, employee morale, and financial performance. Thus, both types are important for a company in navigating toward sustainable development. Much of the efforts to date in developing sustainability metrics, however, have focused on measuring the operational dimensions of sustainability.

6.1.1.2 Development of Sustainability Metrics in the Chemical and Manufacturing Industry. Efforts have been made to develop sustainability metrics for various purposes and industry sectors, such as urban planning (e.g., Shane and Graedel, 2000), green building (e.g., Olgay and Herdt, 2004), and sustainable tourism (e.g., Hughes, 2002; Li, 2004). Manufacturing companies, including chemical manufacturers, have also kept track of the various elements that make up sustainability metrics. This includes resource uses that carry economic costs as well as

emissions and wastes, as mandated by regulation. Over the past decade, the industry has also begun to view their resource consumption and environmental release data in terms of broader sustainability.

Evolution Towards Sustainability Metrics. The development and use of environmental indicators and metrics in the manufacturing industry followed a learning curve summarized in the five evolutionary levels described by Veleva and Ellenbecker (2001):

- *Level 1 – Facility compliance/conformance metrics*, which are compliance-focused (e.g., number of reportable spills and violations, amount of fines, and injury rates);
- *Level 2 – Facility material use and performance metrics*, which are efficiency measures, usually expressed as amounts of inputs and outputs (e.g., energy use, TRI emissions, and so on) normalized by mass of products, monetary revenue, and value added;
- *Level 3 – Facility environmental and human health effect metrics*, which go beyond the quantities of inputs and outputs and look into their potential effects on human health and the environment (e.g., human toxicity potential, global warming potential, acidification potential, and so on);
- *Level 4 – Supply chain and product lifecycle metrics*, which extend Level 3 indicators to consider impacts along the supply chain and the entire lifecycle of the products; and
- *Level 5 – Sustainability metrics*, which further extend the scope to cover long-term issues such as material depletion and the use of renewable resources.

Further integration with social and economic dimensions of sustainability is also commonly considered a criterion of sustainability metrics.

As compliance has historically been a primary driver, conventional EHS metrics typically focus on compliance-driven *Level 1* metrics. While ensuring conformance with laws and regulations, these metrics offer little business competitive advantage to companies implementing them. *Level 2* to *Level 5* metrics, on the other hand, begin to address efficiency and value generation while minimizing undesirable impacts and related costs to the company, the community, and the environment. Early work in this evolution toward sustainability metrics includes the development of eco-efficiency metrics by Canada's National Round Table on the Environment and the Economy (NRTEE, 1999) and World Business Council for Sustainable Development (WBCSD) (Verfaillie and Bidwell, 2000). Supported by participation from their member companies, the organizations recommended eco-efficiency measurements that are defined as ratios, with resource use and environmental impacts in the numerators, and value generation in the denominator, or vice versa.

Informed by the above efforts, eco-efficiency metrics were further refined for application at the operational level through the "sustainability metrics" work of the American Institute of Chemical Engineers (AIChE, www.aiche.org/cwrt/

projects/sustain.htm). In collaboration with BRIDGES to Sustainability, a nonprofit group based in Houston, heuristics and decision rules were developed and tested in industry pilots and for over 50 chemical processes based on the respected Process Economic Program (PEP) data from SRI International (Menlo Park, CA) (Schwarz *et al.*, 2000, 2002). Effort to develop sustainability metrics was also undertaken by Britain's Institution of Chemical Engineers (IChemE, 2002), which resulted in an expanded set that includes economic and societal metrics in addition to those focused on eco-efficiency.

While development of these metrics sets typically began by focusing primarily on the amounts of material and energy flows, measures of potential impacts, such as toxicity potential, were soon recognized as more meaningful and rigorous than simple quantity measures, and thus became part of standard metrics. Extensions to supply chain, lifecycle, and broader sustainability considerations, however, have been more limited, especially in progress tracking and management decision-making, where such data are more difficult to obtain relative to the company's own resource use and environmental release data. Nevertheless, these more extensive sustainability considerations are increasingly applied in assessing products and processes during research and development (R&D). Corporate case studies highlighted in Sections 6.1.2 and 6.1.3 provide such examples.

Criteria for Selecting Sustainability Metrics. Consensus on the criteria for effective sustainability metrics has also emerged from the above efforts (Beloff and Beaver, 2000; Schwarz *et al.*, 2000, 2002; Bakshi and Fiksel, 2003; Sikdar, 2003). It is now generally accepted that sustainability metrics should satisfy the following (Beloff and Beaver, 2000):

- simple and understandable to a variety of audience;
- reproducible and consistent in comparing different time periods, business units, or decision alternatives;
- robust and nonperverse, that is, "better" metrics must indeed indicate more sustainable performance;
- complement existing regulatory programs;
- cost-effective in terms of data collection – making use largely of data already collected/available for other purposes;
- useful for decision-making;
- stackable along the supply chain or the product lifecycle;
- scalable for multiple boundaries of analysis; and
- protective of proprietary information.

6.1.1.3 Defining the Metrics: Eco-efficiency. As described earlier, the development of sustainability metrics requires one to first identify the key aspects that need to be managed and included in the metrics. Equally important, however, one must define how the metric is calculated. For example, when expressed as a

ratio, one must identify aspects that are to be included in the numerator and in the denominator. Furthermore, one must decide how the metric is bounded (e.g., which supporting processes are included) and any heuristics and decision-rules to be used in the calculation.

In terms of eco-efficiency, the critical aspects that need to be captured by the sustainability metrics have been largely recognized. They are categorized usually into a few basic areas based on resource uses and impacts to the environment, which are:

1. *Material consumption.* The use of materials, especially nonrenewable and finite resources, affects the availability of the resources and results in environmental degradation both in the extraction of the materials and when they are converted to wastes.

2. *Energy consumption.* In addition to being another important area of resource use and availability, energy use results in various environmental impacts. The burning of fossil fuels, specifically, relates to impacts such as global warming, photochemical ozone oxidation, and acidification. Renewable energy, such as hydropower, relates to a different set of impacts.

3. *Water consumption.* Freshwater is essential to life and almost all economic activities. With increasing anthropogenic demands and shortage in many water-stressed parts of the world, water consumption becomes an increasingly important consideration.

4. *Toxics and pollutants.* Toxics and other pollutants released in the forms of air emissions and wastewater effluents may result in damages to human health and the environment, and financial liability to pollutant sources.

5. *Solid wastes.* Reducing solid wastes is important for certain industries, especially in regions with no or very limited landfill capacity, such as Europe, Japan, and northeastern United States.

6. *Land use.* Land is another finite resource that provides a variety of ecological and socioeconomic services.

Basic Eco-Efficiency Metrics. The above eco-efficiency categories can be captured by a small number of metrics. For illustration, let us consider the set of "basic metrics" adopted by BRIDGES to Sustainability (Schwarz *et al.*, 2000, 2002), shown in Table 6.2. Informed by the work of NRTEE (1999) and WBCSD (Verfaillie and Bidwell, 2000), the metrics were chosen as ratios. Impacts are placed in the numerators and a measure of output is in the denominator. The denominator can be mass or other unit of product, functional unit, sales revenue, monetary value-added, or a certain measure of societal benefits. Expressing the metrics as ratios allow them to be compared and used in weighing decision alternatives and comparing operational units. Defined in this manner, the lower metrics are better as they reflect lower impacts per unit of value generation.

A metric's robustness can be very much affected by how decision rules, boundaries, and heuristics are set. The energy metric, in particular, may become perverse when defined solely in terms of direct energy use, that is, the energy content of

TABLE 6.2. BRIDGES' Basic Sustainability Metrics

Output:	**Material Intensity:**
Mass of Product	$\dfrac{\text{Mass of raw materials} - \text{Mass of products}}{\text{Output}}$
or	
Sales Revenue	**Water Intensity:**
or	$\dfrac{\text{Volume of fresh water used}}{\text{Output}}$
Value-Added	
	Energy Intensity:
	$\dfrac{\text{Net energy used as primary fuel equivalent}}{\text{Output}}$
	Toxic Release:
	$\dfrac{\text{Total mass of recognized toxics released}}{\text{Output}}$
	Solid Wastes:
	$\dfrac{\text{Total mass of solid wastes}}{\text{Output}}$
	Pollutant Effects

fuel, electricity, steam, and other energy sources consumed within the boundaries of a company, a facility, or a chemical process. A facility that includes an efficient on-site cogeneration of heat and power, for instance, would fare worse than a similar facility that purchases the heat and power off site. In the latter, generation of steam and electricity occurs outside the facility boundary, and thus its inefficiency in fuel use is not considered. Thus, the basic energy metric in Table 6.2 is reported in the equivalence of primary fuels, such that losses in generation and transmission of secondary energy sources, such as electricity and steam, are accounted for regardless of whether they are generated on site or purchased. This also makes the basic energy metric more representative of the energy cost and the amount of greenhouse gas emissions associated with energy use.

Land use is not shown among the basic metrics in Table 6.2 and represents a category that is less developed and may not fit into the impact-per-unit-output format of the other eco-efficiency metrics. IChemE (2002) proposes a land use metric expressed in terms of total area of land occupied and affected per annual revenue. However, heuristics and decision-rules for the land use metric, especially in relation to impacts on biodiversity and the ecosystem, require further development.

Complementary Metrics. Some of the basic metrics shown in Table 6.2 should be accompanied by other metrics. The toxic release metric, defined in terms of total release of toxics recognized by regulatory systems such as the Toxic Release Inventory (TRI), may lead to inaccurate representation when used by itself, as lower quantity of toxics does not necessarily correlate with less impact. The value of this metric lays primarily in its broad recognition and long-time use in the industry. Certain measures of potential toxicity, however, are necessary to complement this basic toxic release metric. For other pollutants not designated as toxic, rather than having a less useful basic metric in terms of total mass quantity, they are reported similarly in terms of various potential impacts.

Therefore, in addition to the basic metrics, complementary metrics may be developed to address specific needs. They include:

- Metrics that emphasize certain elements of the basic metrics, for example, toxic raw materials metric emphasizes an element of the basic material metric;
- Metrics that cover elements not usually included in the basic metrics, for example, transportation energy metric complements basic energy metric; and
- Metrics that weigh the inputs and outputs with respect to their impact potential, for example, toxicity and global warming potential.

Examples of the complementary metrics under the different basic metric categories are provided in Table 6.3.

Impact Assessment. Those developing sustainability metrics have increasingly made use of methodologies developed for lifecycle impact assessment (LCIA, see Chapter 5). This includes methodologies to assess potential human and ecosystem toxicity as well as other impacts of resource use and pollution. Udo de Haes *et al.* (2002) provide an extensive discussion on these methodologies. Each impact is estimated by multiplying individual material flow with a "characterization factor."

LCIA differentiates the impact assessment methodologies into "end-point" and "mid-point." The end-point methodology estimates the magnitude of impacts at the end of environmental cause-and-effect mechanisms. For example, Eco-Indicator 99, a widely used impact assessment methodology developed by the Pré Consultants (Goedkoop and Spriensma, 1999), calculates the impacts of resource uses and

TABLE 6.3. Examples of Complementary Metrics Under Each Metric Category

Material	**Toxic release**
• Packaging materials	• Toxic release under each TRI category
• Nonrenewable materials	• Human toxicity (carcinogenic)
• Toxics in product	• Human toxicity (noncarcinogenic)
• Toxics in raw materials	• Ecosystem toxicity
Water:	**Pollutant effects**
• Rainwater sent to treatment	• Global warming potential
• Water from endangered ecosystem sources	• Tropospheric ozone depletion potential
• Water use relative to water availability	• Photochemical ozone creation potential
Energy:	• Air acidification potential
• Energy consumed in transportation	• Eutrophication potential
• Nonrenewable energy	

environmental releases on three separate end-points: human health (in terms of disability-adjusted life years, or DALY), ecosystem health (in terms of species diversity), and resource availability (in terms of energy required for future mining efforts).

Mid-point analysis, however, appears to be gaining broader acceptance for use in sustainability metrics, partly due to the less complex modeling and analytical requirements compared to the end-point analysis. Impacts are calculated at certain "mid-points" before end-points of environmental mechanisms such as human and ecosystem health are reached. TRACI, or Tool for the Reduction and Assessment of Chemical and other environmental Impacts, developed by the U.S. EPA (Bare *et al.*, 2003), is an example of a mid-point approach. In TRACI, impacts are characterized in terms of impact categories that include human toxicity (cancerous and noncancerous), ecosystem toxicity, ozone depletion, global warming, acidification, smog formation, eutrophication, and so on. An earlier work by ICI group in Britain (Wright *et al.*, 1997), called the ICI Environmental Burden System, is another example and includes a similar set of impact assessment categories.

Perhaps the simplest approach for assessing toxicity potential is to weigh different toxic substances according to their permissible exposure limits, such as the Threshold Limit Values (TLV) defined by the American Conference of Governmental Industrial Hygienists (ACGIH) or the Occupational Exposure Limits (OEL) set by UK Health and Safety Executive. The TLV, specifically, constitute the threshold concentration at which a worker may be exposed to a substance over a 40-hour workweek with no adverse effects. Thus, lower TLV denotes greater threat to human health. In this case, the inverse of TLV (or OEL) serves as the characterization factor in weighting the human toxicity potential of different substances. Owing to its simplicity and foundation on standard human health and safety regimes, this approach has gained relatively broad acceptance in companies and is incorporated as part of the ICI Environmental Burden System, among others.

Exposure factors are also included in many of the more recently developed characterization factors for human toxicity. These characterization factors are obtained through more complex models that incorporate factors such as persistence, pollutant fate, and exposure pathways. Although involving greater complexity in development, end-users may simply employ the more complex characterization factors already calculated by existing impact assessment models such as TRACI.

Similarly, characterization factors have been calculated through complex models for ecosystem toxicity (based on toxicity to representative plant and animal species) as well as other pollutant impacts. International consensus exists on the characterization of global warming potential (Houghton *et al.*, 2001) and stratospheric ozone depletion potential (World Meteorological Organization, 1999). Competing methodologies, however, exist for other impact categories, including human and ecosystem toxicity.

In addition to the LCIA methodology, one may report substances of priority concern, such as those identified as persistent and bioaccumulative toxics (PBTs), as a separate metric. Both approaches are in fact used in companies, as illustrated in the case studies of GlaxoSmithKline (Section 6.1.2) and BASF (Section 6.1.3).

6.1.1.4   Aggregating Metrics. Aggregating metrics into a single performance index is often desirable to facilitate easy comparison. By converting all metrics into the same unit, one may assess the relative importance of different impacts and, hence, determine the focus of future impact reduction efforts.

Economic and panel approaches are arguably the most common methodologies in aggregating metrics. In the economic approach, metrics are weighted in terms of their associated monetary costs and benefits, both internal and external to the company. AIChE's Total Cost Assessment methodology (TCA, see Section 6.2) and the "Environmental Priority Strategy in product development" (EPS) (Steen, 1999) are tools that may be used to assess environmental and sustainability costs and benefits.

The panel approach, on the other hand, involves normalization and weighting, as outlined in the ISO 14042 standards on LCIA. The total impact from each category (e.g., total energy use or total global warming potential) is first normalized to a reference value. The reference value can be the impact within a geographical boundary (i.e., local, regional, or global), or other benchmark values (e.g., company-wide total, industry average, and so on). Weighting values (usually in percent) may be determined by a panel of stakeholders and applied to the normalized impacts to produce a single-value impact score. A slightly modified version of the LCIA normalization and weighting methodology is used in generating aggregate indices in the eco-efficiency analysis by BASF.

Other aggregation methodologies have also been proposed in the literature. One approach is based on thermodynamics: materials, fuels, and wastes can all be expressed as the total consumption or loss of exergy, or energy available to do work (Bakshi and Fiksel, 2003). Alternatively, the metrics may be expressed in terms of land use, or "ecological footprint." Land use impacts of energy sources and common household materials have been reported (Chambers *et al.*, 2000). It is unclear, however, whether proxies such as exergy and land use adequately represent overall sustainability impacts.

6.1.1.5   Integrating the Socio-Economic Considerations. As discussed earlier in this section, eco-efficiency considerations are tightly linked with the social dimension of sustainability. Greater efficiency in the use of natural resources would allow them to be more equitably distributed among nations and between current and future generations. Reduction in environmental releases may also relieve the loads placed on human health and safety. The panel approach discussed above, which incorporates community's priorities by assigning values to the different metrics through a panel of stakeholders, provides one way to integrate the social dimension more explicitly into eco-efficiency metrics.

While eco-efficiency is certainly an important element of sustainability, it is by no means a representation of the whole picture. Although not as well developed as eco-efficiency metrics, separate sets of social and economic metrics of sustainability have been developed, such as by IChemE (2002). Table 6.4 lists examples of social metrics developed by IChemE, which capture a variety of workplace

TABLE 6.4. Social Metrics Developed by the Institution of Chemical Engineers

Workplace:	Society
• Employment	• Number of stakeholder meetings
• Benefits as percentage of payroll expense (%)	per unit value-added
• Employee turnover (%)	• Indirect community benefit per
• Promotion rate (%)	unit value-added
• Working hours lost as percent of total	• Number of complaints per
hours worked	unit value-added
• Income + benefit ratio (top 10%/bottom 10%)	• Number of legal actions per unit
	value-added
Workplace:	
• Health and safety	
• Lost time accident frequency (number, per	
million hours worked)	
• Expenditure on illness and accident prevention	
relative to payroll expense	

Source: IChemE (2002).

and safety issues as well as societal issues, such as number of stakeholder meetings, complaints, and so on, expressed per unit value added.

6.1.1.6   Automated Sustainability Metrics Management Tools. As sustainability metrics have developed to include the assessment of potential impacts and lifecycle considerations, the challenges in calculating and managing the metrics is increased. However, this challenge can be reduced through the use of automated sustainability metrics management software tools, such as BRIDGESworks™ Metrics (Tanzil *et al.*, 2004). The software tool incorporates the basic and complementary set of metrics shown in Tables 6.2 and 6.3, along with their heuristics for calculation, as well as provides the capability to construct a customized set of metrics. In addition, a robust set of impact assessment methodologies is included for use in identifying the potential effects of toxics and pollutants. An example screen capture from the tool is shown in Figure 6.2.

The availability of such automated tools will further facilitate the incorporation of sustainability metrics for management and progress tracking, especially when integrated into a company's internal management information system. The tool will help companies collect data at the process or facility level as well as aggregate metrics from process, facility, all the way to the corporate level. In addition, with the capability to stack the metrics along the product's lifecycle, it encourages lifecycle thinking in management decision-making.

6.1.1.7   Using Sustainability Metrics. While sustainability metrics have become increasingly sophisticated in content and methodology, they continue to consolidate disparate data into meaningful measures that provide insights for decision-making. Metrics can be used to support various decision-making activities, including:

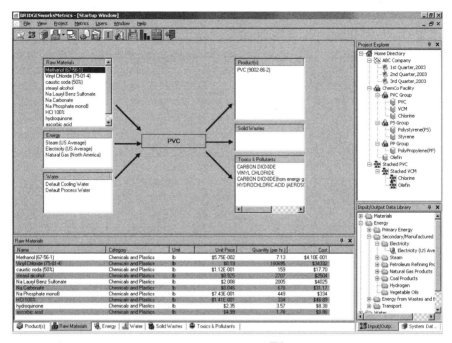

Figure 6.2. Data input screen of BRIDGESworks™ Metrics, an automated metrics management tool.

- Evaluating alternatives, including:
  - technical alternatives, for example, different raw materials and process improvement options, and
  - business alternatives, for example, different supplier and acquisition options;
- Comparing facilities/business units;
- Identifying environmental aspects and impacts of an industrial operation; and
- Tracking performance over time.

When desired, the metrics may also be employed to communicate with stakeholders, such as in external reports.

In Section 6.1.2, Constable *et al.* describe the use of sustainability metrics in product and process development at GlaxoSmithKline (GSK), a pharmaceutical company. To fit the metrics to their own needs, GSK has further developed specific green chemistry metrics, such as mass productivity, reaction mass efficiency, and total solvent recovery energy, for evaluating chemical synthetic processes. To evaluate chemical processing technologies, GSK uses green technology metrics that includes mass efficiency, percent purity, heating and cooling energy, in addition to those similar to basic sustainability metrics shown in Table 6.2.

In Section 6.1.3, Ernst Schwanhold describes an eco-efficiency analysis methodology developed at BASF. Based on standard lifecycle assessment, the analysis involves the use of metrics for various resource uses and environmental and health and safety impacts. The use of normalization and weighting methodologies to generate a single-value environmental performance score is also illustrated in great detail here. Furthermore, the section also presents discussion on the extension of the eco-efficiency analysis to "socio-efficiency," by including various social aspects of sustainability. While the methodology is also primarily used in products and process development, it has also been applied to strategy development as well as to communication with industrial customers and other value-chain partners of BASF.

The use of sustainability metrics in product and process development follows the rule of consecutive refinement in design. This is illustrated in a recently completed pilot on the use of sustainability metrics in the design of steel components (Beloff *et al.*, 2004). In the collaborative work between BRIDGES to Sustainability and Caterpillar Inc., it was found that, in the early stages of design, energy use and its associated greenhouse gas emissions can be reasonably estimated from simulation and lifecycle data from literature. On the other hand, only rough estimates for wastes and emissions are available. Better estimates of wastes and emissions require knowledge of equipment and operating conditions that are not yet identified in the early design stages. Similar conditions exist for chemical processes, highlighting the need for consecutive refinement of the metrics with each design stage.

Obviously, sustainability metrics may also play an important role in management decision-making, especially in progress tracking. As noted by the National Research Council (1999), "The ability to gauge improvement in any endeavor is critically dependent on establishing valid methods of measuring performance. Tracking progress toward an established goal serves to influence behavior by providing continual feedback, and it requires reliable and consistent metrics against which performance can be compared." Many of the eco-efficiency and socio-economic metrics described above are used in companies (see, for example, NRC, 1999; GEMI, 1998) and tracked on a quarterly, annually, or other periodical basis. Metrics used for routine progress tracking, however, tend to be simpler than the more sophisticated ones used in product and process development. The use of impact assessment, in particular, is only gradually gaining acceptance in management use, facilitated by the use of information technology in metrics management.

In developing a metrics program for a company, it is important to realize that "one size doesn't fit all" (GEMI, 1998). Each company has its own distinct set of concerns due to its unique combination of operational, geographical, technological, and community factors. Therefore, an ISO 14001 type of "aspects and impacts" analysis is a good place to start in developing a metrics program. This can be performed by mapping a company's operations and identifying environmental and other sustainability concerns for each operation. Then, a larger set of metrics can be piloted at representative facilities or business units to identify the most relevant ones. For example, a pilot was recently conducted using BRIDGES' sustainability metrics at The Stanley Works, a worldwide manufacturer and provider of tools, hardware, and security products (Nelson, 2003). Comparison of metrics for similar manufacturing facilities within The Stanley Works led to the identification of improvement opportunities in facilities that perform less efficiently than others. When applied to stages in the value chain (production of raw materials, manufacturing processes, and transportation), the metrics provided insights on the relative contribution of the different stages to impacts such as energy use, waste generation, and greenhouse gas emissions. Issues regarding data sources and reliability were also identified in this pilot. Once the most important metrics are identified, one may construct a system for periodic tracking and establish measurable goals at the corporate, business, and facility levels to drive improvement towards sustainability.

6.1.1.8   Summary and Challenges. In the past decade, we have witnessed considerable progress in the development of sustainability metrics, especially in the chemical industry. Metrics, especially those focused on eco-efficiency, have gained acceptance for use in management, progress tracking, marketing, as well as in the assessment of a company's products and processes. Progress has also been made in extending the eco-efficiency analysis to workplace and community well-being as well as in identifying separate social and economic metrics of sustainability. Availability of automated metrics management tools should further facilitate the use of sustainability metrics, especially when integrated into a company's management information system.

Several issues, however, need further exploration. While considerable efforts have been made in developing performance metrics, leading metrics such as management metrics still require additional development. Such metrics, however, may differ considerably from one company or organization to another due to the varying cause-and-effect relationship within each company or organization. Moreover, while anecdotal evidence exists that links various impacts to a company's financial performance, further studies are needed to more explicitly link sustainability metrics to a company's exposure to risks and opportunities.

6.1.2   The GSK Approach to Metrics for Sustainability

DAVID J C CONSTABLE, ALAN CURZONS, AILSA DUNCAN, CONCEPCION JIMÉNEZ-GONZÁLEZ AND VIRGINIA L CUNNINGHAM
GlaxoSmithKline, USA

A considerable number of publications have been written about the use of metrics to drive business, government, and communities towards more sustainable practices. The reader is referred elsewhere for a discussion of what metrics have been proposed. There has also been much written about the characteristics of metrics, or what constitutes a good metric (Bennett and Jones, 1999; NRC, 1999; Corporate Environmental Performance, 1999). It is generally agreed that metrics must be clearly defined, relevant and meaningful (easily understood and accepted), easy to reproduce, few in number but sufficiently diverse, cost-effective, measurable, and objective rather than subjective. They must also yield nonperverse results, should use existing data systems for their collection, be able to be benchmarked and monitored, and must ultimately drive the desired behavior (CWRT, 1998).

The main driver for metrics might be summarized in a common problem faced by a typical research chemist and/or chemical engineer in the chemical processing industries. These individuals are routinely expected to select materials or technologies to carry out various chemical operations and are faced with an array of possibilities that need to balance competing demands for quality, yield, throughput, cost, EHS impact, and the need to provide material for safety and clinical testing. In recent years, scientists are more likely to be asked "which of the options under consideration is the most sustainable?" For example:

1. Several different chemicals used in a chemical synthetic process can serve the same purpose or function; which of these chemicals is the most sustainable?
2. The same chemical, intermediate or Active Pharmaceutical Ingredient (API), can be produced by several different chemical synthetic reactions and/or routes; which of these reactions or routes is the most sustainable?
3. The same chemical, intermediate or API, can be produced by several different processes composed of different unit operations and/or processing aids and solvents; which of the processes is the most sustainable?
4. The same chemical or API may be formulated in several different forms; which presentation is the most sustainable?

Because these questions, and others similar to these, were routinely asked for a number years, it was the goal of the EHS Product Stewardship group at GlaxoSmithKline (GSK) to help in providing an answer. Beginning in 1997, in collaboration with 10–15 chemical and allied industry companies associated with the American Institute of Chemical Engineers' Center for Waste Reduction Technologies (AIChE/CWRT), GSK worked to develop a set of core and complementary sustainability metrics. GlaxoSmithKline used the CWRT metrics to organize our thinking about what is sustainable and adapted them for our work on several programs. A suite of tools was developed as an outgrowth of our metrics work and analysis, and these tools are discussed in greater detail in a later chapter of this book.

Metrics have been used by GlaxoSmithKline in several important program areas that are part of our sustainability initiatives. These program areas include Green Chemistry, Green Technology, Total Cost Assessment, Life Cycle Inventory/Assessment and Green Packaging. Each of these program areas will be discussed to illustrate how metrics have been used.

6.1.2.1 Green Chemistry. As a manufacturer of comparatively large numbers of very chemically complex materials in small volumes, it is somewhat difficult to intuitively or immediately answer questions about what might constitute "green" or sustainable practices. However, by employing the general metrics categories of mass, energy, and pollutants/toxics dispersion developed by the AIChE/CWRT sustainability group and combining these with a few additional considerations around safety and solvents, GSK was able to develop a specific metrics set that could be used to objectively evaluate GSK chemical synthetic processes. An example of the kinds of metrics that have been used by GSK is shown as Table 6.5. This work is more extensively described elsewhere (Curzons *et al.*, 2001; Constable *et al.*, 2001).

It should be noted that developing a metrics set is sometimes far easier than applying them to an existing set of data, especially for the kinds of materials and processes used by the pharmaceutical industry. The chemical complexity and the general lack of data for many materials makes the collection of data difficult, and the interpretation of results sometimes questionable. In addition, our work has indicated that in general, and at a high level, mass and energy metrics are very good surrogates for many of the pollutants/toxics metrics. For example, there is an obvious correlation between greenhouse gases and energy, or between acid gases and energy, or eutrophication potential and energy. By reducing the material and energy intensity of processes, many sustainability issues can be addressed. Following on from material and energy intensity, improvements in sustainability will be driven by the kinds of material and energy that is used, not just the amount (i.e., the inherent hazard and where it comes from).

Exploration of a variety of metrics has revealed the following:

- Pursuing a metric such as yield, a ubiquitous metric that chemists utilize to evaluate reaction efficiency, will not by itself drive business towards sustainable practices. However, from an economic standpoint, yield remains a very good metric, especially for high value-added materials such as pharmaceuticals.

- Reaction mass efficiency combines key elements of chemistry and process and represents a simple, objective, easily derived and understood metric for use by chemists, process chemists, or chemical engineers.

- Mass Productivity appears to be a useful metric for focusing attention away from waste towards the use of materials. As such, it is more likely to drive chemical and technology innovations that will lead to more sustainable business practices.

- Mass productivity seems to be broadly understood by business managers.

6.1.2.2 Green Technology. It is no secret that the chemical processing technology used by the pharmaceuticals industry is largely unchanged from the 50s; that is, the 1850s. While it is true that there have been significant advances in the materials of construction, methods of analysis, and automation of many processes, the basic batch operations have not changed. While this fact is beginning to be recognized by regulatory agencies as a significant contributor to product quality problems, it also represents a significant hurdle for advancing sustainability initiatives in the pharmaceutical industry. Batch operations, by their nature, are inherently inefficient in many cases because they are carried out in a multipurpose chemical plant whose equipment is optimized for its flexibility for general reactions, not specific chemical processes.

TABLE 6.5. GSK Green Chemistry Metrics

Category	Units
Mass	
Mass Productivity	
$\dfrac{\text{Mass of product (kg)}}{\text{Total mass in (kg)}} \times 100\%$	%
Reaction Mass Efficiency (RME)	
$\dfrac{\text{Mass of isolated product (kg)}}{\text{Total mass of reactants used in reaction (kg)}} \times 100$	%
Energy	
$\dfrac{\text{Total process energy (MJ)}}{\text{Mass of product (kg)}}$	MJ/kg
$\dfrac{\text{Total solvent recovery energy (MJ)}}{\text{Mass of product (kg)}}$	MJ/kg
Pollutants/Toxic Dispersion	
Persistent and Bioaccumulative	
$\dfrac{\text{Total (mass persistent} + \text{bioaccumulative) (kg)}}{\text{Mass product (kg)}}$	kg/kg
Human Health	
$\dfrac{\text{Total (mass of material [for all materials]) (kg)}}{\text{Permissable exposure limit (ACGIH) (ppm)}}$	
POCP (Photochemical Ozone Creation Potential)	
$\dfrac{\text{Total (mass of solvent [kg]} \times \text{POCP value} \times \text{Vapour Pressure (mm)])}}{\text{mass of product (kg)} \times \text{Vapour Pressure [toluene]} \times \text{POCP [toluene])}}$	kg/kg as toluene
Greenhouse Gas Emissions	
$\dfrac{\text{Total (mass of greenhouse gas from energy [as kg } CO_2 \text{ equiv])}}{\text{mass of product (kg)}}$	kg/kg as CO_2
$\dfrac{\text{Greenhouse gas, kg } CO_2 \text{ equivalent, ex energy for solvent recovery}}{\text{kg product}}$	kg/kg
Safety	
Thermal hazard	Highlight
Reagent hazard	Highlight
Pressure (high/low)	Highlight
Hazardous byproduct formation	Highlight
Solvent	
Number of different solvents	No
Overall estimated recovery efficiency	%
Energy for solvent recovery	MJ/kg
Mass intensity net of solvent recovery	kg/kg

Given this context, if there is to be any progress towards more sustainable processes, there must be a paradigmatic shift in the way APIs are synthesized. If this shift is to be made, however, one is still faced with the question of whether or not replacement technologies are or are not more sustainable than existing technologies. GSK has therefore developed a systematic, case-scenario methodology for ranking technology that utilizes lifecycle thinking and is, not surprisingly, based on a metrics approach. Details about this approach and the methodology that has been employed are reported elsewhere (Jiménez-Gonzalez *et al.*, 2001, 2002).

Very briefly, the methodology considers a framework of four aspects to evaluate technology including energy, environment, safety, and efficiency. To evaluate these four aspects, specific metrics were chosen depending on the unit operation, and a scoring system was developed to provide a relative ranking of the various technology options under consideration. An illustrative set of metrics is shown in Table 6.6.

6.1.2.3 Lifecycle Inventory/Assessment. The investigation and application of Life Cycle Inventory/Assessment (LCI/A) methodology afforded GSK a unique perspective that has been applied across many of the programs and tools developed for our Design for Sustainability initiatives. Obtaining fundamental LCI data has taken a number of years to assemble, analyze, and apply more broadly, but GSK is now in a unique position to routinely used the data and the insights we have gained to effect focused, prioritized, and meaningful changes in the business.

GSK used the sustainability metrics set developed by the AIChE/CWRT metrics group (CWRT, 1998) to organize or "roll-up" LCI data into impact categories. These impact categories can then be used in a variety of applications to provide relative rankings or scores for materials. This approach was applied to a lifecycle ranking for solvents in the GSK Solvent Selection Guide, and to a lifecycle scoring

TABLE 6.6. Metrics Used to Evaluate Technology

	Units
Mass metrics	
Mass intensity (not including water)	kg/kg usable product
Added solvent intensity	kg added/kg usable product
Wastewater intensity	kg/kg usable product
Specific pollutants released	g/kg usable product:
(e.g., CO_2 emissions, etc.)	
Efficiency	%
Quality	% purity
Energy metrics	
Heating	MJ/kg usable product
Cooling	MJ/kg usable product
Electricity	MJ/kg usable product
Cooling (refrigeration cycle)	MJ refrigeration/kg usable product
Safety considerations	Process safety, OH considerations, etc.
Operational considerations	Heat/mass transfer, continuous, etc.

methodology for synthetic routes in the GSK Fast Lifecycle for Assessment of Synthetic Chemistry or FLASC^{TM} tool. In each case, sustainability metrics were used to develop a practical tool to guide bench-level scientists and engineers in the selection of more sustainable choices.

6.1.2.4   Total Cost Assessment. Another related AIChE/CWRT focus group was convened to develop a Total Cost Accounting tool about the same time as the sustainability metrics focus group was meeting (AIChE, 1999). These two efforts enabled a certain synergistic outcome for both projects given their very complementary discussions on appropriate metrics and valuation. Total Cost Assessment/Accounting provides a very convenient form of valuation for Life Cycle assessments, especially in a business context, and this will hopefully be applied in the future for GSK tools. This is discussed further in Section 6.2.

6.1.2.5   Green Packaging. Packaging issues are of considerable importance, especially in Europe, and manufacturers are increasingly pressured to decrease packaging impacts. GSK has maintained an interest in packaging since the early 1990s and has consistently sought to influence packaging designers and marketing groups to consider the impacts of GSK packaging. Through a thorough knowledge of information gathered for European and UK packaging levies, GSK has been able to develop a considerable understanding of its packaging and potential impacts. These data have been combined with LCI information and traditional EHS considerations to design a metrics-based green packaging guide for its Consumer and Nutritional Healthcare businesses, and they will soon have one for its

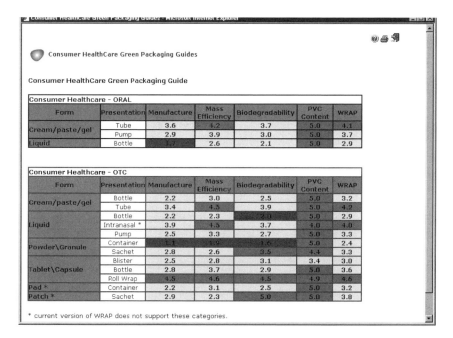

Figure 6.3. GSK Consumer HealthCare Green Packaging Guide.

pharmaceuticals packaging. The guide for Consumer Healthcare is shown in Figure 6.3. GSK has also developed a user-friendly interactive corporate intranet site for packaging designers that is described in Section 8.7.

6.1.2.6 Conclusion. It has been said in a sports context that "if you are not keeping score, you are only practicing." This truism has been applied to business and roughly paraphrased as "if you are not measuring, you cannot be managing effectively." Metrics of many kinds are being used on a continuing basis at GSK to effectively manage, benchmark, and challenge performance in all areas of its business.

6.1.3 The Eco-Efficiency Analysis Developed by BASF

ERNST SCHWANHOLD

BASF Aktiengesellschaft

6.1.3.1 Overview. The eco-efficiency analysis is one of our most important tools for implementing the principles of Sustainable Development in BASF. The key question is "What should the products of the future look like?" For this purpose, BASF, together with an external partner, began to develop the instrument of eco-efficiency analysis as early as 1996. In the meantime, the method has matured and around 200 products and production processes have been analyzed to date. Eco-efficiency analysis makes it possible to consider economy and ecology side by side in the development and optimization of products and processes and to select the most eco-efficient alternative. The objective is to identify products combining optimum application properties and good environmental performance at lowest possible cost. The analysis yields clear indications of possible improvements in the products and processes employed. Scenarios and variants enable us to show the distance between potential new products and existing solutions. Eco-efficiency analysis is a strategic instrument for BASF. Strategies are developed for sustainable analyses of our product lines and support is provided for decisions on capital expenditures.

Fifty percent of the eco-efficiency analyses conducted to date have been used for internal strategy and research decisions. The other half of the analyses has been carried out in cooperation with external partners such as customers, NGOs, and governmental institutions. Right from the beginning the eco-efficiency analysis was discussed in public. This has led to many valuable contributions from external parties and has put the method at a high level of acceptance throughout various industries and stakeholders.

6.1.3.2 The Methodology. The first step of an eco-efficiency analysis is the definition of a specific customer benefit (Saling *et al.*, 2002). The analysis then compares economic and ecological advantages and disadvantages across several product or process solutions that can fulfill the same function for customers. This means that products are not compared with one another in overall terms but rather their application performance such as "painting a square meter of furniture front" or "reductive cleaning for 1000 kg dyed polyester."

The eco-efficiency analysis focuses on each phase of a product's lifecycle "from cradle to grave," beginning with the extraction of raw materials from the Earth and ending with recycling or waste treatment after use. The basis is a lifecycle analysis according to standard ISO 14040 and the following.

In this way, the environmental impact of the products used by BASF as well as of the starting materials produced by others is measured. The usage behavior of the final consumers together with the various possibilities for reuse and disposal are also analyzed. In addition, a comprehensive economic assessment is performed, including all costs incurred in manufacturing or use of a product. The economic analysis and the overall environmental impact are then combined to evaluate the eco-efficiency. Thus, all relevant decision factors are analyzed with specific customer benefits always being the focus of attention.

Eco-efficient solutions to the problems are those that provide a better customer benefit from a cost and environmental point of view.

6.1.3.3 An Example: Reductive Cleaning of Dyed Polyester Fibers to Improve Wash Fastness. An innovative process for the reductive cleaning of polyester was assessed by an eco-efficiency analysis (Steenken-Richter *et al.*, 2002). A major problem in dyeing polyester fibers and polyester-containing blends with disperse dyes is the poor wash fastness. High wash fastness is one of the main requirements for clothing throughout the world. This problem is currently tackled in various ways. In addition to selecting dyes that exhibit very low thermal migration, dyers employ reductive agents in the cleaning process. In reductive post-cleaning, disperse dye molecules adsorbed onto the surface of the fiber are broken down into smaller, often colorless and more readily water-soluble fragments. In the eco-efficiency analysis the customer benefit was defined to reductive cleaning of 1000 kg dyed polyester. The analysis compares two alternatives for reductive cleaning of dyed polyester:

- the traditional hydrosulfite process, and
- a new process based on Cyclanon® ECO, a liquid reductive agent.

In the hydrosulfite process, the dye bath is discharged after dyeing and hydrosulfite is added in an alkaline solution. After reductive cleaning two rinsing steps are necessary. A neutralizing step finishes the process.

When using Cyclanon ECO, the reductive cleaning is performed in the exhausted dye bath itself without changing the pH. After completion of the reductive step one additional rinsing bath is sufficient to complete the process.

6.1.3.4 The Indicators. The representation of a multiplicity of individual results from the actual lifecycle assessment is frequently opaque and difficult to interpret. To improve the interpretation of results, BASF has developed a method that combines the ecological and economic parameters, plotting them as a single point in a coordinate system.

Ecology. The environmental impact is described with reference to six categories, resulting in what is called the "ecological fingerprint":

- consumption of raw materials;
- consumption of energy;

- area requirement;
- emissions into air, water, and soil (wastes);
- toxic potential of the substances employed and released; and
- potential for misuse and hazard potential.

Each of these categories covers a large number of detailed individual criteria.

In the ecological fingerprint the alternative with the highest impact is assigned a value of one; all others are evaluated relative to this – the further inward an alternative is located, the better it is. The ecological fingerprint of the study about reductive cleaning of polyester clearly shows the low environmental impact of the new process. In all environmental categories the new process shows enormous improvements (Fig. 6.4). The advantages are based on reduced consumption of water, reduced consumption of chemicals, and reduced water emissions.

The overall environmental impact of a product or process is given by the combination of the individual data sets. After normalization as described above, the next step is to combine the normalized values via a weighting scheme to form a single value for the ecological impact. This weighting scheme is made up of societal weighting factors and so-called relevance factors.

Societal weighting factors (Fig. 6.5) account for the importance society attaches to the different forms of ecological impact. They are based on public opinion surveys. The relevance factors are based on actual data and indicate the relative importance of individual environmental impacts (e.g., emissions to air) for a particular eco-efficiency analysis. For each specific impact variable, first the impact due to the investigated product or process is divided by the total environmental impact of this variable in the country in question. These numbers are generally very small but the relation between them yields the necessary weighting and thus, the relevance

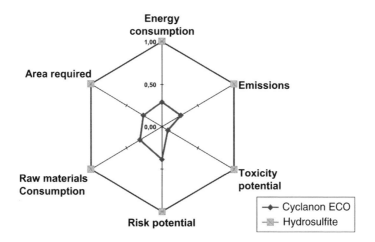

Figure 6.4. The ecological fingerprint.

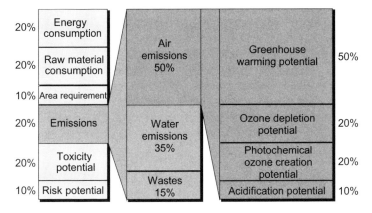

Figure 6.5. Societal weighting factors in eco-efficiency analysis.

factors. This way, both qualitative factors influenced by society as well as quantitative factors based on actual environmental data are used.

Economy. The economic data are compiled over the entire lifecycle. For this purpose the material and energy flows, including all relevant incidental flows, are taken into consideration. The total costs are normalized with respect to the average of all alternatives. This helps in identifying cost drivers and areas offering potential for cost reductions.

Looking at the total costs of the example (Fig. 6.6), the advantages of the new process are obvious. The new process is more cost-effective. Cost reduction can be achieved due to:

- significantly shorter process time;
- saving of rinsing water;
- reduction of effluent load (much less acidic and alkaline reagents);
- fewer chemicals, reduced number of controls, and as a consequence fewer possibilities for errors.

The savings of water and the reduced process costs dominate and easily compensate for the higher costs for chemicals.

6.1.3.5   The Portfolio.   The data on relative costs and environmental impacts are used to construct an eco-efficiency portfolio, which clearly shows the strengths and weaknesses of each product or process. The relative costs are plotted on the x-axis of the portfolio, and the relative environmental impacts on the y-axis. The closer a product or process is to the upper right-hand corner, the more eco-efficient it is. Multiplying the normalized values of the economic and ecological assessment results in a number for the eco-efficiency. Products or processes characterized by the

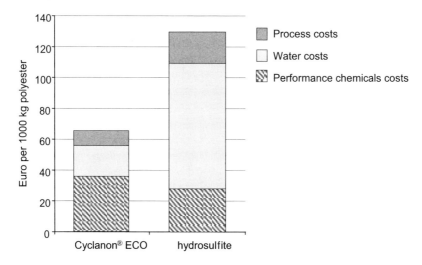

Figure 6.6. Comparison of total costs.

same number are deemed equally eco-efficient. This can, however, be achieved by different combinations of the individual economic and environmental numbers.

The eco-efficiency portfolio of the example clearly presents the advantages of the new process (Fig. 6.7). The process is the more eco-efficient alternative and is positioned in the upper right-hand corner. Costs and environmental impact are lower than for the process based on hydrosulfite.

If all polyester dyers worldwide employed the new process, the residential energy use (heating and electricity) of a city like Frankfurt and the water consumption

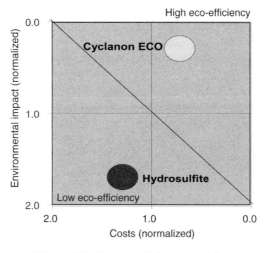

Figure 6.7. The eco-efficiency portfolio.

of a city of six million inhabitants would be saved each year (Dittrich-Krämer *et al.*, 2002).

6.1.3.6   The Value of Eco-Efficiency Analyses. The value of the eco-efficiency analysis tool, apart from its description of the current state, results from the recognition of dominant influences and the illustration of "what if...?" scenarios. The stability of the results is verified by means of sensitivity analyses in every project. Underlying assumptions as well as system boundaries and societal weighting factors are varied and checked within realistic ranges. The results of an eco-efficiency analysis make it possible to identify weaknesses in products and processes over the entire lifecycle. This allows us to identify factors with significant optimization potentials.

Eco-efficiency analysis permits a condensed representation of complex inter-relationships, which, as a result, are easier to grasp and to understand. Often, the facts and implications of a case are only vaguely understood. The results of an eco-efficiency analysis allow the visual presentation of these facts and provide a sound basis for discussion. Moreover, the detailed information gained from the investigations enables a more targeted technical, ecological, and economic development of the products and processes. From a user's perspective, the understanding of costs and various environmental aspects of a product is often of a relatively subjective nature. The eco-efficiency analysis illustrates these aspects and makes them more readily accessible for discussion.

BASF also values the method as a tool for cooperation with customers. Joint projects can reveal improvement potentials, which will in turn realize a multiple-win situation; the winners being the customers, the retailers, the end-users, BASF, health, and environment.

With regard to the markets of the future, we believe it is important to consider the environmental impact of products. We are convinced that eco-efficient products and processes give our partners and us a competitive edge.

6.1.3.7   Further Enhancing the Eco-Efficiency Analysis by Including Social Aspects. Whereas numerous instruments are used in practice for the ecological assessment of products and processes, social lifecycle assessment procedures are still lacking. BASF is cooperating with Karlsruhe University and the Öko-Institut e. V., Germany, in developing a method to include and measure social aspects of sustainability in a lifecycle assessment. The activities are integrated in the Project "Sustainable Aromats Chemistry" of the German secretary for education. The new method, the so-called SEEbalance®, examines and compares three main aspects of different product or process alternatives: costs, environmental impact, and social effects. Socio-eco-efficient solutions combine a good environmental performance with high social benefit and low costs for the customer. For the SEE-balance, the well-proven principles of the BASF eco-efficiency analysis were enhanced by a product-related specification of the social dimension of sustainable development.

Various organizations (several UN organizations, national governments, non-governmental organizations, scientific institutes, private enterprise organizations) have already formulated specific social development goals and, in many cases, also indicators to measure them. We first reviewed the existing social assessment criteria by a check of literature and other references. In a second step, it was examined which of these social goals are relevant for and applicable to product and process assessment. Viable social indicators need to be suitable for the lifecycle assessment and for building up specific databases. Ecologically relevant inputs and outputs are usually easily related to one product unit (usually 1 kg or 1 MJ) and can therefore be cataloged and found in special lifecycle assessment databases. This, so far, has not been the case for social aspects. Correspondingly, no such databases exist from which useful data could be extracted. A possible approach for developing social indices (e.g., the number of occupational accidents or the number of employees) is based on an assessment of entire industrial sectors. For this, statistical data on social indices from various branches of the economy are linked to the amount of manufactured goods in the respective branches, and thus can be related to product units. Analog to the ecological and economical dimension, the social factors are also based on a lifecycle view.

The indicator classification that was developed for the SEEbalance is based on the "business management stakeholder approach" (Schmidt *et al.*, 2005). The following groups were identified as typical stakeholders that may be affected by social effects of production, uses, and disposal of products:

- employees;
- suppliers/business partners;
- end customers/consumers;
- neighborhood and society at large;
- future generations;
- international community.

In order to generate the final assessment, the individual results for the relevant social indicators must be aggregated. Owing to the varying importance of the different indicators, each indicator has to be weighted to determine its influence on the result. Similar to the existing eco-efficiency analysis, weighting of the indicators includes "relevance weighting factors" and "societal weighting factors". The result is then presented as social fingerprint (Fig. 6.8).

Finally, the results of the social, ecological, and economic assessments are combined in the three-dimensional portfolio, the so-called SEEcube® (Fig. 6.9).

The instrument described here is to be regarded as "work in progress," that is, it needs to be constantly and critically reviewed and adjusted to the ongoing developments in society and the discussions in politics, society, and science.

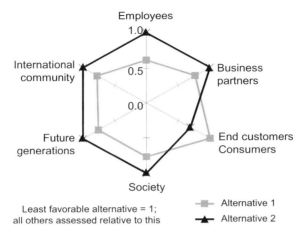

Figure 6.8. The social fingerprint.

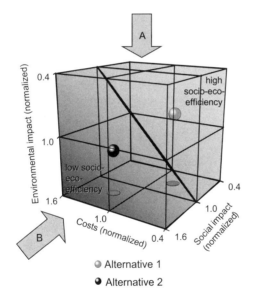

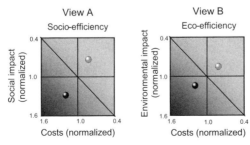

Figure 6.9. The SEEcube.

6.2   ASSESSING VALUES: COSTS AND BENEFITS

6.2.1   Overview

BETH BELOFF

BRIDGES to Sustainability

A critical aspect of integrating sustainability into business is in the area of valuation. How do we know the financial impacts of our decisions? How can performing more sustainability help to reduce costs, avoid future costs, increase the company's value proposition, support growth, and so on? What are the links between intangibles related to social and environmental performance and a firm's overall performance? What contributes to value creation in the chemical industry? How can we reconcile the fact that we may incur costs for performing more sustainably in the short term while the benefits may not be captured until the long term; similarly, how do we justify costs associated with our business decisions when the benefits might accrue to others in the value chain? We know the value of ecosystem services is both invaluable – that is, our life depends on them – and of no real market value; how then should we value ecosystem services in order to protect them for our survival and that of our business institutions? This section attempts to offer points of view on these topics.

Section 6.2.2 is titled "Intangibles and Sustainability." There is a growing body of evidence suggesting that the drivers of wealth creation for business and the economy are less about physical and financial assets and more about intangibles. While conventional accounting and financial metrics yield some insight into a firm's market value, forward-looking sustainability indicators – anything from confidence in a company's management to research leadership, to the management of environmental liabilities – are becoming more relevant to a business's overall value proposition. The link between intangibles related to sustainability and company performance is explored in this section.

Section 6.2.3 is called "Total Cost Assessment: Looking at All the Costs Involved with a Decision." Traditional decision-making typically focuses on direct and indirect costs that appear on the balance sheet. The TCA model defines three additional cost types, contingent liabilities and internal and external intangibles. A methodology for developing total costs related to environmental impacts is discussed.

In Section 6.2.4, "Societal Costs", we learn that societal costs constitute an important element of the costs of environmental impacts from industrial activities, although they often fall outside the private calculation of costs and benefits to a company. This entry gives a perspective on the importance to a company of considering these externalities as a sustainability indicator in both management decision-making and facility design and optimization.

Section 6.2.5, "Valuing Ecosystem Services", describes how development of an economic valuation framework for ecosystem services remains a critical challenge in sustainable development, particularly in balancing the needs of social and economic development, and ecosystem conservation. Without it, ecosystems will remain undervalued and under threat. Ecosystems are capital assets that are being liquidated at humanity's peril, and this should, in turn, create impacts for business, which depend on the health of natural capital for survival. This piece provides an understanding of how to consider the value of ecosystem services.

6.2.2 Intangibles and Sustainability

KARINA FUNK

Massachusetts Renewable Energy Trust

PAMELA COHEN KALAFUT AND JONATHAN LOW

Predictiv

> There is no question in my mind that business and the free enterprise system are essential to making sustainability work. Our focus at Dow is on hard-wiring it into our company in the same way we have fully institutionalized environment, health and safety into our culture and into our work and people processes. Our challenge is to make sustainability sustainable. Ultimately, the world will judge our commitment to sustainability not by what we say, but by what we do.
>
> — William Stavropoulos, CEO of The Dow Chemical Company, quoted in Fiksel *et al.* (2004)

A recession in 2001, accounting scandals in 2002, record-breaking unemployment in 2003, and conflicting evidence of economic recovery in 2004 – all these recent contexts and events in the United States have led investors and other corporate stakeholders to re-think their position on just what is "fundamental" to the valuation of a company. Both institutional and retail investors have learned some painful lessons, re-examined their assumptions about what constitutes tangible and intangible value, and broadened their scope to consider characteristics that can lead to longer-term financial success. One area that has begun to capture the attention of investment professionals is the mounting evidence of financial risks associated with corporate liabilities, including accountability for the environment and for human rights, as well as global problems such as climate change. Indeed, a look at the websites of the Fortune 100 companies reveals that companies such as DuPont, Alcoa, and General Electric think it is smart to address environmental or social responsibility[1] right on their Investor Relations home page. But why would investors care?

Because of stock price, for starters. The link between shareholder value and environmental and social performance is a phenomenon that has spawned a large body of research, literature, and investment activity in recent years. Insurers claim that in the next decade, the annual cost of global warming will rise to $150 billion a year (Webber, 2002). The stakes are increasing as multinationals in the finance community band together to buttress their arguments for sustainability

---

[1]"Social responsibility" is used here as a general term to encompass the community relations information that appears under different labels on the companies' websites (corporate citizenship, charitable work, community service, etc.).

considerations in their finance portfolios. For example, ten leading banks from around the world announced in 2003 a set of voluntary guidelines called the "Equator Principles," whereby they intend to meet the International Finance Corporation's environment, health and safety guidelines in their projects in developing countries. This is an interesting and unprecedented expectation: banking clients must adhere to these principles, and this is relevant to *all* corporations. Principle #8 states that if a project goes out of environmental or social compliance, this constitutes grounds for a default on the loan. The cost of capital will simply increase for companies that fail to address these concerns.

There are now over 200 mutual funds, run by over 800 portfolio managers and analysts, dedicated to socially or environmentally responsible investing. In sum, socially screened portfolios now represent over $2 *trillion* in assets, over 10 percent of the $19.9 trillion assets currently under management in the United States (Social Investment Forum, 2001). Different investment styles have emerged among funds using socially responsible, ethical, or environmental criteria.[2] The majority of the $2 trillion figure consists of screened investments, but credible organizations in the past several years have been developing scoring and ranking tools that rate companies according to environmental, social, and economic criteria. The Dow Jones Sustainability Index scores companies based largely upon their responses to extensive questionnaires (see http://www.sustainability-index.com/), while the FTSE4Good Index analyses Environment, Health and Safety (EHS) and social responsibility activities, with the stated intent of promoting a stronger business commitment (see http://www.ftse.com/ftse4good/index.jsp). These indexes have generally performed in line with or have outperformed the broader market averages.[3]

Researchers at Cap Gemini Ernst & Young take this correlation as further testimony that intangibles – that is, nonfinancial or nontraditional financial capital – matter. The Ernst & Young Center for Business Innovation (CBI) Value Creation Index (VCI) provides insights into these intangible "value drivers" that are strongly correlated with market value. These may vary according to industry, and among durables and nondurables manufacturing we have found consistently that measures related to intangibles such as management credibility, innovativeness, ability to attract talented employees, and research leadership are highly correlated with market value.[4]

"Environment," as measured by environmental, social, and community service scores, also ranks consistently among the top ten value drivers in our VCI model,

---

[2]Distinct investment styles include environmentally effective investing (e.g., Winslow Management), socially responsible investing (e.g., companies screened by FTSE4Good Index), and sustainability investing (e.g., companies screened by Dow Jones Sustainability Group Index or ranked by Innovest EcoValue'21).

[3]Research reveals that there was no statistically significant difference between the risk-adjusted returns of socially responsible or ethical funds in the United States, Germany, and the United Kingdom and those of conventional funds during the time period of January 1990 through March 2001 (Bauer *et al.*, 2002).

[4]Our analysis of actual nonfinancial performance as correlated with market value revealed the following value drivers: Innovation, Quality, Customer, Management, Alliances, Technology, Brand, Employees, Environment. Multiple, statistically independent measures are used as inputs for each driver in order to ensure a robust model.

which explains up to 90 percent of variability in market value.[5] Encompassing a much broader suite of measures than environmental performance, this model touches upon what many would argue are components of sustainability. Certainly, sustainability means different things to different stakeholders; for now let us define "sustainability" as embodying a "desirable" future state for any given stakeholder. For investors, this desirable future state would surely include sustained revenue growth over the long term.

6.2.2.1   *Over the Long Term.* Traditional financial data is historically ... well, historical. A desirable future state demands forward-oriented indicators. Companies have the opportunity to use nonfinancial variables as bellwethers: just as sensors can monitor pollution levels in real time, helping to warn of impending problems and, when combined with databases and analytical techniques, they can help to establish causality.

Without deciphering the links between intangibles and a firm's performance, companies are abrogating a significant opportunity for value creation. Using findings from our research, as well as industry literature and conversations with business and academic researchers, we developed a list of the most critical categories of nonfinancial performance that determine corporate value creation (Table 6.7).

In the chemicals industry, our model showed that innovation, alliances, leadership, and sustainability were the four most significant drivers of value. Companies are increasing both their global reach and their innovative capabilities by acquiring smaller companies and creating new alliances. And as knowledgeable chemical industry insider Calvin Cobb has written, "The fact that leadership and social responsibility and the environment placed third and fourth, respectively, on the list of intangible drivers speaks to the 'chemophobia' that is prevalent in society today. Consumers love the products chemicals companies produce. But they hate the chemistry, if you will, behind those products. Due in part to the negative image chemicals corporations must fight, and in part to regulation, chemicals companies need strong leaders who can communicate effectively about what they are doing to preserve the environment" (Cobb and Hunter, 2002)

Doing good is not only the right thing to do; it also makes smart business sense. Many would argue that environmental spending does nothing more than improve a company's reputation for environmental management, and mitigate liability or regulatory risk down the road. And most investors would argue that there is no proven correlation between environmental responsibility and shareholder wealth. Both fair points.

On the other hand some advocate that companies should bear the full "social" costs of their operations, thereby "internalizing" these externalities. But Wall Street bestows rewards for companies that externalize as much of their cost as possible. In some very specific cases, externalities can be addressed through nonregulatory means such as shareholder advocacy and public relations (Vogel, 1977). But

---

[5]Drivers and results are for durables and nondurables manufacturing sectors.

TABLE 6.7. The Measures That Matter

Customer: The ability to develop customer relationships, satisfaction, and loyalty.
Leadership and Strategy: Management capabilities, experience, and leadership's vision for the future.
Transparency: Does management communicate honestly and openly? Are its communications believed and trusted? Does it hold itself accountable?
Brand Equity: Strength of market position. The ability to expand the market, perception of product/service quality, investor confidence.
Environmental and Social Reputation: How the company is viewed globally, such as: environmental concerns, community concerns, regulators' concerns, inclusion in "most admired company" lists, triple bottom line.
Alliances and Networks: Supply chain relationships; strategic alliances; partnerships.
Technology and Processes: Strategy execution; IT capabilities; inventory management; turnaround times; flexibility; reengineering; quality; internal transparency.
Human Capital: Talent acquisition, workforce retention, employee relations, compensation; what makes a "great place to work."
Innovation: The R&D pipeline; effectiveness of new-product development; patents, know-how, business secrets.
Risk: The ability to effectively manage the balance between potential liabilities and potential opportunities.

Source: Fiksel *et al.* (2004).

let's not mince words. Externalities are, by definition, market failures and are unlikely to be addressed without regulatory intervention.[6]

However, proactive investing in environmental measures beyond that required by law can be good for the bottom line, if for no other reason than to limit downside risk (Rienhardt, 1999). And despite the many studies that debunk statistical correlations (many of the positive correlation studies have been limited in analytical rigor; McWilliams and Siegel, 2000), there has been a proliferation of "green" funds in response to investor demand, and researchers find a paucity of published data implicating a negative correlation between environmental performance and share price among industry peers.[7] Please see Section 8.3, "Sustainability and Performance," for additional discussion on intangible value drivers for the chemical industry.

6.2.2.2 The Beginnings of a Sustainability Model.
Many of the indicators used for the Value Creation Index studies mentioned above were limited to the availability of public information. But one could imagine some of the additional kinds of metrics – beyond that which is publicly available – that would constitute a firm-level model of sustainability. By mining what they already know from a host

---

[6]Externalities used here in the sense of market externalities (positive externalities and pecuniary externalities notwithstanding).

[7]There exist reports that "either criticize current regulatory regimes as overly costly, ineffective, or both. In addition, there are many studies that argue, on cost–benefit grounds, that some environmental standards are unjustified in a public policy sense. However, none of these studies examined financial impacts at the level of the firm or the shareholder" (Earle, 2000).

of qualitative evidence and quantitative measures, and seeking to identify indicators of sustainable business that cut across all their business functions (procurement, supplier relations, product design, and so on), managers can obtain a more comprehensive view of how to encourage sustainable growth.

As a complement to the broad suite of intangible value drivers that make up the industry-wide Value Creation Index models (Table 6.7), Figure 6.10 presents the leading indicators discussed in this article as inputs to an *illustrative* firm-specific sustainability model. Qualitative research and particulars of the industry context can also be used to create a model of sustainability specific to a firm. A survey instrument is designed based on a hypothetical model and given to representatives within the firm and business stakeholders. The results and the quantitative measures are used to test the model empirically. Each driver is made up of multiple indicators to ensure robustness and accuracy. The indicators are used to create scores for each driver (in the circles), which can be scrutinized statistically (checked for multicolinearity, for their relationship to the prescribed intangibles, how well they group together, for independence and statistical significance). These in turn are weighted relative to each other and according to their impact on sustainability. The impacts (Is in the squares) represent the expected effect of a one-unit change in any of those drivers on a normalized score for sustainability, the intangible in question. Most interesting, the impacts leading to performance measures (on the right) suggest the expected effect of a one-unit change in sustainability on the various performance measures.

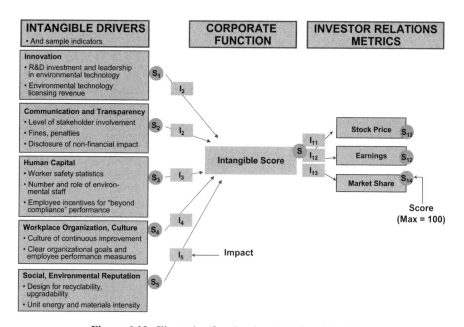

Figure 6.10. Illustrative firm-level model of sustainability.

Companies have an opportunity to develop forward-looking tools like the Value Creation Index that link the strategic contributions of their sustainability efforts to enhanced market valuation. And with long-term success in mind, the importance of sustainability as a shareholder value driver can no longer be ignored. In the words of Gary Pfeiffer, senior vice president and chief financial officer at DuPont: "Every corporation is under intense pressure to create ever-increasing shareholder value. Enhancing environmental and social performance are enormous business opportunities to do just that" (Fiksel et al., 2004).

6.2.3   Total Cost Assessment: Looking at All the Costs Involved with a Decision

LISE LAURIN

EarthShift

GREG NORRIS

Sylvatica

Total cost assessment (TCA) is a decision-making methodology that enables cost-based decision-making based on traditionally unquanitified factors such as the risk of an accident, potential changes in regulations, community relations, and employee satisfaction. It was developed and validated by an industry collaboration assembled by the American Institute of Chemical Engineers' Center for Waste Reduction Technology (AIChE CWRT). This multidisciplinary, scenario-based costing methodology complements traditional cost models by facilitating the examination of all costs associated with a decision.

6.2.3.1   Additional Cost Categories Add Relevant Information for the Decision-Making Process. Traditional decision-making typically focuses on direct and indirect costs that appear on the balance sheet. The TCA model defines three additional cost types, contingent liabilities, and internal and external intangibles. Table 6.8 gives examples of each cost type. While contingent liabilities and internal intangibles (Type III and IV costs) are not typically accounted for on a balance sheet, they are very real costs to a company. External intangibles (Type V) are not direct costs to the company, but it is helpful to include this category, as often the analysis of this category will influence the analysis of Type III and IV costs. One of the most important methodologies in TCA enables companies to estimate Type III, IV, and V costs. This methodology is consistent with sound business decision-making processes, yet allows incorporation of variables that do not directly impact the manufacturing process.

Six Steps to a New Source of Critical Data. When implementing the TCA methodology, a company needs to first assemble a team that has the ability to assess the types of costs shown in Table 6.8. This might include a health and safety professional to help with the cost of a potential accident, a brand marketing specialist

TABLE 6.8. Environmental, Health, and Safety Cost Types in TCA Model

Cost Type	Description	Examples
I. Direct costs (recurring and nonrecurring)	Manufacturing site costs	Capital investment, operating and maintenance costs, labor, raw materials, and waste disposal costs
II. Indirect costs (recurring and nonrecurring)	Corporate and manufacturing overhead costs; costs not directly allocated to product or process	Reporting costs, regulatory costs, and monitoring costs
III. Future and contingent liability costs	Potential fines, penalties, and future liabilities	Fines and penalties caused by noncompliance; clean-up, personal injury, and property damage lawsuits; natural resource damages; industrial accident costs
IV. Intangible internal costs (company-paid)	Difficult-to-measure but real costs borne by the company	Cost to promote consumer acceptance; maintaining customer loyalty, worker morale, worker wellness, union relations, corporate image, and community relations
V. Intangible external costs (not directly paid by company)	Costs borne by society	Effect of operations on housing costs, degradation of habitat, effect of pollution on human health

to assess the cost of loss of brand loyalty, and a product specialist to assist with the direct and indirect costs associated with any option. The team then walks through six steps of planning, research, and analysis, with a final feedback loop into the company's decision process. The team participates in a series of brainstorming and reality checking activities. Experience has shown that the insight generated by the team during an interactive TCA process far exceeds that which could come from the individuals working separately.

Steps in the TCA Methodology
1. *Goal Definition and Scoping.* In the first step of the process, the team agrees upon the options or alternatives to be assessed. The assessment may be across two existing processes or products; it may be between an old way of doing things and a new way of doing them; or it may be between two possible new ways of doing things. The assessment, however, does not need to be limited to two options. The team also agrees upon the purpose of the TCA analysis and the goals of the project result. In nearly all cases there will be

corporate, division, or site goals to consider. In addition, the group may wish to consider sustainability, environmental and health goals, and social impacts. It is important at this phase to determine whether Type V costs will be used, and if so, how they will be incorporated into the decision.

2. *Streamline the Analysis.* In this second phase of the process, the team places limits on the goal and scope to reduce ambiguity about the results. The team may decide to focus on a subset of the lifecycle: gate to gate for example, instead of cradle to grave. Where data gathering may expand the project beyond reasonable expectations, the group may decide to use more accessible data from a similar process as good approximation. The team may decide to focus on a certain set of goals. Some goals may be so important that they can be assigned a "go–no go" limit.

3. *Identify Potential Risks.* In this phase of the analysis, the team identifies the risk scenarios associated with each of the alternatives. Identifying the sources of uncertainty makes it easier to assign an uncertainty to the scenario. The team then identifies the cost drivers (e.g., compliance obligations and remediation costs) for each scenario. This phase can be the most difficult in the evaluation, as the team is asked to put costs and probabilities on situations they have not had to evaluate in the past. The team leader or facilitator must guide the discussion away from unusable answers – "I have no idea!" – to workable boundaries, such as "the probability is less than 10 percent." For each Type III, IV, or V cost the group must agree upon probability, frequency of occurrence, and timing of occurrence where relevant data are available.

4. *Conduct Financial Inventory.* In this step, the team assigns a value to each cost in all five categories. Type I and II costs would be the same costs used in a traditional cost comparison. For the remaining costs, the team reviews the risks identified in Step 3 and assigns costs or a range of costs to each cost driver. Table 6.9 gives some suggestions for sources of cost data. The team then reviews the costs to determine which are the most significant, and to assess how that information can best be incorporated into the decision-making process.

5. *Conduct Impact Assessment.* Once all the costs are available, the data is entered into spreadsheets (available through the AIChE website) or into the automated software program TCAce and the results can be calculated. It is important that the team document all assumptions and results for each scenario and cost decision, especially for important potential impacts that are not currently feasible to cost. The team may choose to change some assumptions and reassess the results. This step is especially beneficial for costs whose probability is not known with any certainty. Refining and rerunning the calculations gives the team an understanding of the robustness or weakness of the results.

6. *Feedback to Company's Decision Loop.* In this step the results are fed into the company's decision-making process. Total cost assessment is not designed to

TABLE 6.9. Sources of Cost Data for Use in TCA

Cost Type	Example	Data Source
III: Future and contingent liability	• Civil and criminal fines and penalties	• EPA's Integrated Data for Enforcement Analysis (IDEA) database; National Compliance Database
	• Cost of accidents	• EPA ARIP database
IV: Intangible– internal	• Staff (productivity/ morale; turnover; union negotiating time)	• Industry-specific studies estimating medical costs and lost wages from workplace injuries
	• Market share (value chain perception, public perception, consumer perception)	• Studies regarding the costs associated with loss of market share due to changes in public perception associated with industrial accidents
V: Intangible– external	• Pollutant discharges to groundwater	• Natural resource damage (NRD) settlements for groundwater contamination
	• Natural habitat impacts: local community	• Published literature on willingness-to-pay scales, related to preservation of natural habitat or to protection of a particular species. Also, data on costs of restoring habitats or species.

replace an organization's traditional accounting system but rather to provide cost information for internal managerial decisions. Each company will have its own policies, principles, and values that will guide how the TCA model is applied within the company. This final step recognizes that the TCA is only one consideration in an overall process that needs to include many types of information.

Tools. There are two main tools to support the implementation of the TCA methodology. The first tool is an MS Excel Workbook with spreadsheets for each type of cost, available from the AIChE. These spreadsheets provide typical costs to be considered under each category and provide simple calculation capability.

The spreadsheets, however, may give an oversimplified result. In the case of a $100,000 accident that has a 5 percent probability of occurrence in each year, for example, the spreadsheets will show an annual cost of $5000. The actual cost will never be $5000 – it will be either $0 or $100,000 in any given year. A software program called TCAce was developed to properly accommodate uncertainties. Users enter data into TCAce, similarly to the way it is entered into the manual spreadsheets, but with more flexibility. TCAce can then perform an analysis similar to the simple analysis provided by the spreadsheets. It can also perform a Monte Carlo analysis, resulting in more realistic probable costs. Figure 6.11 shows the inconclusive result of an analysis between installing a water recovery system and

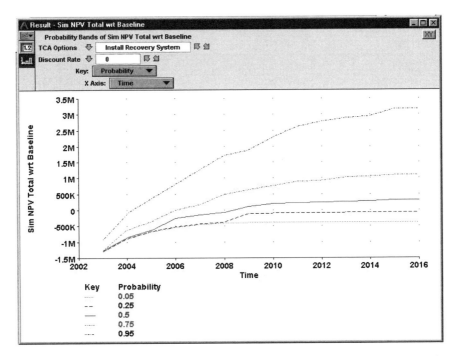

Figure 6.11. Using Monte Carlo analysis, the software program TCAce calculates the probability that an option will cost less than the baseline.

the baseline using Monte Carlo analysis in TCAce. In this case, the recovery system has a 50 percent chance of costing less than or equal to the baseline, but may just as well cost more. Refining the data may provide a more conclusive result.

A Methodology by Industry for Industry. The TCA methodology provides an industry-standard costing framework for assessing process development, product mix, waste management, pollution prevention, facility location and layout, outbound logistics, and other business-wide issues. It allows businesses to quantitatively include financial risk in their decision-making process. The methodology aids managers in making informed decisions about environmental, health, and safety opportunities and impacts, contributing to improved long-term competitiveness.

6.2.4 Societal Costs

BETH BELOFF, DICKSEN TANZIL AND MATTHEW RETOSKE
BRIDGES to Sustainability

Societal costs constitute an important element of the costs of environmental impacts from industrial activities, although they often fall outside the private calculation of costs and benefits to a company. Most of these costs, such as health and property damages due to pollution, usually belong in the category of societal costs (or external costs) borne by society at large. In this chapter, we look at the nature of the societal cost element of Total Cost Assessment (TCA), how it is a cogent indicator of future internal corporate costs, and how it can be estimated.

6.2.4.1 Evolution of Costs. From a sustainability perspective, the TCA methodology should account for both internal corporate costs and societal costs. In the conventional business model (Fig. 6.12a), a company invests on a project when the projected revenues are greater than the projected costs to the company. However, to reflect environmental stewardship and social responsibility, a truly sustainable company should make considerations beyond the conventional business model

(a) Conventional business model

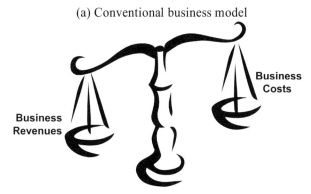

Invest when business revenues > business costs

(b) Sustainability model

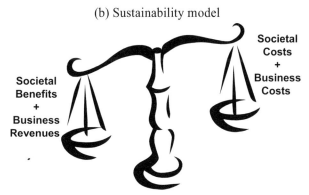

Invest when **business revenues > business costs**
and **total** benefits > **total** costs

Figure 6.12. To reflect environmental stewardship and social responsibility, a truly sustainable company should make considerations beyond the conventional business model and take into account the societal (external) costs and benefits. (©BRIDGES to Sustainability, 2002.)

and take into account the societal (external) costs and benefits. In this sustainability model (Fig. 6.12*b*), a company would weigh total costs and benefits, which include societal costs and benefits, in its investment decisions.

While the external costs do not affect a company's bottom line at the time of impact, they eventually manifest themselves in the internal company costs, as illustrated in Figure 6.13. Previous work on the external costs of "harmless" odors by BRIDGES to Sustainability (Beloff *et al.*, 2000) provides a perfect example on how a seemingly "harmless" external intangible effect can translate into substantial internal company costs in the forms of fines, capital and operating costs for abatement equipment, as well as other more intangible costs associated with loss of goodwill and reputation, job productivity declines due to lowered employee morale, and so on. Similarly, societal benefits – both negative costs and positive value-add – can also be conceptually viewed as providing a flowback to the company creating the benefits for the community. This is further addressed in Section 6.2.2.

The study also found that internal company costs of odors rose markedly when a certain critical mass of affected population is exceeded, in this case at approximately 30 to 50 households affected, as shown in Figure 6.14. It should make intuitive sense that, when there are enough people in a community who are experiencing the direct effects and related costs from a company's environmental impacts, they will find one another, perhaps even be identified by NGOs, organize, consider actions that are within their control, and act. This, in turn, can lead to a disproportionate level of costs to the company as it attempts to respond to the challenges created. For these reasons, societal costs become a cogent indicator for management of future internal

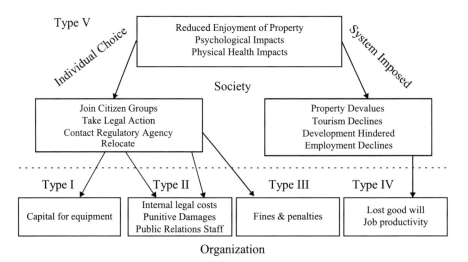

Figure 6.13. Evolution of costs. External or societal costs eventually manifest themselves in the internal company costs (Beloff *et al.*, 2000).

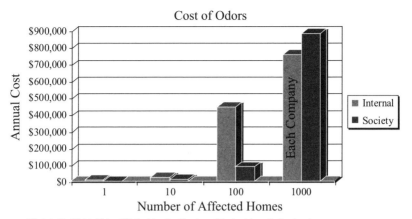

Affected = Real Estate Value + Medical, legal and/or air conditioning + investigational costs

Figure 6.14. The internalization of external costs rises rapidly once a critical mass is exceeded (in this case, approximately 30 to 50 households) (Beloff *et al.*, 2000).

company costs, as they represent societal concerns that are likely to be translated into real business impacts to the company.

6.2.4.2   Societal Costs: Global, Regional, and Local. Industrial activities may result in environmental impacts of different scales: including global, regional, and local. Global warming and stratospheric ozone depletion are examples of global impacts. Impacts and damages in this category, such as natural calamities due to global climate change attributed to greenhouse gases, can be experienced thousands of miles from the source. Regional and local impacts, in contrast, are experienced only within certain regional or local boundaries. Air and water acidification, eutrophication, and urban air pollution problems are examples of impacts in this category.

Tying the societal costs of global impacts to a company's environmental performance requires one to examine the impacts at the global level. Assessment that is limited to regional or local boundaries may result in skewed, if not inaccurate, estimates. The societal costs of regional and local environmental impacts, on the other hand, can vary considerably with region or location and should be assessed only within their corresponding regional or local boundaries. Regional and local variables such as local population, ambient pollution level, climate, topography, and available pathways affect the extent of the environmental damages and their associated costs to the society.

6.2.4.3   Data Sources. A societal or external (Type V) cost database was included in the original TCAce™ (AIChE, 1999). This database contains some literature estimates on the societal costs of pollutant discharges to air, water, and land, and on the economic values of natural habitats. While covering a wide range of impacts, the database is far from comprehensive. For example, for the damage costs of air emissions in the United States, the database is based solely on estimates from Minnesota Public Utility Commission. These cost estimates, although rigorous and valid, were established only for one U.S. Midwestern state and may not be representative of the cost of industrial air pollution elsewhere in the United States. As pointed out in the TCAce™ manual (CWRT, 1999), many of the other reported societal cost estimates are also location-specific and do not represent general U.S. or global estimates. Furthermore, except for the damage costs of air emission, many of the reported costs are not tied to process flows (such as pound of emission) and therefore difficult to use in the design and optimization framework.

Chalmers University's "Environmental Priority Strategy in Product Development" (EPS) (Steen, 1999) is perhaps currently the most comprehensive external damage cost database, especially for use in design and optimization. It includes databases on the willingness-to-pay economic costs of pollution per unit weight of emission and of depletion per unit weight of resource use. The costs of human health, agricultural, forestry, and resource damages are incorporated into their estimates. However, these costs are estimated only in the "average sense" for OECD countries (developed economies in North America, Europe, and the Pacific Rim). Costs in each specific country, region, or locality are likely to differ, and will be higher in particular in industry- and/or population-intensive urban areas.

Societal or external cost data are also available elsewhere in the literature. The Environmental Valuation Resource Inventory (EVRI; available at http://www. evri.ec.gc.ca/evri/) includes a large database of studies of the economic valuation of ecosystems and ecosystem damages. Studies in this database provide primarily "raw data" on the economic values of the damages. Considerable work is necessary to tie the cost of damages to design-related variables such as rates of emissions and resource use. A large amount of literature is also available on the economic costs of global warming and urban air pollution. An overview of the costs of these two impacts is presented as examples in the following sections.

6.2.4.4 Example: Costs of Global Warming. Global warming is an example of a global environmental impact. Economic damages due to climate change in one region may be attributed to CO_2 and other greenhouse gas emissions halfway around the world. The facts that the effects of global warming are uncertain and exist mostly as a future liability add to the complexity of estimating the societal costs of global warming.

One challenge in understanding the future liabilities associated with climate change is that scenarios can be divided into two separate yet interrelated categories. The first is greenhouse gas emissions scenarios, and the second is climate change impact scenarios. Because of the indirect nature of atmospheric carbon pollution, it is important to understand the complex relationship of the two and the limitations of defining that relationship.

Greenhouse Gas Emission Scenario. In determining greenhouse gas emission scenarios, there are four factors that need to be considered: population growth rate, growth rate of per capita GDP, energy intensity of economies, and the carbon intensity of energy. The most common method for estimating future CO_2 emissions levels and the one used by the Intergovernmental Panel on Climate Change (IPCC) is the Kaya Method, based on the work of Professor Yoichi Kaya of Keio University.

Kaya's approach incorporates several demographic factors to estimate the growth rate of CO_2 emissions. It assumes a continuing decrease in both energy intensity of GDP and carbon intensity of energy. The baseline global growth rate of emissions can be assumed to be 3.4 percent if past trends and future predictions hold true. With these assumptions, IPCC's fifteen- (2015) and fifty- (2050) year scenarios demonstrate that even with a decrease in energy intensity of GDP and carbon intensity of energy, emissions will increase at a steady rate.

Climate Change Impact Scenario. The potential effects of global climate change cover a large range of impacts and cost considerations, which are summarized in Table 6.10. Currently, direct internal climate-related costs are not allocated to carbon emissions. For instance, damage due to an extreme weather event or increased insurance premiums would not be identified with greenhouse gas emissions. There is some justification for not doing this – greenhouse gases dispersed in the atmosphere are not like local pollutants where an impact can be tied directly to the pollution. One may fall into an epistemological game of trying to determine cause and effect that is irreconcilable given the nature of greenhouse gas emissions. It is a practical impossibility to allocate damages or costs from one particular event to the greenhouse gas emissions from one location.

Economic studies on the societal costs of climate change present a wide range of dollar amounts per ton of carbon equivalent. Designers and decision-makers should therefore consider a number of scenarios when determining the future costs associated with greenhouse gases. It is a practical impossibility to know the full extent to which these costs are associated with anthropogenic climate change, but given the solid scientific foundation established by the IPCC, one can establish reasonable ranges of potential costs.

One option is to base decisions on the four impact scenarios shown in Table 6.11. These scenarios are based on research primarily from the IPCC, and were developed by BRIDGES to Sustainability (2002) to assist decision-makers when considering

TABLE 6.10. Societal Climate Cost Considerations

Agricultural	Heat stress
Human health	Coral
Food production	Mangrove
Drought	Coastal
Flood	Tundra
Population displacement	Wetlands
Diminished food security	Forests
Fresh water availability	Glacial retreat
Infectious diseases	Threats to fisheries
Desertification	Soil salinization
Infrastructure stress	Coastal erosion
Loss of biodiversity	Tropical cyclones
Sea level rise	Thermal water pollution

TABLE 6.11. BRIDGES' Guestimates of the Per-Ton Cost of CO_2 Based on IPCC Impact Scenarios

Impact Scenario	Description	Carbon Cost Estimate (2000 US$)
Standard	Small carbon costs incrementally increasing, limited legislation, CO_2 increases of 1–4% annually, impacts low–moderate	$25
Technological surge	New technologies accelerate reductions in energy and carbon intensity, adaptive capacity high, impacts low, decrease in per capita CO_2 emissions	$10
Carbon constrained	Increasing internalization of costs to arrest emissions, higher energy prices, energy efficiency mandated, sequestration mandated, impacts moderate–severe	$95
End of hydrocarbons	Serious impacts drive austerity in energy use and carbon emissions, heavy internalization of cost, high carbon tax, severe losses from climate change impacts	$275+

the societal costs of greenhouse gas emissions. The quoted costs are best guess estimates for the future based on the literature estimates. The chosen dollar amounts are a reflection of where data points "bunch" in a variety of different greenhouse gas emissions cost studies depending on assumptions and scale of those studies. These figures are offered here not as precise estimates, but to illustrate the possible range in order-of-magnitude.

6.2.4.5 Example: Costs of Regional and Urban Air Pollution. Industrial air pollution results in regional problems, such as acidification, as well as local environmental problems, especially in densely populated, industrialized urban areas. Numerous studies have been performed to estimate the societal costs of regional and urban air pollution, most of them focusing on the costs of damages to human health. The works invariably entail complex procedures, comprised of multiple steps. The procedures determine the marginal costs of air pollution typically through the following steps:

1. Relating the effect of a reduction or increase in pollutant emission to the regional/local air quality level;
2. Relating the effect of the change in regional/local air quality level to the magnitudes of health damages and/or other impacts;
3. Estimating the economic value of the damages; and
4. Normalizing the economic value of damages per unit mass of reduction/increase in pollutant emission.

Large uncertainties are involved in each of the above steps. Consequently, they result mostly in order-of-magnitude estimates of societal costs.

The step of economic valuation (Step 3) usually uses the willingness-to-pay (WTP) approach, which measures how much the society values a noneconomic resource or how much the society is willing to pay to avoid damages. The WTP may be estimated using carefully developed questionnaires (contingent valuation) or through direct estimates of the monetary costs of the damages, for example, how much people pay to alleviate a case of headache or to extend life through medical procedures.

Another approach used in economic valuation is the hedonic pricing method. By this approach, economic valuation is obtained through estimating the direct effect of environmental impacts on market prices, usually housing prices. The hedonic pricing method is often used to estimate the economic value of more intangible damages due to air pollution, such as reduced visibility and aesthetics.

Figure 6.15 shows some of the external damage cost estimates for NOx. The costs are shown based on the reported best estimates, or the geometric average of lower and upper bound estimates when no best estimates were reported in the original publications. The magnitudes of the estimates differ by over three orders of magnitudes. With such a large range, these estimates are occasionally criticized as unreliable. However, the variations can be understood in terms of local variables. In general, the estimates are higher in densely populated urban areas such as Los Angeles, and lower in more rural and less populated areas such as rural Minnesota.

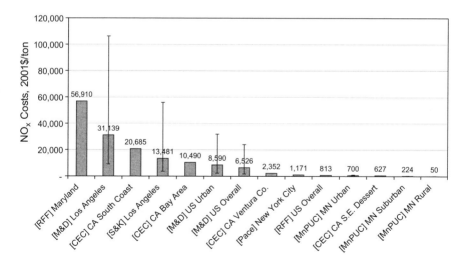

Figure 6.15. Damage cost estimates per ton of NOx emission. Best estimates or geometric average of upper and lower bound estimates, inflated to 2001$ based on Consumer Price Index.
Sources: (Pace) Ottinger *et al.*, 1990; (CEC) California Energy Commission, 1993; (S&K) Small and Kazimi, 1995; (MnPUC) Minnesota Public Utility Commission, 1997; (M&D) McCubbin and Delucchi, 1999.

Differences in assumptions, however, also contribute to some discrepancies. Quite expectedly, the overall U.S. estimates tend to fall in the middle range, less than the high estimates for the highly populated and highly polluted urban areas but greater than the rural estimates.

Empirical correlations may be developed to relate the costs calculated from different studies to key regional or local variables. Wang and Santini (1995) reviewed five studies and 15 sets of cost data to relate the estimated per-ton costs of NOx (as well as VOC, CO, PM_{10}, and SOx) emissions to regional ambient air quality measurements and the size of affected population. Figure 6.16 compares the direct estimate of the costs of NOx emission from various studies with the estimates from a correlation modified from that of Wang and Santini (1995), incorporating results from more studies. Generally, the various NOx cost estimates correlate well with ambient ozone level and the size of the population in the affected metropolitan area or air quality region, although there are a few outliers, most likely due to differences in assumptions and methodology.

6.2.4.6   Weighting Environmental Impacts. As mentioned previously in Section 6.1, societal costs, either alone or in combination with other types of environmental costs, may be used as weighting factors to "value" different environmental impacts. For illustration, Table 6.12 compares damage cost estimates for different air pollutants from different studies. The monetary figures listed are based on those reported

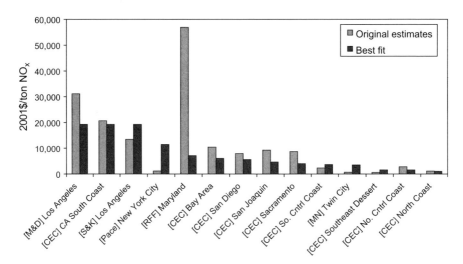

Figure 6.16. Comparison between the original direct estimates of the damage costs per ton of NOx emission with correlation best fit. $\ln(Cost) = 0.514 \cdot \ln(ozone) + 0.790 \cdot \ln(population) + 5.99$ where *ozone* is the 2nd 1-hour maximum in ppm; *population* expressed in 1000s.
Sources: (Pace) Ottinger *et al.*, 1990; (CEC) California Energy Commission, 1993; (S&K) Small and Kazimi, 1995; (MN) Minnesota Public Utility Commission, 1997; (M&D) McCubbin and Delucchi, 1999.

TABLE 6.12. Using Societal Costs to Weigh Environmental Impacts

Source	Geographic Reference	Effect	Unit	PM$_{10}$	SOx	NOx	VOCs	CO
McCubbin and Delucchi (1999); Delucchi et al. (2002)	United States	Health	\$/kg	46.96	30.25	6.53	0.49	0.04
Small and Kazimi (1995)	Los Angeles	Health	\$/kg	126.39	138.85	13.48	4.47	—
CEC (1993)	CA South Coast	Health and other	\$/kg	68.01	10.60	20.68	9.87	0.00
EPS (Steen, 1999)	OECD	Health and other	ELU/kg[a]	36.00	3.27	2.13	2.14	0.03
EI-99 (Goedkoop and Spriensma, 2000)	Europe	Health (respiratory)	DALY/kg	3.75E-04	5.46E-05	8.87E-05	6.46E-07	—
			\$/kg[b]	25.00	3.64	5.91	0.04	—

[a]1 ELU (Environmental Load Unit) ≈ 1 € (1998).
[b]Assuming an economic value of \$66,667 per DALY (disability-adjusted life year) based on a statistical value of life (SVL) of \$5 million divided by a life expectancy of 75 years.

in the original studies, not inflation-accelerated. The EPS (Environmental Priority Strategy in product development) damage estimates (Steen, 1999) are expressed as ELU (Environmental Load Unit) per kg, where one ELU is approximately equal to one 1998 ECU (European Currency Unit, €). Note that EI-99 weighting factors (Goedkoop and Spriensma, 2000) for health damage were reported not in monetary units, but in terms of loss in disability-adjusted life years (DALY) per kg of emission. However, the EI-99 health damage factors were also converted to dollar figures in Table 6.12, to facilitate comparison with monetary damage estimates from other studies, by assigning a literature-based monetary value on their original damage estimates.

Cost estimates and weighting for the different air pollutants vary among the different studies in Table 6.12, which can be attributed primarily to the differences in geographical reference and assumptions. As in our earlier discussion, broad-based cost estimates (e.g., for overall United States, Europe, or OECD member states) tend to be lower than specific urban estimates (e.g., for Los Angeles).

There exists a certain degree of consistency among the studies. All sets of estimates clearly point to airborne particles (PM_{10} is the common categorization among the different studies in Table 6.12) as having among the highest marginal damage costs per kilogram emitted. Carbon monoxide (CO) is also consistently ranked the lowest in terms of cost per kilogram emitted. However, the order of weighting differs among the different studies for the other pollutants, as are the weightings themselves. Thus, in using societal costs in weighting various environmental impacts, one must select one or a few sets of estimates that best match the project's geographic settings as well as one's corporate philosophy. Uncertainties in the estimates must be properly taken into account through sensitivity analysis.

6.2.4.7 Conclusions. Estimations of societal costs (and benefits) can be meaningful sustainability indicators for management in predicting where future costs may arise, and they can be useful in facility design and optimization with respect to "rolling up" environmental impact metrics to include these potential future costs. The process of estimating societal costs, while methodologically complex, is most important not in the acceptance of a final numerical value; that can always be challenged. Instead, they should be viewed in terms of order-of-magnitude and direction of societal concerns. Understanding *this*, looking differently at how business decisions impact the communities from which they are granted license to operate and to grow, and engaging those who are making business decisions to recognize that there are real costs to the company associated with externalities, is an important step in ensuring sustainability.

Acknowledgements. We gratefully acknowledge Gulf Coast Hazardous Research Center (GCHSRC Project 051LUB2810 Extension), which partially funded this work.

6.2.5 Valuing Ecosystem Services

MITCHELL MATHIS

Houston Advanced Research Centre (HARC)

What is the value of nature? Philosophers, ethicists, theologians, and a wide spectrum of social and even natural scientists have responded to this question with a variety of answers based on an equally varied set of frameworks and approaches. These responses are often categorized as either "anthropocentric," in which value is derived from some form of interaction with humans, or "biocentric," in which nature and the organisms of which it is comprised have intrinsic value apart from humans, and where humans are no more or less valuable than any other organism. The response by the field of economics falls within a specific anthropocentric approach known as "utilitarianism." From the perspective of the utilitarian approach, nature has value insofar as it yields benefits or satisfaction to humans. Thus, the notion of ecosystem services – that is, the wide array of processes, functions, and resources provided by the ecosystem that benefit humans – fits squarely within the utilitarian approach to value.[8]

Ecosystem services are closely linked with the existence of "natural capital," discussed in Section 4.5, in much the same way that interest income is linked to a financial asset. In either case, a given amount of the asset, sometimes generically referred to as "the stock," generates a certain flow of benefits, while increasing or decreasing the amount of the asset, which, in turn, changes that asset's capacity to generate benefits.

As the many beneficial functions provided by the ecosystem become better understood, and as the nonmarket benefits the ecosystem bestowed upon humans become more widely appreciated, significant research attention has begun to focus on the economic value associated with ecosystem services (Costanza *et al.*, 1997; Daily, 1997).

As with markets, economics generally defines the benefits that nature provides in terms of goods and services. Some of these goods are inanimate, such as water, rock, oil, and minerals; others are the result of living ecosystems, often referred to collectively as "ecosystem services." These beneficial ecological functions include the provision of renewable natural resources such as fish and timber, as well as a host of other valuable services ranging from erosion and flood control, to pollination, to the mitigation of a wide range of pollutants, to the provision of scenic beauty. The flows of services provided by the ecosystem can be categorized into four types: (1) raw material used as production inputs, (2) life support, (3) recreational amenities and aesthetics, and (4) waste assimilation (Freeman, 1993). Many of the goods, and most of the services that nature provides, are conferred directly to their human beneficiaries by nature, without passing through markets.

---

[8]DeGroot *et al.* (2002) establish categories of ecosystem functions that yield ecosystem services. These include regulation of ecological and natural processes, habitat, production, and information.

Because markets for many, if not most, ecological benefits do not exist, no market prices exist to reflect their use or nonuse values. The absence of a price is not inherently undesirable; however, the lack of a price to indicate the economic value (or loss of value) of such ecological benefits makes it difficult to incorporate these benefits into planning and decision-making processes. Consequently, many of the valued natural resources and ecological services provided by nature "for free" are overlooked and are undervalued in both the private and public sectors.

Over the past half century, the field of economics has developed a rich and extensive literature that advances the theory and methodology of estimating values associated with the myriad unmarketed yet economically valuable benefits provided by nature. While the literature is vast, the number of approaches available to estimate the economic value of ecosystem services remains limited to only a handful, each with its own strengths and weaknesses (for comprehensive reviews, see Mitchell and Carson, 1989; Freeman, 1993; Smith, 2000; Mathis et al., 2003). These include replacement costs and avoided costs analysis, factor income analysis, hedonic pricing, contingent valuation, and travel costs analysis. A discussion of these approaches follows later in this section. Although there is often considerable debate about the methods used and the accuracy of the values derived, there is little doubt that the unpriced economic value of ecosystem services is enormous.[9]

6.2.5.1 Economic Value.

Considerable confusion often exists regarding the terms "economic value," "market value," and "market price." Each of the terms refers to quite different concepts, but they are nonetheless often mistakenly used interchangeably. Thus, before moving forward with the discussion of the economic value of water to sustain the ecosystem, it is useful to discuss what is meant by the term "economic value." Economic theory contains a number of concepts pertaining to value. Some of the most fundamental of these notions are represented in Figure 6.17, which presents a simple depiction of a "market" in which the consumer and the producer interact to generate an equilibrium price, P^*, and a quantity, Q^*. This familiar graph contains a variety of important information regarding economic value, derived from certain characteristics of consumers and producers, and the relationship of these characteristics with the market.

Marginal Costs and Marginal Willingness to Pay. Consider first line segment AB, which maps out the consumer's "marginal willingness to pay" for each successive unit of the goods in question to form the "demand curve." Its downward slope reflects "decreasing marginal utility," that is, the notion that the value to the consumer of each successive unit of the goods decreases the more units of the goods the individual obtains. This line is also often referred to as the "marginal benefits" curve. Turning now to the producer (the firm), consider line segment CD. This line

---

[9]Costanza et al.'s work on "The value of the world's ecosystem services and natural capital" in 1997 sparked an array of debate over valuation methodology as well as its usefulness. *Ecological Economics*, April 1998, was dedicated primarily to the valuation of ecosystem services and a response to Costanza's article that had appeared in *Nature*.

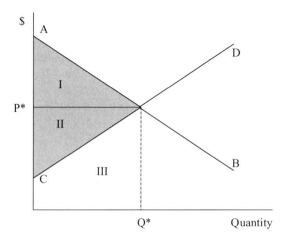

Figure 6.17. Framework of economic value.

represents the firm's "marginal costs," that is, the costs associated with producing each successive unit of the good. The marginal cost curve's upward slope reflects the idea that firms will utilize "low hanging fruit" first, and as more units of the goods are produced, accessing "fruit higher in the tree" increases the cost of production. The marginal cost curve is also the foundation of the familiar "supply" curve.

Market Price. When the consumer and producer come together with "mutual coincidence of wants" to form a market, the consumer will accept a price no higher than his willingness to pay, and producers will accept a price no lower than their cost of production (willingness to sell). The equilibrium market price P^* occurs at the number of units Q^* where the marginal benefits to the consumer equal the marginal costs to the producer. In other words, price simultaneously matches the consumer's willingness to pay and the producer's willingness to sell, thus "clearing the market"; that is, at (P^*, Q^*) there is no gap between marginal willingness to pay, marginal willingness to sell, and the market price. No trades for additional units of the goods will occur because the price exceeds the consumer's willingness to pay, and falls below the producer's marginal cost. It is important to recognize that the market price P^* reflects *only* the value of the last unit Q^* exchanged in the market, and this value is identical for both the consumer and the producer.

Because the market price increases or decreases according to changes in supply and demand – for example, a decrease in supply or increase in demand will both lead to a higher price – price is relied upon as an indicator of economic scarcity, and is sometimes even referred to as "scarcity value." The scarcer the goods, the higher the price, and vice versa.

Economic Surplus. For each unit exchanged in the market previous to Q^*, a gap exists between the consumer's willingness to pay, the producer's marginal costs of production, and the market price at which those units were exchanged. The difference between the consumer's marginal willingness to pay curve and the producer's marginal cost curve for each successive unit of the goods is referred to as "marginal surplus" (or marginal net benefits). This marginal surplus is divided between the consumer and the producer. For each unit, the consumer surplus is the amount by which his willingness to pay exceeds the market price, while the producer surplus is the amount by which the market price exceeds the marginal cost. Total consumer surplus is found by summing up the marginal consumer surplus for each unit of the goods up to Q^*, and is represented in Figure 6.17 by shaded triangle I. Similarly, total producer surplus is represented by shaded triangle II. Total surplus is the combination of both consumer and producer surplus, represented by both shaded triangles, and can be viewed as the total net benefits to society achieved by the production and consumption, sale and purchase, of Q^* quantity of the goods. In the standard neoclassical economic model presented here, it is total surplus that is typically used as the representation of economic value. Total benefits, that is, economic value, are maximized at the point where marginal benefits (willingness to pay) are equal to marginal costs, and the market outcome is said to be "efficient" because no further trade can be made that increases these total benefits. The maximization of total benefits is one of the fundamental objectives of the economic decision-making process.

Observe that economic value, defined as total surplus, is *not* the same as market value. Market value is simply the producer's revenue – the number of units sold, Q^*, multiplied by the price, P^*. While market value includes the producer surplus, represented in Figure 6.17 by triangle II, it also includes the cost of producing Q^*, represented by area III. Moreover, it does not include any surplus that accrued to the consumer (triangle I).

It is also important to distinguish between economic value and "economic impact." Economic impact is a broad concept often used to understand the "pros" and "cons" of changes associated with implementing a project or policy, in terms of the economic system that is affected. Analysis of economic impact can include changes in economic value, as described above, but economic impact is *not* the same as economic value. It typically includes estimates of "direct" effects (e.g. changes in direct revenues and income, changes in the number of jobs) and may also include estimates of secondary effects that result as direct effects work their way through the economic system. For example, a policy or project that yields direct negative impacts on the local fishing industry may also yield negative secondary impacts on the coastal community that provides the fishing industry with a variety of goods and services. When considering increases or decreases in income/revenue, a "multiplier" is often used to capture some of these secondary effects.

6.2.5.2 *Approaches for Estimating the Economic Value of Ecosystem Services.* As described above, to estimate the value of ecosystem services that pass through

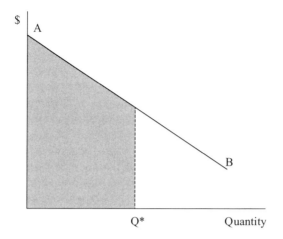

Figure 6.18. Economic value of nonmarket ecosystem services.

markets, it is necessary to have information about both producers' marginal costs and the consumers' marginal willingness to pay.[10] However, because most ecosystem services are nonmarket benefits conferred directly to the consumer, no human producer is necessary. Consequently, there is no marginal cost curve (in the traditional sense described previously) associated with nonmarket ecosystem services. Thus economic value depends only on consumers' willingness to pay for such services. Indeed, most of the approaches used to estimate the economic value of nonmarket ecosystem services are based on information about the willingness to pay for those particular services. In particular, economists seek to estimate the area under the willingness-to-pay (demand) curve, as a measure of total net benefits (see Fig. 6.18). Two important points are worth making here. First, consumers in this context include any entity that derives benefits from the ecosystem service, whether it is an individual or a firm. Secondly, willingness to pay does not imply that the beneficiaries of ecosystem services actually pay for them; in fact, in most cases we do not. Instead, it represents the value to consumer of the benefits they receive.

Estimating the economic value for nonmarket ecosystem services presents a unique challenge because no *direct* evidence of value can be observed in *actual behavior*. Thus, methods that seek to estimate economic value based on willingness to pay generally fall into one of two categories: those that use indirect evidence from observed actual economic behavior, and those that use direct evidence from hypothetical behavior.[11]

- *Travel Cost (TC)*: Service demand may require travel, whose costs can reflect the implied value of the service; recreation areas attract distant visitors whose

[10]Producer costs include the costs of converting the ecosystem service into a marketable good, as well as bringing the good to market.
[11]More technically, these two approaches are respectively based on "revealed preferences" and "stated preferences."

value placed on that service may be at least what they were willing to pay to travel to it. Willingness to pay for the ecosystem service is revealed indirectly through actual expenses incurred during the trip.

- *Hedonic Pricing (HP)*: Service demand may be reflected in the prices people will pay for associated goods; housing prices at beaches exceed prices of inland homes. Willingness to pay for the ecosystem services is revealed indirectly through the willingness to pay higher prices for property that receives a higher level of ecosystem services.

- *Contingent Valuation (CV)*: Service demand may be elicited by posing hypothetical scenarios that involve some valuation of alternatives; people would be willing to pay for increased fish catch or deer bag. Willingness to pay is obtained directly from the consumer in response to a hypothetical context.

Another type of approach is based on estimating the costs of "replacing" the ecosystem service if that service were not available. This approach has two variations: avoidance cost and replacement cost.

- *Avoided Cost (AC)*: Services allow society to avoid costs that would have been incurred in the absence of those services; flood control avoids property damages, or waste treatment by wetlands avoids health costs.

- *Replacement Cost (RC)*: Services could be replaced with man-made systems; natural waste treatment can be replaced with costly treatment systems.

A third type of approach – referred to as the production or factor income method – recognizes the important role ecosystem services often play in the production of market goods. In particular, the ecosystem service is viewed as an intermediate production input. Thus, the ecosystem service contributes a certain portion of the final market good.

In addition to values derived from various utilitarian benefits of nature, that is, ecosystem services, economics can also incorporate concern for the survival or well-being of other species and various other aspects of nature, with which no direct benefits occur. Within the framework of economic value, such concerns are generally expressed as "nonuse" values, because the individual does not consume or otherwise directly benefit from the "good." Sources of nonuse values might include the existence of rare or diverse species of animals, unique natural environments, areas of pristine habitat, or even a way of life, such as family farms or indigenous villages. These values may seem less tangible because they derive from the satisfaction the individual obtains from knowing that such aspects of nature exist, and/or will continue to exist, without actually experiencing them at that moment. However, they can be an important component of nonmarket value associated with nature.[12]

---

[12]Nonuse values can be divided further into "existence," "bequest," and "option" values. Option values are motivated by the desire to preserve the option of the individual to enjoy the environmental benefit at some point in the future, even though the individual may not currently be doing so. Bequest values reflect the desire to leave an environmental legacy, that is, to preserve the option for others in the current

6.2.5.3   Value of Ecosystem Services and the Chemical Industry. Estimates of the value of ecosystem services are important information that can help to incorporate the benefits nature provides into a variety of decision-making processes. Two relatively well-known examples of how such estimates of economic value can be useful (discussed in greater detail elsewhere in this book) are costs–benefits analysis and environmental accounting. Thorough costs–benefits analysis should include estimates of changes in "intangible" nonmarket economic values, in terms of losses or gains associated with resulting ecological impacts and changes in ecological benefits. Economists widely recognize that the omission of nonmarket economic values can result in assessments of economic net benefits that are highly skewed and inaccurate.[13] To implement environmental accounting, whether at the global, national, state, or local level, the need for estimates of economic value of both assets (natural capital) and flows of benefits (ecosystem services) is obvious.

With regard to the chemical industry in particular, certainly many basic production inputs to the chemical industry are derived from inanimate or living natural resources, and are often processed and exchanged as market goods. Thus, certain ecological services may be linked to the chemical industries through the market forces of supply and demand. In such cases, the market price provides important information about the economic value of these services, which is fundamentally important to decision-making processes throughout the industry. However, a variety of nonmarket, hence unpriced, ecosystem services are tied to the chemical industry either as beneficial inputs to the production process, or through negative ecological impacts as a result of byproducts of the production process or of the chemical product itself.

As with firms within any industry, chemical firms that are pursuing sustainable development goals must contend with two generally opposing forces: (1) the firms' drive to reduce costs and (2) the necessity, from a societal perspective, of incorporating important unpriced ecosystem costs and benefits into their decision-making process. Devoting the necessary resources toward understanding the value of ecosystems services, both those that provide benefits to the firm as well as those that are negatively impacted, is an important step. However, this must be carried out in tandem with another, and perhaps more difficult, step: to recognize and incorporate those values into the firms' internal decision-making process.

---

generation, and/or those in future generations to enjoy a given environmental benefit. Finally, existence value is based on the satisfaction derived from simply from knowing that a given element of nature exists.

[13]In general, the economic impacts of a change in the allocation of water will make some people better off and others worse off, and these losses and gains are not distributed equally across neighborhoods, cities, counties, regions, and states. It is therefore important to examine where the economic impacts of a project or policy occur. Although economic theory can contribute extensive insights regarding efficiency and the objective of maximizing total benefits, it provides relatively little guidance as to whether one distribution of losses and gains is better than another.

6.3 AUDITING SUSTAINABILITY PERFORMANCE

6.3.1 Introduction

KAREN L COYNE

CoVeris, Inc.

A key component of developing and implementing a sustainability program within an organization will entail a periodic review of that program focusing on "How are we doing?" and "Where do we need to make improvements?" These are two basic tenets behind the management system function known as auditing, which helps an organization determine how they are doing in their journey along the path to sustainability.

Before embarking on a sustainability audit, the support and involvement of the CEO is essential to ensure alignment of their direct reports over the audit's business value (Waddock and Smith, 2000). There are also a myriad of questions an organization must first address, such as:

- What standard shall we audit against?
- Who is driving the need for a sustainability audit?
- What is the role of third-party assessors?
- How can management be supporting a sustainability evaluation process to convert core values into leading and profitable operating practices?
- What are some EHS management system strategies for conducting Responsible Care Management System®, RC14001 and ISO 14001 audits?
- What is an example of a U.S. state-level program?

In this chapter we will address these considerations on sustainability auditing, presented from a variety of perspectives.

Section 6.3.2, "Sustainability Auditing," from the perspective of an independent third-party verifier, introduces the basic components of a sustainability auditing program, as well as the role of external, third-party auditing and how that may impact an organization with a sustainability program. Section 6.3.3 covers "Corporate Responsibility Auditing: Assuring What Companies Say to the Public is Truthful," while Section 6.3.4 tackles "Auditing Responsible Care® Worldwide." Section 6.3.5 "EHS Management System Audit Strategies for RCMS, RC14001 and ISO 14001," includes information from three EHS consulting companies who conduct auditing, and looks at various management strategies for auditing, including EHS management approaches to conducting Responsible Care Management Systems®, RC14001 and ISO 14001 audits, and how Responsible Care® is audited worldwide.

In Section 6.3.6, "New Mexico's Green Zia Environmental Excellence Program: Third-Party EMS Performance Auditing," one state's innovative management systems program will be reviewed: New Mexico's Green Zia Environmental Excellence Program.

The "Directory of Standards and CSR-Related Organizations," as compiled by The Future 500, may be found in Appendix 2, and is a summary of corporate social responsibility standards and organizations compiled by the Future 500 organization.

6.3.2 Sustainability Auditing

KAREN L COYNE, CPEA[14]

CoVeris, Inc.

A key component of developing and implementing a sustainability program within an organization will entail a periodic review of that program focusing on "How are we doing?" and "Where do we need to make improvements?" These are two basic tenets behind the management system function known as auditing, which helps an organization determine how they are doing in their journey along the path to sustainability. This section will describe what the audit process is and how an organization may audit against a broad-reaching concept such as sustainability. Specifically, we will address the following questions in this section:[15]

1. What is a sustainability audit?
2. What standards or guidelines are being used to audit against?
3. How and why are sustainability audits different from environmental, health and safety (EHS) audits or management system audits?
4. What are the drivers for conducting sustainability audits?
5. What are the impacts of the Sarbanes–Oxley Act of 2002 and corporate governance forces?
6. What is the role of third-party verifiers[16] and how does that impact companies with implemented sustainability programs?
7. Is there a professional certification for sustainability auditors?
8. What are the steps to implement a sustainability audit process?

6.3.2.1 What is a Sustainability Audit? Before specifically addressing sustainability audits, it will be helpful to review some basic terminology,[17] such as what is an audit? Simply put, the audit function is a systematic evaluation process. The audit activity is directed at verifying an organization's status with respect to explicit criteria. An audit is distinct from other types of evaluations that involve conclusions

---

[14]CPEA is the certification acronym for Certified Professional Environmental Auditor.

[15]The closely related topic of audit reports and sustainability reporting is covered in Section 6.4 of this chapter, entitled "Reporting Sustainability Performance."

[16]In this section, the terms auditor, verifier, assessor, and assuror are used interchangeably and have the same meaning.

[17]Each organization will utilize different terms for the same basic function. In implementing an audit process for a sustainability program, the following terms may also be utilized to describe the audit function, dependent upon regional, cultural, and/or linguistic differences: assurance, verification, measurement, assessment, or review.

based on a limited evaluation or, simply, professional opinion (The Auditing Round-table, 1993). It is an event in time (a "snapshot") measuring performance (or lack thereof) against an established set of audit criteria.

The audit function has been defined by the professional association for EHS auditors as "a systematic, documented process of objectively collecting and evaluating factual information in order to verify a site or organization's [. . .] status with respect to specific, predetermined criteria" (The Auditing Roundtable, 1997). The audit criteria are "specific measures or requirements against which the auditor tests and evaluates the information collected as a part of the audit process" (The Auditing Roundtable, 1997). Those criteria may be regulations, management standards, performance standards, performance indicators and metrics, principles, guidelines and/ or corporate directives, to name a few.

The audit process follows a prescribed approach involving the following major steps (The Auditing Roundtable, 1997):

- Define the audit program objective and scope.
- Define the audit program organization. This should include defining the entity conducting the audit, the staffing plan, the authority of those staff, the minimum qualifications of audit program personnel, and the selection criteria for the audit team.
- Select the sites to be audited. This should include the frequency of the audits and the methodology behind site selection.
- Select the audit protocols, checklists, or guides to be utilized during the audit. This should include how the selected document was developed, approved, updated, and retained for future records.
- Define the pre-audit activities. This should include planning on-site activities for meetings, file reviews, site tours, and interviews.
- Establish procedures for audit reporting and document management. This should include defining the audit report content, timing, and distribution, as well as how procedures on how documents generated during the audit will be managed.
- Establish procedures for post audit corrective action and tracking of findings. This should include responsibilities and timing for follow-up and mechanisms to remedy any nonconformance identified.
- Define quality assurance processes to be built into the audit process.

A sustainability audit builds on these basic definitions of the audit and audit process and requires measurement and tracking targeted to the three dimensions of any sustainability program. That is, the sustainability audit process measures value, and subsequent progress, in the three dimensions of sustainability: social, environmental, and economic. The values are measured through performance metrics against a set of performance indices (or criteria), or against a set of sustainability guidelines (both are discussed later in this section). These values can often be difficult to quantify. A sustainability expert in Australia succinctly addressed the core issue of measuring sustainability by stating "I think of sustainability as a broad index or set of indices. That is, I judge an action or an activity on a continuum between non sustainable and sustainable. We all embrace sustainability at the macro level, but we are having trouble applying it at the micro level" (Greenfield, 2000).

6.3.2.2 What Standards or Guidelines are Being Used to Audit Against? At the time of publication, there is not yet a global performance standard or management system scheme for what constitutes "sustainability" or "sustainable development." The International Organization for Standardization (ISO) has decided to develop a related Social Responsibility (SR) guidance (ISO, 2004); however, it is not clear the extent to which this may address the three dimensions of sustainability.[18] As noted in the press release regarding this news, the following statement was made:

> In deciding to develop an SR guidance standard, ISO emphasizes that it is intended to add value to, and not to replace, existing inter-governmental agreements with relevance to social responsibility, such as the United Nations Universal Declaration of Human Rights, and those adopted by the International Labour Organization (ILO), and other UN conventions. Furthermore, it recognizes the need to develop an agreement with ILO on cooperation between the two organizations in the area of social responsibility.
>
> —ISO, 2004

Owing to the lack of a common external standard, the rapidity with which the movement towards sustainability has been developing, and a confusion regarding overlapping terminology, there is a growing proliferation of a broad range of performance standards, guidance documents, principles, guidelines, and frameworks related in one way or another to sustainability (Coyne, 2002). These have been developed by various stakeholders such as governmental, nongovernmental, or advocacy organizations; business or trade associations; or industry groups (BSR, 2000; Coyne, 2002). Some focus exclusively on the social and ethical portions of sustainability, while others include a broader focus including labor, human rights, environment, and corporate governance. Some of these contain principles applicable to an organization's sustainability program, and others are targeted towards the external reporting of these programs. As previously mentioned, none yet combine all three areas of sustainability under a single standard or management system scheme.

The list of these efforts is growing almost monthly; however, a partial, alphabetical listing of the most well-known principles, guidelines, and standards is as follows:[19]

[18]The decision was made by ISO at a senior management meeting held 24–25 June following the input provided before and during the international conference on social responsibility held 21–22 June 2004 in Stockholm, as well as through the efforts of the ISO SR advisory group's analysis and resource report.
[19]See Appendix 2 for a Directory of Standards and CSR-Related Organizations by Future 500. For a useful comparison document outlining the major guidelines and standards in this area, please see *"Comparison of Selected Corporate Social Responsibility Related Standards"* (BSR, 2000).

- *AA1000S Assurance Standard* (AccountAbility, 2003) is an assurance standard developed through AccountAbility, an international membership organization.[20] "The standard is designed primarily for assurance providers in guiding the manner in which they provide assurance. The AA1000 Assurance Standard covers the principles that define a robust and credible assurance process, the essential elements of a public assurance statement, and the independence, impartiality and competency requirements for assurance providers. The Assurance Standard can be used for stand-alone assurance, but can be best understood and used in conjunction with the rest of the AA1000 Series and particularly the AA1000 Framework" (AccountAbility, 2003). *AA1000 Framework* (AccountAbility, 1999) is the initial foundation and general management framework upon which the rest of the AA series is built.

- *Caux Round Table Principles for Business* (Caux Round Table, 1994) is a global set of aspirational principles developed by business leaders for responsible corporate behavior based on the two ethical ideals of kyosei and human dignity.[21] There is a greater focus on business conduct, community involvement and corporate governance, and a lesser focus on environmental and human rights (as distinguished from human dignity) issues (Coyne, 2002).

- *Coalition for Environmentally Responsible Economies (CERES) Principles* (CERES, 1989) includes ten principles for voluntary organizational endorsement, the purpose of which is to "not only formalize their dedication to environmental awareness and accountability, but also actively commit to an ongoing process of continuous improvement, dialogue and comprehensive, systematic public reporting." These are one of the oldest sets of principles and were originally named the Valdez Principles (CERES, 1989).

- *Corporate Responsibility Audit*^{TM} (SmithOBrien, 1995) is a systemic, cross-functional assessment of corporate culture and operating practices, applies over 100 performance indicators in corporate governance, quality management systems, environmental performance and energy conservation, human resources, including labor relations, human rights, community involvement, and stakeholder collaboration to evaluate a company's effectiveness at adhering to its core values, industry or international standards, and stakeholder expectations.

- *The Global Sullivan Principles* (Sullivan, 1999) is an aspirational standard, which includes eight principles on labor, ethics, and environmental practices for global multinationals (BSR, 2000). An annual public pledge by endorsing organizations is required.

---

[20]AccountAbility is "an international, not-for-profit, professional institute dedicated to the promotion of social, ethical and overall organizational accountability, a precondition for achieving sustainable development" (AccountAbility, 2003). The AA1000S is available on the Institutes' website through a no-charge download.

[21]"The Japanese concept of kyosei means living and working together for the common good enabling cooperation and mutual prosperity to coexist with healthy and fair competition. 'Human dignity' refers to the sacredness or value of each person as an end, not simply as a mean to the fulfillment of others' purposes or even majority prescription" (Caux Round Table, 2004).

- *Organisation for Economic Development and Cooperation's Guidelines for Multinational Enterprises* (OECD, 2000) incorporates voluntary principles and standards endorsed by over 33 countries for responsible business conduct by multinationals covering human rights and environmental issues.

- *The Principles for Global Corporate Responsibility: Bench Marks for Measuring Business Performance* (ICCR, 2003) is a set of performance standards and expectations for corporate behavior covering a wide range of over 60 issues that ICCR feels are fundamental to responsible corporate actions, including environment, child labor, and corporate governance.

- *American Chemistry Council (ACC) Responsible Care® Management System* (ACC, 2004) "replaces the current practice of applying six [Responsible Care] Codes (e.g., community awareness and emergency response, distribution, employee health and safety, pollution prevention, process safety and product stewardship) with a combined 106 management practices. Instead, relevant aspects of the existing Codes are subsumed into a RCMS that is based on benchmarked best practices of leading private sector companies, initiatives developed through the Global Environmental Management Initiative, International Standards Organization and other bodies, and requirements of national regulatory authorities" (ACC, 2004).

- *Social Accountability (SA) 8000* (SAI, 1998) is a certifiable, auditable standard containing nine principles focused on labor and human rights for international workplaces. Many of the International Labor Organization's workplace standards were incorporated into SA8000.

- *Sustainability Reporting Guidelines* (Global Reporting Initiative, 2002) are a set of voluntary sustainability reporting guidelines, based on (and with similarities to) financial reporting.

- *United Nations (UN) Global Compact* (UN, 1999) is a voluntary corporate citizenship compact covering nine principles in the areas of human rights, labor, and the environment. It includes specific practices for endorsing organizations to enact in both internal corporate practices and complementary external public policy initiatives.

The proliferation of guidelines and standards, and their inconsistent coverage of the core principles of sustainability, remains a key reason why many organizations have opted to create their own guidance or principles. In a recent survey of 107 multinationals conducted through a study commissioned jointly by the World Bank and its financial arm, the International Finance Corporation, survey participants were asked to identify what guidelines their organization used to determine corporate social responsibility. Fifty-one percent of survey respondents identified their own organization's code of conduct, while one-third adhered to an external code or standard (WBCSD, 2004a). Until a globally accepted, consensus-based standard is developed, such as that which may form out of the efforts of the International Organization for Standardization, there will likely be little in the way of uniform practice or approach.

In addition to the proliferation of guidelines and standards, much research has also occurred in the (different, yet related) field of sustainability performance indices and associated metrics,[22] both within organizations as well as externally through governmental and nongovernmental organizations (NGOs). Performance indices for measuring sustainability or assessing sustainability performance may be developed by each organization, based on organizational-specific sustainability goals, risks, opportunities, and/or commitments. Alternatively, the performance indices and associated metrics may be developed by using external indices with company-specific customization.

A well-known example of the externally derived sustainability indices and metrics may be found as part of the standardized sustainability reporting process under the Global Reporting Initiative (GRI). In the most recent set of sustainability reporting guidelines, a set of hierarchical, core indicators has been developed, whose purpose is "to provide information about the economic, environmental, and social impacts of the reporting organization in a manner that enhances Comparability between reports and reporting organizations" (GRI, 2002).[23] These indicators were developed as one way to measure economic, environmental, and social performance of organizations. In the context of developing and implementing a sustainability auditing program with explicit, predetermined criteria, the GRI list of indicators may be a place for organizations to start the process of defining their own indicators and setting up the data collection efforts needed to track those indicators.[24]

Of the three categories of sustainability performance measurement, the social area may be the most difficult to address. Where financial and environmental indicators lend themselves easily to quantitative performance measurement, "many of the social issues that are the subject of performance measurement are not easily quantifiable" (GRI, 2002). Social indicators must often rely on qualitative measures of the organization's operations. This is an area where further research will be needed. An example of the core social indicators and metrics from GRI's Sustainability Reporting Guidelines is illustrative of this difficulty:

GRI Core Indicators for Social Performance (GRI, 2002):

Area: Labor Practices and Decent Work – Aspect: Health and Safety

Core Indicator: Practices on recording and notification of occupational accidents and diseases, and how they relate to the ILO Code of Practice on Recording and Notification of Occupational Accidents and Diseases

---

[22]See Section 6.1, "Assessing Impacts: Indicators and Metrics."

[23]The GRI Core Indicators are defined in a hierarchy of category (highest level), aspect, and indicator (the most detailed level). The detail of the indicators for each of the listed aspects may be found in GRI's 2002 *Sustainability Reporting Guidelines* (GRI, 2002).

[24]The GRI Core Indicators are heavily weighted towards the social indicators, and less so towards environmental and economic indicators, respectively. Therefore, the GRI list of indicators may simply be a starting point for some organizations, and the environmental and economic indicators enhanced through the application of other indicators developed elsewhere.

Metric: Qualitative description

Core Indicator: Description of formal joint health and safety committees comprising management and worker representatives and proportion of workforce covered by any such committees

Metric: Qualitative description

Metric: Quantitative data – percent of workforce covered

Core Indicator: Standard injury, lost day, and absentee rates and number of work-related fatalities (including subcontracted workers)

Metric: Quantitative data

Core Indicator: Description of policies or programs (for the workplace and beyond) on HIV/AIDS

Metric: Qualitative description

Once the financial, environmental, and social indicators and associated metrics have been developed, they may be used to assess an organization's sustainability performance through a sustainability audit, either internally by the organization, or by external stakeholders such as shareholders or environmental nongovernmental organizations (NGOs).[25]

6.3.2.3 How and Why Are Sustainability Audits Different from Environmental, Health And Safety (EHS) Audits or Management System Audits?

In the United States, professional environmental auditing had its beginnings nearly 25 years ago when in the late 1970s a handful of organizations began to review and assess their compliance with specific environmental regulatory programs such as the hazardous waste management requirements promulgated under the Resource Conservation and Recovery Act (RCRA). The continued growth of regulatory programs and other industry-specific initiatives led to numerous, independent, single-focus audits covering health and safety, environment and transportation, and Responsible Care®. In the 1990s, a transition began where the health and safety auditing function was integrated with the environmental auditing function, and was consolidated under one organizational EHS auditing entity in most corporations.

Dow Corning Corporation's audit history is illustrative of this process. "In the beginning, Dow Corning had corporate values but no auditing" (Kimball, 2002). Health and safety audits were begun in the early 1980s; environment and transportation audits in the late 1980s; and commitment to Responsible Care® in 1988. Integrated environmental, health, safety and transportation audits began in 1999–2000, and in 2002 Dow Corning began piloting integrated audits with Responsible Care® (which included EHS&T, Emergency Response, Community Awareness, Product Stewardship, Process Safety, Loss Prevention, and TSCA); business process (including financial); information technology; and quality (Kimball, 2002). This

---

[25]The impact of third-party auditors is addressed later in this section.

maturation process of an integrated internal audit function is characteristic of many organizations' internal audit programs today.

With the arrival of the ISO 14001 Environmental Management System Specification in 1996 (ISO, 1996), some corporate EHS auditing functions took on the additional responsibility of auditing management systems beyond those that may have already been developed under programs such as Responsible Care®. While organizations in Europe had traditionally approached EHS auditing primarily through a management systems approach under earlier-generation standards such as BS7750 (and not a regulatory focus, which developed later), this was a new slant for U.S.-based organizations. The audits were now based on a holistic systems viewpoint and not a single issue regulatory performance viewpoint. What in fact was occurring was "a gradual evolution from a regulation-based, compliance-driven regime to an environmental stewardship process [...]" (STP, 2004).

With the advent of the concept of sustainable development and integration of that concept into the core of an organization's business practices, internal and/or external assurance or verification (i.e., auditing) of implemented sustainability programs became an important task. Sustainability auditing is similar to EHS compliance and management systems auditing in that they are all part of an overarching approach of systematically managing the environmental, health, safety and social aspects and impacts of the organization. That is, they are part of a larger management system even where the specific sustainability aspects have not been labeled as such. A technical report by Det Norske Veritas (DNV) described this relationship well by reporting

> Most companies already have at least one management system, and many companies have well defined activities which actually are CSR-activities but have not been systemized under a CSR perspective. Thus [the plan-do-check-act management system approach] must be adapted to the actual company and must be coordinated with other management systems in the companies.
>
> —DNV, 2002[26]

Despite the similarities that sustainability audits have to compliance and EHS management systems verification processes, there are some important yet subtle distinctions to note. Those distinctions include: the scope or boundaries of the assurance/verification activities, standards of practice, assurance criteria and evidence, and content of the assurance report.

Boundaries or Scope. Compliance audits, management system audits, and early sustainability audits all generally measured activities based on the traditional

---

[26]There are many different ways of looking at the relationship between Corporate Social Responsibility (CSR) and Sustainable Development and the impact on the role of business and industry. However, there are, in our opinion, no major substantial differences between CSR as we have described it in this report and sustainable development as it originally was presented in Our Common Future. Or to put it another way: every aspect of CSR should already be covered by sustainable development' (DNV, 2002).

boundary criteria used in financial reporting. That is, the verification and reporting activities were focused on an organizational slice (facility, business unit or corporate), or on perspectives based on legal ownership and direct control. Perhaps due in part to the influence of environmental management systems developed under ISO 14001, companies began to look beyond the "fence line" and started expanding the definition of the organizational sphere of influence of the organization, activities, and/or products. Defining those boundaries in a sustainability context is a challenging exercise, and the GRI Sustainability Reporting Guidelines recommend extensive interaction with organizational stakeholders to help determine the appropriate boundaries.

Standards of Practice. For EHS compliance and management system audits, standards of professional practice have been developed by various EHS professional audit organizations. These have defined how to perform an EHS audit (The Auditing Roundtable, 1993; BEAC 1999), and also how to design and implement an EHS audit program (The Auditing Roundtable, 1997). For environmental management systems auditing under ISO 14001, ISO created a set of guidelines and audit procedures, initially in ISO 14011:1996, and then later as a unified guideline in ISO 19011:2002 for environmental and quality auditing (ISO, 2002). For sustainability auditing, the only standard currently in place for audit practices is AA1000S. This open, nonproprietary document is "a generally applicable standard for assessing, attesting to, and strengthening the credibility and quality of organizations' sustainability reporting, and their underlying processes, systems and competencies" (AccountAbility, 2003).

Assurance Criteria and Evidence. Assurance criteria and evidence refer to the "specific measures or requirements against which the auditor tests and evaluates the information collected as a part of the audit process" (The Auditing Roundtable, 1997). As outlined earlier in the discussion on guidelines and standards, the assurance criteria for an EHS compliance audit would include regulations or statutes. For a sustainability audit, it could include recognized performance indicator protocols or reporting guidelines. In either case, an objective and evidence-based approach should be used to gather information (either qualitative or quantitative) to determine whether the audit subject (system, organization, activity, event, condition, or information) conforms to the audit criteria. For sustainability audits, the information being evaluated may include more qualitative, than quantitative, evidence.

Content of Audit or Assurance Report. Compliance audit and management system reports are developed based on the standards of practice referenced earlier, and are directed to whoever is the audit client (defined as the organization commissioning the audit; STP, 2004). Sustainability assurance report content also varies depending on the audience for whom the report is intended. The entity conducting the assurance review must determine whether the report is intended for (1) the organization itself for internal use (such as using the indicators and targets in the audit report for strategic and operational decision-making tools and processes), or (2) external stakeholders of the organization for use in interpreting and using the information in a way that is relevant to their specific needs (such as comparing one organization's report with those from other organizations and against relevant standards) (Account-Ability, 2003). Finally, AA1000S outlines three important principles pertaining to report content. These are materiality, completeness, and responsiveness (Account-Ability, 2003). The levels in which these principles are applied are directly related to the degree of assurance being sought.

6.3.2.4 What Are the Drivers for Conducting Sustainability Audits? Owing to

the growing focus today on corporate governance,[27] stakeholders representing a number of perspectives are requiring increased corporate accountability and transparency. Demands are coming from shareholders, employees, and financial institutions to "disclose and discuss a wider and deeper range of sustainable development issues related to [an organization's] activities, products and services" (WBCSD, 2003a). Additional stakeholder communities pressing for issuance of nonfinancial performance reports include competitors, peers, customers, potential investors, the media and lobby groups (AICPA, 2002).

Although organizations have released sustainability reports, "evidence suggests that information contained in sustainability reports is rarely used by either stakeholders (including investors) or by management to inform judgments and actions – the key test of credible and useful communication" (AccountAbility, 2003). There has been a credibility gap, and one way to narrow that gap is through sustainability audits conducted through the third-party assurance process. "Independent assurance is perceived as a key factor in building the quality and credibility of nonfinancial reporting" (PWC, 2004). The move towards independent verification has grown, precipitated both by requirements for such from Sarbanes–Oxley, as well as from the call from nongovernmental organizations for nonbiased, independent verification of corporate performance. There has been some movement in the marketplace to establish companies focused solely on provision of corporate performance verification without apparent conflict of interest from consulting activities. The degree of true separation varies, however, from firm to firm, and the client would be well advised to investigate the corporate structure and degree of independence. A few of these independent verification companies currently exist, such as CoVeris, Inc.,[28] SmithOBrien, and ERM Certification and Verification Services (ERM CVS),[29] and more may soon develop.

In addition to the factors discussed above, there is also a pull from the financial markets to conduct sustainability audits. The call to organizations to more fully provide information on their environmental and social performance is due to developing evidence that shows good performance in the environmental and social arenas often

---

[27]This movement is in reference to the recent United States corporate governance scandals and the resulting law addressed at curbing those abuses, the Sarbanes–Oxley Act of 2002. For full text of this act, see the following link: http://frwebgate.access.gpo.gov/cgi-in/getdoc.cgi?dbname=107_cong_bills&docid=f:h3763enr.txt.pdf.

[28]For contact information, see the link: http://www.coveris-cpv.com.

[29]For contact information, see the link: http://www.ermcvs.com.

means better financial performance. To illustrate this point, a review showed that the Dow Jones Sustainability Index outperformed the Dow Jones Global Index over a five-year period from January 1997 to April 2002 (WBCSD, 2003). "Over half of all companies listed in the German DAX-30 index view sustainable business practices as 'key to the long-term success of the company', according to a recent survey by Susanne Bergius, Sustainability Expert of the Handelsblatt" (WBCSD, 2003a).

A number of large institutional investment bodies also now take into account sustainability reports of organizations when making investment decisions. This comes from the demand for socially responsible investing from church groups as well as from the growing number of mutual funds formed around investments in companies screened for certain social responsibility criteria (AICPA, 2002). Calvert,[30] a large mutual fund company leading the field in socially responsible investing, is seeking greater transparency and disclosure regarding environmental and social performance. It recently requested that a number of publicly traded companies issue a sustainability report based on GRI Guidelines (Calvert, 2004). A review of press releases suggests that this request is not limited to Calvert, and is in fact is a widely covered subject with numerous examples to cite.

For some organizations, the driver for implementing a sustainability program and related reporting comes from an internal management need. The drivers may be to manage business risk and opportunity, support internal decision-making, and/or build trust within various stakeholder communities. Lastly, although the concept of sustainability has largely been a voluntary corporate initiative, there is now a driver from a regulatory perspective. Mandatory requirements pertaining to aspects of sustainability have been implemented in France, Germany, and several Nordic countries (WBCSD, 2003).

6.3.2.5   What Are the Impacts of the Sarbanes–Oxley Act of 2002 and Corporate Governance Forces? The Public Company Accounting Reform and Investor Protection Act of 2002 (also known as the Sarbanes–Oxley Act, Public Law 107-204) was passed after the well-publicized corporate governance failures starting in 2001.The intent of the act was to strengthen corporate governance and financial disclosure rules under the Security and Exchange Commission's (SEC) rules for publicly traded companies.

There are a number of sections in the Sarbanes–Oxley Act that may affect the general world of EHS auditing, including sustainability auditing. Although the legislation and the ensuing Securities and Exchange Commission's (SEC) regulations apply to publicly held companies in the United States, a number of other organizations have also been following these sweeping changes. The act changed what information must be disclosed, how that information must be developed and disclosed, and the consequences for not disclosing fully and properly (Wagner and Cannon, 2003). While the act made no substantive changes in the disclosure requirements that specifically address environmental reporting, it did add new requirements

---

[30]Calvert is the nation's largest family of socially responsible mutual funds, with over $9.7 billion in assets under management (Calvert, 2004).

for certification, attestation, and disclosure that will make EHS audit reports subject to new levels of scrutiny (Wagner and Cannon, 2003). In particular, the following sections of the Act are important to note:

- *Section 404 on Management's Report on Internal Control over Financial Reporting.* Management must provide a statement that they are responsible for establishing and maintaining adequate internal controls over financial reporting. In addition, there must be an annual assessment of the effectiveness of the company's internal control structure and procedures for financial reporting. This information must be certified by the top management of the company.
- *Section 409 on Real Time Issuer Discloser.* This requires disclosure to the public on an urgent basis of any "material" changes in the company's operations and/or financial condition. This information should be any thing that could affect a company's market valuation.

The practical effects of the Sarbanes–Oxley Act on companies are as follows (Wagner and Cannon, 2003):

- Ensure that adequate internal procedures exist to estimate and disclose environmental costs and liabilities;
- Identify and document events and emerging trends in environmental regulation or enforcement that could have a material financial impact on the company's operations;
- Implement an internal reporting system focused on reporting to appropriate personnel; and
- Establish procedures to evaluate and quantify potential environmental liabilities.

6.3.2.6   What Is the Role of Third-Party Verifiers and How Does That Impact Companies with Implemented Sustainability Programs? The EHS assurance and audit function has matured into verification of a broader range of corporate performance and social responsibility criteria, with demands for assessment by independent and objective third parties. Stakeholder expectations further necessitate that there is no perceived conflict of interest between a third party's verification and other consulting assignments. In addition, with an increase in sustainability reporting, such as that outlined under GRI's Sustainability Reporting Guidelines, the need and demand for independent assurance of those reports has been growing. In order to provide independent, third-party assurance as to content, veracity, and accuracy of those reports, a growing industry has been developing providing those independent assurors. These assurors come from a wide range of firm types: consulting and engineering firms, public accounting firms, and certification organizations (AICPA, 2002). Some of those firms have established distinct and separate units dedicated to independent verification of an organization's performance in the areas of environmental compliance and stewardship, health, safety, and social responsibility.

Until very recently, there has been no standardized approach for conducting those assurance audits or for how to account for sustainability; however, several approaches for these have recently been issued. The approaches are very similar to the EHS audit process described earlier in this section. In addition to AA1000S, which was previously described, a second standardized approach to sustainability accounting was issued in September 2003 entitled "Sigma Guidelines – Toolkit, Sustainability Accounting Guide" (The Sigma Project, 2003). Targeted toward the finance function within an organization and sustainability practitioners, the guide provides "information on the variety of approaches that are known to be available for sustainability accounting; alternative frameworks to improve understanding of how these approaches may fit together; information on each approach and how organizations can start to use them; and suggestions of areas that require further development" (The Sigma Project, 2003). Although embryonic in their development, these two standardized approaches form the framework by which independent, third-party verifiers or interested stakeholders will view an organization's sustainability program.

The role of third-party verifiers, therefore, is to provide an independent, non-biased assessment of the content, veracity, and accuracy of an organization's sustainability report or sustainability program. Because stakeholders are requiring increased corporate accountability and transparency, an organization embracing the sustainability perspective should be prepared for scrutiny and assessment of where their sustainability program actually stands. Impacts on an organization from having a sustainability program may vary depending on the industry sector, spheres of operation, stakeholder communities, financial markets and/or geopolitical events; however, a few common impacts may include the following (GRI, 2002; WBCSD, 2004b):

- Having to provide forums for stakeholder consultation;
- Responding to specific stakeholder requests and expectations;
- Expanding, improving, and/or auditing data collection and information systems;
- Securing external expert witness testimony;
- Conducting issue-specific audits;
- Engaging suppliers;
- Deciding how "open" or transparent an organization should be with stakeholders and with competitors;
- Debating the future legal implications and potential for lawsuits for publicizing their adherence to codes of conduct;
- Incurring significant costs (both human and financial resources) to develop, implement, and report on sustainable development management systems.

Once a sustainability audit process has been established, either internally or externally, there is an additional important consideration for the reporting process and third party verifiers that GRI has outlined. This is regarding the auditability of the data. The auditability principle states: "Reported data and information should be recorded, compiled, analyzed, and disclosed in a way that would enable internal auditors or external assurance providers to attest to its reliability" (GRI, 2002). This principle relates to the degree to which data collection, information management and communication systems are able to be examined for accuracy, internally and externally.

6.3.2.7 Is There a Professional Certification for Sustainability Auditors? There currently is no accredited, professional certification scheme specific to sustainability auditors. In the United States, the Board of Environmental, Health and Safety Auditor Certifications (BEAC)[31] offers a Management Systems Certified Professional Environmental Auditor (CPEA) certification. This certification is not specific to sustainability, but does include the GRI Sustainability Reporting Guidelines in its exam for the Management Systems CPEA certification (BEAC, 2004).

There is also voluntary guidance that has been issued with respect to the desirable attributes for a sustainability assurance provider (auditor). In GRI's Sustainability Reporting Guidelines, a number of issues are recommended for consideration by organizations as they select their assurance provider (GRI, 2002):

- Degree of independence, and freedom from bias, influence, and conflicts of interest;
- Ability to balance the consideration of interests of various stakeholders;
- No involvement with the design, development or implementation of the organization's sustainability management system;
- No involvement with the organization's sustainability monitoring efforts or reporting;
- Due professional care, including sufficient time to effectively carry out the assurance process; and
- Competence, as demonstrated through an appropriate level of experience and professional judgment, to meet the objectives of the assurance process.

Finally, there is an accreditation process for social accountability audits under Social Accountability, Inc. This entity accredits organizations – known as certification bodies – to conduct audits, certifying workplaces as complying with SA8000. Accreditation must be earned before a firm can have its staff perform SA8000 certification audits (SAI, 2004).

This is currently an unmet need in the marketplace. Once developed, a universally accepted certification scheme for true sustainability auditors and assurance providers will greatly enhance the credibility and consistency of assurance efforts in the future.

---

[31]"BEAC is an independent, nonprofit corporation established in 1997 to issue professional certifications relating to environmental, health, and safety auditing and other scientific fields. BEAC was originally created as a joint venture between The Institute of Internal Auditors (The IIA) and The Auditing Roundtable (Roundtable). [...] BEAC is a member of the Council of Engineering and Scientific Specialty Boards (CESB), a third-party accreditation board. CESB has granted full accreditation to BEAC's Certified Professional Environmental Auditor (CPEA) certification" (http://www.beac.org/about.html).

6.3.2.8 What Are the Steps to Implement a Sustainability Audit Process? "In 1987, the World Commission on Environment and Development (Brundtland Commission) called for the development of new ways to measure and assess progress towards sustainable development" (IISD, 1997). The refinement of a sustainable development audit process has been ongoing since that initial call to action from the Brundtland Commission; yet, to date, a fully mature audit process for assessing sustainable development has not yet emerged. Several notable efforts have been undertaken, however, which have provided a leap forward in the maturation process.

The earliest, notable effort in this area was the issuance of the "Bellagio Principles for Assessment" through the International Institute for Sustainable Development (IISD, 1997). The document provided a consensus review and synthesis of ideas and insights from then-current, ongoing efforts for undertaking and improving the sustainable development assessment process. Ten basic principles were issued and formed the "Bellagio Principles for Assessment," named after the site of where the meeting took place (Bellagio, Italy). The principles were intended to provide advice in selecting indicators, measuring progress, and interpreting and communicating results.

A white paper produced by the NGO Business for Social Responsibility provides another useful template for the implementation of a verification program (BSR, 2003). Based partially on SA8000 and the GRI Guidelines, it includes the following verification steps (which would be applicable towards the development of a verification process):

- Establish audit principles;
- Define the role of the verifier;
- Assist the verifier in understanding the company and its stakeholders;
- Plan the audit;
- Facilitate data collection and information;
- Seek expert opinions due to the different spheres of activity and the complexity of measurements;
- Publish the verification statement from the assessor; and
- Request that the verifier provide a letter to management separate from the published verification statement.

Finally, a recommendation for implementing a sustainability audit process would be to incorporate or integrate the sustainability efforts into the existing EHS management system scheme. In this way, the sustainability audit process can dovetail into the existing EHS audit process and subsequent management review process. The DNV report entitled "CSR Management Systems"[32] provides very useful guidance in Chapter 4 of that report on establishing a sustainability management system process based on the plan-do-check-act principle (DNV, 2002). In

---

[32] As noted earlier, the CSR term in the context of this report has the same meaning as sustainable development.

addressing the routines for the review of such management systems, the following excellent guidance is provided (DNV, 2002):

> The intention of the review is for management with executive responsibility to confirm the continuing suitability and effectiveness of the system. The CSR Management System including the indicators and the corresponding targets should be evaluated and improved periodically. In addition the Value Statement, the Code of Conduct and the high level CSR goals need to be reviewed regularly. Typical inputs are monitoring records, measurement results, possible deviations and recommendations for improvement. Conclusions from these reviews should be followed up in action plans to ensure continual improvement of the system. Management reviews of the CSR Management System will typically be part of the management review of a complete management system if such exists (e.g. as required in ISO 9000/14000). In addition, the routines must ensure that conclusions from the management review are followed up in and linked to the existing management system activities where appropriate.

6.3.2.9 Conclusion. Successful implementation of a sustainable development mindset in an organization requires true integration of the values and principles behind sustainability, and not a shallow, cursory, stand-alone "project." That is, sustainability is only one small piece of a larger approach to organizational responsibility in the global environment. The audit or assurance process is an important and valuable step in this approach that cannot and should not be overlooked.

6.3.3   Corporate Responsibility Auditing: Assuring What Companies Say to the Public is Truthful

NEIL SMITH AND PAUL SCARBROUGH

SmithOBrien

At a time of unprecedented cynicism about corporate ethics, a company's explicit commitment to corporate responsibility is central to its success, and is increasingly seen by employees and investors as a proxy for general managerial competence. Yet, rigorous examination of how things get done, even in the best of firms, reveals large gaps between what a company tells its stakeholders and how it operates. Thus, corporate leaders, more than ever before, want assurance that the company is actually doing what it reports to the public.

Corporate responsibility auditing, a recognized management discipline first applied at APT Associates in the 1970s and later refined by The Body Shop and other companies in the mid-1990s to evaluate their environmental and social impacts, can help corporations assess how well they do what they say they do, and proactively convert rhetoric into reality.

The premise of corporate responsibility auditing is the recognition by executives that the information provided to shareholders, as well as other stakeholders, on a corporation's performance needs to be truthful and reliable. Inherent in the auditing process is the provision of crucial knowledge that will sustain the public's trust and sharpen managements's collective focus on key nonfinancial value drives (e.g., employee retention, innovation, customer satisfaction, and reduced pollution) and the events that can affect them both positively and negatively. These elements – determinants of future financial results – are easy enough to describe but not always easy to practice.

At a minimum, corporate responsibility auditing can provide a complete picture of how well a company performs against its values, industry or global standards, peers, and the expectations of essential stakeholders. Many companies use the results to put in place, both internally and in partnership with major suppliers, credible and measurable process improvements that *responsibly* drive efficiencies, reduce risk, seize new business opportunities, and strengthen relationships with stakeholders. For many companies, corporate responsibility auditing also provides a verifiable and credible process for reporting to stakeholders on their social and environmental impacts and the effects on profitability.

Advancements in corporate responsibility auditing have also led to its use in risk management, by identifying certain risks, with damage potential, that typically lie unidentified and unmanaged until they are brought to the attention of senior officials by activist groups or the media. Risks, such as employee discrimination, high turnover, lost productivity, and suppliers' labor and environmental practices – which can quickly erode a company's integrity and shareholder value – are rarely on the business agenda of executives, and when they surface, catch company officials unaware.

Corporate responsibility auditing can also identify how capital has been allocated across departments and business units of the company and the level and types of risks to which the capital is exposed. An experienced audit team will advise on whether risks in one part of the company may exacerbate or ameliorate those in another and, in consultation with employees and managers, recommend process improvements, which the audit will undoubtedly yield, where risks are too high. This information can provide management with a clearer understanding of acceptable tolerance levels for key value drivers, and reliable data with which to respond proactively to shifts in the work environment, key markets, and in society.

6.3.3.1 Getting Started

Laying the Groundwork. The typical corporate responsibility auditing process (Table 6.13) has a dozen or so steps to ensure it is thorough and conducted efficiently. At the start, the CEO's evident commitment and involvement are a must. Setting the right tone for what can be learned from a rigorous review, she or he can ensure support from their direct reports on the audit's business value, essential in order for the audit team to freely drill down deep and wide into the organization's practices.

After adopting an auditing framework, companies often form a steering committee of employees and managers to guide and manage the audit process. Some committees also include representatives of key external stakeholder groups. While hiring an independent CSR auditing services firm to conduct the audit, the steering committee's responsibilities include:

TABLE 6.13. Steps in Conducting a Corporate Responsibility Audit

1. Gain CEO commitment, and the support of direct reports.
2. Appoint a steering committee to guide auditing process.
3. Select a qualified, external CSR auditing services firm.
4. Diagnose the readiness of the corporate culture and leadership to adopt more progressive and responsible practices.
5. Evaluate efficacy of governance structure (BOD and senior management) and operating practices, such as environmental practices and energy conservation; quality management systems and supply chain management; human resources and labor relations; human rights; community involvement; and stakeholder engagement/collaboration.
6. Complete internal interviews or focus groups and collect internal operational data, available benchmarking data (e.g., performance information on comparable noncompeting companies, industry, and international standards).
7. Interview external stakeholders.
8. Identify fundamental or underlying reasons why operating performance and corporate/operating goals are not aligned.
9. Compare internal performance data with the performance of peer companies and the expectations of stakeholders.
10. Audit team presents preliminary findings and recommendations to steering committee and CEO for feedback.
11. Audit team completes financial analyses of top three to five recommendations.
12. Steering committee prepares Final audit Report and presents to senior management team and Board of Directors.

- Defining the audit scope, including the key performance indicators against which current practices are assessed;
- Setting desired outcomes and learning;
- Identifying employees to be interviewed, internal documents to be reviewed, and external research of best practices and the websites of watchdog groups;
- Establishing realistic timelines, and internal reporting procedures.

It is critical that key senior officials actively participate on the steering committee and perhaps in the audit process itself. This will help ensure their buy-in of the process and motivate them to implement the resulting recommendations. Involving external stakeholders assures an increased level of objectivity in the process and can begin to build mutually beneficial relationships where they did not exist.

The audit team must include content experts, who have extensive industry experience in the business disciplines being audited, as well as organizational development specialists. The most useful audits rely on interviews and focus groups to ensure depth of information on the company's practices, and perceptions. Participants typically include employees and managers, influential suppliers, customers, regulators, and activist groups. These types of discussions, generally, unveil telling information that survey tools cannot.

6.3.3.2 The Auditing Process. A corporate responsibility audit needs to be systemic in approach and cross-functional in scope. In addition to assessing the readiness of the company's culture and leadership to adopt and sustain more pro-gressive practices, the audit typically focuses on one or more of six key business functions: environmental practices and energy conservation; quality systems (including supply chain management); human resources and labor relations; human rights; governance; and community involvement. Opportunities to collaborate with stakeholders are sought throughout these areas.

A corporate responsibility audit can provide a system for managing change. Thus the audit team should start by assessing the corporate culture and leadership to deter-mine early whether or not the organization has the capacity to change and how best to address obstacles (e.g., people and systems) before assessing the efficacy of the operating practices.

When risks or operating deficiencies are later uncovered, the audit team must think about the underlying causes and associated costs and how best to reduce or eliminate them. Replacing one toxic with another "alternative" one is not enough. Rethinking the process to remove the need for all toxic inputs in the target, and is frequently achieved. Additionally, the team will want to consider the costs and effects of its findings and recommendations across different operating areas.

The most efficient approach to conducting a corporate responsibility audit is to assess some practices company-wide, such as governance, human resources, energy conservation, and community relations. Other business functions, including environmental, health and safety (along with quality systems and production prac-tices), human rights, and supplier relations, should be assessed at a business unit or even plant level, where the audit team, along with these employees, can consider various scenarios for improving performance and reducing costs that may also be replicable later in other facilities. Additionally, in order to accurately analyze the financial effects of making any substantial improvements, the audit team will need to review the raw direct and overhead costs of current practices. These costs are usually readily available at lower-level operating units.

In analyzing the business case for its recommendations, audit teams employ a number of accounting techniques. Because the effects of most corporate responsibil-ity issues cross departmental and divisional lines and because *activities drive costs*, it is necessary to understand how certain activities and practices impact financially and strategically the entire organization.

For these reasons, Activity-Based Costing (ABC), the most accurate method of cost analysis, is particularly helpful in identifying all of the individual costs of an operating practice or an event, such as a chemical spill. In analyzing a spill, for example, the audit team can attach a dollar value to waste of materials and labor, system changes, unused capacity, and the spill's myriad overhead costs, as well as corporate legal, PR, and other administrative support costs, all of which need to be carefully understood in considering process improvements.

For the most part, few of these costs will appear in traditional cost estimates. Although ABC has proved, in most cases, too complex and time-consuming to fully replace most traditional (financial) accounting systems, it can be very success-fully used for special studies to break down activities and estimate their associated costs, provoke change, and make decisions. Even in the most sophisticated companies, the full costs of events, such as chemical spills, discrimination lawsuits, product liability lawsuits, or other litigation and regulatory risks, are very poorly understood. A cross-functional analysis of the problem, including its costs, is normally not done, except by a corporate responsibility audit, in most companies.

Real Options is another valuation method used by audit teams for nonfinancial value drivers, by estimating how the sustainability of the driver will affect the company's financial performance. Based on the fact that variability has an important effect on economic value, particularly in trying to protect value drivers from worst-case scenarios, Real Options infuses a more global and market-focused thought process to quantifying how improvements in the company's social and environmental performance will affect the bottom line. The approach draws upon the knowledge and a consensus of opinions from key experts inside the company (e.g., HR, EHS, or community affairs managers) to forecast the probability of certain events, such as a product breakthrough, or the probable positive and negative financial effects of current or improved practices, including the risk of litigation and available reserve capacity for emergencies. A natural consequence of this type of data review and interviewing is that the internal experts also become more careful and dedicated to the quality of their input.

Although Real Options uses a sophisticated mathematical model to make the numerical calculations, the analysis, an extension of classic financial options theory, is based on the data and opinions provided by the managers themselves instead of having been parachuted in by a consultant-provided ABS study. This method captures the expert judgment of managers and allows it to become a source of competitive advantage over time.

6.3.3.3   Presenting the Audit Results. Once the audit has been completed, the audit team then presents a Preliminary Report to the entire steering committee and the CEO for feedback on its findings and recommendations. The team will want to especially highlight *low hanging fruit* – areas in which simple, more responsible practices would quickly improve the bottom line and begin to demonstrate the business case for adopting more complicated changes later.

These changes can be as simple as adopting practices to reduce paper use, or as complex as suggesting that the HR module of an ERP application be purchased to track diversity issues more cheaply.

The Final Audit Report includes the team's detailed cost–benefit analyses of the three to five most value-added recommendations. The steering committee then presents the audit results to the entire senior management team and the Board of Directors.

It is important to recognize that managers are traditionally reluctant to undertake a rigorous examination of their operating practices for fear that any negative findings will be seen only as an indication of failure. Additionally, good social and environmental performance is often perceived as too costly. Yet, example after example demonstrate that operating responsibly saves money and, in some cases, even creates profitable new opportunities. According to companies that have completed corporate responsibility audits and implemented the recommended improvements, the payback has ranged from six to twenty times the cost of the audit over six months to three years. For example, Dupont, which has been conducting similar reviews for years, reports having saved more than a half billion dollars.

6.3.3.4 Conclusion. For many leading companies, operating responsibly is implicit in their core business strategy, and corporate responsibility auditing is routinely used in maximizing values drivers and in reporting honestly to stakeholders on the company's entire performance.

Others, however, remain unconvinced, and still see operating responsibly as an external extraction of money and management's time.

Experience in corporate responsibility auditing suggests that it can be part of a company's core business model, or an expensive and useless external exercise. When built into the fabric of the business as part of creating shareholder value, it reveals value opportunities not observable otherwise. When tacked on to the fringes of business operations, corporate responsibility auditing functions in a similar fashion to business fads; it becomes an expensive and pointless tactic. So, just like everything else a business does, there is a big, big difference between doing it and doing it well.

6.3.4 Auditing Responsible Care® Worldwide

BRAD VERRICO

Verrico Associates

6.3.4.1 Current International Audit Programs. Currently, the International Council of Chemical Associations (ICCA) does not endorse a single method of Responsible Care® implementation. Nor does it require or endorse a single method to audit Responsible Care®.

American Chemistry Council (ACC) sponsored Responsible Care® certification standards, RC14001 and RCMS Certification, are not limited to the United States, or to chemical companies. There is already an example of a non-ACC member receiving RC14001 certification. A number of other countries have expressed interest in RC14001 and RCMS certification. These include India, Thailand, Brazil, Argentina, Canada, China, Malaysia, and Japan. Many other companies are continuing to express interest. Typically, American-based companies look to include their international facilities in overall Responsible Care® certification programs, although they are not required to by ACC membership obligations.

The Canadian Chemical Producers' Association (CCPA) has the longest-standing Responsible Care® verification program in the world. Its Responsible Care Management System Verification (MSV) preceded the U.S. MSV process, and the current ACC Responsible Care® certification program. The CCPA Responsible Care® verification process has three purposes:

- To help the companies improve;
- To help the industry improve;

- To improve credibility.[33]

Every three years, CCPA member companies are verified by teams of industry experts, public advocates, and local citizens, who write a consensus report summarizing the verification process and players, opportunities for improvement, findings requiring corrective action, and successful practices.

New companies are verified three years after joining CCPA using a comprehensive Responsible Care-in-Place Verification protocol, with subsequent verifications following a more focused protocol. Verification reports are made available to their local communities and other interested parties by the verified company.

6.3.4.2 The Future of International Responsible Care® Audits.

While the CCPA and the ACC are the only current trade-association sponsored Responsible Care® auditing programs, all ICCA-recognized trade associations have requirements for procedures to verify that Responsible Care® is being implemented. Many associations have vigorous self-assessment programs. Some are developing external verification and certification programs.

One notable example is in the United Kingdom. There, the Chemical Industries Association (CIA) has Responsible Care Management System Guidance and a mandatory self-assessment process for its members. On the CIA website, there is a discussion of the current system of self-assessment and the CIA's aims:

> Over 80 per cent of member sites now operate health, safety and environmental management systems. Self assessment against the Association's systems is now something all of our members carry out. A small number of sites also have these systems verified by external organisations: our aim is to increase the level of external verification.[34]

Many U.S.-owned organizations with sites overseas have committed to Responsible Care® certification – RC14001, RCMS, or a combination of these specifications. These companies are typically members of the ACC. These certification commitments should affect facilities in Canada, South America, Europe, and Asia-Pacific. It is projected that this initiative by the United States and other organizations will have an effect on the global Responsible Care® program, eventually making Responsible Care® certification an accepted – even expected – practice worldwide.

6.3.5 EHS Management System Audit Strategies for RCMS, RC14001, and ISO 14001

RAINER OCHSENKUEHN

First Environment, Inc.

As risk management has evolved in organizations, so has the audit strategy. Assessments of internal control structures are common, and more organizations are moving

---

[33]From CCPA website, http://www.ccpa.ca/print/ResponsibleCare/verification.aspx.
[34]From CIA website, http://www.cia.org.uk/newsite/responsible_care/iop.htm.

towards management system auditing instead of the more traditional compliance audits. Traditional internal auditing involves identifying the main risks in a business unit and developing audit programs to test controls that mitigate those risks. Compliance audits are still conducted, but now they are part of a more overarching audit including all elements of a management system. Management system auditing has become more popular with CEOs and the Board of Directors since the Sarbanes–Oxley Act is also applicable to environment, health and safety (EHS).

There are lots of good advice, articles, how to books, directions, and so on about "Audit strategy and how to perform a management system audit," but they focus mostly on audit techniques and approaches; rarely are they addressing the essentials to an environmental, health and safety management system, and the audit of such a system. As with most corrective actions that we see as auditing professionals, the current management systems deal mostly with the symptoms, not the real root cause of the situation, AWARENESS, ATTITUDE, and COMMITMENT. This means the employee awareness related to EHS issues, the organization's EHS attitude, and top management commitment.

So, what are the conclusions looking at established management system audit results such as ISO 14001, new approaches, and even the inclusion of security as part of the recent certification audits to the American Chemical Council's (ACC) Responsible Care Management System (RCMS) and RC14001 Technical Specifications?

First of all, an internal audit on one's own management system needs to be a solid and comprehensive approach in order to determine the effectiveness of management systems. Currently, the development shows an increased mismatch between the maturity of management systems and the stagnation of audit programs put in place to verify the system implementation, maintenance and effectiveness, thus not leading to any value-added auditing and even worse not to be able to address potential nonconformances.

An example is Organization A, which had utilized their initial questions originally developed for a gap analysis in their internal audit program. These questions are completely useless. After the system is implemented, it is not giving them any value to ask something like "Is the policy documented? Yes/No." On top of it, the audit checklist is only reflecting checks and no comments whatsoever to indicate what and who was audited in the first place. This ineffective audit leaves Organization A fairly blind – not being able to evaluate the actual system, but merely its surface. Why is this not detected? This may be because the audit of the audit program is often just limited to the existence of an audit schedule and whether this schedule is met. What is missed is to evaluate the ability of the audit program to ensure that the system is working properly and that the changes to the organization, processes, and/or products are incorporated into the system.

This is also related to the second issue found frequently – the root causes of nonconformances are not always determined correctly or, in other words, a formal problem solving of any kind is not existent. Instead of coming up with the real issues leading to the problem, organizations blame individuals (operator error) or training for most of the problems identified. It is not required to come up with a root cause analysis for each and every problem; however, where it is deemed necessary by the organization, it has to be done correctly and not just as a cursory exercise. If the root cause was relating the nonconformance back to the underlying problem, the preventive action is a breeze, ensuring that the problem does not come up again.

Auditing as a company fear factor is usually based on company culture in the form of a traditional command and control approach to EHS. Companies with no previous experience in management systems tend to show the same attitude performing management system audits as for regulatory-driven compliance audits or, even worse, the typical command and control attitude "I'm here to getcha!" It takes training, an open mind, and quite some time for auditors who are used to this regulatory approach to see and understand the difference in the approaches. The emphasis in performing a management system audit is to establish objective evidence that the system is implemented, maintained, and effective. Unfortunately, some auditors still think they need to find nonconformances in order to justify their job. The sum total of the audit should be to provide the organization with a value-added audit that includes the evaluation of the strength of the organization's management system, commitment to EHS compliance, and most importantly a prediction of the ability of the organization to continually improve. On the other side, management makes the mistake of judging the performance of the management system based on the number of negative audit findings, without looking behind the scenes. The part of evaluating the management system by the number of nonconformances is really insignificant and these numbers can be highly misleading to management. Compared audits might have had different scope and emphasis. Furthermore, the number of audit findings is subjective based on who performed the audit, in what area, and related to what elements of the management systems. Remember, an audit is a sample. It is not a complete assessment of the management system in all its details.

6.3.6   New Mexico's Green Zia Environmental Excellence Program: Third-Party EMS Performance Auditing

JEFF WEINRACH

New Mexico Environment Department

6.3.6.1   Introduction. Several years ago, New Mexico embarked on a strange, yet wonderful, odyssey – the Green Zia Environmental Excellence Program. The Program (Green Zia) was designed to take the successful elements of the Malcolm Baldrige Criteria for Performance Excellence and state Quality Awards Programs such as Quality New Mexico, which utilize the Baldrige Criteria, coupled with the principles of pollution prevention (P2) and Environmental Management Systems (EMS) and develop an effective set of EMS tools and methodologies designed to provide organizations in New Mexico, especially small businesses, with a long-term viable approach to sustainable development through environmental excellence. At the time, we were the only state that was formally adopting the Baldrige Criteria, albeit modified, as the cornerstone to our P2 initiatives. After five years of implementation, the Green Zia Program has become a model for other state P2 programs and is well aligned with many federal and international environmental management initiatives.

One of the principal elements of Green Zia that will be the focus of this paper is the third-party assessment and feedback report process. Like our Baldrige brethren at both the national level and at the state Quality Award Program level, Green Zia employs a number of volunteers to read applications that are submitted to our state Environment Department, provide *nonprescriptive* feedback on strengths and opportunities for improvement, determine a "score" for the application based on a 1000-point scale, and possibly conduct a site visit to verify/clarify the application. This paper will describe all aspects of the third-party examination process and will provide some personal insight as to the strengths and opportunities for improvement of our own process.

6.3.6.2 Background. The Malcolm Baldrige Criteria for Performance Excellence and the application processes, both at the national and state levels, typically include the following elements:

- Organizational development of an approach to quality utilizing the Baldrige Criteria as a framework;
- Submittal of an application (upwards of $\sim$50 pages) that describes the organization's approach to quality and the applicant's key results;
- Review of the application by a team of examiners, who prepare a feedback report including both strengths and opportunities for improvement and a score based on a 1000-point scale;
- Determination by a panel of judges as to whether the applicant, based on the feedback report and score, will receive recognition and/or an award by the sponsoring organization; and
- Improvement of the organization's approach to quality, based, in part, on the feedback report.

The Baldrige Criteria contains seven "categories" of quality that include leadership, planning, information management, and results. Recognizing the inherent similarities between quality and environmental management, we developed the Green Zia Program to use a similar set of seven categories to address various aspects of an organization's EMS. Six of the seven categories address approach and deployment of EMS processes and systems: How do the various components of the EMS work and to what degree are they being deployed throughout the organization? The seventh category specifically addresses the results obtained through the deployment of the EMS. For organizations that have an EMS in place and are looking for opportunities for improvement, each category is divided into "items" that allow the applicant to describe aspects of their EMS in greater detail. For example, the leadership category is divided into two items: organizational leadership and community leadership. For organizations that do not have an EMS in place yet, we recommend that their application be written to the category level (~10 pages) instead of to the item level. The feedback reports are written at the same level as the application.

Each category (or each item) has a number of points associated with it. The total number of points that an applicant can receive is 1000 points. (Note: Baldrige winners typically will score around 700 points.) The leadership category for Green Zia is worth 125 points out of the total of 1000 points. The results category for Green Zia is worth 325 points (or approximately one-third of the total points available!). This implies to the applicant that the results section is of a higher "weight" than any of the other categories in terms of an overall score. The score for a particular approach/deployment category (how many points awarded compared to the total number available) is based on a number of factors such as whether a particular EMS approach is systematic, well-deployed across the organization, aligned with environmental and organizational goals and objectives, and so on. The score for the results category is based on the EMS performance levels, trends, and comparisons to other organizations as well as the relevance of the results to the overall EMS deployment. If an applicant were to reapply to Green Zia or to a state Quality Award Program on a regular basis, a different group of examiners would be used each time to review the application. This is to assure that each review is objective and is based solely on the merits of the current application. Likewise, examiners who have profound knowledge of the applicant or who have a conflict of interest with the applicant would not serve on the particular examination team.

In spite of these precautions, there is an inherent subjectivity to the application review. Teams may be comprised of four to seven people, typically of diverse backgrounds. Each team member reviews the entire application. Consensus is used to ensure that the team agrees on all aspects of the review process. Because of the subjectivity in the process, scores are usually not provided to the applicant directly. The applicant may be provided with a scoring "band" to let them know where the examination team thought the applicant's system resides (100–200, 300–400, and so on). The thinking is that if an application scores a 350 one year and a 325 the next year (with a different review team), the applicant may think that their EMS has not improved when, in fact, it may have improved but was evaluated slightly differently by the second review team. Many organizations that utilize Baldrige as their quality framework conduct an internal self-assessment using the Criteria, in which case the score is known by the applicant and used to assess levels of performance.

Baldrige and Green Zia are built upon a foundation of performance. That is, there is no "bar" that examiners look for relating to some particular level of effectiveness. Performance-based systems rely on measuring effectiveness as it relates to competitors, industry standards, or benchmarks, or other approaches relevant to the applicant. Since these programs are not "certification" programs, we are not looking for a particular scoring threshold to be reached to tell an applicant that they have "succeeded" in their EMS implementation. On the contrary, the focus of these types of approaches is to focus on continuous improvement through analysis and assessment, not through certification.

6.3.6.3 *Choice of Examiners.* As was stated in the previous section, examiners should not have a conflict of interest with the applicant to maintain objectivity. The feedback reports do not prescribe solutions, nor do they use language such as "ineffective, inadequate, wonderful, excellent," and so on. Examiners, who have an interest in the applicant, whether it is positive or negative, have a difficult time not using this type of language. Effective examiners should have intimate knowledge of the criteria but do not have to have intimate knowledge of the applicant. For example, schoolteachers or administrators may be examiners for manufacturing applicants. As long as the examiners understand the criteria that the application is based upon, knowledge of the applicant is not necessary. The only exception is if the applicant uses a lot of lingo and terminology that examiners who are not familiar with the particular sector would not understand. In this event, it is helpful to have at least one member of the examination team who is familiar enough with the applicant's type of business that the lingo is understood.

As mentioned earlier, examination teams usually are comprised of four to seven members to assure diversity of viewpoint. Consensus is used to make sure that the team is agreeable to all aspects of the feedback report. Consensus is often a difficult process, but it is necessary to make sure that the feedback is objective and as reliable as possible. Teams are often comprised of a combination of new and experienced examiners, again to ensure diversity of viewpoint and to minimize subjectivity.

6.3.6.4 *Application Review.* Once the examination team has been identified, their primary job is to read the application and to prepare the feedback report. Reading a Green Zia application is not like reading a novel or even a proposal. It is more like trying to read a road map. There is not an obvious beginning or end; it is more where the application takes the reader. Perhaps the most important section of the application is the organizational overview, which describes for the examiners what the organization does and why it does it. It also describes the organization's primary environmental impacts and why the organization is interested in minimizing or eliminating these impacts. Ideally, the rest of the application tells the story of how the organization is addressing these impacts logically and systematically. If anything, the review of the application needs to be thorough, especially considering that the applicant may have never described all aspects of its EMS to this level of detail before. As examiners, we are looking for consistency and context. Are approaches sound and systematic? Are they deployed in all appropriate work units? Is there a "system" in place or are activities carried out in a random manner? Are the results aligned with the processes that generated them? These are some of the questions that we consider during an application review.

After reading the application *thoroughly*, each examiner generates comments for each category of the application. The comments are nonprescriptive complete thoughts about the applicant's approach, deployment, and/or results relating to components of their EMS. For example, in the planning category, an individual examiner may comment that the applicant's EMS planning process does not include input from the organization's key customers and stakeholders (an opportunity for improvement), or the applicant's EMS information management process collects relevant data from all available sources (a strength). The comments usually include a statement illustrating the "context" of the comment; that is, why is the comment relevant to the organization. These "so what" statements usually reflect something that the applicant stated in the organizational overview as being important to them, such as their position in the market or the demographics of their customers. This is one of the major differences between Green Zia and other EMS approaches: Green Zia views every organization that applies as different – different issues, different approach to their EMS, and so on. The leadership system for a local machine shop may look quite different from the leadership system of General Motors. Each can be effective in its own right, but we should not expect the same approach to be followed. The organizational overview often provides the examiner with that critical context to determine whether the approaches used by the applicant "make sense."

Typically, we use what we call "category champions" in the consensus process. Even though each examiner needs to review the entire application and provide comments for each category or item, we typically employ category champions who are team members responsible for the consensus discussion for that particular category. For example, I may be the category champion for category 2 (strategic planning), which means that I will lead the consensus discussion for category 2 even though all team members have written comments for category 2. I should probably know the category 2 section of the application well if there are any issues that come up during consensus that the group needs to resolve.

After consensus has been reached for all seven categories of the application, the team then determines the score of the application based on the team's consensus. Scoring must be done after consensus. I have seen many examples where one team member was totally smitten with a section of the application and another team member thought that the section was inadequate. The consensus discussion for this section must allow for these two examiners to come to an agreement about the strengths and opportunities for improvement for the section. If they were to score prior to consensus, the team member that liked the section would score it high (70 percent for example) and the other team member could score it low (20 percent for example). This does not mean that the "real" score for that section should be 45 percent. It could be that one of the examiners saw something that the other examiner overlooked. In that instance, consensus would determine what the overall team impression of that section was and subsequently what the score of the section should be.

One note regarding consensus – it is not compromise. It is not "I'll give you 10 percent on leadership if you give me 10 percent on planning." Consensus means that each team member can "live" with the decision of the team. All points of view were heard. If one team member cannot agree with the direction that the team is moving regarding the review, then consensus has not been attained and the group will need to do more work to achieve consensus.

6.3.6.5 *Scoring.* In the scoring process, which is also condensed after each category has been agreed to by the team, we are looking for an overall sense of the application by section. Scoring is divided in scoring bands, both for approach/deployment categories and for results categories. Each scoring band describes to the examiner what a section would or would not have to score in that particular band. For example, a scoring band of 30 to 40 percent in the approach/deployment categories has the following attributes:

A sound, systematic approach, responsive to the basic purposes of the item;

Approach is deployed, although some areas or work units are in early stages of deployment;

Beginning of a systematic approach to evaluation and improvement of basic item processes.

A particular item of an application may have two of these three but not the third. We would still score it in the 30 to 40 percent band even if all three of the bullets have not been recognized by the examining team. It is basically the best "fit" that we use to determine scores for the application. The scoring bands for the results section are quite different than for the approach/deployment sections since we are looking for the quality of results in that section as opposed to how sound and systematic a process is or to what degree the process is being deployed.

After the scoring has been completed, the team prepares the feedback report for the applicant, which includes the consensus comments for each section of the application plus an executive summary, which highlights the most significant strengths and opportunities for improvement for the applicant. It is essentially like an abstract. Typically, the category champions write their particular sections since they know that section intimately. After the feedback report has been written (and edited!), the team usually performs a sanity check looking at both the feedback report and the score to see if they are commensurate. We also look for inconsistencies in our feedback report between various sections. We do not want to say that a process has strength in one section of the application and say that the same process is an opportunity for improvement in another section of the application. Organizations who receive feedback reports from Malcolm Baldrige, state Quality programs, or Green Zia put a lot of attention into the feedback reports as an effective approach for continuous improvement. We owe our best effort to these applicants in order for them to make the best decisions about improvement.

6.3.6.6 Site Visits. After the feedback report and score are submitted to a panel of judges, who make recognition and award decisions, the judges may ask the examination team to conduct a site visit to verify or clarify the application. This is typically done if the judges feel that the applicant might be "worthy" of an excellence award, and the judges want to make sure that the application is stating the truth. If the team is asked to conduct a site visit, they are not going to be answering many questions about the Green Zia process or what the team thought about a particular process, system, or result. The team is there to verify and clarify what the application stated. Does the process exist? Is it being used across the organization? Are the results what the application says they are? Site visits have two primary aspects to them: interviewing people and looking at data and reports. Interviewing people is certainly an interesting aspect to the site visit. It is not unusual to get a different response to a particular question if you interview senior leadership versus the maintenance department or the third-shift workers. You can really see if deployment is occurring if everyone you talk to has a similar answer to a question. It is often helpful to speak to people randomly so that management cannot "brief" people on how to answer particular questions.

Site visits for Green Zia last two days. Site visits for the Malcolm Baldrige Award take up to a week. The work is tiresome but fulfilling for the examination team. They often meet during the site visit (team members may split up during the site visit to talk to as many people as possible) to compare notes and discuss what they have observed and learned.

After the site visit, the team usually will revise the feedback report accordingly. Comments may be removed or modified to address what was discovered during the site visit. Opportunities for improvement that started with "It is not clear," which is a common preface to an opportunity for improvement may be replaced with "The applicant does not have." Sometimes strengths become opportunities for improvement or vice versa. The team does not necessarily need to rescore the application but they do need to convey to the judges whether their impression of the applicant after the site visit was better, worse, or stayed the same. This will allow the judges to make any final decisions regarding award or recognition level.

6.3.6.7 Exit Interviews. After the applicants have received their feedback reports (they do not receive scores but may be notified as to what overall scoring band they were at), the applicant may ask for an exit interview with the examination team. This is where the applicant can ask the team "What did you mean by this comment?" If the applicant asks: "What should we do about this?" or "How can we improve this?", the team usually does not answer those types of questions. It is meant more to clarify what the feedback report was addressing. Programs such as Green Zia and Malcolm Baldrige provide applicants with a great deal of training and technical assistance to help them improve their organizations. It is more appropriate for applicants to use these vehicles than to ask the examination team how they can improve a particular process.

6.3.6.8 Conclusions. The time commitment to serve as an examiner for a program such as Green Zia is significant, but the process is extremely valuable and a lot of fun. It is a great learning experience for those who want to see how others address their EMS issues. We often recommend that organizations who are thinking of applying to Green Zia have someone in their organization become an examiner first just so they can see how the process goes. As programs like Green Zia continue to grow and evolve, we will continue to learn much from our examiners, who are our primary eyes and ears.

6.4   REPORTING SUSTAINABILITY PERFORMANCE: LATEST TRENDS IN CORPORATE REPORTING, NEW TOOLS, AND PRACTICES

STEPHANIE MEYER

Stratos Inc.

6.4.1   Sustainability Reporting is Becoming a More Common Practice

Since the early 1990s, there has been a trend toward greater voluntary public disclosure of corporate environmental management and performance information throughout most sectors in North America and Europe. In recent years, companies have begun to expand the scope of information they disclose to include a greater range of social and economic issues and impacts. While companies may use a range of vehicles to make this information available, a growing number are realizing benefits from publishing a hard copy or electronic report that provides detailed information on their environmental, social, and economic performance. While the issues coverage of these reports varies substantially (see Table 6.14), these reports can be broadly grouped under the heading "sustainability reports."

Internationally, sustainability reporting is becoming increasingly prevalent. At the end of 2002, a study conducted by the University of Amsterdam and KPMG estimated that close to 2000 companies worldwide were producing sustainability

TABLE 6.14. Types of Information That May be Included in a Sustainability Report[a]

Environmental Performance	Economic Performance	Social Performance
Energy inputs	Key financials	Human resource management and employee relations
Water and material inputs	Investment in intellectual capital	Health and safety
Solid wastes and hazardous wastes	Employee compensation	Workplace diversity
Effluent and spills	Taxes and royalties	Labor rights
Air emissions	Direct economic contribution	Human rights
Greenhouse gas emissions	Community development	Business ethics and integrity
Land use, biodiversity, habitat and species	Customer satisfaction	Indigenous peoples
Fines and noncompliances	Fines and noncompliances	Fines and noncompliances

[a]Good quality sustainability reports provide the reader with contextual information on the company, its products and services, and its business lines and facilities. They also provide information on corporate policies, organizational structure, and management systems. While each company determines the specific issue areas to cover within its report (based on their significance to the company and its stakeholders), the following columns identify the types of issues that a number of companies are addressing in their reports.

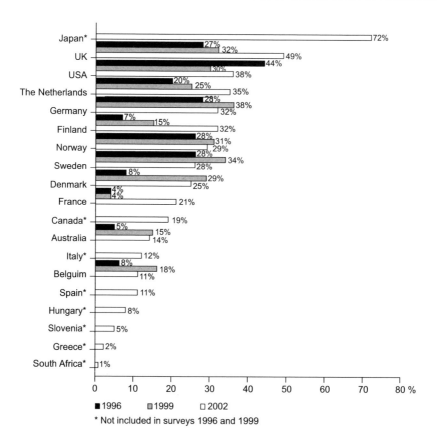

Figure 6.19. Corporate reporting by country, top 100 in 19 countries (1996, 1999, 2002). (*Source*: KPMG, 2002.)

reports (KPMG, 2002). That study looked at the percentage of top 100 companies by revenue that publish sustainability reports in 19 countries. Figure 6.19 illustrates the growth in reporting over the past decade in most countries.

As is the case in other sectors, a growing number of chemical companies are publishing sustainability reports. A recent benchmark survey of corporate sustainability reporting in Canada identified nine Canadian chemical companies that publish some form of sustainability report, out of a total of 100 identified companies that provide such information on their Canadian operations (Stratos, 2003).

In an effort to improve the consistency, credibility, and comparability of sustainability reports, the Global Reporting Initiative (GRI) has been working since 1997 to develop a set of widely-accepted principles and guidelines for sustainability reporting.[35] The second version of the GRI's *Sustainability Reporting Guidelines* was published in 2002. By October 2004, 571 companies had informed the GRI that they used and referenced the 2002 *Sustainability Reporting Guidelines* in preparing their reports.[36] Of these, 24 were in the chemical sector.[37]

Apart from stand-alone sustainability reports, many chemical companies make some information on their environmental, health and safety and social performance publicly available through other mechanisms. These mechanisms can include a combination of the following:

- *Association-level reports.* In many countries, it is a Responsible Care requirement that participating facilities provide key performance data to the national association, which publishes an annual Responsible Care report. These data are usually presented at the association level, and are not made publicly available at the facility level.[38]

- *Community newsletters.* The Responsible Care program requires that participating chemical facilities demonstrate that "an effective, complete and responsive community dialogue has been established." As part of this dialogue, chemical facilities are expected to communicate company performance data to key stakeholders, including employees, communities, and governments. For example, a number of chemical companies distribute community newsletters to local residents several times a year. This targeted communications mechanism helps keep local residents abreast of new developments and issues of local concern.

- *Inclusion in the annual report.* Some companies provide information on their sustainability goals and performance in their annual reports. In many cases, this information is limited to statements of sustainability objectives and information on environmental risks and liabilities. However, a small but growing number of

[35]The Global Reporting Initiative is a multistakeholder process and independent institution whose mission is to develop and disseminate globally applicable *Sustainability Reporting Guidelines.* The Guidelines have been informed by the active participation of representatives from companies, accountancies, investment organizations, and environmental, human rights, research and labour organizations from around the world. For more information, see www.globalreporting.org.

[36]GRI website (www.globalreporting.org), October 20, 2004.

[37]These companies are: AECI, South Africa; Agrium, Canada; AVEBE, The Netherlands; BASF AG, Germany; Cognis Deutschland GmbH & Co. KG, Germany; Dow Chemical Company, United States of America; Dow Corning Corporation, United States of America; DSM, The Netherlands; DuPont, United States of America; Hitachi Chemical, Japan; Imperial Chemical Industries, United Kingdom; Johnson Matthey, United Kingdom; Kuraray, Japan; Kyowa Hakko Group, Japan; Methanex, Canada; Mitsubishi Plastics, Japan; NOF Corporation, Japan; Potash Corporation of Saskatchewan (PotashCorp), Canada; SASOL, South Africa; Scandiflex, Sweden; Sekisui Chemical, Japan; Sumitomo 3M, Japan; T. Hasegawa, Japan; Toyo Ink, Japan.

[38]According to the International Council of Chemical Associations (ICCA, 2002), "each national association signed up to Responsible Care is required to make an annual progress report on implementation of the initiative. Worldwide, 28 countries have published the required codes or guidelines for implementation of Responsible Care, with 29 countries reporting a range of performance indicators such as emissions, incidents and injuries; 20 of these are making the indicators public and discussing them with interested parties."

companies are producing a single, integrated report that includes both all relevant financial disclosures as well as substantive information on the company's sustainability performance. The most fully integrated of these reports are premised on a vision that links corporate sustainability with ongoing financial success.

6.4.2   The Organizational Scope of Reporting Varies Substantially

Sustainability reports have been written to cover a company's

- Global operations;
- Operations within one country;
- Operations within one business line; or
- Facility-level operations.

The scope of the report should reflect the interests and information needs of the intended target audience(s). For example, investors will likely be most interested in a report that covers a company's entire operations and focuses on key financial risk issues. Local residents will likely prefer to see information presented at the facility level, allowing them to consider the performance of their local facility relative to other facilities owned by the company.

As public reporting evolves, a number of companies are experimenting with different mechanisms to allow the reader to focus on the organizational scope of greatest interest to them. Interesting approaches include:

- *Provision of company-wide and facility-specific information within one report.* A number of companies include facility-level performance information within the body of their reports or in an appendix (e.g., DSM, Kuraray).
- *Publication of multiple reports.* Within large, multinational corporations, it is not unusual to see the parent company publish a global report, which is then supplemented by country-level or subsidiary reports. This is the approach used by Sekisui Chemical. Similarly, Dow Chemical supplements its country-level report with facility-level reports.
- *Provision of additional, interactive web-based information.* As more companies move to e-reporting or hybrid reporting (in which a hard copy report is supplemented by additional information on the company's website), they are able to provide interactive features that allow the user to customize information to suit their needs. For example, Novo Nordisk (an international pharmaceutical company) produces an electronic report that includes a section titled "Interactive charts." Here, users can generate custom data charts for various social and environmental indicators either by site or by year.

6.4.3   There are Sound Business Reasons Why Companies are Reporting

Effective corporate reporting can play an important role in attracting investment and protecting markets. In the current era of highly mobile capital, companies are using sustainability reports to help secure investor confidence. Such reports can demonstrate the company is managing all relevant risks and positioning itself to address emerging opportunities. Reporters are responding to increased demands for transparency and to the benefits of disclosure. They are realizing that reputation, as a key corporate asset, can be affected by performance in a range of environmental, social, and economic areas (Table 6.15). Investors are starting to learn that companies that manage environmental, social, and economic issues effectively tend to be better managed overall and represent less risky and higher value investments. Reporting companies are also benefiting from improved relationships with employees and communities.

6.4.4 This is a Time of Experimentation: To be Effective, Reports Should Evolve to Address a Number of Emerging Issues

This section discusses three key issues reporters in the chemical sector currently face, and suggests specific ways companies can improve their reports:

1. Provision of high-quality, credible information;
2. Expanding the scope of their issues coverage to move beyond Responsible Care reporting; and
3. Employing effective communications mechanisms to reach their target audience.

6.4.4.1 High-Quality, Credible Information. Research confirms that corporate sustainability or responsibility can have an important impact on a company's reputation, and that credibility lies at the core of reputation (see, for example, Ipsos Reid, 2003). Regardless of the form it takes, to be effective in enhancing reputation and in supporting management decision-making, corporate sustainability reporting must therefore be credible – it must provide information that is relevant to the intended audience in a trustworthy manner.

TABLE 6.15. Why is Sustainability Reporting Important?

Business Value	Societal Value
Meet stakeholder expectations	*Promote transparency*
• Demands for performance information	• Increases credibility
• Build relationships	*Enhance corporate accountability*
Internal benefits	• Public demand
• Attract/retain high-quality employees	
• Staff motivation	
• Strengthen systems and data collection	
• Board integrated corporate vision	
Business opportunities	
• Licence to operate	
• Maintain/access new markets	
• Attract investment	
• Attract new business partners	

As shown in Table 6.16, reputation can be seen as a function of effective performance and credible communications. To be effective, sustainability reports need to provide high-quality performance information in a way that the intended target audience perceives to be credible. While this may be easier said than done, companies can improve the quality of their sustainability reports through improving their performance information and employing various mechanisms to enhance the credibility of their reports.

Performance Information. On its own, performance information is almost meaningless. Part of the value in publishing a sustainability report is that it allows the company to tell a story about its performance: where it has been, how its performance compares to relevant government standards and sectoral norms, why performance has changed over time, and where it is heading in the future. Many reports fail to provide this important contextual and directional information around their performance data, resulting in a missed opportunity. Surveys of Canadian and international reporting (Stratos, 2001, 2003) suggest that many companies can substantially improve their performance reporting by

- Referencing sectoral or other *benchmarks* to allow judgments to be made about the adequacy of the performance reported;
- Presenting *trend data* and *explanations for significant changes* in performance to enable the reader to develop a reasonable understanding of whether and how performance is improving and how effectively it is being managed;
- Including *targets* to enable the reader to understand and form an opinion about the company's future intentions;
- Presenting *qualitative* as well as *quantitative* data in *absolute* and *normalized* formats where appropriate; and
- Restating assumptions for data measurement when corporate structures change.

Credibility Mechanisms. A lack of trust in reported information will prevent stakeholders from reading and using the information contained in sustainability reports and may also lead to a loss of trust in the company. Companies can employ a range of mechanisms to enhance the credibility of their communications, as identified in Table 6.16.

TABLE 6.16. Reputation Equation:
Reputation = Performance + Credible Communications

Performance Information	Credibility Mechanisms
• Absolute and normalized data	• Transparency in issue selection
• Trend information over time	• Balanced reporting of issues
• Explanation of changes	• Data reliability
• Future targets	• Verification or assurance
• Benchmarks	

Transparency in Issue Selection. One of the most significant challenges facing reporting companies is determining what information to include in a sustainability report. There are trade-offs to be made between completeness (an approach encouraged by the GRI) and a focus on the issues that are of greatest importance to a company, its shareholders and other stakeholders, and the audience for whom the report has been prepared. Whatever approach a company uses, transparency of process can enhance the report's credibility.

Reports should provide the reader with information on how the company has assessed the relative importance of the environmental, social, and economic issues and how it has selected the issues to discuss within its report. Readers will be interested to know if and how stakeholder input has been solicited and considered in this assessment, as this may provide an indication of the value the company places on the interests and input of its key stakeholders, particularly if those same stakeholders are the intended audience for the report.

Where companies choose to exclude issues from their reports, the reasons for the exclusion – for example, lack of relevance to their operations – should be stated. A report that fails to address all issues of obvious relevance to the sector will not enhance corporate reputation.

Balanced Reporting. Many sustainability reports appear to overemphasize the good news, skimming over or omitting the bad news. In discussions with stakeholders, many indicate they have more confidence in reports that discuss both the company's achievements and its shortcomings within the reporting period – what can be called "warts and all reporting." A 2003 survey of 56 U.S.-based NGOs revealed that "the most important approach a company can take to improve a report's credibility is to acknowledge noncompliance, poor performance, or significant problems" (Burson-Marsteller, 2003).

While the tendency to emphasize good news is understandable, good news should be accompanied by a candid discussion of shortcomings and challenges. Key areas of the report that should provide a balanced perspective of company achievements, shortfalls, and challenges include:

- *The CEO Statement.* This statement sets the tone for the report and, as the first section that many read, heavily affects first impressions of the report's credibility. These statements should include a discussion of key performance indicators for the reporting period, identifying strengths and weaknesses in performance. They should also be forward-looking, acknowledging key business and sustainability challenges the company faces, and demonstrating how the company plans to address these challenges.

- *Performance Information.* A company should report on all issues that are material to the company and its key stakeholders. The report should include meaningful performance indicators for each of these issues, along with a candid discussion of performance relative to targets. Where a company has not achieved a desired result, the report should acknowledge the shortcoming and explain how the company plans to improve its performance in this area.

- *Discussion of Key Challenges.* In addition to providing a review of historical and current performance, sustainability reports should be forward-looking. They should identify key challenges and opportunities facing the company, and discuss how the company is positioning itself to respond to them. Major issues of relevance to the sector should not be overlooked or glossed over. For example, reports from the chemicals sector should acknowledge and discuss the company's position relative to emerging debates such as those surrounding endocrine-disrupting hormones and the use of chlorine as a feedstock. The Dow Chemical Company's report includes discussion of a number of debates and dilemmas, including vinyl, animal welfare, and biotechnology.

Data Reliability. Stratos' 2003 benchmark survey of sustainability reporting in Canada found that the majority of reports reviewed in the chemical sector – indeed, in all sectors – presented performance data with little to no discussion on data quality and reliability. At a minimum, reports should explain how the company obtains and compiles its data, and the limitations and uncertainties associated with the data and the compilation process. Knowledgeable readers understand that there are challenges associated with collecting and aggregating data from multiple sources. It would be refreshing to see more reporters acknowledge these challenges and describe how they are addressing them.

In particular, reports could be strengthened by providing the reader with information on:

- How the data and information contained in the report have been collected and rolled up;
- The methods used to estimate, monitor, or measure performance and associated confidence levels;
- Significant changes in measurement or estimation methods from previous years and if and how historical data have been restated to ensure data comparability;
- Significant changes in the scope of the reporting organization and if and how historical data have been modified or restated to enable data comparability; and
- Steps the company has taken to enhance the reliability, accuracy, and comparability of reported data and information.

Verification and Assurance. A small but growing number of reporters are experimenting with verification and assurance mechanisms in an effort to provide independent assurance about the reliability, accuracy, and completeness of their reports. These assurance mechanisms can be broadly grouped into three categories (Stratos, 2003):

- *Testimonials.* Those that include some form of statement from community members that attests to the company's record in a particular area.
- *Assurance on data and information accuracy and management.* Those that provide assurance related to the accuracy of reported data and information

and the reliability of underlying data collection, compilation and management processes.

- *Assurance on completeness, relevance and accuracy.* Those that provide an opinion on the completeness, relevance, and accuracy of all information presented in the report.

The key test of effectiveness is whether any of these assurance mechanisms enhance the perceived credibility and reliability of reported information in the eyes of a particular stakeholder group. Additional work needs to be done in this area to determine which of these mechanisms meets the needs of different target audiences, including internal and external audiences. At this stage, companies may benefit most by focusing their efforts on pointed discussions and engagement with key stakeholder groups to understand current "trust barriers" and to identify what companies can do to increase their stakeholders' level of trust and confidence in reported data and information – and in the company itself.

6.4.4.2 Moving Beyond Responsible Care Reporting. Many of the reports produced by the chemical sector have been prepared in line with Responsible Care principles and codes. In a 2001 benchmark survey of sustainability reporting in Canada, Stratos found that most chemical company reports emulate the Responsible Care focus on ensuring that products are produced in a safe and environmentally appropriate manner: "The leading reports [from the chemicals sector] provide good information on environmental performance related to manufacturing, building community awareness, emergency response, and efforts to ensure the safe transportation of their products." This study also found that few chemical company reports cover material and energy inputs as thoroughly as environmental emissions. However, here has been a slight improvement in this area based on findings from the 2003 study. More fundamentally, few chemical company reports venture beyond safety and environmental production performance and community outreach to discuss the full range of social issues relevant to their operations in communities and to their businesses more broadly. For example, not many reports discuss the company's positions on the important ongoing public debates that may determine the social and environmental acceptability of some of their existing products. The willingness of chemical companies to address and report on emerging debates such as those surrounding biotechnology, endocrine-disrupting hormones, and the use of chlorine as a feedstock is seen as an important litmus test of their stated commitment to openness and accountability.

Chemical companies can strengthen their sustainability performance and reporting by actively managing and publishing information related to a broader range of environmental, social, and economic impacts. For example, companies could provide information on:

- How they address sustainability considerations in the selection and design of new products;
- The company' positions on important ongoing public debates that may determine the social and environmental acceptability of some of their existing products;

- Their work with suppliers and customers to influence their environmental and/
  or social practices and performance;
- The company's economic impact on local and regional economies, including
  through provision of jobs, payment of taxes, and purchase of goods and services;
- Information on their customer satisfaction program, including indicators;
- Information on their employee relations and workplace practices; and
- Information on their corporate governance structure and business ethics and
  integrity practices.

6.4.4.3 Effective Communications that Reach Target Audiences. Companies
can provide sustainability information to stakeholders in a number of ways. For
example, they can incorporate it into their annual report, provide a stand-alone
sustainability report, include it on the company's website, or distribute focused
reports to specific target audiences through magazine or newspaper advertisements,
inserts, CD-ROMs, or newsletters. Although most reporting companies continue to
produce paper-based sustainability reports on an annual basis, a growing number are
providing electronic and web-based versions of their reports or are supplementing
the information included in paper reports with online information. This development
is occurring worldwide. In 2003, over 1200 reports were provided in electronic
format – HTML and PDF (2004 ACCA and CorporateRegister.com report).

As companies gain experience with hybrid and electronic reporting, we are
seeing a number of innovative practices that harness the abilities of web-based
communication. Examples of such innovative approaches from within and outside
the chemical sector include:

- *Novo Nordisk* and others provide interactive graphs that enable the reader
  to customize the presentation of data to suit their needs. The user can view
  company-wide or facility-specific data, selecting the performance indicators
  and years of greatest interest to them. *DSM's* electronic report provides
  users with drop-down menus from which they can select and view environmental performance data by site.
- *BT* produces a web-based report that allows the reader to organize and view
  the company's sustainability information according to different frameworks
  or norms, including alignment with the GRI *Sustainability Reporting Guidelines*, BT's statement of business conduct, or the nine principles of the
  United Nations *Global Compact*.
- Some companies, such as *VanCity* and *TransCanada*, use symbols (Fig. 6.20)
  in their paper reports to indicate that their website provides more information
  on the subject.

Figure 6.20. Symbols used in paper reports to refer the reader to a website.

- *Rio Tinto* provides an interactive map, enabling the user to focus on a facility of interest.

Additional experimentation with electronic reporting will continue to yield new ways to make information more accessible and usable to target audiences. To use e-reporting effectively, however, companies need to adopt the benefits of the Internet while retaining core features of good paper reporting, including clearly identifying relevant performance information, indicating the time period the information covers and when it was last updated, and delineating the overall structure of the "report."

6.4.5 Moving Forward on Sustainability Reporting

This chapter has suggested a number of ways in which current reporters can strengthen their reporting practices to enhance the quality and credibility of their public sustainability reports. To make these reports more useful, it is equally important to encourage more companies to make sustainability management and performance information publicly available, thus allowing users to compare approaches and results across companies.

Figure 6.21 provides some guidance to first-time reporters on how to begin public sustainability reporting.

6.4.6 Sustainability Reporting Case Study: Conducting Today's Business with Tomorrow in Mind (*PotashCorp's 2002 Sustainability Report*)

THOMAS C PASZTOR

PotashCorp

The transparency inherent in sustainability reporting goes directly to the issue of trust in public/corporate relationships at a time when this issue is front and center in everyone's mind.

Many sustainable practices have long been ingrained in PotashCorp's culture and values. However, before the Company published its first Sustainability Report in 2003, there was no central place or common voice for communicating its efforts, progress, and remaining challenges in sustainable development.

Researching and writing the report helped to provide clarity and direction to the Company's sustainability initiatives. Through the reporting process, we were able to better gauge our strengths, as well as identify those areas where we can improve.

The reporting process also enabled us to take inventory of our performance measures, our current sustainability initiatives, and our strategic framework for continuous improvement in corporate transparency.

We wrote our inaugural triple-bottom-line report to satisfy the needs of the broadest audience possible. Industry jargon was replaced with common words and terms. And for the more knowledgeable stakeholder we provided a substantial amount of facts and documentation. Finally, to ensure ease of use, our report begins with a brief "How to Use This Report" section.

How to decide whether to report?

You will need to consider a number of factors when making your decision, but the bottom line will be: *do you expect sustainability reporting will be beneficial for your company?* You can inform your decision by identifying the types of benefits and costs you anticipate will be associated with a public sustainability report.

How to begin reporting?

You will need to make a number of initial decisions to inform and guide the report scope and content, and the reporting process itself. These include determining the:

- Issues coverage – What scope of issues will you include within the report? Many companies begin by reporting on one focus area (e.g., environment); others start by compiling and presenting the data and information they already have available.
- Boundary of the report – What business units, what facilities, what countries, and what time period will be covered within the report?
- Level of integration – Will the report be a stand-alone report or integrated into the annual report? Will it address economic, environmental, and social factors in discreet sections, or will it attempt to link and/or integrate a number of these factors to present a full picture of company performance and direction?
- Medium of reporting – Printed or electronic? Some companies begin by gradually making more sustainability information available on their corporate website; others begin by producing a printed report, supplemented by additional information on the website.
- Theme and key messages – Communications messaging and report design are important considerations. The report should be consistent with corporate branding and image, and should involve communications personnel from the beginning.
- Level of verification or assurance to be used – while few companies engage external verifiers to verify their first reports, it is helpful to engage stakeholders in the reporting process from the outset to identify their expectations for the report.

Figure 6.21. First steps in sustainability reporting.

In the months following the publication of our initial Sustainability Report, we have actively used the document with customers, employees, and investors. We also have made it available to the public, and mailed copies to targeted community, government, and media audiences.

Customer surveys conducted in December 2003 showed that about a third of our customers were aware of the report. They noted that they weigh economic, social, and environmental performance when determining whether to do business with a company. One large, international customer requested several additional copies of the report for forwarding to all of its division presidents as "must reading."

6.4.7   Tales from the Trenches

An essential step in producing a credible Sustainability Report is to ensure that you have across-the-board support from senior management. This was not an issue at PotashCorp. The directive to publish a public document that reported on the Company's sustainable development initiatives had, in fact, come from senior management.

To help guide the reporting process and provide assistance with data gathering, four members of senior management were named to a Sustainability Committee. Their contribution was invaluable. Working with the report's production team, the Committee provided insight to areas ranging from corporate core values to specific achievements and metrics that warranted inclusion in the report.

The structure of the report separates our performance measure data from those sections devoted to telling the PotashCorp story. The report gives the readers facts, figures, and support data. Interpretation is left to the reader without influence from the Company. The report also explains the company's operations and how they fit into a strategy based on sustainability.

With this reporting structure now established, subsequent reports will likely merge performance measures with commentary on the Company's progress toward meeting goals, targets and the overall sustainability mission of the Company.

PotashCorp's initial Sustainability Report represented its first foray into this type of reporting. Based on the feedback we have received, our initial effort gives us a solid platform for future reports and will serve as a benchmark for making Potash-Corp's commitments to sustainability explicit and transparent.

6.5 SECURITY AND SUSTAINABILITY

SCOTT BERGER

American Institute of Chemical Engineers (AIChE)

Given the greatly increased focus on industrial and public security in the years since the September 11, 2001 terrorist attacks, few would disagree that sustainability and security are closely linked. Without much imagination, one could put forward the hypothesis that a sustainable manufacturing process that consumes few, if any, material or energy resources and that uses more benign materials would contain nothing usable by a terrorist. At best, however, this hypothesis is only partially true. At worst, the hypothesis not only misses the point of sustainability almost entirely, it also distracts from the true mission of security.

This section will provide a broad overview of security from the sustainability perspective, going well beyond the subjects of fences and cameras, in order to give the reader an appreciation of the subject. Additional references, to help the reader learn more about security, are provided at the conclusion of this section.

In order to understand security in a sustainability context, it is necessary to treat security the way you treat any other sustainability issue – by evaluating security from cradle-to-cradle. It is important to consider:

- the roots of terrorism;
- terrorist goals and strategies;
- terrorist tactics;
- exploitable consequences;

- security vulnerability;
- security countermeasures;
- the link between security and sustainability; and
- review and continual improvement.

6.5.1 The Roots of Terrorism

Step #1: Understand What Motivates the Terrorist. Over the course of history, mankind has undergone enormous political and social changes. While some changes have occurred peacefully, it has been more the norm for changes to be precipitated by violence. War, coup d'etat, revolution, riot, and terrorism are all violent means to achieve social change.

Of course, the way one perceives a given change depends largely on one's position relative to the conflict. We consider those who cause death and anxiety in our society to be terrorists. These same people consider themselves revolutionaries or martyrs, and they believe as much in the change they wish to make in the way we live, as we believe in keeping our way of life on its present course.

When we talk about terrorism today, we generally mean individuals who intend to kill themselves, or put themselves at significant risk, in the process of implementing their attacks, although this is not necessarily the case. Suicide terrorists must believe intensely in their causes, but also must be at a position that their life prospects have become so bleak that death is an attractive option. Religious fervor often is employed to sell the terrorists' vision of societal change and make the prospect of death more attractive. The mere existence of suicide terrorists is terrorism in itself.

However, we should not forget terrorists with no interest in suicide, such as Timothy McVeigh, nor should we forget the manual laborers who, during the French Revolution, threw their wooden shoes ("Sabots," in French) into their machinery, such "sabotage" impacting the basic infrastructure of the country. However, both suicidal and nonsuicidal terrorists commonly have come to the conclusion that they need to take their extreme actions because society is not meeting their basic needs. As other authors in this publication have argued, meeting societal needs is a basic tenet of sustainability. In other words, the absence of sustainability is at the very roots of terrorism.

6.5.2 Terrorist Strategies and Tactics

Step #2: Understand What Potential Terrorists Wish to Accomplish, and How They Operate. While there are would-be terrorists who are content to create nuisance events – hoax bomb threats for example – serious terrorists aim to cause major damage to people, property, and society, in four basic ways.

- *Inflicting mass casualties by explosion, fire, or toxic release:* for example, detonating a bomb, attacking a chemical storage tank to release its contents, releasing a biologic agent, or arson in a public building.

- *Contamination:* for example, adding a pathogen to drinking water, introducing an agricultural disease.

- *Disruption of basic infrastructure or society:* for example, initiating a computer virus attack, blowing up key electrical substations or factories, attacking government officials and offices, and attacking key cultural icons such as sporting events, tourist destinations, and national monuments.

- Theft or improper acquisition of materials for various purposes including to conduct the types of attacks described above, for example, stealing an ingredient with which to manufacture illegal drugs (which finance terrorist activities), buying fertilizer from an agricultural distributor for use as a bomb, hijacking an airplane.

Actions may fit into multiple categories. For example, the 9/11 terrorists misappropriated airplanes, crashed them into the World Trade Center and Pentagon, causing fire, mass casualties, and disruption of communications and financial market activity, not to mention hundreds of billions of dollars in military spending and the lives lost in subsequent military action.

Depending on the extent of the result the terrorist seeks, a terrorist or a terrorist group may plan its attack well in advance, in some cases for many years before launching the attack. Planning can include obtaining publicly available information about the target, infiltrating the target organization, conducting surveillance, and colluding with an employee inside the target organization, either forcefully or with that employee's knowledge.

The advance planning and the associated surveillance provides the best opportunity to deter, detect the attack during the planning stages so that the attack can be stopped before it happens, or to delay an attack to lessen its impact or provide time for military or police response.

Different groups of terrorists tend to organize their attacks in common patterns. For example, Al Qaeda favors broad-scope attacks that are technically advanced, but do so much less frequently than Palestinian terrorists who act more frequently – by some accounts more than 20,000 attacks on Israel from 2000 to 2004 – with a single operative, on a relatively smaller scale.

Therefore, by being aware of the results certain terrorist groups wish to achieve and the tactics used by these groups, a facility owner can better understand how the facility could be used by the terrorist.

6.5.3 Security Vulnerability Analysis

6.5.3.1 Introduction

Step #3: Decide Which Approach to Vulnerability Analysis to Use. Security vulnerability analysis (SVA) is the activity of identifying how potential terrorists can breach security at your site in order to impact an asset and cause a terrorist event. In this context, asset can mean a piece of equipment, a store of product, a key building, a computer system, a person, or anything else of importance to your company or organization. Security is vulnerable when three factors coexist: (a) an identified terrorist threat, (b) an asset that terrorist can possibly exploit, and (c) insufficient security measures to deter, detect, or delay the terrorist, and protect the asset from attack. There are numerous approaches to security vulnerability analysis, some of which are described in Table 6.17. The main factors that differentiate these methods are customizations relative to a particular sector.

All vulnerability analysis methods fall within a spectrum ranging from qualitative ("Asset-based") to quantitative ("Scenario-based"). In general, one will choose an SVA method on the scenario-based end of the spectrum when

- the asset or the consequences resulting from attacking the asset is particularly attractive to a terrorist;
- when you simply cannot afford the consequences of an attack; or
- when little prior experience exists for analyzing that asset's vulnerability.

By contrast, one will choose an asset-based SVA when the consequences are relatively less or when considerable experience exists for analyzing the asset.

For example, the U.S. Secret Service uses an exhaustive scenario-based approach to protect the President of the United States. Likewise, the nuclear industry uses a rigorous scenario-based approach. In the former case, the symbolic value of a U.S. President makes him or her a very attractive target. In the latter, the political

TABLE 6.17. Security Vulnerability Analysis Approaches

Method	Developer	Basis	Reference
ACC Tier 4 SVA	American Chemistry Council	Asset-based approach for low-hazard, low-impact sites	www.responsiblecare toolkit.org
CARVER	U.S. Department of Defense	Asset-based approach with wide general applicability	
CCPS SVA	Center for Chemical Process Safety (American Institute of Chemical Engineers)	Practical scenario- and asset-based approaches for fixed manufacturing sites	www.aiche.org/ccps/ sva
SOCMA SVA	Synthetic Organic Chemical Manufacturers Association	Asset-based approach for small, specialty chemical manufacturers	www.socma.org
VAM and RAM (various versions)	Sandia National Laboratories	Scenario-based approaches tailored for specific sectors (chemical, water, dams, etc.)	

and actual consequences of a nuclear loss of containment are significant enough that we cannot afford to let it happen.

On the other hand, asset-based SVA approaches tend to be used in situations where many assets are similar, like public buildings. Once the symbolic value of the asset, the potential consequences of attack, and other factors are evaluated, security measures needed are determined based on prior experience with similar types of buildings.

In reality, almost all SVA methods lie somewhere between asset-based and scenario-based. This is simply practical: to the degree that shortcuts can be applied without sacrificing results, the faster countermeasures can be identified, and the more money remains available for protections.

The following description of SVA follows the treatment presented in "Guidelines for Analyzing and Managing the Security Vulnerabilities of Fixed Chemical Sites," Center for Chemical Process Safety (AIChE, 2002).

6.5.3.2 Portfolio Screening

Step #4: Prioritize the Work. Unless your company needs to consider only one easily demarcated facility, it is important to prioritize your efforts to analyze vulnerability and implement security countermeasures. Prioritization should take into account the attractiveness of the asset or target, the difficulty with which an attack could be carried out, and the potential damage that could result. In the simplest form, attractiveness, ease, and damage could be ranked on a qualitative scale (e.g., 1–3), then the three scores summed. Screening approaches with more detailed scales may also be used if finer detail is needed.

The first factor to be considered is the damage that could be inflicted. Consider loss of life as well as economic loss, and try to envision the worst-case possibilities. In the case of chemical facilities, the worst case resulting from a terrorist attack may be worse than the so-called "worst-case scenario" developed for chemical facilities covered by the EPA's Risk Management Plan rule (EPA, 1996).

The second factor to be considered is the target attractiveness. Terrorists tend to consider national monuments, major cultural, political, and sporting events, and the financial sector to be particularly attractive, as an attack on such a target is viewed as an attack on their enemy's entire way of life. Likewise, key infrastructure components such as key bridges, tunnels, highways, and railways are more attractive. Finally, the public's fear of chemical and petroleum facilities may make these more attractive targets, more so if materials in the facility have potential off-site consequences if released.

To really understand what makes a target attractive to a terrorist, search the Internet for "The Al Qaida Manual." This document includes a sobering discussion of attractive targets. Another useful resource is the book "American Jihad."

The third factor is difficulty of attack. Consider the manpower, other resources, and planning that would be required in order to mount an attack that would cause significant damage.

6.5.3.3   Identify Assets

Step #5: Determine What You Need to Protect. An asset is anything the facility owns or employs that could possibly be exploited by a terrorist. In a chemical plant, physical assets include tanks, reactors, and warehouses. Bridges, trains, power lines, herds of cattle, and assembly lines are examples of assets in other sectors. In all sectors, people are assets, as are computer infrastructures. In the asset identification step, you need to identify everything under your control that you may need to protect.

6.5.3.4   Set the Scope of the Study

Step #6: Determine What You Cannot Protect (What Someone Else Needs to Protect). Before starting, decide the boundaries of your study. For example, are in-bound rail shipments to a facility considered only inside the facility gate? 100 yards outside? When it leaves the shipper? It should be clear that depending on where the boundaries are set, the problem of vulnerability analysis can become quite large. Regardless of where one sets boundaries, it would be prudent to identify the parties responsible up to your boundaries, and confirm that their boundaries line up with yours, so that areas are not neglected. Outside-boundary parties to keep in mind include rail, truck, and marine transportation, utilities, pipelines, and near-neighbors. It is also important to identify the kinds of antiterrorist activities you can realistically undertake, and distinguish these from those for which you must rely on local law enforcement and military. For example, you may decide that it could be appropriate to have unarmed guards, but not armed, and you will rely on the military to protect against an attack by air.

6.5.3.5   Estimate Potential Consequences

Step #7: Determine What Kind of Impact a Terrorist Can Have. For each asset, determine the potential consequences of a successful attack, including fatalities, injuries, economic impacts, and social impacts. For chemical releases, conventional release modeling techniques may be used – however, be sure to include consideration of toxic materials that are not on regulatory lists if significant consequences are possible. Consider both personal and regional/national economic consequences. When looking at personal economic consequences, include replacement costs, lost business, and clean-up costs. When considering regional/national economic consequences, ask if your plant might be one of many in the country that makes a product critical to public health or the military, or provides a material that is used in such a product.

6.5.3.6   Analyze Threats

Step #8: Identify the Types of Terrorist Attacks You Need to Consider. Find out about different terrorist groups, who each group targets, whether they are active in your sector or region, whether they may be targeting operations like yours, and what types of strategies they use.

In almost every case, you will not have the experience and current knowledge of terrorist activities to be able to conduct the threat analysis yourself. Involve local law enforcement, and discuss with them whether to involve regional, state, and national law enforcement and intelligence. If you do not have security expertise on staff, you should consider engaging a professional security consultant. It is also possible to subscribe to security alert services that provide updates on terrorist activities. These services are useful, but should not replace establishing a good relationship with law enforcement.

The final result of this step is a set of design basis threat statements that you can use to develop attack scenarios in the next step. Some examples are

- vehicle-borne improvised explosive device (VBIED);
- armed assault;
- infiltration to place fixed explosives;
- stand-off assault, for example involving rocket-propelled grenades (RPGs);
- cyber-attack;
- theft.

6.5.3.7 Asset–Threat Pairing

Step #9: Match Threats to Potential Consequences to Identify Possible Attack Scenarios. In this step, you put together the information obtained in the previous three steps to establish scenarios for potential terrorist attacks using the design basis threats against your assets to produce adverse consequences. For example, you might identify that a terrorist could drive a VBIED close enough to an anhydrous ammonia storage tank that upon detonation, the tank will collapse and release ammonia, with resultant impact on the nearby population.

You should strive to identify all reasonable scenarios that terrorists that could be interested in your facility might use, without being unnecessarily duplicative. For example, if you have two ammonia tanks, you do not really need to consider a VBIED attack on each tank and on both tanks simultaneously. One scenario should be sufficient to represent all three possibilities.

In identifying scenarios, you should involve a team representing diverse backgrounds, including security and law enforcement, process, operation, and business knowledge, and use a brainstorming approach. Team members should attempt to place themselves in the minds of a potential terrorist, and consider how they would attack specific assets via the design basis threats. Such an approach is often called "Red-Teaming."

6.5.3.8 Evaluate Countermeasures.
In this step, you evaluate whether the security measures you have in place are adequate to deter, detect, delay, or respond to an attack. In conducting this phase of the vulnerability analysis, again involve law enforcement and also include security experts. As a result of this step, you will identify action items to implement over time. You may also identify scenarios for which conventional security measures may not be adequate. In such situations, your relationship with law enforcement is critical, as a regional or national security solution may be required.

6.5.4 Security Countermeasures

Step #10: Determine if Existing Countermeasures are Sufficient to Address Possible Attack Scenarios. Security countermeasures envisage a four-tier approach, involving deterrence, detection, delay, and response. These four components are described briefly below. Persons wishing a detailed discussion of security countermeasures should obtain a copy of the "Protection of Assets Manual," published by the American Society of Industrial Security (www.asis.org).

Deterrent countermeasures either discourage terrorists from considering your facility or stop attacks in progress. For example, pop-up bollards in roadways may be used to stop an intruding vehicle. Large earthen berms around storage tanks serve double duty to stop a vehicle and contain a chemical spill. Real or dummy video cameras can cause a terrorist to consider another target.

Delay countermeasures slow the progress of a terrorist attack, or delay its onset. Many people believe that fences are a deterrent countermeasure. Not so: a fence takes but a few seconds to get through. Rather, fences are a delay countermeasure, which cause a terrorist to conduct surveillance from a distance, making it take longer to establish plans. This gives more time for facility personnel to realize they are being watched and to involve law enforcement. Tall, thick, thorny bushes may be even more effective then fences (or can supplement them), because they are much harder to get through and see through.

Detect countermeasures allow facility personnel to identify surveillance and incipient attack. Security cameras play an important role in detection, and recent developments in software make it possible to pick unusual behavior out of crowds for more thorough investigation. Entry alarms and proximity alarms are additional detection countermeasures. In extremely sensitive cases, detection countermeasures may be set a distance away from the facility to identify a possible attack before it arrives at the boundary.

Military or submilitary response to a terrorist attack is something best prepared for well in advance of an attack, and such plans should be the result of involvement of law enforcement in cases where significant consequences coincide with a vulnerable target. However, in many plants using toxic materials, a military response may be inappropriate, as it may be possible for the military response to cause the same kind of consequence that the terrorist intended to cause. For example, if it becomes necessary for armed personnel to protect chemical facilities, they must be well trained to avoid such events.

6.5.5 Long-Term Security Management

Step #11: Periodically Review Security and Implement Improvements. Like any other component of a business operation, security must be managed to ensure that it continues to function properly, to adjust to changes in the nature of the assets and the threats, and to implement opportunities to improve efficiency and effectiveness. Security programs should involve ongoing discussions with law enforcement to monitor changes in the threat as well as interaction with company technical and business efforts to monitor changes in operations and assets. In addition, formal security vulnerability analyses should be repeated periodically.

Since late 2001, draft chemical security bills have been on the docket in the U.S. Senate. The two key bills, submitted by Senators Corzine and Inhofe, would require security analyses to be conducted and various kinds of security measures and plans implemented depending on the results of the analysis. The Senate is not acting aggressively on either bill at this time. While the bills are similar, the Corzine bill requires a technology assessment step aimed at actively driving inherently safer technologies. These bills are assigned different reference numbers as they proceed through committees, so check www.senate.gov for current status.

Even if security bills are passed in the current congress, implementation through the Department of Homeland Security and other agencies will very likely take several years. In the interim, many trade organizations have implemented security management requirements for their members. One organization that should be highlighted is the American Chemistry Council (ACC). Members of the ACC must implement a comprehensive environment, health, safety, and security management system following a template similar to ISO 14001 called Responsible Care® 14001 (RC14001), or a somewhat less comprehensive system called Responsible Care® Management System (RCMS). Most trade organization members following their organization's requirements should have little trouble meeting the Inhofe bill requirements (www.responsiblecaretoolkit.com).

6.5.6 Can Sustainability Prevent Terrorist Attacks?

By this point, it should be readily apparent that even the most sustainable product or service, made by the most efficient and green manufacturing process, can still be subject to a terrorist attack. Indeed, if we accept that sustainability includes meeting societal needs, then a process or product that can best meet the needs of society but is not available to all may well be the biggest target of a terrorist who does not have access to that process or product. In this case, even though the process or product may be completely absent of explosive energy or toxic materials that a terrorist might use, the terrorist could still add a contaminant or toss his "sabot" into the manufacturing or distribution works.

This does not diminish in any way the importance of striving for sustainability, nor does it mean that a sustainable manufacturing process makes no contribution to security. In fact, by virtue of striving to meet societal needs, sustainability addresses terrorism at its roots. Therefore, it is vitally important when proceeding down the road to sustainability to consider the basic needs of those who would wish us harm, because meeting these needs may well turn potential foes into, if not friends, people we can coexist with.

6.6 BUILDING CORPORATE SOCIAL RESPONSIBILITY

NEIL SMITH AND JOAN BIGHAM
SmithOBrien

In Tomorrow's Company's paper, "Redefining CSR," the London-based think tank offers one of the simplest and clearest definitions of "corporate social responsibility

(CSR)." "Social means 'towards society.'" The word that really matters and often provokes the most reaction is "responsible." Corporate social responsibility, according to the paper's author, Mark Goyder, is "the responsibility we expect companies to show in being part of society. In short, we expect... that companies not only behave responsibly towards their shareholders, but in all their relationships with people, the natural environment, and society generally." The paper goes on to identify two types of corporate social responsibility: "compliance" CSR, driven by the expectations of external stakeholders and the pressure of public reporting; and "conviction" CSR, which is led by a company's vision and values. The most successful industry leaders, such as Dow or DuPont for their air and water quality initiatives, are those who have a vision that goes beyond the profitable delivery of products or services to affect the quality of people's lives. They also ensure that the company does what it says it stands for. In doing so, these business leaders insist that the company's values are part of every business decision (Goyder, 2003).

The best companies do not confine their practices only within laws and regulations that govern their business, but instead prefer to encourage employees to stretch their thinking to go beyond the minimum legal and regulatory requirements. To these companies, explains Goyder, "CSR is a natural and practical expression of values for which they have always stood and will continue to earn the public's trust. Every behavior flows naturally from the company's purpose and values."

6.6.1   The Business Case for Corporate Social Responsibility

Before investing the time and resources required to adopt proactive and more responsible practices, corporate managers often want to know the business case – how such changes will increase shareholder value. To some managers the business case is intuitive, for others it is thought to rely more on the vagaries of intangible factors and is therefore elusive.

Dupont CEO Chad Holliday no longer views environmental, health and safety (EHS), and social responsibility as separate functional areas, but instead has woven them into three corporate priorities: creating products and services that deliver greater value to customers and shareholders with fewer inputs; improving operating efficiency and capital utilization, while reducing the supply chain environmental footprint; and seeking technological innovations that advance the quality of life (Holliday and Pepper, 2001).

Companies make money in essentially two ways – they either increase revenue and/or reduce costs. On the revenue side, the business case for CSR is more difficult to quantify reliably, for too many factors influence a company's performance. The general economic climate, industry conditions and market demands, volume discounts to distributors to clear out inventory, and acquisitions all affect revenue, but none has anything do with a company's CSR. Similarly, with stock price, while CSR companies may be more attractive to socially responsible investors (SRI), their corporate responsibility is only one of a half dozen or more factors that influence the buying decisions of money managers and stock analysts.

On the cost-savings side, the business case is clearer and more measurable. Here the return on investment (ROI) on pollution controls, for example, can be measured in fewer fines and penalties. A more welcoming and inclusive workplace that values differences may avoid employee discrimination suits and accompanying litigation costs. Companies that are proactive in their dealings with stakeholders enjoy fewer costly management distractions that come with reacting to problems as they occur – the old "it could never happen to us" scenario.

Every company needs to create economic value to return to shareholders. But the delivering on that value happens in large measure by dealing openly and honestly with essential stakeholders and delivering sustainable value to society at large as well.

The chemical business, despite its contributions to people's day-to-day lives, still remains one of the most revered and often misunderstood industries. It is continually plagued by people's doubts over the safety of many of its products and manufacturing processes. The Bill Moyer's 2001 television special, "Trade Secrets," on what the chemical industry knew, but failed to inform the public, about the harmful effects of certain products only exacerbates the problem. Since 1988, the industry's Responsible Care voluntary regime of safety, management, and environmental impacts has attempted, though with limited success, to allay the public's fears over the health and environmental effects of making and using chemicals. The consequences of this perception (see Chapter 3), whether real or imagined, are manifested in the resistance companies face when they try to build new facilities and the proliferation of restrictive and costly government regulations.

To address the concerns of its diverse stakeholders, the chemical industry has formed hundreds of community advisory councils and partnerships with community and environmental activists in an attempt to hear their points of view. German chemical giant BASF, for example, has numerous collaborative projects with customers and NGOs. The company is working with the U.S. Agency for International Development (USAID) to promote the use of mosquito nets impregnated with BASF's insecticide FENDORA® to protect people from malaria in African countries. In another example, BASF works together with the World Health Organization, UNICEF, the German Society For Technical Cooperation, and the Carter Foundation to use BASF's larvicide ABATE® to fight parasites in Ghana, Nigeria, and Sudan (BASF, 2001).

By effectively engaging stakeholders, responsible companies have grown to expect the following:

- Improved employee job satisfaction, resulting in higher productivity, greater innovation, problem solving and personal ownership for outcomes. If employees respect what their employer stands for, they will stay longer, leading to lower turnover and recruitment costs, absenteeism, and illness.
- Improved quality systems resulting in fewer returns, loyal customers, and more repeat business.
- Reduced waste and toxic emissions that lower the depletion of natural resources and improve air quality, resulting in healthier employees and

communities, and lower legal exposure from fines and penalties, local disputes, bad press, and management distraction costs.

- And, finally, companies can enjoy higher profits and shareholder value, resulting from lower operating and overhead costs, for example, administrative and sales, regulatory reporting, and cost of capital and greater access to new capital markets.

6.6.2 Managing Responsibly

Recent corporate scandals, which have led to the near demise of some of the world's largest corporations and have ended the careers of numerous CEOs, illustrate a strong correlation between enforcing ethical behavior and business results. Empirical research also confirms a link between CSR, risk ratings, reputation standing, and various economic indicators over a number of years. However, it would be foolish to think of the adoption of CSR practices as a predictor of business success. CSR, an indicator of a well-run company, along with an effective overall strategy, high-quality products and effective branding, smart investments and innovation, and an ability to learn and adapt faster than competitors to changing conditions combine to make a successful company (Waddock and Smith, 2000).

Managing an organization responsibly is not in itself complex or entirely new. For centuries smart managers have recognized that employees are more productive when they are treated with respect, and valued and rewarded for what they know. Successful companies recognize that before they can deliver reliable high-quality product or service they must spend time with their customers to understand their known and, more importantly, latent needs.

However, what is new is that corporate responsibility now takes place in a context in which business is more

- global in reach,
- powerful and influential,
- under greater scrutiny through the media and the Internet,
- under pressure for more than financial results from investors,
- under pressure from powerful activist groups, and
- mistrusted by large parts of the general public (Goyder, 2003).

Corporate leaders who acknowledge these conditions ensure that in all of their business relationships there is a clear and consistent idea of why the company exists and what it stands for. This is how trust and loyalty are created. They also recognize the competitive advantages of building integrated organizations, in which all stakeholders are networked in an effort to build a truly open and responsive company. Their leadership styles are visible and consistent, while allowing others to assume greater responsibility.

Successful corporate leaders have found ways to break down the walls of functional silos that inhibit accountability and communication. Their senior management teams have bought into some simple, smart business practices – which avoid polluting and accompanying fines; which respect and value employees; and which contribute resources to the community in ways that not only grow their business, but the community as well. These executives are also effective at breaking through the intricate web of managers whose self-interests and internal politics can undermine a company's directions, success, and ultimately, its public trust.

Today's responsible company seeks a clear understanding of stakeholder expectations, yet does not raise the bar too high in delivering on them. The company's values do not only rest on the wall for all to see, but are integral to strategic planning and decision-making.

Lastly, but no less importantly, successful leaders admit where their companies face significant challenges or have failed. They also admit what they do not know about running a sustainable and socially responsible company.

Having successfully made the transition from chemicals to bioscience, Monsanto pioneered the development of genetically modified crop seed that reduces the need for pesticides and herbicides. Company officials, however, never anticipated the major opposition from environmental groups, consumers, and antiglobalization activists, and the public fears of genetically modified foods. In 2000, then CEO Robert Shapiro admitted as much: "We learned that there is often a fine line between scientific confidence, on the one hand, and corporate arrogance on the other. . . . We didn't listen very well to people who insisted that there were relevant ethical, religious, cultural, social and economic issues as well." Today, Monsanto continues to struggle with a tarnished image and lackluster financial performance (Vidal, 1999).

6.6.3 Drivers of Corporate Social Responsibility

Credibility and trust are *everything* in business. Corporate social responsibility first caught the attention of the public in the late 1960s and early 1970s, following a host of moral failings including neglecting consumer safety, bribing government officials, polluting the environment, exploiting child labor, and even toppling governments in developing countries (Paine, 2003). Angry Americans, whose mistrust in corporations was fueled by the belief that corporations only listen to complaints when forced to do so, began boycotting major brands, throwing up picket lines outside corporate headquarters and, aided by a 1970 federal court ruling allowing shareholders to use the proxy process to raise concerns about corporate behavior, bought stock in major companies to ensure a say at shareholder meetings. Throughout the 1980s and 1990s, numerous exposes of exploitative labor practices in global supply chains put additional pressure on companies to adopt codes of conduct and demand compliance from their suppliers. The fall of Enron in 2001 – adding fuel to the fire of corporate malfeasance – renewed calls for greater corporate integrity, accountability, and transparency.

Transparency is fundamental to business. Investors need to be confident that reported profits are real, that executives will not use their positions to enrich themselves, and that accountability systems are in place to expose and punish abuses. Workers have to believe in a company's commitment to build value if they are to put their careers, and the security of their families, into their hands. Customers assume the integrity of the transaction; once that trust is broken, they rarely give the company a second chance. In the absence of transparency, all of these relationships are at risk (Batstone, 2003).

6.6.3.1   Pressures from Shareowners.
Investors expect a competitive return from their investments in companies. In addition, they expect transparency, reliable forecasting, timely information, and opportunities to be heard. The quality and legitimacy of EHS, workplace and board diversity, and corporate governance and management's leadership in these areas are increasingly seen as a proxy for shareholder value. An Ernest & Young survey of 300 Wall Street analysts found up to 86 percent of oil and gas analysts confirmed that regulatory compliance, employee health and safety, community relations, and lawsuits can materially affect a company's value (Cap Gemini Ernest & Young, 1996). These performance pressures are today a normal part of corporate life.

There are also some growing investor pressures that are expanding the definition of corporate responsibility and demanding a greater say in how companies are run. There is mounting evidence, for instance, of the financial risks of global problems, such as climate change. European insurers claim that in the next decade, the annual cost to industry and governments of global warming will rise to $150 billion a year (Webber, 2002).

The social investment movement has become a significant source of pressure by investors on companies to manage more responsibly and to be more responsive to shareowner concerns. By 2001, more than $2 trillion, one out of eight professionally managed investment dollars in the United States, were part of a socially responsible portfolio. New share indexes have also emerged, such as the Dow Jones Sustainability Index and FTSE4Good, launched by the London Stock Exchange and the *Financial Times* publishing company, which track companies whose strategies integrate environmental and social considerations into their financial performance.

The long-held assumption that operating responsibly is too costly no longer holds up. For example, the Domini Social Index, a stock index created to track the performance of socially screened companies against other nonscreened indexes, has generally outperformed the S&P 500 on a total-return basis and on a risk-adjusted basis since its inception in May 1990, although it trailed the S&P 500 during 2000 (SRI, 2003). Furthermore, academic studies in finance and accounting have consistently found either positive or neutral performance differences between socially screened and unscreened investments.

Groups of shareholder activists, representing institutional investors, such as the Interfaith Center on Corporate Responsibility (ICCR), a coalition of 275 religious institutional investors, submit numerous shareholder resolutions annually calling on management to change their company practices. Among the issues are sweatshops and human rights abuses, global warming, equal opportunity, executive compensation, and the election of board members. In 2003, chemical and pharmaceutical companies faced 24 shareholder resolutions, filed by ICCR and other activist shareowners whose concerns ranged from eliminating PVCs in manufactured goods to phasing out the use of dioxin.

In Europe, calls for increased disclosure got a boost in 2001 when the Association of British Insurers (ABI), a 400-member trade association of Britain's insurance industry, issued new guidelines. The ABI members account for more than 20 percent of stockmarket investment in London. These guidelines ask companies to disclose information about the social, environmental, and ethical risks and opportunities they face and how they plan to handle them. ABI officials say the guidelines "represent an important opportunity for investors and companies to work together both to protect shareholder value and improve their understanding of corporate social responsibility" (see www.abi.org.uk).

6.6.3.2 Pressures from Employees. Employees want to be paid well, with competitive wages and salaries, benefits, and increasingly, stock. They want the resources to do their jobs well. And, equally important, they want an employer who treats them fairly and with respect and values their knowledge and different life experiences. Employee perceptions about how a company accepts and manages its social responsibilities are also increasingly part of employee decisions about where to work. Furthermore, unions and related institutions put pressure on companies to reform their labor practices to meet global labor standards.

6.6.3.3 Pressures from Customers. Commercial customers, especially those who have a long-term relationship with or are dependent on a company, have a vested interest in the company's future as a going concern. They want suppliers who are reliable, produce quality products and services at a fair price, and, like employees, customers want to be respected and treated fairly. Most customers also want to know about their suppliers' safety and environmental records. With this in mind, customers want to know about the values and attitudes that underpin their suppliers' business dealings and the risks associated with their business practices, products, and services. This shift towards greater accountability can be seen in the rising number of companies seeking ISO certification and the adoption of nonfinancial auditing standards such as SA8000 and the Ethical Trading Initiative.

Consumers too are increasingly pressuring companies to accept and manage more responsibly through their purchasing power. Many want to know that the products they are buying are environmentally safe and that companies recognize that their "license to operate" is a privilege, not a right. Studies by reputation experts point out that consumers admire most companies that consistently demonstrate their concern for employees, the environment, product quality and reliability, the communities in which they operate, as well as financial performance.

6.6.3.4 Pressures from NGOs and Activists. Aided, in part, by the speed and ease of using the Internet, global activists and NGOs have emerged and are increasingly vocal in their demands that companies adhere to high expectations regarding safety, labor standards, the environment, human rights, and corporate governance. The capacity of activists to mobilize their own resources, disseminate negative information about companies, and take concerted action against practices they find offensive or problematic has never been greater (Waddock *et al.*, 2002).

6.6.3.5 *Pressures from Communities and Governments.*

Communities and even nations, many of which fiercely compete for foreign investment by companies, are becoming aware of the negative consequences of eroding tax bases and a general lack of long-term commitment by companies to where they operate. While still offering generous incentives to induce companies to bring jobs and tax revenues to their communities, municipalities also expect companies to obey the laws, protect the environment, and help solve community problems. Consequently, companies find it increasingly necessary to act or become a "neighbor of choice," by partnering with local citizen groups and living up to high standards of excellence with respect to communities (Burke, 1999).

Within the chemical industry are hundreds of local citizen advisory panels. Company representatives meet several times a year with interested residents, who represent schools, hospitals and health care specialists, businesses, environmental associations, community groups, and municipal authorities, to discuss immediate concerns such as emissions and traffic problems.

The chemical industry is one of the most regulated industries in the developed world. In general these regulations primarily relate to worker safety, air, soil and water quality, groundwater contamination, and waste handling and disposal. In response to the acceleration of a global marketplace, regional and country governmental entities are raising the safety bar for chemical manufacturers and users. One legislative proposal, which has far-reaching implications for the makers and users of chemicals, is REACH (Registration, Evaluation, Authorization of Chemicals). Proposed by the European Commission in May 2003, the REACH System – created to allow users to choose safer alternatives – would require producers, importers, and users of chemicals to provide public information on the risks associated with their use. Registration of chemical substances, an integral part of the new system, would involve the release of extensive information on the intrinsic properties and hazard of each substance, intended use, and potential exposure and risks to people and the environment (see Section 3.4 for a more detailed discussion of REACH).

6.6.3.6 *Pressures from Institutional Developments.*

A number of institutional developments have led to pressures for more responsible corporate behavior, creating a need for greater transparency of and accountability for corporate actions.

A major source of pressure on company performance is the numerous ratings and ranking schemes from prominent business publications in recent years, as well as highly visible awards for best practices. In contrast to traditional corporate rankings that have largely evaluated companies on financial criteria, size and growth, for example, the Fortune 500, these new ratings and rankings evaluate companies' performance with respect to their treatment of a variety of stakeholder issues. *Fortune* magazine's widely recognized "Fortune's Most Admired Companies" has been ranking companies on multiple criteria other than financial since the early 1980s. Employee issues are now covered, for example, in *Working Women* magazine, and *Black Enterprise* magazine. These rankings compete with *Business-Week*'s "Best Companies for Work and Family," rankings. Management quality is covered by *Fortune*'s rating as well as *Industry Week*'s "100 Most Admired" company ratings. Further, global "most admired" rankings of businesses are now published in Europe and Asia (Waddock *et al.*, 2002).

These rankings are more than a name on a page. At 3M, ranked among the top ten companies on The Harris Annual Reputation Survey, company officials believe that its high scores for quality of management and innovation translate into shareholder value by enhancing brand preference amongst consumers and attracting and retaining a diverse and talented workforce (Low and Kalafut, 2002).

Global standards and principles are another source of institutional pressure. The UN's Global Compact represents one prominent example to promote values-based practices in global corporations. The Global Compact is only one of over 400 different initiatives related to codes of conduct, principles, and standards globally, according to the International Labor Office. This proliferation of standards suggests that there are certain baseline expectations or responsibilities to which companies are increasingly expected to advance by a wide range of stakeholders. Codes related to corporate social policy encompass employment issues, training, working conditions, labor relations and child labor, anticorruption, as well as environmental performance and sustainability. Codes of any sort, whether internally or externally generated, will be respected and credible only when they are consistently reported. Public reporting of corporate activities provides the transparency necessary for codes to be implemented and monitored (Waddock *et al.*, 2002).

Standards, principles, and codes are only useful if they are implemented and to the extent that companies can assure stakeholders that they are living up to them. To establish credibility with stakeholders, particularly activists and critics, companies are beginning to engage in more transparent reporting practices, many of which are now emerging from international multistakeholder coalitions. According to one study, some 54 percent of the world's largest companies now disclose some type of social and environmental information on their websites. Perhaps the most prominent reporting and accountability initiative, which is linked to the implementation of both standards and codes, is the Global Reporting Initiative or GRI. Like the Global Compact and many other initiatives, the GRI is voluntary and companies are not currently required to provide external assurance as to the accuracy of the information reported (Waddock and Bodwell, 2002).

For an inventory of CSR standards, codes, guidelines, and organizations, refer to Appendix 2.

6.6.4 Managing Supply Chains Responsibly

For many companies, the supply chain is their greatest challenge, requiring new systems of communication and evaluation. While demanding more from suppliers in the form of higher quality at lower prices and just-in-time delivery, companies are finding their brand inextricably linked to the behavior and practices of their suppliers. Consequently, product brands have often been the target of negative media and watchdog reports on abuses within company supply chains in recent years.

The chemical industry, along with others, has been seriously affected by negative publicity surrounding the operating practices of suppliers, especially with regard to worker safety, labor rights, and the environment. Companies that source from developing countries have been particularly hard hit by labor and environmental activism, low ratings in various corporate rankings and public opinion surveys, consumer activism directed at their products and manufacturing processes, shareholder activism on labor, human rights, and ecological issues, and the antiglobalization movement generally.

Managing supply chains responsibly is complex. Working collaboratively with suppliers, while increasingly more the norm, can require lengthy negotiation and dialog to understand differences and improve relationships and trust. It also takes a functioning relationship and effort of company managers to align the supply chain with the company's values and the full participation of its operations and procurement decision-makers.

Chemical companies own some of the most valuable brands in the world and face a barrage of negative publicity when an accident occurs at their own plants or those of their suppliers. In only a one-month period, June 20–July 20, 2000, 14 significant accidents were reported globally (see AcuSafe™ website). These ranged from a fire at a Canadian acid transformation plant, forcing the evacuation of 3000 nearby residents, to a gasoline pipeline explosion in Nigeria that killed 200 villagers. Whenever accidents like these occur, public interest focuses on the safety conditions at the raw materials producers – the closest link in the supply chain to the brand owner – and the potential harmful effects on surrounding communities. In turn, businesses make even greater demands for improvements in the management and operating practices of their suppliers.

At BASF, suppliers and business partners are expected to observe the International Labour Organization's "Declaration on Fundamental Principles and Rights to Work." The Declaration calls on governments, companies, and worker organizations to uphold employees' right to organize and bargain collectively and to eliminate child labor, forced labor, and discrimination in the workplace. BASF suppliers found to violate these labor standards can be terminated "without notice if necessary" (BASF, 2001).

In negotiating to purchase raw materials, BASF uses a safety matrix to evaluate the dangers posed by their physical and chemical properties. Suppliers are also subject to environmental and safety audits by BASF employees to minimize pollution and to assure their use of safety standards are in accordance with Responsible Care. The company's environmental criteria are supplemented with "minimum social standards" that suppliers are expected to meet, including the prohibition of child labor and the use of forced or bonded workers.

Since the 1990s, many companies have produced social and environmental codes of conduct for their suppliers as a solution to managing their supply chain and averting what otherwise could become a catastrophe. Many companies have also dramatically reduced the number of suppliers to have greater influence over their practices. However, assurance of compliance with sourcing codes remains patchy.

Choices need to be made about the best way of evaluating a supplier's practices. Raw materials and new technologies, for example, may demand tough economic choices between a lower-cost, higher-polluting energy producer or one with cleaner-burning plants. The blanket assumption that every company has the same business model, understands your terminology, or has the means to adhere to what are often multiple standards imposed by their customers no longer applies.

What seem like basic issues of environmental or social standards control are often challenging when applied to global supply chains. For example, companies need to appreciate the nuances of differing local cultures and norms, along with knowing which environmental standards or regulatory requirements a supplier can realistically apply and successfully practice. In addition to knowing where the knowledge gaps exist, companies and suppliers need to negotiate how to close them and where the resources will come from to make the necessary changes.

Furthermore, the cost of adopting a company's code of practice is not always warmly welcomed by most suppliers, while their customer boasts of draining every last cent of value out of the supply chain. It has been proven that the adoption of environmental management systems (EMSs) almost always leads to cost savings. Yet what if a supplier does not have the resources to install one or the knowledge to use it effectively? In addition to taking a leap of faith that savings might exist, suppliers are expected to be patient, because of the time lag between installing systems and getting the expected returns.

Verification of supplier compliance with a company's code of conduct is particularly important to international supply chains. And who will pay for introducing new methods of verification is again a big issue to suppliers.

In addition to using their own internal assessment processes, companies employ external local auditors as well, not only to add credibility to their monitoring process, but also because local auditors have on-the-ground knowledge and languages. Local auditors can uncover important issues much faster than overseas auditors and will do so with a better understanding of how to diplomatically and permanently resolve them. Similarly, because of the evident cultural differences, companies have to be careful not to impose Western values on other cultures. Some companies have gone even further to include local activists in the verification process. By engaging activists, companies can focus the concerns of these groups, while mitigating the reputation threats associated with their criticisms.

Meeting product quality and environmental specifications as well as employing safe and fair working conditions are for some companies the end purpose of supplier auditing so that consumer concerns over product integrity and threats to health and workers' rights and the wider environment are allayed. Having reliable data makes sales and customer relations stronger and gives employees real pride in the company for which they work.

Ashland Distribution Company, a division of Ashland, Inc., offers customers a one-source "closed-loop" process to not only supply chemicals, but also to manage hazardous and nonhazardous wastes. It offers a range of processing and treatment options and compliance assurance throughout North America (Low and Kalafut, 2002).

6.6.5   Protecting and Contributing to the Community

When chemical companies lack a commitment to the communities in which they operate, local people, the environment, and the companies themselves suffer the negative consequences. In Anniston, AL, Monsanto hid its knowledge of the health risks of PCB pollution from the local community for 20 years until the *Washington Post* exposed serious health problems affecting people, animals, and water. One individual lawsuit alone was settled for $43 million and many others are pending.

In a small town in Ohio, household tap water was polluted with C8, a contaminant that stays in the human body for years. Dupont, owner of the Teflon factory across the river, knew about the toxic pollution for decades but kept it secret from the local communities. Now several thousand area residents are suing Dupont for polluting their tap water. And the question lingers that if Monsanto and Dupont hid what it knew from their closest neighbors, what else are these companies hiding from other communities?

Developing nations, many of which compete fiercely for foreign investment by chemical companies, can also suffer from a lack of local community involvement. Chlorine containers that ruptured and then dumped in a pit near Qiqihar, China, recently resulted in 130 people being sent to hospital. In the same month, 23 people were killed and 74 injured in an explosion at a petrochemical complex in Algeria where company officials had been warned about an outdated and faulty boiler. The explosion destroyed three of the refinery's LNG plants and caused the shipping port to close.

Rather than risk this kind of exposure in both reputation and expense, companies will find it increasingly important to become a "neighbor of choice," living up to high standards of excellence in their relationship with their communities both locally and abroad. A two-year collaboration between Dow Chemical and the Natural Resources Defense Council and local activist groups led the chemical producer to voluntarily shift from traditional environmental compliance to pollution prevention. Through a multistakeholder participatory process, community leaders also gained a better understanding of Dow's decision-making process and found areas of common ground with Dow managers (Low and Kalafut, 2002).

If companies do not take the initiative to collaborate with affected stakeholders, they will come under increasing pressure from communities and governments to protect and inform local people. One initiative aimed at strengthening community relationships was the U.S. Emergency Planning and Community Right-to-Know Act (EPCRA) enacted in 1986 after the Bhopal India chemical disaster. A chemical accident is reported in the United States an average of 21 times a day. The EPCRA is designed to inform communities about chemicals and chemical hazards present and transported in the community and to involve the community in developing emergency planning and response. A typical example: a Pawtucket, Rhode Island delivery truck dumps sodium hydroxide into an outdoor tank of hydrogen peroxide and triggers a noxious cloud. Because of EPCRA, the tanks were clearly labeled and the nature of the cloud was immediately known so that schools and residents could be quickly evacuated.

The chemical industry has received a lot of recognition for its Responsible Care program. Originating in 1988, Responsible Care was implemented as a voluntary program to achieve environmental health and safety performance beyond levels required by the U.S. government. Since then the program has resulted in significant performance improvements, including a code of conduct called the Community Awareness and Emergency Response Code of Management. This code helps to achieve some of the original Responsible Care guiding principles:

- To recognize and respond to community concerns about chemicals and our operations;
- To report promptly to officials, employees, customers, and the public, information on chemical-related health or environmental; hazards and recommended protective measures;
- To participate with government and others in creating responsible laws, regulations, and standards to safeguard the community, workplace, and environment;
- To promote the principles and practices of Responsible Care by sharing experiences and offering assistance to others who produce, handle, use, transport, or dispose of chemicals.

The code demands a commitment to openness and community dialog and through community outreach it promotes an open, ongoing dialog with employees and the community. Information is provided about such activities as waste minimization, emission reduction, health effects of chemicals, and efforts to ensure the safe transport of chemicals.

Box 6.1 presents an example of chemical company endorsement of Responsible Care.

BOX 6.1   MILLENNIUM CHEMICALS' COMMITMENT TO SAFETY, HEALTH, AND ENVIRONMENT

Responsible Care®, program developed by the American Chemistry Council, is one of the key drivers of Millennium's achievement of the *vision* to "be the most value-creative chemical company in the world."

Responsible Care reminds us that we have a moral obligation to protect our employees, our neighbors, and the environment. As a critical component of who we are and what we wish to be, Responsible Care is prominently included as its own section of our website.

> Millennium Chemicals is committed to attaining Responsible Care excellence. As a global company, we have developed and implement our own *global standard of excellence* for Responsible Care so that all our operations – in five countries on four continents – have common *performance goals* and metrics for measuring progress.

Historically, the chemical industry has managed its relationship with the public and the community on a voluntary basis. In 1998, the EPA and the chemical industry launched a much-ballyhooed voluntary testing program for high production volume (HPV) chemicals. At the time, 43 percent of the chemicals produced in volumes of one million pounds or more per year had no toxicity data. Nine hundred chemical companies were "invited" to participate in the EPA voluntary testing program. Three years later, about half the companies had responded, but 25 percent of the chemicals identified for testing remained without any commitment from their manufacturers. Voluntary programs are difficult to implement and, by definition, impossible to enforce. But as company behavior becomes more transparent and communities increase pressure on their neighborhood chemical companies, compliance will become more important.

In 2002, the industry reviewed the performance of the Responsible Care program and decided to make endorsement of a new set of management principles mandatory for its members (90 percent of U.S. chemical manufacturing capacity). The objective was to realize greater business value from improved EH&S performance, higher product yield, enhanced operational efficiencies and better community and stakeholder relationships.

Carus Chemical, a manufacturer of chemicals to treat drinking water and waste water, invites local teachers to work in their chemistry laboratories during summer breaks as paid interns. The knowledge they gain is transferred to the classroom to show how science is useful and to upgrade teaching programs.

Today, the USDA is putting more emphasis on whether proposed bio-crops are safe for the environment and human health, a change that might delay or even block the commercialization of some crops genetically engineered to make drugs and industrial chemicals. But if these genetically engineered crops might harm people when mixed with crops intended for food, companies need to recognize that the interests of the community and human health need to come before the commercial progress of science. Chemical companies have the opportunity to take a leadership position with regard to human health and safety and their corporate social responsibilities.

Another area of community impact is the movement of toxic chemicals through areas of human population. Access to toxic and dangerous chemicals poses a threat under normal circumstances and a heightened threat under conditions of terrorism alert. In the District of Columbia passenger trains and freight trains share the same tracks and the presence of graffiti on rail tankers proves how accessible they are to human intervention. Responsible companies are increasing their inspection of tracks and monitoring hazardous material shipments to help protect against terrorist attack. The CEO of Eastman Chemical recently explained, "Each year we report the number of accidents, the amount of hazardous wastes we produce, and how efficiently we're using energy ... we add to that how much money we're spending on environmental protection." All of this and more go onto the Eastman web site. He concludes, "We don't publish this information so people will think we're wonderful or because we're trying to make ourselves feel good. We publish it because it makes good business sense. The communities where we operate are vitally interested in making sure we operate safely and without harm to the environment."

REFERENCES

ACCA and CorporateRegister.com, *Towards Transparency: Progress on Global Sustainability Reporting 2004*, 2004.

AccountAbility, the Institute of Social and Ethical Accountability, *AA1000S Assurance Standard*, March 25, 2003. Available at http://www.accountability.org.uk/aa1000/default.asp?pageid=52, March 13, 2004.

AccountAbility, the Institute of Social and Ethical Accountability, *AA1000 Framework*, November 1999. Available at http://www.accountability.org.uk/uploadstore/cms/docs/AA1000%20Framework%201999.pdf, March 14, 2003.

AIChE Center for Chemical Process Safety, *Guidelines for Analyzing and Managing the Security Vulnerabilities of Fixed Chemical Sites*, American Institute of Chemical Engineers, New York, 2002.

AIChE, *Total Cost Assessment Methodology*, American Institute of Chemical Engineers Center for Waste Reduction Technologies, New York, NY, July 1999.

American Chemistry Council, *Responsible Care® Management System Technical Specification*, Document No. RCMS 101.01, 2004a. Available at http://www.responsiblecaretoolkit.com/pdfs/RCMSTech_012504.pdf?opendocument&Login.

American Chemistry Council, *Responsible Care® ToolKit*, 2004b. Available at http://www.responsiblecaretoolkit.com/rcms.asp, October 25, 2004.

American Institute of Certified Public Accountants, *FAQs on Sustainability Reporting*, October 4, 2002. Available at http://www.aicpa.org/innovation/baas/environ/faq.htm, March 15, 2004.

ASIS, *Protection of Assets Manual*, American Society of Industrial Security, 2004.

B. R. Bakshi and J. Fiksel, "The Quest for Sustainability: Challenges for Process System Engineering," *AIChE J*, 49(6), 1350–1358 (2003).

J. C. Bare, G. A. Norris, D. W. Pennington and T. McKone, "TRACI: The Tool for the Reduction and Assessment of Other Environmental Impacts," *Journal of Industrial Ecology*, 6(3–4), 49–78 (2003). Software available at http://epa.gov/ORD/NRMRL/std/sab/iam_traci.htm#download.

BASF, *Values Create Value*, BASF Social Responsibility Report 2001, BASF, Ludwigshafen, Germany, 2001.

D. Batstone, *Saving The Corporate Soul & (Who Knows?) Maybe Your Own*, Jossey Bass, San Francisco, 2003.

R. Bauer, K. C. G. Koedijk and R. Otter, "International Evidence of Ethical Mutual Fund Performance and Investment Style," Linberg Institute of Financial Economics Working Paper No. 02.59, 7 March 2002.

B. Beloff and E. Beaver, "Sustainability Indicators and Metrics of Industrial Performance," Paper SPE 60982 presented at the SPE International Conference on Health, Safety, and Environment in Oil and Gas Exploration and Production, Stavanger, Norway, 26–28 June, 2000.

B. Beloff, E. Beaver and H. Massin, "Assessing Societal Costs Associated with Environmental Impacts," *Environmental Quality Management Journal*, 10(2), 67–81 (2000).

B. Beloff, D. Tanzil and L. Chuzhoy, "Assessment of Potential Life Cycle Environmental Impacts of Microstructure-Level Simulation for a Steel Component: An Exploratory Research," final project report to the National Science Foundation, May 2004.

M. Bennett and P. James, Eds., *Sustainable Measures*, Greenleaf Publishing Ltd., Sheffield, 1999.

Board of Environmental, Health and Safety Auditor Certifications, *Standards for the Professional Practice of Environmental, Health and Safety Auditing*, 1999. Available at http://www.beac.org, March 14, 2004.

Board of Environmental, Health and Safety Auditor Certifications, *BEAC CPEA Management System Examination Study Guide*, 2004. Available at www.beac.org/FinlRevBEAC%20CPEA-MS%20Study%20Guide%209%20dec.pdf, March 19, 2004.

E. Burke, *Corporate Community Relations: The Principles of Neighborhood Choice*, Praeger, Westport, CT, 1999.

Burson-Marsteller, *2003 Building CEO Capital*, Burson-Marsteller, New York, 2003.

Business for Social Responsibility, *Comparison of Selected Corporate Social Responsibility Related Standards*, November 2000, p. 3 footnote.

Business for Social Responsibility, *Verification*, 2003. Available at http://www.bsr.org/BSRResources/WhitePaperDetail.cfm?DocumentID=440, September 9, 2003.

California Energy Commission (CEC), "1992 Electricity Report," CEC, 1993.

Calvert Group, *Calvert Announces 2004 Shareholder Resolutions*, February 9, 2004 press release. Available at http://www.calvert.com/pressindex_newsArticle.asp?article=3796&image=&keepleftnav=Press + Releases, March 15, 2004.

Cap Gemini Ernest & Young, *Measures That Matter*, A survey of 575 analysts as well as interviews with portfolio managers, Cap Gemini Ernest & Young, Paris, 1996.

Caux Round Table, *Caux Round Table Principles for Business*, 1994. Available at http://www.cauxroundtable.org/documents/PrinciplesforBusiness.doc, March 14, 2004.

Caux Round Table, *Introduction to Caux Round Table Principles for Business*, 2004. Available at http://www.cauxroundtable.org/principles.html, March 14, 2004.

N. Chambers, C. Simmons and M. Wackernagel, *Sharing Nature's Interest. Ecological Footprints as an Indicator of Sustainability*, London, Sterling, VA, Earthscan, 2000.

Coalition for Environmentally Responsible Economies, *Coalition for Environmentally Responsible Economies (CERES) Principles*, 1989. Available at http://www.ceres.org/our_work/principles.htm, March 14, 2004.

C. B. Cobb and W. Hunter, "Measuring What Matters: The Value Creation Indices for the Energy, Utility, and Chemical Industries," in *Perspectives on Business Innovation Special*

Issue: *The Adaptive Enterprise: Energy, Utilities & Chemicals*. Cap Gemini Ernst & Young Center for Business Innovation, Paris, 2002.

Committee on Industrial Environmental Performance Metrics, *Industrial Environmental Performance Metrics, Challenges and Opportunities*, Committee on Industrial Environmental Performance Metrics, National Academy of Engineering, National Research Council, 1999.

D. J. C. Constable, A. D. Curzons, L. M. Freitas dos Santos, G. R. Geen, R. E. Hannah, J. D. Hayler, J. Kitteringham, M. A. McGuire, J. E. Richardson, P. Smith, R. L. Webb and M. Yu, *Green Chemistry*, 3, 7–9 (2001).

Corporate Environmental Performance 2000, Vol. 1, Strategic Analysis, Haymarket Business Publications, Ltd., 1999.

R. Costanza, R. d'Arge, R. de Groot, S. Farber, M. Grasso, B. Hannon, K. Limburg, S. Naeem, R. V. O'Neill, J. Bruello, R. G. Raskin, P. Sutton and M. van den Velt, "The Value of the World's Ecosystem Services and Natural Capital," *Nature*, 387, 253–260 (1997).

K. L. Coyne, "Overview: Corporate Social Responsibility and Sustainable Development Standards." Presentation made to The Auditing Roundtable, New York Metro–New England Regional Meeting, March 6, 2002.

A. D. Curzons, D. J. C. Constable, D. N. Mortimer and V. L. Cunningham, *Green Chemistry*, 3, 1–6 (2001).

CWRT (Center for Waste Reduction Technologies), "Sustainability Metrics Interim Report No. 1," American Institute of Chemical Engineers, New York, 1998. Available at http://aiche.org/cwrt/pdf/smwi.pdf.

CWRT 1999

G. Daily, *Nature's Services: Societal Dependence on Natural Ecosystems*, Island Press, Washington, DC, 1997.

R. S. DeGroot, M. A. Wilson and R. M. J. Boumans, "A Typology for the Classification, Description and Valuation of Ecosystem Functions, Goods and Services," *Ecological Economics*, 41, 393–408 (2002).

Det Norske Veritas and Vestlandsforskning, "Technical Report: CSR Management Systems," Report Number 2002-1072, Revision 1, p. 26, DNV, Hovick, Norway, 2002.

B. Dittrich-Krämer, P. Saling, N. Külzer, A. Bastian, H. Heissler and K. Siemensmeyer, "Kurs Nachhaltigkeit," *UmweltMagazin*, 12, 24–25 (2002).

R. Earle, "The Emerging Relationship Between Environmental Performance and Shareholder Value," The Assabet Group White Paper, 2000, p. 4.

S. Emerson, *American Jihad: The Terrorists Living Among Us*, The Free Press, New York, 2002.

EPA (Environmental Protection Agency), §40 CFR 68, EPA, Washington, DC, 1996.

M. J. Epstein and P. S. Wisner, "Using a Balanced Scorecard to Implement Sustainability," *Environmental Quality Management*, 11(2), 1–10 (2001).

J. Fiksel, J. Low and J. Thomas, "Linking Sustainability to Shareholder Value," *EM Magazine*, June 2004.

J. Fiksel, K. Funk, P. C. Kalafut and J. Low, *Clear Advantage: Building Shareholder Value. Environment Value to Investors*, Global Environmental Management Initiative, Washington, DC, February 2004.

A. Freeman, *The Measurement of Environmental and Resource Values: Theory and Methods*, Resources for the Future, Washington, DC, 1993.

Global Environmental Management Initiative (GEMI), *Measuring Environmental Perform-ance: A Primer and Survey of Metrics In Use*, GEMI, Washington, DC, 1998.

Global Reporting Initiative, *GRI and Other Initiatives*, 2004. Available at http://www.global reporting.org/about/iniaa1000.asp, March 5, 2004.

Global Reporting Initiative, *Sustainability Reporting Guidelines 2002*, 2002. Available at http://www.globalreporting.org/guidelines/2002.asp, March 5, 2004.

Global Reporting Initiative, *Sustainability Reporting Guidelines*, GRI, 2002. Available at www.globalreporting.org.

M. J. Goedkoop and R. S. Spriensma, *The Eco-indicator 99, Methodology Report. A Damage Oriented LCIA Method*, VROM, The Hague, The Netherlands, 1999.

M. Goyder, *Redefining CSR*, Tomorrow's Company, July 2003.

P. F. Greenfield, "The Way Ahead," The Australian Academy of Technological Sciences and Engineering, Sustainable Australia, Academy Symposium, November 2000. Available at http://www.atse.org.au/publications/symposia/proc-2000p20.htm, March 4, 2004.

C. Holliday and J. E. Pepper, *Walking the Talk: The Business Case for Sustainable Develop-ment*, World Business Council for Sustainable Development, Geneva, Switzerland, 2001.

J. T. Houghton, Y. Ding, D. J. Griggs, M. Noguer, P. J. van der Linden, X. Dai, K. Maskell and C. A. Johnson, Eds., *Climate Change 2001: The Scientific Basis*, Cambridge University Press, Cambridge, UK, 2001.

G. Hughes, "Environmental Indicators," *Annals of Tourism Research*, 29(2), 457–477 (2002).

ICCA, *On the Road to Sustainability – A Contribution from the Global Chemical Industry to the World Summit on Sustainable Development*, ICCA, August 2002.

Institution of Chemical Engineers (IChemE), *The Sustainability Metrics: Sustainable Devel-opment Progress Metrics Recommended for Use in the Process Industries*, IChemE, Rugby, UK, 2002.

Interfaith Center on Corporate Responsibility, "The Principles for Global Corporate Respon-sibility: Bench Marks for Measuring Business Performance," *ICCR, 2003 Corporate Examiner*, 31(4–6) (2003). Available at http://www.bench-marks.org/downloads/ Bench%20Marks%20-%20full.doc.

International Institute for Sustainable Development, *Assessing Sustainable Development: Principles in Practice*, International Institute for Sustainable Development, Winnipeg, Manitoba, Canada, 1997, p. 1.

International Organization for Standardization, *Environmental Management Systems-Specification with Guidance for Use* (ISO 14001:1996(E)), International Organization for Standardization, Geneva, Switzerland, 1996.

International Organization for Standardization, *Guidelines for Quality and/or Environmental Management Systems Auditing (ISO 19011:2002(E))*, International Organization for Stan-dardization, Geneva, Switzerland, 2002.

International Organization for Standardization, *ISO to Go Ahead with Guidelines for Social Responsibility*, June 29, 2004 press release. Available at http://www.iso.org/iso/en/ commcentre/pressreleases/2004/Ref924.html, October 1, 2004.

Ipsos Reid, "Reputation Management," Presentation at IQPC Conference: Communications and Media Relations for Corporate Social Responsibility, August 27, 2003, Toronto, Canada.

C. Jimenez-Gonzalez, A. D. Curzons, D. J. C. Constable, M. R. Overcash and V. L. Cunning-ham, "How Do You Select the 'Greenest' Technology? Development of Guidance for the Pharmaceutical Industry," *Clean Products and Processes*, 3, 35–41 (2001).

C. Jimenez-Gonzalez, D. J. C. Constable, A. D. Curzons and V. L. Cunningham, "Developing GSK's Green Technology Guidance: Methodology for Case-Scenario Comparison of Technologies," *Clean Technology Environmental Policy*, 4, 44–53 (2002).

K. Kimball, "EHS Auditing Tools: Past, Present, and Future: the Industry Representative's Perspective," Presentation made to The Auditing Roundtable 2002 Winter Meeting, January 7–9, 2002, Phoenix, AZ.

D. Koch, "Dow Chemical Pilot of Total 'Business' Cost Assessment Methodology: A Tool to Translate EHS 'Right Things To Do' into Economic Terms," *Environmental Progress*, 21 (1), 20–28 (2002).

KPMG and Universiteit van Amsterdam, *KPMG International Survey of Corporate Sustainability Reporting 2002*, KPMG, The Netherlands, June 2002.

W. Li, "Environmental Management Indicators for Ecotourism in China's Nature Reserves: A Case Study in Tianmushan Nature Reserve," *Tourism Management*, 25(5), 559–564 (2004).

J. Loh and M. Wackernagel, Eds., *Living Planet Report 2004*, WWF – World Wide Fund For Nature (formerly World Wildlife Fund), Gland, Switzerland, 2004.

J. Low and P. Kalafut, *Invisible Advantage*, Perseus Press, Cambridge, MA, 2002.

M. Mathis, A. Fawcett and L. Konda, "Valuing Nature: A Survey of the Non-Market Valuation Literature," HARC Discussion Paper VNT-03-01, Houston Advanced Research Center, The Woodlands, Texas, 2003.

D. R. McCubbin and M. A. Delucchi, "The Health Costs of Motor-Vehicle-Related Air Pollution," *Journal of Transport Economics and Policy*, 33(3), 253–286 (1999).

A. McWilliams and D. Siegel, "Corporate Social Responsibility: Correlation or Misspecification?" *Strategic Management Journal*, 21, 603–609 (2000).

Minnesota Public Utilities Commission (MN PUC), "Order Affirming in Part and Modifying in Part Order Establishing Environmental Cost Values," Docket No. E-999/CI-93-583 before the Minnesota Public Utilities Commission, July 2, 1997.

R. Mitchell and R. Carson, *Using Surveys to Value Public Goods: The Contingent Valuation Method*, Resources for the Future, Washington, DC, 1989.

National Research Council (NRC), *Industrial Environmental Performance Metrics*, National Academy Press, Washington, DC, 1999.

National Round Table on the Environment and the Economy (NRTEE), *Measuring Eco-efficiency in Business: Feasibility of a Core Set of Indicators*, Renouf Publishing Co., Ottawa, 1999.

K. Nelson, "Applying Sustainability Metrics at The Stanley Works," presented at the 11th International Conference of the Greening of Industry Network, San Francisco, October 12–15, 2003.

V. Olgay and J. Herdt, "The Application of Ecosystems Services Criteria for Green Building Assessment," *Solar Energy*, 77(4), 389–398 (2004).

Organisation for Economic Cooperation and Development, *OECD Guidelines for Multinational Enterprises*, June 27, 2000. Available at http://www.oecd.org/dataoecd/56/36/1922428.pdf, March 13, 2004.

R. L. Ottinger, D. Wooley, D. R. Hodas, N. A. Robinson, S. E. Babb, S. C. Buchanan, P. L. Chernick, E. Caverhill, A. Krupnick and U. Fritsche, *Environmental Costs of Electricity*, Oceana Publications, New York, 1990.

L. S. Paine, *Value Shift*, McGraw-Hill, New York, 2003.

PricewaterhouseCoopers, LLP, *Sustainability Solutions Service: Assurance of Non-Financial Information*, 2004. Available at http://www.pwcglobal.com/Extweb/service.nsf/docid/ 9A1196B7E2D49BC985256D890078A352, March 13, 2004.

F. Rienhardt, "Bringing the Environment Down to Earth," *Harvard Business Review*, 77(4), 149–157 (1999).

SAI, *Accreditation Process*, 2004. Available at http://www.sa-intl.org/Accreditation/ Accreditation.htm, October 1, 2004.

P. Saling, A. Kicherer, B. Dittrich-Krämer, R. Wittlinger, W. Zombik, I. Schmidt, W. Schrott and S. Schmidt, "Eco-Efficiency Analysis by BASF: The Method," *International Journal of Life-Cycle Assessment*, 7(4), 203–218 (2002).

P. Saling, A. Kicherer, B. Dittrich-Krämer, R. Wittlinger, W. Zombik, I. Schmidt, W. Schrott, and S. Schmidt, "Eco-efficiency Analysis by BASF: The Method," *International Journal of Life-Cycle Assessment*, 7(4), 203–218 (2002).

I. Schmidt, P. Saling, W. Reuter, M. Meurer, A. Kicherer and C.-O. Gensch, "SEEbalance – Managing Sustainability of Products and Processes with the Socio-Eco-Efficiency Analysis by BASF," *Greener Management International*, 2005, accepted for publication.

J. Schwarz, B. Beloff and E. Beaver, "Use Sustainability Metrics to Guide Decision-Making," *Chemical Engineering Progress*, 77(4), 58–63 (2002).

J. Schwarz, E. Beaver and B. Beloff, "Sustainability Metrics: Making Decisions for Major Chemical Products and Facilities," final report to the U.S. Department of Energy Office of Industrial Technologies, BRIDGES to Sustainability, Houston, 2000.

A. M. Shane and T. E. Graedel, "Urban Environmental Sustainability Metrics: A Provisional Set," *Journal of Environmental Planning and Management*, 43(5), 643–663 (2000).

S. K. Sikdar, "Sustainable Development and Sustainability Metrics," *AIChE J*, 49(8), 1928–1932 (2003).

K. A. Small and C. Kazimi, "On the Costs of Air Pollution from Motor Vehicles," *Journal of Transport Economies and Policy*, 29, 7–32 (1995).

V. K. Smith, "JEEM and Non-market Valuation," *Journal of Environmental Economics and Management*, 39, 351–374 (2000).

SmithOBrien, *Corporate Responsibility Audit*™, 1995. Available at www.smithobrien.com.

Social Accountability International (SAI), *SA 8000*, 1998. Available at http://www. sa-intl.org/SA8000/SA8000.htm, March 13, 2004.

Social Investment Forum, "Report on SRI Trends in the United States," 2001.

Specialty Technical Publishers, *ISO 14001 Environmental Management Systems*, Specialty Technical Publishers, North Vancouver, British Columbia, 2004, pp. Intro-1, 1–14.

(SRI) Social Investment Forum, *Report on Socially Responsible Investing Trends in the United States*, SRI, Washington, DC, 2003.

B. Steen, "A Systematic Approach to Environmental Priority Strategies in Product Development (EPS). Version 2000 – General System Characteristics," Report 1999:4, Centre for Environmental Assessment of Products and Material Systems (CPM), Chalmers University of Technology, Sweden, 1999.

I. Steenken-Richter, R. Wittlinger and U. Baus, "Eco-Efficiency Analysis – A Tool for the Future," in *Proceedings of the 29th International Aachen Textile Conference*, Aachen Textile Press, Aachen, Germany, pp. 104–111 (2002).

Stratos Inc., *Stepping Forward: Corporate Sustainability Reporting in Canada*, Stratos Inc., 2001. Available at www.stratos-sts.com.

Stratos Inc., *Building Confidence: Corporate Sustainability Reporting in Canada*, Stratos Inc., 2003. Available at www.stratos-sts.com.

R. L. Sullivan, *Global Sullivan Principles for Social Responsibility*, 1999. Available at http://www.globalsullivanprinciples.org/.

D. Tanzil, G. Ma and B. R. Beloff, "Automating the Sustainability Metrics Approach," presented at the AIChE Spring Meeting, New Orleans, April 25–29, 2004.

The Auditing Roundtable, Inc., *Standards for the Performance of EHS Audits*, 1993. Available at http://www.auditing-roundtable.org/members_area/standards/ehs_audits. html, March 14, 2004.

The Auditing Roundtable, Inc., *Standard for the Design and Implementation of an Environmental, Health and Safety Audit Program*, 1997. Available at http://www.auditing-round table.org/members_area/standards/ehs_program.html, March 13, 2004.

The Sigma Project, *Sigma Guidelines* – Toolkit, Sustainability Accounting Guide, September 2003. Available at http://www.projectsigma.com/Toolkit/SIGMASustain abilityAccounting.pdf, March 15, 2004.

H. A. Udo de Haes, G. Finnveden, M. Goedkoop, M. Hauschild, E. G. Hertwich, P. Hofstetter, O. Jolliet, W. Klöpffer, W. Krewitt, E. Lindeijer, R. Müller-Wenk, S. I. Olsen, D. W. Pennington, J. Potting and B. Steen, Eds., *Life-Cycle Impact Assessment: Striving Towards Best Practice*, Society of Environmental Toxicology and Chemistry (SETAC), Pensacola, FL, 2002.

United Nations Commission on Sustainable Development, "Indicators of Sustainable Development: Guidelines and Methodologies," United Nations, 2001. Available at http://www.un.org/esa/sustdev/natlinfo/indicators/isdms2001/isd-ms2001isd.htm.

United Nations, *UN Global Compact*, 1999. Available at http://www.unglobalcompact.org/Portal/Default.asp, March 13, 2004.

V. Veleva and M. Ellenbecker, "Indicators of Sustainable Production: Framework and Methodology," *Journal of Cleaner Production*, 9, 519–549 (2001).

H. A. Verfaillie and R. Bidwell, *Measuring Eco-efficiency: A Guide to Reporting Company Performance*, World Business Council for Sustainable Development (WBCSD), Geneva, 2000.

J. Vidal, "We Forgot to Listen, Says Monsanto," *The Guardian*, October 7, 1999.

D. Vogel, *Lobbying the Corporation*, Basic Books, New York, 1977.

M. Wackernagel, N. B. Schultz, D. Deumling, A. C. Linares, M. Jenkins, V. Kapos, C. Monfreda, J. Loh, N. Myers, R. Norgaard and J. Randers, "Tracking the Ecological Overshoot of the Human Economy," *Proceedings of the National Academy of Sciences USA*, 99(14), 9266–9271 (2002).

S. Waddock and C. Bodwell, "From TQM to TRM: Emerging Total Responsibility Management Approaches," *Journal of Corporate Citizenship*, Summer 2002.

S. Waddock and N. Smith, "Corporate Responsibility Audits: Doing Well by Doing Good," *Sloan Management Review*, 41(2), 75–83 (2000).

S. Waddock, C. Bodwell and S. Graves, "Responsibility: The New Business Imperative." *Academy of Management Executives*, 16(2), 2002.

F. Wagner and H. Cannon, "Distilling the Sarbanes-Oxley Act for EHS Auditors," Presentation made to The Auditing Roundtable 2003 Fall Meeting, Baltimore, MD, September 3–5, 2003.

Q. W. Wang and D. J. Santini, "Monetary Values of Air Pollutant Emissions in Various US Regions," *Transportation Research Record #1475*, Transportation Research Board, Washington, DC, 1995.

M. Webber, "The Economic Impact of Global Warming," *World Business Review*, 14 October 2002.

World Business Council for Sustainable Development, *Sustainable Development Reporting: Striking the Balance*, June 1, 2003a, p. 10. Available at http://www.wbcsd.org/DocRoot/GGFpsq8dGngT5K56sAur/20030106_sdreport.pdf, March 15, 2004.

World Business Council for Sustainable Development, "Sustainable Management is Key to Long-term Success," December 9, 2003 press release, 2003b. Available at http://www.wbcsd.org/plugins/DocSearch/details.asp?type=DocDet&DocId=MzMzMg, March 15, 2004.

World Business Council for Sustainable Development, *Accountability and Reporting*, 2004a. Available at http://www.wbcsd.org/templates/TemplateWBCSD1/layout.asp?type=p&MenuId=ODg, March 14, 2004.

World Business Council for Sustainable Development, "The WBCSD Identified as 'Influential Forum' by Recent World Bank Survey," WBCSD press release, 2004b. Available at http://www.wbcsd.org/plugins/DocSearch/details.asp?type=DocDet&DocId=-MzkONQ, March 14, 2004.

World Meteorological Organization (WMO), "Scientific Assessment of Ozone Depletion: 1998," Report No. 44, WMO, Geneva, Switzerland, 1999.

M. Wright, D. Allen, R. Clift and H. Sas, "Measuring Corporate Environmental Performance: The ICI Environmental Burden System," *Journal of Industrial Ecology*, 1(4), 117–127 (1997).

7

FUTURE DIRECTIONS FOR THE CHEMICAL INDUSTRY

7.1 SUSTAINABLE DIRECTIONS FOR THE CHEMICAL INDUSTRY: A LOOK TO THE FUTURE

KEN GEISER

Lowell Center for Sustainable Production, University of Massachusetts–Lowell

The chemical industry arose in England and Germany during the 19th century as a means of synthetically manipulating natural materials into products that could be sold to customers. The industry was organized around a simple materials flow model that involved extracting minerals, biomass, or fossil fuels from the earth and processing them into useful substances that could be bought by customers interested in making products or services. The customers of the chemical industry went on to manufacture a broad array of valuable products that have been used and dispersed, or used and then disposed back into the environment. Either way, these chemicals have often been returned into the different media of the environment in ways that have compromised the balances of the ecological processes. The simple linear material flow model has inadequately accounted for environmental effects. The model assumes that the natural environment has an endless supply of raw materials for extraction and an unlimited capacity to assimilate wastes.

Such a material flow model and such a system of chemical manufacturing is not sustainable. The natural systems of the planet are not linear, one-pass systems. Instead, the planet's systems are remarkably sophisticated cyclical systems in which materials and energy constantly flow through repeating cycles. The homeostatic equilibrium of ecological processes is resilient up to a point, but the torrent of synthetic chemical products and wastes that have been produced by the chemical

Transforming Sustainability Strategy into Action: The Chemical Industry, Edited by B. Beloff, M. Lines, and D. Tanzil
Copyright © 2005 John Wiley & Sons, Inc.

industry over the past 60 years now threatens the "natural services" of the planet (Daily, 1997).

The transition to a sustainable chemical industry requires a thorough reconceptualization of the industry and its products. Future generations will continue to need chemicals and the industrial transformation of chemicals, to meet human needs, will continue to require ingenuity and enterprise. However, the types of chemicals and how they are used must be significantly reconsidered. Fossil fuels will need to play a much smaller role, and wastes from production and consumption will need to be managed and recycled in ways that conserve materials and protect the environment.

This transition will require a new mission for the industry that promotes human health and environmental quality as seriously as the market promotes economic efficiency and product effectiveness. Put conceptually, a sustainable chemical industry would be one that optimizes value from the use of chemicals, adds no new risks to everyday life, increases natural capital, minimizes the transfer of risks from one generation to another, respects and enhances the natural functioning of the planet's ecosystems, and assures no net loss of valuable resources. Creating such an industry will require government policy and market incentives that promote sustainability. This will require financial investment institutions and international development programs that are committed to developing an industry that is as ecologically sound and socially sensitive as it is economically productive.

The avenues for this development are already being laid. Leading firms in the industry and thoughtful government leaders are exploring new goals and new directions. Some of the most progressive firms have established corporate sustainability policies and many of these firms publish annual environmental reports. The chemistry and chemical engineering fields have responded with new professional statements, conferences that explore sustainable directions, and educational curricula and texts that integrate environmental considerations into conventional education (Allen and Rosselot, 1997; Allen and Shonnard, 2002).

A brief review of two of these new directions – chemical stewardship (services) and green chemistry (function) – and then a look to the future, offer illustrations of how the industry may be moved towards sustainability.

7.1.1   Chemical Stewardship: Services

Research on new chemicals, new routes of chemical synthesis, new feedstocks, and new chemical services have begun to pay off with cleaner production systems, reduced energy consumption, and products that are more easily recycled or biologically disposed. Indeed, a new conception of the chemical industry is emerging that sees the industry as a service industry as much as a materials industry. There is a growing business in chemical management services that provide management services where compensation is based on performance-based metrics rather than just chemical sales. Large chemical manufacturers such as Henkel, Ashland Specialty Chemical, BP, and PPG Industries have expanded their businesses to include these services. Some of these chemical service contracts involve retaining full or partial responsibility of chemicals while they are used. This is called chemical stewardship (*Chemical Week*, 2003).

Under the principles of chemical stewardship, chemical management services maintain control and even ownership of chemicals while assisting users to use them safely and efficiently and reclaiming them when the customer's use is finished. The chemical manufacturing firm that produces a chemical maintains management responsibility for the chemical either by maintaining ownership of the chemical and "leasing" the chemical to customers or by standing ready to take back the chemical when its useful life is over. For instance, Ashland Specialty Chemicals provides a Total Chemical Management service for product manufacturers that purchases chemicals from suppliers, stores them for just-in-time delivery, manages them during their application, supervises waste chemical collection, and finds recyclers for the chemical wastes. Providing chemical management services to Motorola's semiconductor fabrication facilities, Ashland worked with the production managers to push more wafers through the processes and reduce the amount of chemicals used and the amount of chemical wastes generated.

DuPont Canada has developed a new business model that redefines its relationship to its key customers through managing material supply chains. Working with the Ford Motor Company on processes for applying automobile finishes, the firm found it could reduce the amount of paint used, close-loop the painting system, turn waste streams into useful products, and, thereby, develop new ways to generate revenue (GEMI, 2001).

Moving the chemical industry towards sustainability requires developing new business models, based less on selling chemicals and more on selling the services of chemicals. This means maintaining some form of the responsibility for chemicals throughout their lifecycles and being capable of taking back chemicals that cannot be safely released to the environment, and effectively decomposing them or preparing them for reuse.

7.1.2 Green Chemistry: Materials and Function

The state of chemistry, biology, and physics, and knowledge about physiology and toxicology has advanced dramatically over the past half century. We know far more about what makes materials toxic and how to make safer chemicals than we once did. Here lies the opportunity for the material sciences. Many chemists and engineers today recognize that there is adequate knowledge to design chemicals and chemical processes that pose less risk to human health and the environment. Yet, seldom are material scientists asked to turn their design talents to minimizing or eliminating the hazards of materials. If material scientists and synthetic chemists were challenged to come up with less toxic materials, product designers and process engineers would be better able to produce more environmentally benign products and processes. Already, there is a growing number of chemists and chemical engineers interested in the process of environmentally compatible chemical process design.[1]

---

[1] For an extensive review, see Cano-Ruiz and McRae (1998).

The idea of using existing chemistry knowledge to design more environmentally friendly chemicals and chemical processes has opened up a new specialty in chemistry referred to as environmentally benign chemical synthesis, or "green chemistry." Green chemistry has been defined to mean "the utilization of a set of principles that reduces or eliminates the use or generation of hazardous substances in the design, manufacture and application of chemical products" (Anastas and Warner, 1998).

By taking what toxicologists and ecologists already know, it is possible to generate a set of design criteria that could be used to guide the development of more environmentally friendly substances. Paul Anastas and John Warner in their text on green chemistry lay out 12 design principles for green chemistry (Anastas and Warner, 1998). (See Chapter 5 for more in-depth coverage of the 12 principles.)

There appear to be two complementary trends in green chemistry. The first is based on chemistry and chemical engineering and focuses on developing alternatives to current production technologies and practices. This includes research on alternative feedstocks, environmentally benign solvents, reagents, and catalysts, aqueous processing, and safer and more readily recyclable chemical products. The second is based more on biology, toxicology, and ecology and focuses more on developing chemistries that reduce or eliminate materials or processes that are not compatible with biological integrity and ecological functioning. This involves research on nonpersistent, nonbioaccumulative, ecocompatible materials.

Both trends begin with research on alternative feedstocks. This involves preferring renewable (bio-based) feedstocks over nonrenewable (petroleum-based) sources and seeking starting materials that demonstrate the least hazardous properties (e.g., toxicity, flammability, accident potential, ecosystem incompatibility, ozone-depleting potential). For example, glucose can be used as a raw material rather than benzene in the production of hydroquinone, catechol, and adipic acid, all of which are important intermediates in the production of commodity chemicals. Current research has shown that relatively nontoxic silicon is a useful replacement for carbon as a starting base for the synthesis of some organic chemicals (DeVito and Garrett, 1996).

Additional research focuses on alternative reagents and catalysts. This involves identifying catalysts that function in chemical transformations with minimal environmental harm (e.g., minimize energy inputs, maximize yield, minimize waste outputs, generate the least occupational exposure and accident potential). For instance, addition reactions are preferred over subtraction reactions, because they incorporate much of the starting materials and are less likely to produce large amounts of waste. Alternatives to the heavy metal catalysts are sought, because the common metal catalysts are so often extremely toxic. The use of liquid oxidation reactors replaces metal oxide catalysts with pure oxygen and permits lower temperature and pressure reactions with higher selectivity and no metal contaminated wastes. New catalysis techniques that rely on enzymes, microwaves, ultrasound, or visible light may obviate the need for harsh chemical catalysts.

Research on alternatives to halogenated solvents includes investigations of the aqueous chemistries, ionic liquids, immobilized solvents, and supercritical fluids.

Water has been shown to be an effective solvent in some chemical reactions such as free radical bromination. Supercritical fluids such as liquified carbon dioxide are already commonly used in coffee decaffeination and hops extraction. However, supercritical carbon dioxide can also be used as a replacement for organic solvents in polymerization reactions and surfactant production. Future work may involve solventless or "neat" reactions such as molten-state reactions, dry grind reactions, plasma-supported reactions, or solid materials-based reactions that use clay or zeolites as carriers.

These green chemistry initiatives have received a substantial boost by the U.S. government's sponsorship of an annual Presidential awards ceremony for the best examples of green chemistry applications. Over the past several years, this awards program has recognized Bayer's environmentally friendly synthesis of bio-degradable chelating agents, PPG Industry's use of yttrium as a substitute for lead in cationic electrocoatings, and Rohm and Hass's design of an environmentally safe marine antifouling coating to replace tributyltin oxides. For other examples see the Internet site: www.epa.gov/greenchemistry.

7.1.3 A Look to the Future

This chapter also takes a broad, futuristic, and macroview of the chemical industry. We find here ideas that link chemical production to biological processes, the phase out and substitution of the most dangerous chemicals, and the dawn of new means of transforming chemicals at the molecular and submolecular level.

Europe has long played a patriarchal role to the chemical industry. Many of the roots of the industry and so much of the leading science have arisen there that it is not surprising that Europeans should put forth some of the most far-reaching policies on managing chemical risks. Frustrated with the endless production of dangerous chemicals and the slow pace of innovation that could lead to safer substitutes, the European Union has proposed a massive overhaul of government policy that would shift the burden of testing chemicals and assuring their safety onto the industry itself. In Section 7.2.1, Beverley Thorpe takes a critical look at these new policies and finds that the industry is as likely to resist these overtures as embrace them and there is a whole lot of negotiation that is going to take place before the conventional government policies are transformed.

Joanna D Underwood takes us into the world of commercial products in Section 7.2.2. It is too easy to see the chemical industry as quite distant from the local toy store or furniture maker. However, in a world where commodities are increasingly made from petrochemicals and polymers, which are derived from fossil fuels, the resulting products are often a puree of scores of chemicals, many added to control and tame the reactivity and decomposition of the fuel. Underwood notes how frequently many of these substances are persistent, bioaccumulative, and toxic, a combination of factors that when present in domestic products sets ripe conditions for human and ecological harm. She concludes that many of these substances have no place in a sustainable chemical industry.

Mark Dorfman, in Section 7.3.1, describes an alternative path for chemical processes that occurs under ambient conditions, requires small amounts of energy, generates high yields, and produces limited wastes that are easily consumed by other natural processes because they mimic natural processes. For thousands of years people have been using natural processes such as microbial processing, biodegradation, and fermentation to make chemicals. More recently, biosynthetic routes for making industrial organic feedstocks, acids, and polymers have been developing from the work of organic chemists, microbiologists, polymer scientists, and those involved in the loosely organized subfield of biomimicry. Dorfman presents an encouraging overview of this rapidly developing science and describes a wide range of naturally based substances and products already on the market or soon to be.

While biomimicy seeks to direct chemical production to the most obvious natural processes, Mark Wiesner, in Section 7.3.2, looks ahead at the industry's new fascination with the obscurity of very tiny processes. Nanotechnology offers the possibility of manipulating substances at the molecular and atomic levels with a precision and efficiency never before possible without technologies like the atomic force microscope. While there is much promise in these new processes, there is also much to be concerned about. Here, it is not the technology, but the manner in which the chemical industry will use the technology that raises cautions. It remains to be seen whether this new technology will emerge along the same paths that brought the public to fear atomic energy and genetically modified biotechnology or whether the industry will learn from the past and direct the development of nanochemistries towards safety and sustainability.

Finally, Andrea Larson, in Section 7.3.3, presents a business case for the sustainability of the chemical industry, recognizing that the market is not driven by some invisible hand, but rather by the entrepreneurship and creative capacities of real professionals who daily make decisions about what chemicals to develop and use, what products to make and promote, and what costs will be borne by the public and the environment. It is through guided innovation that chemicals are developed and employed, and where innovation embraces environmental values there are a host of business opportunities for creating a sustainable future.

7.1.4   Moving Towards Sustainability

The challenge of creating a sustainable future requires directly addressing the chemical industry. The products and consequences of this one industry so shape the material basis of our contemporary economy as to be a central factor in determining its future. We will not achieve sustainability unless this industry is reconceived. The linear flow of materials that is largely indifferent to the ingenious and elegant processes of natural systems and overbearing in its disgorge of hazardous wastes cannot be a model for the future.

New directions for the industry are emerging. Chemical stewardship, green chemistry, biomimicy, chemical substitution, and nanotechnology all provide approaches that are potentially cleaner, greener, and more productive. At this point they are marginal initiatives. However, the history of industrial change is full of examples of marginal developments growing to revolutionize conventional practices.

Each of the new directions for the chemical industry described here requires more and better information. In order to phase out the worst chemicals we need better information on the thousands of chemicals that remain on the market with little or no health and hazard information. In order to develop effective chemicals steward-ship infrastructures, we will need to build and maintain a comprehensive chemicals tracking system that permits industries and government to monitor the flow of chemicals. In order to develop greener chemicals and chemical processes we need more information about how chemicals behave in the environment and in our own bodies. Finally, we need information that is more accessible to the public if con-sumers and workers are going to have more of a role in encouraging the industry to manufacture chemicals that are functional and safe.

As pressures mount for a world more respectful of resource limits and the material needs of future generations, the chemical industry must find its own path to sustainability. Much of the groundwork has already been laid. Enormous invest-ments in research, instrumentation, and experimentation over the past half-century have produced a wealth of knowledge about chemicals and their behavior. That information is now as valuable as the chemicals themselves. Knowing how to make materials and how to make them safely and sustainably provides the basis for the changes now needed in the chemical industry.

7.2 RETHINKING PRODUCTS

7.2.1 Safer Chemicals Within Reach

BEVERLEY THORPE
Clean Production Action

As public awareness of hazardous chemical contamination grows, there is increasing demand for the cessation of persistent, bioaccumulative and toxic materials in pro-duction processes and products. Reaction to this increasing public concern by pro-duct manufacturers and governments can be one of risk management or the adoption of the substitution principle. Increasingly, substitution is seen as the preferred alternative.

Indeed the principle has been enshrined in many international forums, particu-larly those that have set the generational goal to achieve the elimination of hazar-dous substances (Thorpe, 2003). The Nordic countries, in particular, have based many of their chemical policies round the principle of substitution. Most notable is the Swedish Chemicals Products Act of 1985, which puts the onus on anyone handling or importing a chemical to avoid chemical products for which less hazardous substitutes are available (Geiser and Tickner, 2003).

In June 2003 the United Kingdom's Royal Commission on Environmental Pollution recommended that the U.K. government adopt substitution as a central objective of the country's chemicals policy because "considerable inherent uncertainty in our understanding of the way that chemicals interact with the environment means that there will continue to be a risk of serious effects..." This requires a precautionary approach to chemicals management, and this is best implemented through substitution (RCEP, 2003).

Sustainable chemical policies are being proposed at regional and national levels. Sweden's new chemical policy has the objective of "a nontoxic environment." Here a variety of goals and strategies include the phase-out of the most harmful substances; increasing the information of chemical content in products; promoting more ecolabelling and product declarations; using public procurement to stimulate the market for safer materials; and establishing ongoing dialogs with companies in various sectors to move towards safer chemical use (Geiser and Tickner, 2003). The new draft European chemical policy is itself a paradigm shift in the way chemicals are regulated. The proposed legislation, called REACH, requires the registration, evaluation, and authorization of chemicals and requires companies to provide environmental and human health toxicity data. This reverses the onus of proof and effectively ensures that by the year 2012, all chemicals must be registered prior to marketing. Chemicals found to be carcinogens, mutagens, or reproductive toxins will be substituted or require strict authorization for ongoing production and use. The European Commission sees this as a way to stimulate the market for safer substitutes, provide more complete information for downstream users of chemicals, and protect human health and the environment in the process.

In particular, downstream chemical users will benefit from a more efficient and streamlined flow of information among the supply chain, allowing a more discerning choice of materials and enhancing the profit potential of new materials.

Skanska, one of the world's largest construction companies, with 75,000 employees and activities world-wide, claims that operating for many years under substitution regulation in Sweden has lead them to

> ... continuously seek less harmful alternatives. This is something that our clients expect from us as a producer of buildings or infrastructure. As we are not experts on the components in our products we have to go back to our suppliers with the requests that our clients put on us. As manufacturers of building components they will have to go back to their suppliers etc. This is the way we want the market to work in order to reduce the environmental impact.
>
> —Skanska, 2003

H&M (Hennes Mauritz) sells over 500 million items a year in its 844 sales outlets in 14 countries. One of Europe's largest retail chains, it is a strong proponent of safer substitutes. They have phased out the use of brominated flame retardants along with alkyl phenol ethoxylates, organotins, azo dyes and all carcinogenic dyes, PVC, bisphenol-A, phthalates, antimony, and a wide ranges of heavy metals as well as chlorinated aromatic hydrocarbons. They stipulate a clear set of criteria to all their suppliers, used testing to ensure compliance, and rely on their suppliers and chemical formulators to provide alternatives. The company states that

We have encouraged our suppliers to be innovative and when we have found a better alternative somewhere among our suppliers we have helped to spread that knowledge to other suppliers and other markets. In doing so we have found that almost anything is possible as long as you set clear guidelines on what is not acceptable. We have not had to compromise on fashion or quality in a way that has harmed our business. Prices may have gone up temporarily but as soon as mass production has started, the prices have gone back to previous levels.

—Schullstrom, 2003

Just as sustainability presents us with the most troubling and complex technical challenges we face, it also highlights the most important technological opportunities crying out to be cracked by today's chemists. Finding the solutions will result in major economic progress (Collins, 2003).

A concerted focus on substitution would be an enormous boost to Green Chemistry initiatives by chemical producers and would allow creative new thinking to tackle the challenge of transitioning to more sustainable materials.

For example, Marks and Spencer, a leading retail chain in the United Kingdom targeted the substitution of alkyl tins in the dyeing and finishing of clothing, along with azo dyes and alkylphenol ethoxylates some years ago. Their product specialists are working with the Green Chemistry department at York University to explore safer alternatives.

Similar challenges of researching and implementing safer alternatives are being dealt with by Boots, the leading cosmetics retailers in the United Kingdom. The company has its own product development department and in 1994 decided to phase out alkylphenol ethoxylates (APEs). A difficulty was first identifying where APEs were used in their 40,000 to 50,000 product range as some of their supply chains were complex. However, by the year 2000 the company had achieved a 90 percent phase out.

Increasing concern over brominated compounds has catalyzed major electronic firms to move out of brominated flame retardants. The German Federal Environment Ministry notes that "It is encouraging that there is a general trend to refrain from the use of halogenated flame retardants in products and to replace them with less problematic flame retardants or to redesign flame retardant systems, e.g. by creating greater distances to potential heat sources" (UBA, 2001).

In response to the German dioxin ordinance of 1994, Sony Europe started investigating safer substitutes for halogen-based flame retardants and was successful in developing halogen-free circuit boards used in European TV sets, VCRs, and DVD players. Sony's engineers adopted a resin structure containing nitrogen to increase heat resistance and the company's goal is to be free of all brominated flame retardants by the end of 2005.

The Restriction on Hazardous Substances Directive and its mandate that new electrical and electronic equipment must be free of PBDEs by 2006 has in large part been a major catalyst for product redesign by electronic equipment producers. However, it can also be assumed that leading manufacturers had realized the feasibility of substitution. For instance, Hewlett-Packard monitor housings typically contain phosphorus-based flame retardants and its computer casing has no brominated flame retardants. Motorola uses a halogen-free flame retardant that is a nitrogen/phosphorus combination for most of their product lines. National/Panasonic began marketing the world's first wide screen television for which halogen compounds had been eliminated for virtually all components.

NEC has a target to phase out the use of all halogenated flame retardants by 2011. In 1999 the company launched a polycarbonate containing a silicone flame retardant that it claims to be "far superior" to conventional flame retardants and is neither phosphorus nor halogen based and can be recycled up to five times for the same purpose.

The above examples demonstrate that substitution is indeed feasible and occurring among progressive companies. However, it cannot be assumed that all chemical users will adapt safer substitutes, or that all chemical suppliers will embark on Green Chemistry initiatives as a priority. Ensuring the development and production of safer materials is facilitated by regulation, which levels the playing field. Clear criteria of what is not acceptable, such as all carcinogens, mutagens and reproductive toxins, must be clearly defined and targeted for substitution within a well-defined time frame.

For example, the electronics industry was adamant that the material phase-outs stipulated within the Restriction of Hazardous Substances (RoHS) Directive for new electrical and electronic equipment must apply equally across Europe. In a joint letter to the Commission sent by leading producers, they urged a "clear legal basis for the RoHS directive, so as to provide a high level of protection for citizens without creating uncertainty for business and undermining the Single Market" (EC, 2001).

Lack of data is a serious and unconscionable situation. We should no longer tolerate the fact that 95 percent of all existing chemicals lack basic environmental and health data, allowing society and the ecosystem to effectively remain a grand chemical experiment. The recent crisis in rising body burdens of PBDEs in North American women has resulted in a scramble by health advocates to propose various state legislations banning PBDEs while the EPA negotiates voluntary restrictions with the bromine industry. We should not be reacting after the fact for every one of the hazardous chemicals yet to be discovered.

Lack of data about chemicals and the lack of information along the supply chain and within the industrial sector also hinder the adoption of safer alternatives.

The German Environmental Protection Agency surveyed 13 flame retardants for toxicity to humans and the environment and their suitability for closed-loop substance management. The aim was to assess the feasibility of substitution with less hazardous flame retardants. They ranked the summary of their findings taking into account the lack of data for many of the alternatives and urged expedient research to fill these data gaps (UBA, 2001). A similar study by the Danish EPA also observed that the amount of data available is often very limited, particularly for important criteria such as degradation (DK, 1999).

The power dynamics along the supply chain appear to put buyers at a disadvantage. Some retailers express frustration that they do not receive adequate information from their chemical suppliers and wish information were more forthcoming, particularly pertaining to new materials being developed (Thorpe, 2003).

The lack of information dissemination within industry sectors is another barrier to the adoption of safer substitutes. A recent study details how cement manufacturers in Scandinavia solved the problem of skin contact with cement containing hexavalent chromium back in the 1980s, but the information was not diffused to other European manufacturers, resulting in needless worker exposure (UBA, 2003).

Substitution has been shown to be feasible; now we need to seriously escalate our efforts. Fundamentally, however, the chemical industry must acknowledge the need to urgently produce data for all the chemicals they market prior to placing them on the market. Then we can accelerate the process to stop producing and using materials that persist and accumulate in our environment. This would pave the way to design and adopt materials that are safe within our technical and biological cycles and ensure for the next generations "a nontoxic environment."

7.2.2 Addressing the Challenge of PBTs

JOANNA D UNDERWOOD
INFORM

Finding solutions to the problem of PBTs in commerce depends on the ingenuity and expertise of chemical industry leaders, and of entrepreneurs in other industries who can devise product alternatives that are PBT-free or as close to that as possible. Examples of equally challenging innovation provide encouragement that the same inventiveness that brought us the chemical advances that contribute so much to today's quality of life can likewise meet the challenge posed by these hazardous chemicals.

Persistent toxic chemicals, because they resist degradation in soil, water, and air, can travel long distances from where they were discharged and have impacts long after their initial release to the environment. Many of these stable substances also bioaccumulate, building up to dangerous levels in living organisms, even when released in very small quantities. Mercury, for example, once it enters a water body, can accumulate in predator fish to concentrations as much as one to ten million times greater than those of the surrounding water, endangering humans and wildlife for which the fish is a food source (U.S. EPA, 2001). Similarly, a single meal's worth of fish from Lake Michigan can expose a person to more polychlorinated biphenyls (PCBs) than a lifetime of using the lake as a primary source of drinking water (EPA).

Even when a PBT has been banned from use, its impacts can persist for many years. Despite a ban on the manufacture of PCBs that went into effect more than two decades ago, hot spots of PCB-contaminated sediments remain in the Kalamazoo River in Michigan.[2] And because these chemicals remain in the environment for so long, they can travel long distances from where they were discharged. PCBs have been transported by migratory animals and by air, water, and ice from the various

---

[2] It is important to note that levels of several persistent toxins (such as lead, DDT, and PCBs) have significantly declined in fish in the Great Lakes Basin ecosystem following restrictions on their manufacture and use (U.S. EPA, 2003).

parts of the world where they are used or were used in the past to as far away as the Arctic, where they now pose a major threat to the health of people and animals living in that remote region (Tenenbaum, 1998).

Because of the exceptionally long-lasting risks posed by PBTs, the only way of safeguarding the environment, wildlife, and human health from their impacts is to use them in a closed-loop system, which would prevent them from escaping into the environment, or to phase out their use altogether. The first option is problematic because fugitive emissions are an issue in even the most well-controlled manufacturing processes, and because PBTs have effects at every stage of their commercial lifecycle – from extraction through use and disposal – they would pose risks both before and after they entered the closed-loop process.

Ultimately, the choice of whether to continue using a chemical depends on weighing the importance of its role in society versus the degree of risk it presents. While scientific opinion is relatively consistent regarding the long-term threats of PBTs, the more subjective political and economic judgments about their value may vary and be debated for years. Nonetheless, the second option – to phase out the use of these chemicals by aggressively pursuing alternatives known to be less persistent, less bioaccumulative, or less toxic, *but equally effective* – is an effective strategy for reducing or eliminating their risks.

Some leading companies are voluntarily rising to this challenge. Swedish retailing giant IKEA has stopped using brominated flame retardants (see IKEA website), numerous building systems companies, including Emerson, Trane, and Robert Shaw, are providing mercury-free controls and HVAC systems (INFORM, 2003), and Fujitsu now manufactures some of its computer components (such as its printed circuit boards) without lead, a PBT (see Fujitsu Siemens Computers website).

The ingenuity shown by these and other companies could be a key to success in the emerging global green products market. In some jurisdictions, using substances without accounting fully for their environmental and health effects will no longer be possible. For example, in February 2003, the European Union passed the Directive on the Restriction of the Use of Certain Hazardous Substances in Electrical and Electronic Equipment (known as the RoHS Directive), which requires member countries to develop legislation banning the use of lead, cadmium, hexavalent chromium, mercury, and two categories of brominated flame retardant (polybrominated biphenyls and polybrominated diphenyl ethers) from all electrical and electronic equipment sold in the EU by 2006. Manufacturers that continue to incorporate these hazardous substances into their products will no longer have access to the European market (European Parliament, 2003a).

Also in February 2003, the European Union enacted the Directive on Waste Electrical and Electronic Equipment (known as the WEEE Directive), calling for manufacturers to take end-of-life responsibility for their products (European Parliament, 2003b). Until August 15, 2005, this directive applies the principle of "collective producer responsibility" to wastes generated by electrical and electronic products. This principle allows companies to pool monies used for managing their wastes. However, for wastes generated by products manufactured after that date, it applies the principle of "individual manufacturer responsibility," with the result that the companies with the most easily and cost-effectively refurbishable or recyclable products may gain an economic advantage over those generating more waste. Similar policy approaches are being adopted in Japan, China, and other countries.

Measures such as these, which give manufacturers direct fiscal responsibility for managing the waste generated by their products, may well create an added incentive for removing hazardous and persistent substances from the products in question, as well as encourage ingenious reuse strategies that can turn potential wastes into raw materials for future products. Ericsson Electronics, for example, has already communicated to vendors the names of chemicals it will not accept as constituents of the products it buys (Fishbein, 2002).

The EU directives may also motivate manufacturers to work toward the elimination of other substances, besides those banned under RoHS, in the products they make. For example, electronics manufacturers, because they are now required to take back their products and meet recycling targets set by the WEEE Directive, may want to eliminate other chemicals (such as other types of flame retardant) that create impediments to recycling or pose risks to their workers' health.

In the United States, interest in purchasing products that are PBT-free or that contain reduced quantities of these chemicals has been growing steadily over the past few years. INFORM initiated a program focused on state purchasing offices in 2000 and is now advising procurement officials in 16 states – identifying the products they purchase that contain PBTs and alternative options that are equally effective. One by one, purchasers are shifting their product choices in favor of products that are performance-equivalent, cost-competitive, and readily available. Low-mercury lighting systems, mercury-free thermostats and other building equipment, bio-based lubricating oils, PVC-free IV bags (items made of PVC can create dioxin, a PBT, when burned in medical waste incinerators), lead-free boat and road paint, mercury-free thermometers, and other hospital equipment are all increasingly in demand. Demand is also growing for government fleet vehicles free of mercury components and for vehicles that burn natural gas instead of diesel fuel, whose combustion releases several PBTs.

U.S. EPA, too, is likely to begin paying more attention to products. While the agency has traditionally placed most of its regulatory focus on wastes and emissions from manufacturing operations (it has a specific goal of reducing PBTs in such wastes and emissions by 50 percent by 2005, compared to 1991 levels), a 2000 INFORM analysis of data from the expanded right-to-know programs in New Jersey and Massachusetts showed that more than 95 percent of the persistent toxins (many of which are PBTs) leaving industrial facilities were going into products, compared to only 5 percent generated as industrial waste.

In light of broadening regulatory concerns and the trend toward analyzing the products purchased by government agencies in favor of the least toxic options, chemical companies can get ahead of the game by examining their operations for PBT uses and impacts and bringing their product planners in on the search for PBT-free feedstocks. They can ask such questions as: If PBTs are used in manufacturing and are leaving the plant in products, can they be replaced by other substances? If PBTs are created during manufacturing, are controls available to prevent releases to the environment? If PBTs such as dioxins, lead, and mercury are generated or released when products are disposed of in incinerators or landfills, is this a reason to pursue the search for new feedstocks?

As awareness of the impacts of PBTs grows, local governments and environmental organizations are going beyond chemical bans such as those imposed by the EU's RoHS Directive, and beyond policies that extend the responsibility of producers for their products when they become waste, such as those applied to numerous product categories in the European Union and elsewhere. Recently, government purchasers such as the New Jersey Purchase Bureau, the Massachusetts Operational Services Division, and the New York State Office of General Services have begun to require disclosure of PBT-containing items in purchasing contracts or the removal of specific PBT-containing items from state contracts. While companies may not be eager to supply such information, those that strive to use fewer PBTs may be able to use these requirements to demonstrate their good faith efforts and gain market share, even before the phase-out of a particular material is complete. Those companies that do eliminate substances of concern will certainly be able to profit from the product evaluations and comparisons that the disclosure requirements permit.

7.3   NEW DIRECTIONS

7.3.1   Biomimicry: How and Why R&D Should Be Driven By Nature's Design

MARK DORFMAN
Independent Environmental Science Researcher

Ironic is the public perception that "chemicals" are essentially man-made, industrial substances that contaminate an otherwise "chemical-free" natural world. The fact is, Mother Nature is the ultimate chemical engineer and manufacturer – all of nature is alive with chemistry, from the colorful chemical contortions of corals, flowers, and insects, to the technical triumphs of deep-sea vent communities. Nature's mastery has led to the creation of a highly sophisticated chemical give-and-take underlying all living things. Out of necessity, nature speaks chemistry with a forked tongue – one prong seeking continued technical advances while the other maintains life-friendly conditions. What better model is there for a sustainable chemical industry? The pace of technological advancements in biochemical analysis and molecular manipulation are poised to take *biomimicry* – emulation of natural materials and processes – from the pages of science fiction to the pages of prestigious scientific journals. While emulation of nature's chemistry may not be the sole solution to all chemical industry-related challenges, her wealth of low-toxicity, low-energy, low-impact technologies behooves us to maximize our knowledge base for potential future industrial application.

7.3.1.1 Biomimicry. While there is no formal definition of biomimicry, for the purpose of this section, biomimicry refers to (1) uncovering the chemical configuration of natural materials and the network of natural processes involved in their production and use, and (2) designing new chemical industry products and processes based on these natural models. Applied, industrial-scale outcomes are likely to be engineered hybrids – a compromise between nature's designs and man-made creations. Ideally, these applications would exploit the environmental and health benefits of the natural chemistry model (Table 7.1).

7.3.1.2 Nature as Mentor. Imagine if instead of a few thousand years since the dawn of man-made technology there had been a few billion in which to perfect our chemical craft and invent new products and processes – where might our chemical technology be today? Nature recently celebrated the 3,000,000,000th anniversary of her invention and mass production of photosynthetic technology (cyanobacteria) (de Duve, 1995). Even if we took the lifetime of every major scientist who ever lived, from Aristotle to Einstein, and strung the years end-to-end, nature would still have a multibillion-year advantage of technological trial and error.[3] Those millennia of experience have resulted in many highly sophisticated and energy efficient chemistries with neutral or beneficial, rather than detrimental, impacts on the environment.

Tools of the Trade Pry Open New Doors. In 1901, Franz Hofmeister described the extent of current knowledge of the living cell as "a vessel, filled with a homogeneous solution, in which all chemical processes take place" (Ball, 2002). A hundred years later, the pace of advancement in tools for chemical analysis and molecular manipulation, including within living cells, is exponential. It seems as though each issue of the top scientific journals (such as *Nature, Science,* the *Journal of the American Chemical Society,* and the *Proceedings of the National Academy of Sciences*) has a first-run research article describing a new device or technique that brings the living molecular world into better focus and/or provides greater dexterity for the chemical engineer.

For example, proteins, the workhorses of living chemistries, perform functions ranging from chemical synthesis (enzymes) to chemical communication (hormones). Understanding the intricacies of protein structure and function is an important stepping-stone towards biomimicry. Indeed, chemical manufacturing has already benefited from improved catalysts based on a working knowledge of enzymatic activity (Ball, 2002). Key to understanding how proteins work is understanding their three-dimensional, folded configurations in the liquid environment and the intermediate steps involved in their biological activities – a challenging feat given folding and unfolding timeframes of micro and nanoseconds. Recent advances in picosecond X-ray crystallography "allows one to literally 'watch' the protein as it executes its function" (Schotte *et al.*, 2003).

[3](3000 years of chemistry) $*$ (500 great minds per year) $*$ (75 years per mind) $= 112,500,000$ years.

TABLE 7.1. General Characteristics of Chemical Production and Use: Man Versus Nature

	Man	Nature
Raw material input	Often toxic raw materials	Nontoxic raw materials (unless toxicity is intentional for defense, predation, or reproduction)
Energy sources	Most often conventional nonrenewable energy sources: coal, oil, natural gas, nuclear fission	Most often renewable solar, thermal, or chemical
Production process conditions	High temperatures and pressures in toxic solvents	Ambient temperatures and pressures in an aqueous medium
Products	Materials with a wide variety of useful characteristics	Materials with a wide variety of useful characteristics
Ultimate "post consumer" fate of products and byproducts	Often disposed of into air, water, or soil, where it can remain for long periods and/or may initiate undesirable reactions in humans or the environment[a]	Often used as raw material input for another chemical process or degrades into substances that serve as raw material for other processes (McDonough, 2003)

[a]According to the U.S. Toxics Release Inventory, the chemical industry released 583 million pounds of manufacturing-related chemical wastes into the environment in 2001, excluding wastes associated with downstream product use and disposal (U.S. EPA, 2003). The industry ranked third after metal mining (2.8 billion pounds) and electric utilities. A portion of the metal mining industry's products is used by the chemical industry – biomimicry might reduce the need for these materials and related wastes.

Total internal reflection fluorescence light microscopy has been used to determine the nanoscale motion of myosin, one of the so-called "motor proteins" that drive muscle contraction and move vesicles around cells (Molloy and Veigel, 2003). Such proteins are likely to be the models for synthetic nanometer-sized motors that may play a key role in biomimetic chemical production units. Solid-state magic-angle-spinning NMR has been used to investigate the structure of certain proteins not amenable to other analytical techniques, particularly certain membrane proteins (Castellani *et al.*, 2002). Membrane proteins and their associated ion channels form critical components of important biological processes, such as photosynthesis and nervous system communications. In 2002, researchers at Cornell University used X-ray crystallography to determine the structure of a potassium ion channel, "opening new windows in cellular function" (Abbott, 2002).

A Wealth of Potential Lessons. Millions of species have yet to be scientifically identified and investigated before their biomimetic potential can even begin to be realized. Coral reefs, for example, are home to millions of aquatic organisms that offer particular promise because of the array of chemicals produced for predation, camouflage, and reproduction – a potential that has only barely been explored (Bryant *et al.*, 1998). Examples of biomimetic potentials that are already being investigated and to some degree mimicked include the following.

- *Adhesives*. Most engineered adhesives require clean, dry surfaces or they simply will not work. Nature has devised a variety of nontoxic ways of attaching surfaces with or without these conditions. They range from barbs on weed seeds (which inspired the invention of Velcro), sea mussel adhesive (works under water on rough surfaces), and gecko toe hairs (bond weakly with the molecules in any surface – recently mimicked as "gecko tape") (Benyus, 1997; Geim *et al.*, 2003).

- *Color*. Certain industrial colors and coatings have been associated with toxic metals and solvents. Living creatures employ a variety of novel methods to create and change color and may serve as models for new technologies. For example, butterflies and peacocks create brilliant colors and patterns by diffraction of light via "a very ingenious and simple strategy" of ordered surface microstructures rather than pigmentation (Sato, 2003; Zi *et al.*, 2003). Squid and octopi expand and contract microsacs of various pigments at the surface of their skins to make rapid and reversible color changes triggered by ambient temperature, light levels, pH, or chemicals (Akashi *et al.*, 2002). Although diffraction colors have been used since Renaissance ceramic "luster-ware," increased understanding of these phenomena has led to preliminary man-made prototypes of more sophisticated color systems.

- *Ceramics*. Ceramics are used in a wide variety of industrial applications from fixtures to electronic components. Conventional production methods require enormous amounts of energy – nature produces incredibly strong yet resilient ceramics at ambient temperatures. The prospect of low-energy manufacturing and the combined characteristics of strength and resilience have led to both academic and industry research programs targeted at mimicking abalone shell nacre (Diop, 2003). The abalone's protein-based method of producing nacre's calcium carbonate structure has also attracted the attention of chemists fighting pipe scaling – the continuous build-up of calcium carbonate on the inside of pipes, which causes millions of dollars in damage and water pollution each year. The same protein signals that tell calcium ions to stop forming the abalone's shell might be harnessed to prevent pipe scaling "thus saving millions of dollars in chemicals and pipe replacements with a simple preventive solution" (Koelman, 2003).

- *Photosynthesis*. Photosynthesis is perhaps the holy grail of biomimicry because command of the process would provide at least three major functions of use to the chemical industry: (1) solar generated power, (2) carbon-based chemical synthesis, and (3) light-based, molecular-level computing. With carbon dioxide serving as the carbon source for solar-driven organic chemical synthesis, could the chemical industry one day be a carbon dioxide sink, consuming more from the atmosphere than it produces?

7.3.1.3   The Path Forward. Making serious inroads into biomimicry in the chemical industry will take substantial inputs of public and private research dollars and the brightest, most creative of scientific minds. The fruits of that research are likely years beyond the timeframe of conventional chemical industry investments (typically one- to two-year payback times or less) and so would require a concerted effort by government and industry. One immediate option might be a biomimetic component created in the recently formalized National Nanotechnology Initiative with its $3.7 billion, four-year budget – given its potential public health and environmental advantages, unraveling and mimicking natural chemistries should be a top priority of nanotechnology research and development.

Distinguishing Biomimetic from Synthetic. Given the spatial and temporal needs of human society versus that of the natural world, future generations of new industrial chemical products and processes will likely be hybrids of natural and synthetic chemistries, rather than exact mimics of natural ones. Therefore, criteria will be needed to determine which of these new chemistries should be considered biomimetic and which are more appropriately deemed synthetic. This distinction is important in order to drive research and development towards biomimetic designs that exploit environmental and public health benefits of natural systems. Eager to ride the tide of biomimicry's growing popularity, attempts may be made to fit largely synthetic designs into the provocative biomimetic appellation. An analogous situation arose in the 1980s when industrial pollution prevention eclipsed waste treatment as the leading environmental strategy – even today, vigilance is needed to ensure that a useful distinction remains between activities that avoid the creation of waste and those that manage waste after its generation, particularly as government and industry policies are modified. The ultimate decision to employ a new technology should depend on how well that particular technology meets a variety of criteria, including on-the-job performance, lifecycle impacts, and economics. Biomimetic technologies should be subject to the same set of criteria. The hypothesis here is that new technologies based largely on natural systems will have a head start in meeting these conditions.

Table 7.2 begins to illustrate the range of actual and hypothetical chemical designs and their relationship to biomimicry. The more a new design mimics the composition, configuration, and production of the natural substance or process,

TABLE 7.2. Examples of Designed Chemistries and Their Relationship to Biomimicry

Desired Outcome	Design	Relationship to Biomimicry
Photosynthesis	Porphyrin-based photosynthetic device	Substantial
Color without pigment	Microstructured film that requires styrene for its production	Partial
Spider silk	Goats genetically engineered to produce a silk protein in milk	Minimal

the more substantial is its relationship to biomimicry. For example, researchers are developing a photosynthetic device that, like nature, uses a light-harvesting center (a porphyrin in place of chlorophyll), electron-relay molecules, and the ATP synthase enzyme embedded in a lipid membrane. This light-induced electron transfer generates a flow of hydrogen ions across the membrane, which drives the conversion of ADP to ATP – a key energy-generating process of cellular metabolism. If coupled, as hoped, to the generation of the coenzyme NADPH, this system could "power all kinds of enzymatic chemistry" (Ball, 2002). Such a system, used to produce proteins, might obviate the need to genetically engineer bacteria to perform that function (Benyus, 1997).

Brilliantly colored feathers and wings owe their magnificence to microstructured surfaces of benign biological materials such as keratin and melanin having the same order of size as the light wavelength – researchers in Japan have reproduced the brilliance of such surfaces, but relied on the use of polystyrene, a substance based on the toxic styrene monomer (Sato, 2003). At the far end of the spectrum is the goat genetically engineered to produce a spider-silk protein in its milk – a technique far from the chemical production process inside a spider's spinnerets (Osborne, 2002). Although a technological feat in its own right, a more appropriate term for this application of genetic engineering might be biomanipulation rather than biomimicry. A more extensive list of new chemical designs and their relationship to biomimicry will help address the challenge of creating a set of criteria that distinguishes biomimetic achievements from the rest of the pack – a distinction that, given nature's life-friendly operating conditions, might ensure the prominent position of environmental and public health protection as research and development into new technologies compete for funding resources.

Ask Questions First, Shoot Later. At the dawn of the 21st century, we should take heed of the lessons learned in the past from succumbing to the enticement of new technologies before we give the green light to mass production and use. In short, expect the unexpected. Had we done so in the early 20th century, the global environmental distribution of DDT, PCBs, CFCs, and other synthetic chemicals borne of the petrochemical age, might not be what it is today – better understanding of the potential environmental and public health consequences of CFCs, PCBs, and DDT may have reduced current levels of tropospheric ozone depletion and bioaccumulation of persistent toxins. The predicted outcomes about how living systems or environmental cycles will respond to new man-made substances and their synthetic pathways should factor into distinctions between biomimetic and nonbiomimetic chemistries. Examples of new chemistries that beg these questions are described below.

Scientists are synthesizing new organic substances that, apparently, do not exist in nature. For example, bacteria have been coaxed to produce "non-natural" amino acids and proteins (Schultz *et al.*, 2002), and synthetic protein folds that are outside the basic shapes found in nature are not only physically possible, but can be extremely stable (Kuhlman *et al.*, 2003). Similarly, researchers at Stanford University have synthesized a new form of DNA that uses the natural backbone but has all four base pairs replaced by new, and larger pairs – the new double helices "are more thermodynamically stable than the Watson–Crick helix" (Liu *et al.*, 2003). Certainly these are exciting outcomes of our growing knowledge of, and increased dexterity with, biology-based systems, but stability of new chemical substances has triggered problems in the past that only came to light after mass production and global commercial distribution.

Researchers at the University of Seattle, Washington, report that polypeptides can be genetically engineered to specifically bind to selected inorganic compounds (Sarikaya *et al.*, 2003). A hypothetical device based on this principle, described by scientists at Rutgers University, would be made up of a carbon nanotube backbone with peptide "limbs" attached at the front end of the tube and a motor protein, such as dynein, serving as a propeller at the rear end (Mavroidis and Dubey, 2003). The device would be small enough to flow within blood vessels where it would seek out specific infected cells and use its chemical-specific peptide limbs to deliver drugs or destroy the cell. Although the device is hypothetical, carbon nanotubes have recently been shown to be able to cross the cell membrane and accumulate in the cytoplasm or reach the nucleus (Pantarotto *et al.*, 2004). If such nanodevices, borne of a marriage between nature's designs and our own inventiveness, were mass-produced and widely used, could they eventually disperse in the environment where their protein limbs might grab onto global chemical detritus, such mercury or PCBs, and effectively deliver it to our living cells?

7.3.1.4   Conclusion.   Albert Einstein is quoted to have said, "Look deep, deep into nature and then you will understand everything better." A concerted, transparent, and publicly debated effort by government, industry, and academia to maximize our understanding of living chemistries and use that knowledge to create a substantially biomimetic chemical industry could pave the way for a world with better products and services from the chemical industry, but a smaller footprint on the environment and public health: a *space-age* society with a *stone-age* impact.

7.3.2   Sustainability, Environment and Nanomaterials

Mark R Wiesner
Environmental and Energy Systems Institute, Rice University

A 2003 estimate by the Nanobusiness Alliance indicates that the largest single category of nanotech start-ups is in the area of nanomaterials. The production, use, and disposal of nanomaterials can be anticipated to engender a wide range of benefits and unintended consequences in social, economic, and environmental terms. It is very likely that applications of nanoscience will lead to new means of reducing the production of wastes, using resources more sparingly, remediating industrial contamination, providing potable water, and improving the efficiency of energy production and use. Nanomaterials will also have an increasing presence in consumer products that we encounter everyday. Commercial applications of nanomaterials currently or will soon include nano-engineered titania particles for sunscreens and paints, carbon nanotube composites in tires, silica nanoparticles as solid lubricants, and protein-based nanomaterials in soaps, shampoos, and detergents. Industrial applications currently being marketed include the use of alumina nanoparticles in the manufacture of propellants, pyrotechnics, and ceramics membranes, nanoparticles in semiconductor manufacture, and numerous biomedical applications. If the current trend in commercial ventures continues, we will soon find ourselves with a relatively large nanomaterials industry. Our vantage point early in the trajectory of this industry confers upon us a particularly promising opportunity to "get this technology right" and ensure that nanotechnologies emerge as a tool for sustainability.

Issues surrounding the environmental applications and implications of an emerging nanochemistry and materials industry were raised for the first time in December 2001, at an international symposium organized by the Environmental and Energy Systems Institute at Rice University. Press coverage of this symposium and follow-up events created considerable public interest and controversy. In one case, a Canadian environmental advocacy group, ETC, released a report largely based on information gathered from the Rice symposium that concluded that there should be an immediate halt to the production of all nanomaterials and ultimately called on delegates to the Johannesburg summit on sustainable development to take up this cause. Although the merits (and practicality) of the ETC conclusion are debatable, it underscores the need for sound research on the environmental implications of nanotechnology now rather than later.

7.3.2.1 *Nanotechnologies as Environmental Technologies.* Nano-engineered materials will find numerous applications that will improve environmental technologies and help protect public health. These applications will include industrial separations, potable water supply, chemical synthesis, energy generation and transmission, groundwater remediation, and air quality control, to name a few. A key strength of a nanotechnology-based approached to maintaining or improving environmental quality lies in the potential to approach the thermodynamic limits of the production and clean-up process. Gains in efficiency produced by a shift away from top-down manufacturing and end-of-pipe remediation will reduce energy consumption per unit of production with associated environmental benefits.

Particularly promising near-term applications of nanomaterials that may yield order-of-magnitude improvements over current generation technologies include (1) membrane separations, (2) catalysis, (3) contaminant sensing, (4) energy production and storage, and (5) contaminant immobilization.

Membrane technologies are playing an increasingly important role as unit operations for environmental quality control, resource recovery, pollution prevention, energy production, and environmental monitoring. Membranes are also key technologies at the heart of fuel cells and bioseparations. Nanoscale control of membrane architecture may yield membranes of greater selectivity and lower cost. Membrane separations based on the placement of nano-bio conjugates within the membrane matrix would yield membranes with high specificity for a given biological agent or biological product. Nanochemistry might be used to obtain "smart membranes" that integrate sensing and separation capabilities in the same structure, allowing membranes to adaptively select compounds for transport, detect a compromised membrane surface, or adapt to changes in the environment such as temperature or pH.

Catalysis is increasingly replacing the use of many hazardous substances in chemical production to achieve cleaner chemical synthesis. Catalysis is also important in reducing downstream pollutants after they are generated as is done in the catalytic converters of automobiles.

Nanostructured substrates for spectrophotometric measurement have the capability of greatly enhancing the sensitivity of measurements for many environmental contaminants. This will allow for few manipulations of samples, direct measurement in the field and potentially more stringent controls on sources of compounds as our awareness of the presence of contaminants in our air, water, soil, and biosphere increases. Molecular electronics-based devices have the potential for dramatically reducing the size and cost of current analytical devices, with a potential to create distributed analytical networks. Distributed analytical networks would have an impact similar to the development of small inexpensive microprocessors, which shifted computing power from centralized facilities to desktops, appliances, and myriad other locations. The capability to perform measurements at many locations will improve both the quality and quantity of environmental information and may affect the way that sources of pollution are regulated and monitored. Distributed analytical capabilities would also improve the ability to perform real-time process control, thereby improving the efficiency of process with associated reduction in costs and wastes generated.

Energy production, storage, and transmission have been priority areas for nanotechnology research. Improvements in the efficiency and cost of solar cells and batteries are achievable with nanotechnology-based methods for the creation of thin films. The storage of hydrogen, particularly for the development of fuel cells, is an important challenge that has been recently focused on the use of carbon nanostrutures. The first reported results for hydrogen storage capabilities in carbon single wall nanotubes (SWNTs) was published in 1997. Although reported efficiencies remain controversial, work on hydrogen storage towards the goal of 6.5 percent weight storage by carbon nanotubes remains a high priority. The use of carbon nanotubes as replacement for copper wires faces many important technical hurdles, but with a substantial potential payoff in terms of reduced losses during energy transmission.

The long-term storage/immobilization of hazardous materials is of great concern for industries ranging from nuclear power generation to chemical wastes. First generation solutions involved embedding these materials in solids such as cement or incorporating materials in vitrified masses. Nanotechnology offers the possibility of constructing highly stable masses for long-term storage at the molecular level.

7.3.2.2 Anticipating Environmental Impacts.

7.3.2.2 Anticipating Environmental Impacts. Production of significant quantities of anthropogenically-derived nanomaterials will inevitably result in the introduction of these materials to the atmosphere, hydrosphere, and biosphere. Key questions to be addressed in future research include "Where will these materials most likely end up in the environment?" "What are the most probable paths of exposure to these materials?" "Are these materials toxic?" "How persistent will these materials be?" "How will these materials interact with other chemical species and with organisms?" and "How do the properties and quantities of anthropogenically-derived nanomaterials compare with those of naturally derived nanomaterials?"

Research is needed to explore the impacts of nanomaterials and nanomaterial production on the environment and public health. One framework for assessing these impacts is that of comparative risk assessment. Applied to an assessment of the production, use, and disposal of nanomaterials, a risk assessment typically considers both the *potential for exposure* to a given material and (once exposed) *potential impacts* such as toxicity or mutagenicity. The need to elucidate both of these components of risk in assessing the consequences of nanomaterials on the environment and public health is essential.

Questions regarding nanomaterial toxicity are extremely speculative. Concerns include the potential of a given material to act as an irritant, its potential for transporting known environmentally hazardous materials that "piggy-back" on the nanomaterial, the role of some nanomaterials in generating oxidizing compounds that may damage cells, and the effect on cellular function when nanomaterials enter cells. It should be noted that the most of the properties of nanomaterials that may create concern in terms of environmental impact (such as nanoparticle uptake by cells) are often precisely the properties desired for beneficial uses such as in medical applications. Nanomaterials may differ from other particulate materials in both size and surface chemistry. Nanomaterials are all surface: a gram of single-walled carbon nanotubes, for example, has over ten square meters of available surface. Control over the chemistry of this interface is essential in developing nanotechnologies. Rarely will bare inorganic solids be present in the solution phase of nanostructures; rather, nanoparticles will likely be coated chemically to give them specific desired properties or simply take up other materials from the environment. Biological functionality may be integrated into a surface to draw nanoparticles into certain cells, or to disperse particles in particular organs.

Equally as speculative as our attempts to foresee possible toxicity issues is the issue of nanomaterial exposure. Exposure will be determined in large part by the chain of production, use, and disposal of nanomaterials. Although the mobility of nanoscale particles in the environment is poorly understood, speculation based on laboratory experience in making nanomaterials and experience garnered from colloid science suggest that the mobility of anthropogenic nanoparticles, and thus the potential for exposure, may often be small. Nonetheless, as in the case of toxicity, the relevance of our current experience in this domain is unclear.

Observations reported by others that nanoparticles are subject to cell uptake suggest that nanoparticle accumulation into cells may contribute to their fate in the environment. Possible impacts of nanomaterial uptake include direct toxicity to cells, persistence within the cell, and the ability to transport other associated materials into cells such as contaminants or scavenged genetic material. Understanding the extent to which uptake and persistence will occur in microorganisms may provide critical insight into one important exposure pathway to higher organisms known as bioaccumulation.

Nanoparticles and colloids may facilitate the transport of otherwise immobile contaminants in groundwater aquifers and surface waters, as well as reducing the removal of particle-associated contaminants by water treatment facilities. For example, colloid-mediated transport of plutonium waste in Nevada led to rates of radionuclide transport through groundwater orders of magnitude larger than predicted. Moreover, significant hysteresis has been observed in contaminant partitioning to fine particles, suggesting that physical–chemical interactions with small particulate phases may also affect the bioavailability of these contaminants.

Carbon nanotubes present an extremely hydrophobic surface that complicates their "dissolution" for subsequent manipulation. This hydrophobicity also makes nanoparticles prime targets for the adsorption and facilitated transort of hydrophobic contaminants. To overcome problems in solubilizing nanoparticles, laboratory preparations of nanoparticles typically include modifications to the particle surface such as polysaccharide wrapping, adding end-group functionality or "mops" to nanotubes, or capping mineral nanoparticles with ionic groups. Such chemical modifications may decrease their propensity to aggregate, or attach to environmental surfaces, possible increasing their mobility in the environment.

In summary, ensuring that the nanomaterials industry is created in the image of sustainability will require an informed basis for managing risks rather than an entire elimination of both the benefits and risks of a given nanomaterial. Not all risks, applications or consequences can be anticipated. However, as the next generations of the world's scientists and engineers embark on their careers, the goal should be to equip them to explore new technologies that will break down barriers to the implementation of technologies that improve the human condition while being mindful of the unintended consequences that may accompany new discoveries.

7.3.3   Entrepreneurship and Innovation

ANDREA LARSON
Darden Graduate School of Business Administration

Many companies, within the chemical industry and outside, now understand that cost reductions and product/process improvements result from efficiency analyses of environmental issues. Documented cost reductions in materials input, waste streams, and energy use are readily available. In recognition of the efficiency gains to be realized, as well as risk reduction and regulatory advantages, most firms acknowledge the benefits that result from taking the environmental management agenda seriously. In addition, companies know they can help avoid the adverse effects of ignoring these issues, such as boycotts and stockholders resolutions, which generate negative publicity.

However, the efficiency improvements and risk reduction sides of environmental concerns and sustainability are just one side of the opportunity. They may be dwarfed by the opportunities for innovation. Innovation goes beyond improvement to the existing systems and ways of doing things to the creation of alternatives. This future-oriented perspective on sustainability thinking – geared toward new processes, products, technologies, and markets – offers prospects for competitive advantage over rival firms. At a deeper level, this mode of strategic thinking holds the potential for transforming the way we think about the management of economic activity over the next few decades.

Sustainability logic and innovation are best understood as a wave of change in the current global economy. This wave has its origins in the entrepreneurial creativity of individuals and organizations that track and anticipate trends. As with any emergent frontier of innovation, certain pioneering firms are actively pursuing the emerging entrepreneurial opportunities. Companies such as Rohm and Haas, Pfizer, Shaw, Cargill-Dow, and Chemecol are creating alternative polymers, materials, and products that incorporate health, environment, and social benefit into the design and application of the product.

It makes sense that a radically different way of thinking – that is, viewing economic activity as necessarily dependent upon and interactive with natural systems and communities (whether human, or living but nonhuman) – would create entrepreneurial opportunities. It is consistent with entrepreneurship researchers' observations and therefore not surprising that the most enterprising among us, the entrepreneurial actors whether individuals or organizations, see and pursue these opportunities.

We use the classical definition of entrepreneurship here. This definition focuses on innovation. Entrepreneurial innovation occurs when individuals and firms act to offer more attractive alternative products, processes, and technologies that ultimately come to supplant, or substitute for, existing offerings. These future products, services, markets, even industries, emerge because human creativity irrepressibly generates new and better ways of doing things as human knowledge expands. They also emerge because some individuals and organizations are more attuned than others to national and international trends, new science knowledge, and shifting buyer values. As economists tell us, inefficiencies in markets and changing knowledge and technology perpetually drive innovation. The astute, informed, and observant discover or create opportunities from this raw material.

Peter Drucker (1985) described categories of opportunity that included technological insufficiencies; major shifts in demographic, social, political, and economic forces; and new knowledge from discoveries and innovations. Economist Joseph Schumpeter, who coined the phrase "creative destruction," drew attention to the disruptive character of entrepreneurial innovation. What he called *adaptive change* by companies made only incremental adjustments to existing ways of designing products and processes. These were minor adaptations to ongoing systems. *Creative change*, however, provided *new* products, processes, technologies, organizing structures, and markets.

Viewed through this lens, innovative activity focused on opportunities for new markets in sustainably designed products presents the chemical industry, and the industries it supplies, with potentially enormous entrepreneurial opportunities. As feedstock providers, chemical companies have the opportunity to shape new competitive space in the near future. They can differentiate their products and strategies in ways that will gain future competitive advantage over those who fail to react. In so doing, they also position themselves advantageously in markets where rising concern over hazardous and dangerous materials drives purchasing decisions.

The activities and trends discussed in this book represent a wave of change, but more importantly, opportunity. Companies that stand to benefit are those that carefully read the forces at work with an eye toward market creation, and cost as well as differentiation. In the face of growing challenges to business as usual – for example, demands for social and environmental accounting, transparency, and corporate responsibility – the chemical industry has the opportunity to benefit a broader set of stakeholders while better serving the financial interests of stockholders.

For many in the chemical industry these claims may sound hollow. Everyone knows that global expansion of the chemical industry, even with good faith efforts to comply with safety and pollution requirements, is assumed to carry with it adverse and unintended effects from chemical production and use. Such is the price of economic growth – correct? Wrong. In fact, our passive acceptance of this paradigmatic view of the chemical industry may pose the most formidable obstacle to change. If so, the challenge may be more in our minds than our capabilities.

The reality is that innovative individuals and companies have already challenged these assumptions with alternatives to conventional chemical design and application, working from the feedstock molecules up, and from the final product down to its material components. For these innovators, new scientific information, future-oriented conceptual maps, new design parameters, and the commitment to benign alternatives all combine to create breakthroughs. The opportunity offered to the chemical industry can be understood if one looks through the lens of entrepreneurship and innovation to see the possibilities for new products and markets.

Innovation, broadly defined, means that concrete and specific product and process innovations are considered as well as more abstract innovations in thinking and the novel organization of ideas that shape innovation. New markets in ideas and new markets in products are similar in their challenge to traditionally accepted approaches. The phenomenon of markets in ideas *and* markets for products and services is particularly important in the arena of sustainable development because innovations are taking place in both market arenas. Conceptual innovations – new ways to think about future product design – some of which have been discussed in Chapters 4 and 5 under Planning Frameworks and Designing for Sustainability – now complement, guide, and thereby help drive product and service innovations. This dynamic simply reflects the fact that just as innovation in the design of products and services creates new markets, so too do innovations in ideas change traditionally accepted ways of thinking. These changes in turn alter the way people act and envision the future, including the way many are now making efforts to redesign commerce. Companies typically neglect this entrepreneurial aspect of change in the way chemicals and the products they comprise are considered. Yet this new markets and future products focus accurately describes what we see occurring in innovative sustainable business activity in the chemical industry and beyond.

To illustrate, new and lucrative markets have emerged from the need to design environmental and health hazards *out* of products. Obviously, creation of these markets is supported by the historically unprecedented convergence of government pressure, NGO and stockholder activism, and growing scientific evidence that our prior assumptions about the economy–natural systems relationship are outdated. As a consequence, we see already existing growth markets and fast emerging markets across a range of the economy: organic foods, wind power, hybrid vehicles, green building materials and design, and environmentally benign chemical compounds offered for a wide variety of applications. Critics may point to the small size of these markets, but the size of these markets is less important than their rate of growth and the increasing number and intensity of factors favoring their continued expansion. Strategic, revenue, and profitability opportunities are the most compelling reasons for corporations to look to these kinds of markets. Note that the markets listed cover food, electric power, transportation, shelter, and materials. They also are closely linked to and have positive spillover effects for air quality, health protection, and climate change. Of course, chemicals are the fundamental building blocks of each of these marketplaces; therefore the chemical industry has enormous opportunities to turn our updated knowledge about environmental impacts into new products, processes, technologies, services, and markets.

Any student of entrepreneurship and corporate strategy will tell you that companies must continuously adapt their strategies to a changing context, always watching for ways to develop capabilities and new products and services that better meet stockholder and stakeholders' needs. The wise executive in the chemical industry will be actively and aggressively seeking information on the key trends converging in his or her industry that will determine the definition of *future* competitiveness. Options thinking should guide that individual's resource allocation process. Applied to sustainability, this is the only way to guarantee the company will be positioned to act quickly as markets incorporate "clean" materials, processes, and technology design requirements. The trends and forces driving markets toward this goal are unlikely to diminish. Business decision-makers ignore this reality at their peril; innovators and entrepreneurs embrace this reality as value creation opportunity.

This is the opportunity-oriented perspective on environmental issues. It can be used and applied in any business context. Yet many companies remain focused on compliance and regulatory issues as their definition of environment. Those entrepreneurial thinkers within corporations who perceive market opportunities for more environmentally responsible products face significant obstacles if senior level managers are ill-informed about current market conditions and societal trends. Companies unable to adapt take the risk of falling by the wayside as their competitors move quickly to pursue new markets for environmentally benign materials and technology.

What many executives fail to understand is that a fundamental change has occurred in the last decade. The environment today is inseparable from biological systems health (including human) and economic growth. This perception remains at odds with conventional understanding. The chemical industry, long under the environmental regulatory microscope, has been more attuned to environmental issues as traditionally defined (compliance), but the fundamental approach of most companies in their core businesses has stayed relatively constant, and narrowly conventional. The environment was understood as "out there," with human impacts seen as local and limited. Visible pollution could be managed and was thought easy to address. Besides, ultimately the responsibility of environmental impact was with the government; the company's responsibility was to comply with regulations and be careful not to engage in activities that carried obvious liability risks.

Businesses defined the issues as external to the firm's core concerns – they were not real business issues – and they responded to government requirements by devising technological fixes to solve what appeared to be discretely bounded problems. Traditional segregation of environmental concerns into environment, health and safety departments assured that "environment" would continue to be viewed as overhead costs and marginal to true business concerns. When understood in these terms, entrepreneurial opportunities could never be associated with environment.

This picture contrasts dramatically with the innovative leaders. While commitments to existing asset investments may prevent wholesale switching over to new operating and product design approaches, strategically savvy companies are gaining expertise and market presence. Sony is using corn-based plastics in components and packaging. Available now only in Japan, the casing for the Walkman stereo, wrapping film, and blister packaging are vegetable-based. "Such products are designed using environmental technology to encourage recycling activities, energy conservation and reduction in the use of hazardous materials. This replacement plastic uses only 45 percent of the oil-derived resources of conventional plastic, with commensurate reductions in carbon dioxide emissions."

Ford has introduced a concept SUV with a hydrogen engine and using biomass-derived materials including Cargill Dow's PLA (polylactide) biopolymer derived from corn. Bio-based materials contribute to the roof, carpet mats, tires, seating foam, and lubricants. Collaboratively with Toyota, Toray Industries has developed PLA material for use in an extensive range of automotive components. Raw materials are sourced through an alliance with Cargill Dow, but Toyota expects to have a raw material facility on line soon. Toray Industries expects to reach revenues of $86 million in 2005.

From the European Union, "eco-technology" that includes innovative processes and products is expected to be one of the fastest growing market arenas in the 21st century. The export market to developing countries for environmental technologies was reported to have grown by 10–17 percent between 1998 and 1999, with a lower growth rate for activity in developed markets. The sector was reported to have created more than 500 jobs between 1997 and 2002.

Reducing manufacturing costs is a big incentive for many. Materials maker DuPont used to make 10,000 tons of a polymer called Sorona each year using a petrochemical process in which oil was the raw material. Sorona was used in swimsuits, slacks, and jackets, since fibers made from it were soft and recovered their shape after being stretched. Such fibers are not biodegradable, however, and they are expensive to make because their manufacture requires high-temperature, high-pressure reactors.

DuPont engineers cloned genes from a bacteria, inserted them into a fermentation process, and created a microorganism that did not exist in nature. A bioengineered organism, the new creature turned glucose – a form of sugar – into Sorona in a single step. This new process debuted commercially in 2002 (Dupont, 2005). As a result, DuPont will not need either its oil-based raw material or the expensive reactors. Not only will production be much cheaper, says DuPont research manager Scott Nichols, but Sorona will be biodegradable, and the process for making it will be more environmentally friendly.

GE Plastics, a subsidiary of General Electric (GE), has developed a technology that by 2006 could make it unnecessary for carmakers to paint vehicles. It costs around $300 million to build paint lines and they take up about half of a car plant's floor space. GE's Lexan SLX product promises to eliminate the expense and the space requirements.

Thus far, the technology has been used only to paint the bumpers of a two-wheel personal transportation system called Segway. A thin film, when heated it can be molded into shapes that correspond to parts of a car. While the film will be more expensive than paint, the process eliminates the need for paint lines and uses a plant's existing molding equipment. GE Plastics expects to announce its first contract with a carmaker in 2004.

Well-worn terms such as pollution prevention have innovative sides. The effort to lessen or eliminate pollution before it is created uses *green chemistry* – preemptive manufacturing measures that companies use to decrease or eliminate future liabilities. For example, the National Environmental Technology Institute – created by the Massachusetts legislature, the chemical industry, and the University of Massachusetts – is working on plant-based materials to replace the toxic ones in foams and adhesives. The Institute is also developing water-based coatings for cedar and yellow pine. They replace oil-based paint, which, as it dries, contributes to smog. There is no shortage of examples of innovative firms moving to translate changing market conditions into opportunities for top and bottom line growth. Drivers of these changing conditions show no slowing signals – in fact, quite the opposite. As more information is relayed from chemistry and its related and off-shoot scientific disciplines to governments and corporations, it seems clear that industry must respond. The question for individual firms is whether they will resist and lose competitive advantage, or proactively respond to capture their share of the new value that will inevitably be created.

REFERENCES

A. Abbott, "Ion Channel Structures: They Said it Couldn't be Done," *Nature*, 418, 268–269 (2002).

R. Akashi, H. Tsutsui and A. Komura, "Polymer Gel Light-Modulation Materials Imitating Pigment Cells," *Advanced Materials*, 14, 1808–1811 (2002); as described by P. Ball, "Nature Inspires Colour Change Gel," *Nature Materials Update*, January 8, 2003.

D. T. Allen and K. S. Rosselot, *Pollution Prevention for Chemical Processes*, Wiley, New York, 1997.

D. T. Allen and D. R. Shonnard, *Green Engineering: Environmental Conscious Design of Chemical Processes*, Prentice Hall, Upper Saddle River, NJ, 2002.

P. T. Anastas and J. C. Warner, *Green Chemistry: Theory and Practice*, Oxford University Press, New York, 1998, pp. 11, 30.

P. Ball, "Natural Strategies for the Molecular Engineer," *Nanotechnology*, 13, 15–28 (2002).

J. M. Benyus, *Biomimicry: Innovation Inspired by Nature*, Quill William Morrow and Co., New York, 1997.

D. Bryant, L. Burke, J. McManus and M. Spalding, *Reefs at Risk: A Map-Based Indicator of Potential Threats to the World's Coral Reefs*, World Resources Institute, 1998, p. 9.

J. A. Cano-Ruiz and G. J. McRae, "Environmentally Conscious Chemical Process Design," *Annual Review of Energy and Environment*, 23, 499–536 (1998).

F. Castellani, B. Van Rossum, A. Diehl, M. Schubert, K. Rehbein and H. Oschkinat, "Structure of a Protein Determined by Solid-State Magic-Angle-Spinning NMR Spectroscopy," *Nature*, 420, 98–102 (2002).

Chemical Week, "Chemical Management Services Gains Traction," *Chemical Week*, June 4, 2003.

T. Collins, Director of the Institute for Green Oxidation Chemistry, 2003.

T. Collins, "The Importance of Sustainability, Ethics, Toxicity and Ecotoxicity in Chemical Education and Research," *Green Chemistry*, 5(4), G51.

G. C. Daily, Ed., *Nature's Services: Societal Dependence on Natural Ecosystems*, Island Press, Washington, DC, 1997.

C. de Duve, "The Beginnings of Life on Earth," *American Scientist*, September–October (1995).

S. C. Devito and R. Garrett, Eds., *Designing Safer Chemicals: Green Chemistry for Pollution Prevention*, American Chemical Society, Washington, DC, 1996.

J. C. Diop, "General Electric's Nanotechnology Program: Tougher Ceramics Based on the Microstructure of Seashells," *Technology Review*, December 2002–January 2003, 65 (2003).

DK (Danish Environmental Protection Agency), *Brominated Flame Retardants: Substance Flow Analysis and Assessment of Alternatives*, Environmental Project #494, DK, Copenhagen, Denmark, 1999.

P. Drucker, *Innovation and Entrepreneurship: Practice and Principles*, 1st edition, HarperCollins, New York, 1985, 277pp.

Dupont, "Sorona" Available at http://www.dupont.com/sorona/ourheritage.html. Accessed April 2005.

EC (European Commission), Joint letter by Electrolux, HP, ICL, Nokia, Ericsson, Agilent, IBM, Gillette, Sun Microsystems, and Intel to the European Commission, 2001, concerning the WEEE and RoHS Directives, 2001.

European Parliament, Directive 2002/95/EC of the European Parliament and of the Council of 27 January 2003 on the restriction of the use of certain hazardous substances in electrical and electronic equipment, 2003a. Available at http://europa.eu.int/eur-lex/pri/en/oj/dat/2003/l_037/l_03720030213en00190023.pdf.

European Parliament, Directive 2002/95/EC of the European Parliament and of the Council of 27 January 2003 on waste electrical and electronic equipment (WEEE), 2003b. Available at http://www.icer.org.uk/WEEE20030127.pdf.

B. K. Fishbein, *Waste in the Wireless World: The Challenge of Cell Phones*, INFORM, Inc., New York, 2002, p. 95.

Fujitsu Siemens Computers, "Environment." Available at http://www.fujitsu-siemens.com/aboutus/company_information/business_excellence/environmental_care/products.html. Accessed April 2005.

A. K. Geim, S. V. Dubonos, I. V. Grigorieva, K. S. Novoselov, A. A. Zhukov and S. Yu. Shapoval, "Microfabricated Adhesive Mimicking Gecko Foot-Hair," *Nature Materials*, 2, 461–463 (2003).

K. Geiser and J. Tickner, *New Directions in European Chemicals Policies: Drivers, Scope, and Status*, Lowell Center for Sustainable Production, University of Massachusetts – Lowell, Lowell, MA, 2003.

Global Environmental Management Initiative (GEMI), *Environment: Value to the Top Line*, GEMI, Washington, DC, 2001, pp. 50–54.

Ikea, "IKEA: Environmental and Social Issues." Available at http://www.ikea-group.ikea.com/corporate/responsible/brochure.html. Accessed April 2005.

INFORM, Inc., "Specifying and Sourcing Mercury-Free HVAC and Building Equipment," fact sheet, 2003. Available at http://www.informinc.org/fact_P3hvac.php.

O. Koelman, *Biomimetic Buildings: Understanding & Applying the Lessons of Nature*, Rocky Mountain Institute Solutions, Snowmass, CO, 2003.

B. Kuhlman, G. Dantas, G. C. Ireton, G. Varani, B. L. Stoddard and D. Baker, "Design of a Novel Globular Protein Fold with Atomic-Level Accuracy," *Science*, 302, 1364–1368 (2003).

H. Liu, J. Gao, S. R. Lynch, Y. D. Saito, L. Maynard and E. T. Kool, "A Four-Base Paired Genetic Helix with Expanded Size," *Science*, 302, 868–871 (2003).

C. Mavroidis and A. Dubey, "From Pulses to Motors," *Nature Materials*, 2, 573 (2003).

W. McDonough, M. Braungart, P. T. Anastas and J. B. Zimmerman, "Applying the Principles of Green Engineering to Cradle-to-Cradle Design," *Environmental Science & Technology*, 37(23), 435A–441A (2003).

J. E. Molloy and C. Veigel, "Myosin Motors Walk the Walk," *Science*, 300, 2045–2046 (2003).

L. Osborne, "Got Silk," *New York Times*, June 16, 2002.

D. Pantarotto, J. P. Briand, M. Prato and A. Bianco, *Translocation of Bioactive Peptides Across Cell Membranes by Carbon Nanotubes*, Chemical Communications, Royal Society of Chemistry, Cambridge, UK, 2004, pp. 16–17.

RCEP (Royal Commission on Environmental Pollution), *Chemicals in Products. Safeguarding the Environment and Human Health*, RCEP, Westminster, London, UK, 2003, p. 165.

M. Sarikaya, C. Tamerler, A. K. Y. Jen, K. Schulten and F. Baneyx, "Molecular Biomimetics: Nanotechnology Through Biology," *Nature Materials*, 2, 577–585 (2003).

O. Sato, *Angew. Chem. Int. Ed.*, 42, 894 (2003), as described in "From Butterflies to New Materials," *Chemical & Engineering News*, February 24, 2003, p. 31.

F. Schotte, M. Lim, T. A. Jackson, A. V. Smirnov, J. Soman, J. S. Olson, G. N. Phillips Jr., M. Wulff and P. A. Anfinrud, "Watching a Protein as it Functions with 150-ps Time-Resolved X-ray Crystallography," *Science*, 300, 1944–1947 (2003).

I. Schullstrom, *H&M Response to the EC's REACH Regulation Public Interest Internet Consultation*, Hennes & Mauritz AB, Stockholm, Sweden, 2003.

P. G. Schultz *et al.*, *Journal of the American Chemical Society*, 125, 935 (2002); as described in "Reinventing Biology," *Chemical & Engineering News*, January 20, 2003, p. 7.

Skanska, *Response to the EC's REACH Regulation Public Interest Internet Consultation*, Stockholm, Sweden, 2003.

D. Tenenbaum, "Northern Overexposure," *Environmental Health Perspectives*, 106, A54–69 (1998).

B. Thorpe, *Safer Chemicals within Reach: Using the Substitution Principle to Drive Green Chemistry*, Clean Production Action prepared for Greenpeace Environmental Trust, London, UK, 2003.

UBA (Umweltbundesamt), "Subchem Case Studies – Product and Process Innovation," 2003. Available at www.subchem.de.

UBA (Umweltbundesamt), "Substituting Environmentally Relevant Flame Retardants: Assessment Fundamentals," 2001. Available at www.umweltbundesamt.de, "Issues."

U.S. Environmental Protection Agency, "*The Great Lakes: An Environmental Atlas and Resource Book*," Chapter 4, 1994. Available at http://www.epa.gov/glnpo/atlas/glat-ch4.html.

U.S. Environmental Protection Agency, "Kalamazoo River Area of Concern," April 9, 2003. Available at http://www.epa.gov/glnpo/aoc/kalriv.html.

U.S. Environmental Protection Agency, "Mercury Update: Impact on Fish Advisories," fact sheet, June 2001. Available at www.epa.gov/ost/fishadvice/mercupd.pdf.

U.S. EPA, *2001 Toxics Release Inventory Public Data Release*, U.S. EPA, Washington, DC, 2003, p. 2-1.

J. Zi, X. Yu, Y. Li, X. Hu, C. Xu, X. Wang, X. Liu and R. Fu, "Coloration strategies in peacock feathers," *Proceedings of the National Academy of Sciences*, 100(22), 12576–12578 (2003).

8

THE BUSINESS CASE FOR SUSTAINABLE DEVELOPMENT

8.1 OVERVIEW

BETH BELOFF

BRIDGES to Sustainability

The key question asked by companies considering pursuit of sustainability practices is what is the business case – a profit or shareholder value case – that can create justification for allocating resources and the attention of corporate boards and executives. As was written in *Walking the Talk* (Holliday *et al.*, 2002), companies tend to get involved in activities long before they can prove the business case for doing so.

But the case *is* emerging and, although the corporate response is fragmented and at many stages of maturity in the United States relative to other places in the world, there are many noteworthy efforts highlighted in this section of the book that are moving us toward that definitive business case (Fig. 8.1). Companies moving up the sustainability learning curve are recognizing that, from a business competitiveness perspective, by moving beyond traditional EHS concerns to include (1) social issues regarding the workforce/workplace and community, and (2) long-term issues, not just short-term ones, they can move from compliance and managing near-term risks to creating new business opportunities. This perspective contributes to the longer term survivability of the enterprise.

We have organized this section into four parts:

1. The 2004 Chemical Industry Sustainability Survey and related Focus Groups Summaries, conducted by PwC and BRIDGES to Sustainability, with the

Transforming Sustainability Strategy into Action: The Chemical Industry, Edited by B. Beloff, M. Lines, and D. Tanzil
Copyright © 2005 John Wiley & Sons, Inc.

support of AIChE, to capture a view of where the industry stands on the issues and practices related to sustainability.

2. An overview of sustainability and performance, which links the intangibles of sustainability to market performance.

3. Business case studies expanding on the areas covered in the survey, and written by sustainability leaders in major companies with chemical operations, including DuPont, Shell, BASF, GlaxoSmithKline as one major pharmaceutical company, and 3M as a major customer of chemicals.

4. Various other important perspectives on the business case for the industry by others – the investor community perspective, an investment analysis piece providing a cautionary tale regarding Dow and Bhopal and potential market backlash, the scientific and NGO community perspective, and a commentary on corporate governance and sustainability.

Again, here we are not attempting to be comprehensive. There are numerous other worthy case studies and perspectives that could well be highlighted. Our goal is to present a prism of various thought-provoking perspectives both from the heart of the chemical industry and from those related to it in one way or another as stakeholders. We hope this stimulates further discussion among industry groups about where to push the business case.

Figure 8.1. Corporate sustainability learning curve. (Adapted by Beloff from a conceptual model developed for The Stanley Works by BRIDGES to Sustainability and Convergence Consulting, 2004.)

8.2 2004 CHEMICAL INDUSTRY SUSTAINABILITY SURVEY (COMPILED BY PRICEWATERHOUSECOOPERS, LLP)

ANDREW SAVITZ, DOUGLAS HILEMAN and MICHAEL BESLY[1]
PricewaterhouseCoopers, LLP

8.2.1 Overview

Every industry in the United States is grappling with sustainability, or corporate social responsibility, and the chemical industry is no exception. During the first quarter of 2004, PricewaterhouseCoopers LLP (PwC), in collaboration with BRIDGES to Sustainability (BRIDGES), conducted a survey, two focus groups, and interviews of senior executives and managers in the chemical industry, which revealed the following.

1. Although certain chemical companies are seen as targets by advocates who believe their products are inherently dangerous and their processes essentially unsustainable, others are considered global leaders in addressing environmental, economic, and social concerns.
2. Although Responsible Care® raised the bar and provided a common framework for chemical companies on environmental stewardship, a large (and possibly growing) gap exists between leaders and laggards within the industry on sustainability.
3. Although Responsible Care® was a highly visible, industry-wide approach, sustainability is seen to be primarily a matter of shareholder value, hence an area for competitive advantage. This raises a question as to whether an industry-wide approach to sustainability is either feasible or advisable. On the other hand, it is possible that a major adverse event, regardless of the company involved, could taint the entire industry or create pressure for a retroactive industry-wide regime. In any case, it seems clear that laggards would benefit from this assistance.

In addition to this dilemma, sustainability, and the multiple driving forces behind it, are creating a wider and more diverse set of responsibilities for the chemical industry. Our research revealed these additional challenges:

1. Numerous respondents expect to use the Responsible Care Management Systems (RCMS) framework to implement and manage sustainability, but believe that no significant changes to existing programs would be necessary. As many analysts and some respondents have noted, Responsible Care® does not address many of the areas covered by sustainability.
2. Most respondents indicated that sustainability touches more on public relations than any other function. However, stakeholder engagement,

[1]The authors wish to thank Beth Beloff of BRIDGES to Sustainability for her assistance.

fundamental to the success of any sustainability program, did not appear to be high on the radar screen of most respondents. One conclusion is that responding companies are focused on the reporting aspect of sustainability, without sufficient emphasis on the engagement aspects. This may be a fruitful area for industry-wide collaboration.

3. The lack of metrics on social and economic elements of sustainability continues to be a challenge. This may also present an opportunity for industry-wide collaboration.

8.2.2   Introduction

"Sustainable Development," "Sustainability," "Corporate Social Responsibility," are terms often used to describe the trend towards enhanced performance and reporting with respect to environmental, social, and economic indicators.

In response to the rapid development of sustainability as a business movement, PwC was approached by BRIDGES and the American Institute of Chemical Engineers (AIChE) to conduct a sustainability survey of the chemical industry in the United States. The results from this collaborative effort were to become an integral part of the publication, *Transforming Sustainability Strategy into Action: The Chemical Industry.*

The purpose of the survey was to:

- provide specific information on current attitudes, concerns, approaches, and activities in sustainable development concepts and practices;
- identify specific sustainability-related trends as they relate to the chemical industry; and
- stimulate a discussion on best practices for incorporating sustainability into the chemical companies' management frameworks.

In February and March 2004, PwC surveyed senior executives and managers with responsibilities for overseeing sustainability at 35 chemical companies with major operations in the United States, to identify key issues and challenges facing the industry and to gain insight into how individual chemical companies are handling "triple bottom line" performance management within the context of Responsible Care®.

Additionally, BRIDGES and AIChE's industry-led center CSTP,[2] together with PwC, organized two follow-up focus groups, consisting of representatives from nine companies and Dr. Terry Yosie from the American Chemistry Council (ACC). The purpose of these focus groups was to discuss our preliminary assessment of survey responses, gain more in-depth understanding on certain issues, and offer a forum for knowledgeable parties to raise relevant issues around sustainability that went beyond the survey instrument itself.

---

[2]The Institute for Sustainability's "Center for Sustainable Technology Practices."

8.2.2.1 Survey Design. To ensure that the respondents addressed the areas of sustainability – economic, environmental, and social – the survey presented three commonly used definitions for sustainability[3] and respondents were asked to consider their responses in terms of those definitions. This was critical given the broad array of definitions and understanding related to the concept of sustainability.

The survey instrument consisted of 19 questions divided into four categories:

Company Specific: Current Attitudes, Opportunities & Risks

Company Specific: Current Policies, Practices & Metrics

Industry Specific: Significant Developments, Influences & Issues

Industry Specific: The Role of Responsible Care® and Other Guidelines

The brief survey was designed to highlight certain top line issues. The closed-end questions were developed by PwC and BRIDGES to elicit uniform, comparable, and measurable responses. However, respondents were given the opportunity to add their own narrative responses throughout. Prior to distribution, the questionnaire was reviewed by industry representatives, a process facilitated through the AIChE, to test relevance and survey mechanics.

8.2.2.2 Targeted Respondents. The target respondents were senior executives and managers with responsibilities for sustainability at chemical companies with major operations in the United States. The companies and contacts within them were identified by BRIDGES and the AIChE. Contacts were asked to forward the survey to the appropriate individual if they believed that they were not the most knowledgeable on the company's sustainability approach and activities. In addition, contacts were informed that the results of the survey would be nonattributable and that all company-specific information provided would remain strictly confidential.

The survey was distributed and administered as follows:

- 225 companies initially invited to participate via e-mail;
- target list narrowed to 75 preferred targets based on subsequent screening;[4]
- 35 responses received via fax or web submission.

8.2.2.3 Response and Limitations. Representatives of 35 chemical companies responded to the survey (Fig. 8.2). As with many surveys of this nature, companies that see themselves as leaders or early movers are more likely to respond. Therefore, the survey likely includes the attitudes and approaches of some of the most active

---

[3](a) The World Commission on Environment and Development (Brundtland Commission) definition: "Development that meets the needs of the present without compromising the ability of future generations to meet their own needs." (b) The "Triple Bottom Line" definition: To minimize harm resulting from business activities and to create economic, social, and environmental value. (c) The Dow Jones Sustainability Group Index definition: "A business approach to create long-term shareholder value by embracing opportunities and managing risks deriving from economic, environmental and social developments."

[4]AIChE member companies with significant chemical operations within the United States.

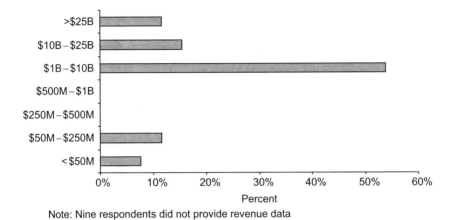

Note: Nine respondents did not provide revenue data

Figure 8.2. Breakdown of responding companies (35 companies responded, representing over US $225 billion in annual revenues). (*Source*: PricewaterhouseCoopers, LLP.)

companies in this area. While every effort was made to reach a diverse sample group, the findings presented here may not be representative of all chemical companies operating within the United States. As such, we have made a conscious effort to avoid drawing broad, industry-wide conclusions, but rather have analyzed, consolidated, and presented the survey results as they were received.

8.2.2.4 Profile of Respondents. The 35 respondents are predominantly publicly traded, multinational companies and most are well known to the media and the public (Table 8.1). The responding companies represent approximately $US 225 billion in annual sales. Over 90 percent of the respondents operate and have assets outside the United States. The survey therefore collected information from companies whose primary or initial experience with sustainability may have been gained in jurisdictions outside the United States.

Finally, the respondents fall into two groups: those whose principle business is the fabrication, processing, and/or handling of chemicals ("Primary Chemicals" companies), and those for whom the chemicals business is one of a number of different businesses ("Specialty Chemicals" companies).

8.2.3 Survey Analysis and Follow-Up

A team of PwC professionals with experience in sustainability and the chemical industry reviewed the survey results and consulted with a panel of experts and practitioners from the industry for their perspective on what the raw numbers and responses indicated. We then developed our preliminary findings. We also identified additional issues where more structured discussion might elicit further insights and conducted two focus groups to better understand some of the findings and validate our overall conclusions.

TABLE 8.1. Survey Respondents

Primary Chemical Companies	Specialty Chemical Companies
3M Company (The)	Arch Chemicals, Inc.
Air Products and Chemicals, Inc.	Bayer Corporation
Burlington Chemical Co.	BP
Chevron Phillips Chemical Company	Cambridge Major Laboratories, Inc.
Dow Chemical	Cardolite Corporation
DuPont	Celanese AG
Eastman Chemical Company	Eastman Kodak
Eliokem	Elementis Specialties
Formosa Plastics Corp.	Lubrizol
EMD Chemicals	Methanex Corporation
Occidental Chemical Corporation	Monsanto Company
Shell Chemical Companies	Pfizer
	PPG Industries, Inc.
	Rohm & Haas
	W.R. Grace & Co

Note: Eight companies declined to disclose company name. (*Source*: PricewaterhouseCoopers, LLP.)

The focus groups were convened in collaboration with BRIDGES and the AIChE's Institute of Sustainability, and involved industry representatives with both sustainability and chemical industry experience. These groups included individuals who had responded to the survey as well as individuals at companies who had not responded.

8.2.3.1 Survey Responses and Initial Observations. Our analysis of initial responses suggests the following insights.

Responses. Of the respondents, 83 percent rated sustainability of "high" importance to the company (i.e., rated 7 or higher out of 10), while only 46 percent indicated a "high" degree of integration of sustainability with business strategy (Figs. 8.3 and 8.4).

Initial Observation: Sustainability is Important, But Not Yet Integrated into the Business Processes. This response is not surprising. Even if senior management has articulated a vision for sustainability, it takes time and effort to convey it to the organization, and embed it into roles, responsibilities, and management systems. The immediate challenge is effectively to communicate the business case for sustainability to all levels of the organization and to develop programs to integrate sustainability into the business.

Responses. Of the respondents, 97 percent have written environmental policies, but only 40 percent have a written policy on sustainability – only human rights was less frequently addressed at 34 percent (Fig. 8.5).

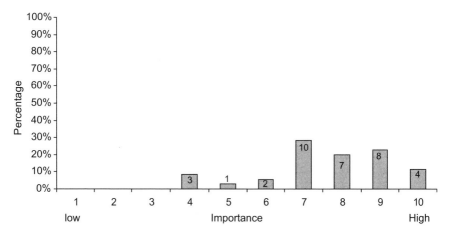

Figure 8.3. How important is sustainability to your company? (Question 1). (*Source*: PricewaterhouseCoopers, LLP.)

Initial Observation: Sustainability Policies Lagging. Owing to prolonged industry-wide focus on the subject, most companies have environmental policies. Sustainability, however, is not clearly defined by many companies, and written policies do not yet exist. Also, many areas covered under most definitions of sustainability may be perceived to be addressed already by existing policies, negating the need for a separate sustainability policy.

Responses. The highest ranked opportunities for sustainability included enhanced reputation, cost savings, and innovation, indicated by 65, 62, and 53 percent of respondents, respectively (Fig. 8.6).

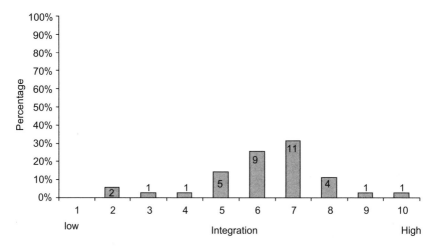

Figure 8.4. How much is sustainability integrated into your company's overall business strategy? (Question 1a). (*Source*: PricewaterhouseCoopers, LLP.)

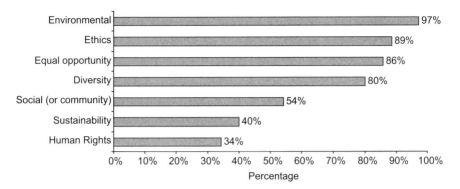

Figure 8.5. Does your company have a written policy that specifically addresses the following areas? (Check all that apply.) (Question 2). (*Source*: PricewaterhouseCoopers, LLP.)

Initial Observation: "Hard" and "Soft" Opportunities. The chemical industry is very familiar with cost savings related to environmental performance improvement. Sustainability offers many similar cost minimization opportunities. Damaged reputation has long been something to avoid, and it can easily be associated with top line performance. Companies are now beginning to appreciate the upside value of company reputation – and that sustainability performance can drive this intangible value.

Responses. Chemical companies use a variety of frameworks to manage sustainability. Of respondents, 62 percent use Responsible Care®, more than any other framework (Fig. 8.7).

Initial Observation: Variety of Management Frameworks. If most companies do not have written sustainability policies and sustainability does not seem well understood

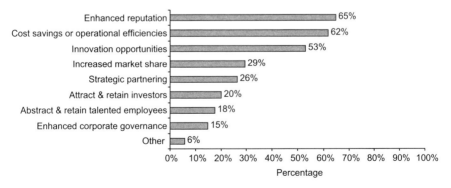

Figure 8.6. What do you see as the greatest sustainability-related opportunities for your business? (Check the top 3.) (Question 3). (*Source*: PricewaterhouseCoopers, LLP.)

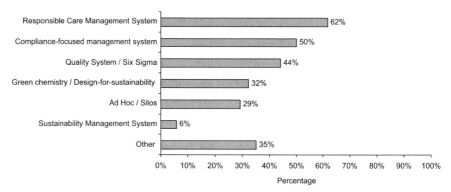

Note: "Other" category includes integration into various existing management systems/processes (including ISO 14001), business planning, local SD champions, SD committees, and Board of Directors.

Figure 8.7. How does your company manage sustainability? (Check all that apply.) (Question 5). (*Source*: PricewaterhouseCoopers, LLP.)

or integrated, it seems odd that 62 percent of the respondents say they manage their sustainability efforts through an RCMS.[5] Despite the fact that Responsible Care® is cited as the dominant approach, companies differ considerably in terms of how they view sustainability, where it is located, how they manage and measure it. The question remains "Are companies managing their sustainability programs in a systematic way?", and the answer is not clear. The focus groups confirmed that Responsible Care® is not an ideal vehicle by which to manage this area. It is helpful for environmental issues, but less so for social issues and even less so for economic issues related to sustainability.

Responses. Respondents indicated that responsibility for sustainability is located in a variety of functions within their organizations. By far, the most frequent response was Public Relations (74 percent), with Environmental, Health & Safety (EHS) a distant second at 44 percent (multiple responses permitted) (Fig. 8.8).

Initial Observation: Possible Confusion on Roles and Responsibilities. These responses seem at odds with the fact that 62 percent of the respondents manage sustainability via Responsible Care®, and provides evidence of program fragmentation, which is typical for all industries in the early stages insofar as sustainability cuts across numerous corporate functions. Coordination and integration are proportional to the maturity of a sustainability program, and "Public Relations" and "Communications" are often the starting point of such programs, so it is no surprise that a high percentage of respondents cited these functions as having primary responsibility. One would expect that the management of sustainability would move out of the public relations – and even the EHS function – once it becomes more integrated within

---

[5]It may be that some companies are using environmental stewardship as a proxy for sustainability, which indicates a lack of understanding concerning the latter.

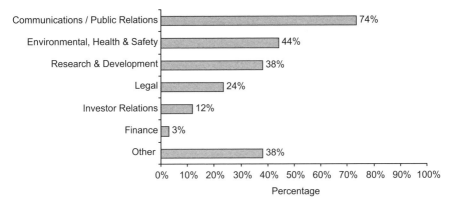

Note: "Other" category includes Board of Directors, Marketing, Human Resources, Government Affairs and Stakeholder Relations.

Figure 8.8. Within which functional area(s) is sustainability currently housed? (Check all that apply.) (Question 6). (*Source*: PricewaterhouseCoopers, LLP.)

the core of the company's business strategy. We note that many companies outside of the chemical industry have established cross-functional sustainability committees.

Responses. Just over half (51 percent) of the respondents now have a senior-level executive or executive committee with dedicated responsibilities to coordinate and promote sustainability. Another 14 percent indicated they expected to have this in place within two years, leaving just over a third (35 percent) with no plans to do so (Fig. 8.9).

Initial Observation: Organizational Placement is Beginning to Reflect Importance of Sustainability. With executives increasing focus on sustainability, companies

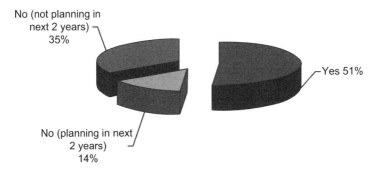

Figure 8.9. Does your company currently have a senior level executive or executive committee with dedicated responsibilities to coordinate and promote sustainability throughout your organization? (Check one.) (Question 7). (*Source*: PricewaterhouseCoopers, LLP.)

are likely to comprehend the extent and depth to which sustainability affects their operations, compliance, risk profile, and competitive position.

Responses. Of survey respondents, 58 percent publish an external sustainability report, or plan to do so within a year. Other survey answers related to reporting the following (Figs. 8.10 to 8.13).

- 37 percent indicated they use triple bottom line metrics;
- 60 percent indicated they would make no changes to their program if Responsible Care® were expanded to include social and broader economic criteria;
- 74 percent indicated that sustainability is housed in Public Relations (highest among response options);
- 74 percent indicated they use continuous improvement metrics to track their sustainability performance. This comports to the 80 percent who indicated they adhere to certified Environmental Management Systems (ISO 14000 or similar) – where continuous improvement is a basic tenet;
- 29 percent indicated that they use stakeholder outreach and evaluation to measure the success of their sustainability program.

Initial Observation: Reporting Making Progress, But Major Challenges Remain. Although almost 60 percent are reporting, or plan to, only 37 percent of the respondents indicated they use triple bottom line metrics to report on sustainability, which raises a question of how the reports can effectively capture basic sustainability information.

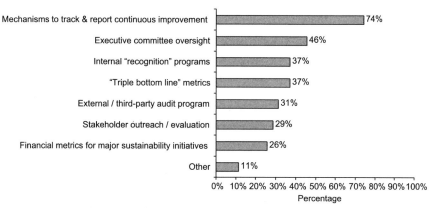

Figure 8.10. How does your company measure (or plan to measure) success with respect to your sustainability program? (Check all that apply.) (Question 8). (*Source*: PricewaterhouseCooper, LLP.)

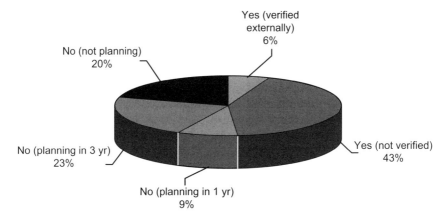

Figure 8.11. Does your company issue a Sustainability Report? If so, is the report externally verified in whole or in part? If not, do you expect that your company will issue a report? (Question 9). (*Source*: PricewaterhouseCoopers, LLP.)

Responses. Most respondents (89%) believed there would be more emphasis on sustainability in five years than today. Over half the respondents expected regulations to be an emerging factor driving sustainability in the chemical sector (Fig. 8.14).

Initial Observation: Sustainability – Here to Stay. To the EHS or sustainability professional, this is not a surprising response, but it may be to others within the organization. However, a survey of over 900 CEOs conducted by PwC in 2003 revealed that 79 percent of CEOs believe that sustainability is vital to the profitability of their companies, an increase from 69 percent in 2002.[6]

Responses. Respondents saw increased public right-to-know (71 percent), changing regulations for processes and products (68 percent) and globalization (62 percent) as the most significant developments that have influenced the chemical industry to consider sustainability (Fig. 8.15).

Initial Observation: Regulatory Requirements are Important Drivers for Sustainability, But Not the Only Ones. Although regulations play a central role in affecting company behavior, so do other factors, for example, increased transparency, competitive advantage, reputation, growth, and innovation. Globalization and regulation may be linked, as many of the biggest changes in regulatory schema are emerging from outside the United States, but will have a significant effect on chemical companies operating within the United States. Furthermore, issues related to regulatory compliance, environmental and sustainability reporting are becoming more complex for companies headquartered in one country, with business units

[6]PwC 2003 Global CEO Survey released at World Economic Forum in Davos, Switzerland.

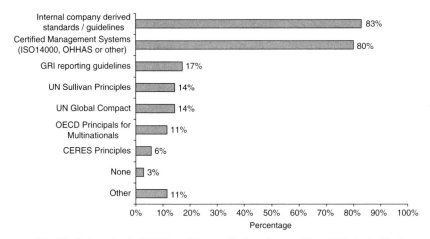

Note: "Other" category includes WBCSD eco-efficiency and Center for Corporate Citizenship Standards of Excellence.

Figure 8.12. Other than Responsible Care, what voluntary standards or guidelines does your company adhere to? (Check all that apply.) (Question 10). (*Source*: PricewaterhouseCoopers, LLP.)

and operations in others. Sustainability offers the opportunity to consolidate programs and reporting.

Responses. New regulations that address chemical risks throughout a chemicals lifecycle (53 percent) and requirements for intensive chemical management risk assessment processes (50 percent) were viewed as the most significant regulatory issues facing the industry over the next five years (Fig. 8.16).

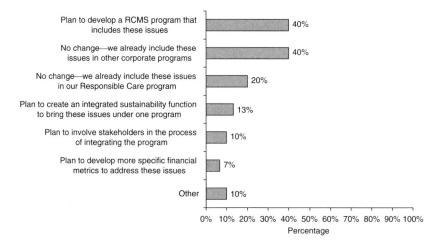

Figure 8.13. If Responsible Care were to include provisions for social and economic responsibility, how would your company modify its current program to incorporate these changes? (Check all that apply.) (Question 15). (*Source*: PricewaterhouseCoopers, LLP.)

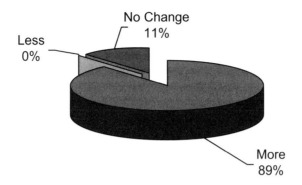

Figure 8.14. In your view, will your company place more or less emphasis on sustainability in 5 years than it does today? (Check one.) (Question 11). (*Source*: PricewaterhouseCoopers, LLP.)

Initial Observation: Risk Management is a Major Issue Going Forward. The extent to which the chemical industry effectively understands and manages risk will greatly depend on the nature of its interaction with key external stakeholders – especially the users of its products. Once again, regulatory oversight will play an important role here and, as such, we noted that this should be another topic of discussion for our focus groups.

8.2.4   Focus Groups' Perspectives

Two follow-up focus groups were convened, consisting of a total of nine companies, five of whom participated in the survey (Table 8.2). Dr. Terry Yosie of the ACC also participated.

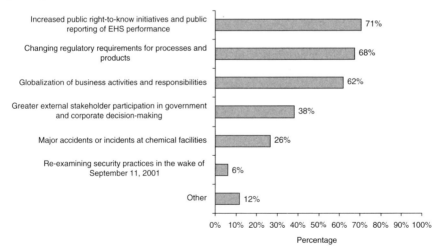

Note: "Other" category includes the commitment of large-scale, benchmark chemical companies to sustainable development, product portfolio growth & innovation and distinct competitive advantage experienced.

Figure 8.15. What are the most significant developments or events that have influenced the chemical industry to consider sustainability practices? (Check the top 3.) (Question 12). (*Source*: PricewaterhouseCoopers, LLP.)

Figure 8.16. Which of the following does your company believe will be a significant issue over the next 5 years? (Check the top 2.) (Question 13). (*Source*: PricewaterhouseCoopers, LLP.)

During each focus group, our initial observations were briefly presented and we discussed questions provided to the focus group participants in advance. In addition to these questions, the groups also discussed other issues related to the survey results, sustainability, and/or the chemical industry in general.

8.2.4.1 Focus Group Perspective #1: Responsible Care® and Sustainability – Cousins, not Twins. Responsible Care® and sustainability are related, but different. Responsible Care® originated, in part, in the chemical industry's desire to reduce its negative public profile, and avoid additional regulations. Responsible Care® focuses on compliance. Responsible Care® also focuses on intra-industry sharing and collaboration on better practices. Sustainability, on the other hand, is broader, involving engagement with an array of internal and external stakeholders, not just on environmental issues, but on social and economic ones too. Sustainability allows individual companies to differentiate themselves in a competitive environment. Even as Responsible Care® expands to include more sustainability parameters, it is not

TABLE 8.2. Focus Group Participants

Company	Survey
Air Products & Chemicals	Yes
BASF	No
BP	Yes
Cytec Industries	No
DuPont	Yes
Eastman Chemicals	Yes
FMC	No
Rohm & Haas	Yes
Shell Chemicals	Yes

sufficient to drive or manage a comprehensive sustainability program. Linkages between Responsible Care® and sustainability exist, but may not be universally understood. Further, Responsible Care® is viewed as a subset of sustainability by many leading companies, but to others it may be viewed as the only path forward into the sustainability arena.

8.2.4.2 Focus Group Perspective #2: Gap Between Leaders and Others. Chemical Industry companies tend to be either advanced when it comes to sustainability, or struggling to understand and adopt sustainability. Leading chemical companies are primarily large, multinational companies, although not all large, multinational companies have yet arrived. Chemical industry representatives acknowledge that robust sustainability programs are lagging in the majority of small- and medium-sized companies, who may not have the necessary resources to fully understand or launch such programs.

8.2.4.3 Focus Group Perspective #3: Little Formal Measurement of Stakeholder Outreach and Engagement. Focus group participants confirmed our initial hypothesis that relatively few companies use formal stakeholder outreach or evaluation to measure success with respect to sustainability programs. Given that survey respondents identified Public Relations most often as a function involved with sustainability, this may be a gap – and may present opportunities – for sustainability programs. Further, focus group members noted that stakeholders may not understand the information that the companies provide. We note that this is not unique to the chemical sector.

8.2.4.4. Focus Group Perspective #4: Greater Understanding of the Costs and Benefits of Sustainability Needed. The focus groups confirmed the need to identify, categorize, measure, and assess costs and benefits associated with sustainability. The groups also identified a number of key impediments, including the fact that costs and benefits may accrue to third parties as well as to the company. The inherent imbalance between costs and benefits hinders faster adoption of sustainability.

8.2.4.5 Focus Group Perspective #5: Need for a Standardized Reporting Framework. Chemical companies are constantly bombarded by investors and other stakeholders seeking sustainability-related information. Most stakeholders do not understand or appreciate the cost of this effort, or the limitations on what companies can provide. Companies, the industry as a whole, and external stakeholders would each benefit from a standardized approach or format for tracking and reporting sustainability related information.

8.2.4.6 Focus Group Perspective #6: Need for Standardized Sustainability Metrics. The absence of standard metrics, especially social and economic, is hindering the development and reporting of sustainability initiatives. Significant difficulties exist in translating sustainability to meaningful, measurable performance metrics and standards.

8.2.5   Comparison to Other Published Surveys

8.2.5.1   SustainAbility Stakeholder Survey: February 2004. The United Kingdom based consulting firm SustainAbility published *External Stakeholder Survey, Final Report for Global Strategic Review of Responsible Care*® in February 2004. SustainAbility's survey was oriented towards stakeholder impressions[7] of the chemical industry, and how the industry currently fares with regard to Responsible Care® and sustainability. The survey results highlight impressions from stakeholders that reinforce many of our findings. Findings and comments presented in SustainAbility's survey include the following.

- Perceived weaknesses of the chemical industry include lack of transparency and engagement, lack of accountability and common metrics, a poor understanding of product risks, and a gap between the performance of large companies and medium-to-small companies.
- Responsible Care® has improved both the reputation and the performance of the industry, but it does not adequately address less technical issues related to sustainability such as communication and accountability.
- Respondents believe the public has a very negative general perception of the chemical industry, although their own perception is only slightly negative.
- Respondents believe that Responsible Care® has had a positive effect on industry reputation.
- Respondents believed that a broad array of issues – including product safety impact, product responsibility, climate change, clean water – are important to the chemical industry, but that the industry was not prepared to deal with these issues.
- REACH[8] legislation in Europe, and the industry's opposition to it, continues to keep the industry center stage with stakeholders, and fuels negative public perception.
- The industry is seen by many stakeholders as reactive and secretive.
- Environmental issues are addressed more effectively than social issues (other than worker health and safety), and economic issues (such as economic development, education, and standard of living) are not addressed at all.

Stakeholders indicated that they want to engage in structured dialog; they do not simply want to be flooded with information. They also suggested areas for improvement to Responsible Care®, including

- more robust requirements for reporting and external engagement;
- requiring third-party verification of Responsible Care® implementation;

---

[7]Forty-one individuals interviewed.

[8]REACH (Registration, Evaluation, and Authorization of Chemicals) is a new regulatory framework that aims to simplify and streamline chemical regulation in Europe.

- improving traditional EHS operational performance; and
- establishing industry performance targets, goals, and timelines.

8.2.5.2 PwC 2002 Sustainability Survey. PwC conducted a sustainability survey in 2002, obtaining responses from 104 companies in many industry sectors. Fourteen of the respondents were from the chemical sector. Although the questions were not identical, responses from the chemical sector indicate similar attitudes towards sustainability.

Listed below are some of the key results of the 2002 survey:

- Forty-three percent had either defined, or planned to define, sustainability, and the same number rated the importance of sustainability a 7 or higher on scale of 1–10.
- Only 29 percent of the respondents had a formal policy on sustainability, although all respondents had environmental policies.
- Factors that influenced the decision to implement sustainability practices included:
  - Enhanced reputation – 100 percent
  - Industry trends – 91 percent
  - Cost savings – 82 percent
  - Competitive advantages – 82 percent
  - Innovation opportunities – 73 percent
  - Maintain license to operate – 73 percent
  - CEO/Board commitment – 64 percent
- Factors that influenced decision not to implement sustainability practices included limited Resources, and the lack of a clear business case.
- One-third of the respondents issued a Sustainability Report.
- One-third had a senior-level committee for sustainability issues/promotion.
- Sustainability activities being pursued included:
  - Pollution prevention & waste minimization – 100 percent
  - Community education and outreach – 93 percent
  - Environmental management systems – 79 percent
  - Corporate philanthropy – 79 percent
  - Employee volunteering – 79 percent
  - Direct investment in communities where company operates – 79 percent
- Almost all respondent (93 percent) believed that more emphasis would be placed on sustainability in five years time.

Reporting, and timing of reporting on sustainability, is about on track with previous results. In 2002, 36 percent of respondents published a sustainability report. Three of these used the Global Reporting Initiative (GRI) as a guide for their report, and one respondent obtained external verification for their report.

Of the six respondents to PwC's 2002 sustainability survey who did not produce a sustainability report in 2002, their responses and progress as of August 2004 are as follows:

- Three expected to publish a report within two years; as of August 2004, one of these had posted sustainability reports for 2002 and 2003 to their websites.
- Two expected to publish a report within five years. As of August 2004, one had published a report in 2002 but not in 2003.
- One had no plans to publish a report, and had not done so as of August 2004.

8.2.6   Conclusions and Path Forward

8.2.6.1   Conclusions. Survey results, the focus group discussions, and the authors' experience suggest several conclusions with regard to sustainability and the chemical industry.

- Sustainability is important to the industry and is likely to increase in importance.
- Stakeholders, chemical companies, and industry associations are all focused on sustainability. Areas of focus include:
  - Transparency
  - Reporting
  - Common metrics
  - Engagement
- Responsible Care® has been successful in elevating environmental performance within the chemical industry. It is a notable example of a voluntary industry-wide program where the industry appears to have gained credibility with stakeholders. Some within the industry are comfortable with this as a working model for sustainability program development.
- However, the linkages between Responsible Care® and sustainability are not universally understood. Many in the industry may misjudge its adaptability to sustainability. Although the majority of respondents reported that they have adapted (or plan to adapt) their RCMS to address sustainability, when the question arose during the focus groups, the consensus was that Responsible Care® and sustainability, although related, completely aligned. Sustainability extends beyond the traditional Responsible Care® focus on compliance into more opportunistic/competitive issues.
- Leading companies have learned how to successfully embed sustainability into their business operations and have been able to achieve significant competitive advantages as a result. Responsible Care® was developed on the basis of industry collaboration and knowledge sharing. As leading companies continue to build sustainability programs – and the competitive edge that comes with it,

– and as laggards move more slowly if at all, this gap will widen. Calls for industry cooperation, as with Responsible Care®, may not be returned.

- Metrics, data assurance, reporting, and communications are all critical aspects of improved sustainability programs. Responsible Care® does not fully address all aspects.

- Sustainability is not widely recognized as a core business process or as an important element of stakeholder commitment. Many key industry stakeholders, notably customers and investors, wield considerable financial impact. Therefore, many aspects of sustainability warrant the same level of organizational structure, oversight, and governance currently afforded environmental compliance or other control functions.

- Social and economic performance, and stakeholder engagement are evolving, from the areas of specific activity, to performance measurement, management and oversight, and communication and reporting of these issues.

- The ACC's suggested direction for modification of Responsible Care® is likely to include many issues of interest to an array of stakeholder groups, beyond environmental issues.

- Rigorous sustainability performance, transparency, reporting, communication, and stakeholder communication will undoubtedly improve company and industry reputation. Significant cost savings are also available, including reduction in the cost of legal compliance and other commitments to external parties. This should continue to be a powerful driver for sustainability program elements, but it may be hindered by a fragmented approach to sustainability, by the lack of complete understanding of the true cost of compliance, or by the inability to recognize the potential for savings aggregated from several functions throughout an organization.

8.2.6.2 *Path Forward.*

The chemical industry will continue to live, as they say, "in interesting times." Our survey results suggest that the industry is definitely on a path forward with regard to sustainability. The path will no doubt include changes to policies, programs, procedures, metrics, reporting, and stakeholder engagement. For leading companies, it will also include a fundamental re-thinking of the way business is conducted in almost every aspect.

The ACC's expanded RCMS will be a useful tool in pursuing sustainability, although it is not a panacea. Top-level commitment, clear vision of company mission, risk assessment, control activities, information and communications, and monitoring is essential. It is important to gain a thorough understanding of how RCMS supports sustainability – and how it does not.

The calls for industry knowledge sharing are likely to continue, particularly from smaller and medium-sized companies. More leading companies, with a full appreciation of the competitive advantages that sustainability can bring, are likely to be reluctant to share all their strategies and appeals. The chemical industry has a history of collaboration – Responsible Care® being a notable example – but later adopters should maintain reasonable expectations. We have suggested some areas where collaboration may be fruitful (see overview at the beginning of this contribution).

With the need for so many functional groups to be involved in the effective management of sustainability programs, coordination and communication within the company is the key to success. Mistakes and failures are inevitable, yet the leaders will persevere and will realize performance improvement and bottom-line benefits. Companies will eventually view sustainability as an integrated business process, and will clarify roles and responsibilities for functional groups. They will roll out more systematic programs, develop monitoring activities, and periodically re-think their approach.

As sustainability continues to grow in importance for the chemical industry, it will become increasingly important to measure the true costs and benefits. Frameworks now in their nascent stages will evolve to help companies focus on areas of greatest importance.

As a collaborative activity, the chemical industry should look to develop a common framework or standard for sustainability reporting and responding to stakeholder inquiries. As part of this effort, the industry should continue to monitor other emerging external reporting frameworks, such as the Global Reporting Initiative, One Report®, and the Corporate Responsibility Exchange, which all seek to standardize reported information and may be helpful to both the industry and its stakeholders.

8.3 SUSTAINABILITY AND PERFORMANCE: AN OVERVIEW[9]

KARINA FUNK

Massachusetts Renewable Energy Trust

The dust of the recent economic implosion has yet to settle. Capital assets and fortunes have been dreamed up, liquidated, and lost more times than any industrial-size paper shredder can handle in a lifetime. Analysts and investors will bear for years the burden of ignoring the high debt loads of many of the now-fallen giants of technology. Former executives and employees of now defunct companies will continue to battle in the courts for years. Fewer CEOs stay with their company for the long term,[10] while the media and the public have hotly debated whether corporate returns have kept up with hyperbolic CEO pay.[11] The number of business failures has been

---

[9]A previous version of this article appeared in *Sloan Management Review*, Winter 2003.

[10]Five years ago 22 percent of Fortune 100 CEOs had been with their companies for over 35 years compared with just 10 percent today (Spencer Stuart, 2004).

[11]For example, Forbes.com reported that 36 percent of CEOs at large companies in the United States, who have held the position for three years or less, have delivered annualized total returns of 15 percent or more to shareholders (Gillies, 2002). By contrast, the Institute for Policy Studies and United for a Fair Economy report that the CEOs of the 23 large companies under investigation for accounting irregularities earned 70 percent more from 1999–2001 than the average of CEOs among large companies, while shares of these companies lost 73 percent of their total value (Klinger *et al.*, 2002).

increasing – over 73,000 per year on average in the last 10 years, compared to 19,000 per year in the previous 30 years.[12] And corporations seem less able to predict their true earnings.[13]

Not surprisingly in the current economic environment, investors have begun to watch closely information related to how sustainable a business is, and many companies are voluntarily stepping up to this demand by issuing sustainability reports and engaging with a growing group of socially and environmentally responsible investors. Yet the word "sustainability" is quite politically charged, particularly within the lexicon of business. When, as is commonly the case, the term is limited to encompass environmental management or social equity, sustainability is often perceived to be at odds with fiduciary responsibility and unlinked to business strategy. This narrow view of the term, however, ignores the role that innovation, human capital development and strategic execution have to play in creating sustainable businesses.

This article attempts to make a business case for sustainability by adopting a broader view of the term, defining it as organizational characteristics and actions that are designed to lead to a "desirable future state" for any given stakeholder. For investors, this desirable future state would surely include sustained revenue growth over the long term. However, stakeholders alongside the investing public include consumers, employees, and regulators. All of these stakeholders value sustainability in some sense – whether it is diversity in the workforce for stakeholders in the talent market, innovation and risk management for stakeholders in the capital markets, or pollution prevention for stakeholders in the community. On the firm level, a desirable future state would include maintaining viability and profitability as well as consistent environmentally and socially responsible practices. In this broader view of sustainability, the convergence of innovation, adaptation, and unpredictability makes for a fertile environment for growth, for the bottom line, and for the evolution of sustainable businesses. Companies that actively manage and respond to a wide range of sustainability indicators are better able to create value for all stakeholders over the long term.

---

[12]From the 1950s through the mid-1980s, the number of annual business failures in the United States, was fairly constant at fewer than 20,000 firms a year. Beginning in 1982, the rate has increased to the point that the rate of business failures exceeds 80,000 firms a year. One might argue that this is because there are simply more businesses being launched. Yet, the failure rate per 10,000 listed businesses has also dramatically increased, rising from 43 from 1950–1979 to 91 from 1980–1997 (Mankin, and Chakrabarti, in press; Council of Economic Advisors, 2000).

[13]As evidenced by the number and amount of special items being reported in the S&P 500 over the past 18 years. The number of firms taking "special items" for restructuring charges, inventory write-downs, and asset write-downs, has risen significantly. The number of S&P 500 firms taking special charges has increased more than fourfold, from 68 in 1982 to more than 300 in 2001. (*Source*: Cap Gemini Ernst & Young Center for Business Innovation analysis, August 2002.)

BOX 8.1   DOES "SUSTAINABLE" MEAN FOREVER?[14]

The concept of sustainable development has received widespread attention (if not agreement) in the international community in the past decade. Early momentum for the concept was supplied by governments (especially in developing economies), NGOs, multilateral organizations, and has since begun to attract serious consideration from the private sector.

In the business world, the most relevant elements of sustainability have to do with environmental and social responsibility – within which lie issues such as reducing materials and energy intensity, concern for consumer and employee health, and being a "good steward" in the community in which a company is involved. But just as organisms survive not by living forever but by propagating their species, sustainable business must be a dynamic state. The opportunities of sustainable growth must be emphasized as businesses evolve and adapt according to economic realities. Limiting our scope to merely a time dimension assumes that sustainability is a zero-sum game. The convergence of innovation, adaptation, and unpredictability make for a fertile environment for growth, for the bottom line, and for the evolution of sustainable business. Proactive response to regulation, building customer loyalty, as well as attention to the relevant intangible indicators are all things that, through one feedback loop or another, may be translated into value creation (Reinhardt, 1999).[15]

8.3.1   Intangibles and Value Creation

Since the year 2000, the European Commission (Clark, 2000), the U.S. Securities and Exchange Commission (Blair and Wallman, 2001), and the U.S. Financial Accounting Standards Board (Upton, 2001) have all commissioned studies on the importance of intangibles in the economy. All conclude in no uncertain terms that the drivers of wealth creation for business and the economy are less about physical and financial assets and more about intangibles such as intellectual property and employee talent. While conventional accounting and financial metrics yield some insight into a firm's market value, forward-looking sustainability indicators – anything from confidence in a company's management to research leadership to the management of environmental liabilities – are becoming more relevant to a business's overall value proposition.

As a result, a good deal more effort is being paid to codifying the nonfinancial and intangible aspects of businesses. In France, their *Nouvelles Regulations Economiques* mandates (among other things) reporting on human resources, community, and labor standards. And in the United Kingdom, the government requires ethical, social, and environmental information of occupational pension funds' investment policies.

---

[14]Funk, K. "Performance – and Sustainability – Over the Long Haul," Cap Gemini Ernst & Young Center for Business Innovation white paper, June 2001. See Funk (2001).

[15]Reinhardt argues that working with regulators can also be used as a competitive weapon (Reinhardt, 1999).

A Value Creation Index (VCI), created by Cap Gemini Ernst & Young, attempts to not only quantify nonfinancial performance, but also to link key intangibles to a company's valuation in the market (see Section 6.2.2 for a discussion of the VCI). The results clearly reveal the importance of business assets that most companies do not measure, manage, or disclose. Social and environmental responsibility were among the top intangible factors identified, but the VCI offers a far broader picture of the elements of sustainable business. In the chemicals industry, a VCI model (Box 8.2) reveals that the top four indicators that are highly correlated with the market value of equity of these companies are related to innovativeness, strategic alliances, leadership, as well as to environmental and social responsibility.[16]

Wall Street is gradually becoming aware of the importance of measurement and disclosure of nonfinancial elements of a business. Although analysts may not speak of sustainable development, 50 percent of oil and gas industry analysts surveyed by Cap Gemini Ernst & Young confirmed that regulatory compliance on environmental issues, community service, and lawsuits do indeed impact the value of a firm.[17] In this case, it is clear that attention to environmental and socially responsible performance can mitigate the downside risk of corporate liabilities. Of those same oil and gas analysts, 68 percent believe that intangibles related to employees impact the value of a firm (http://www.cbi.cgey.com/) – and insofar as things such as employee satisfaction, diversity of workforce, workplace turnover, and productivity have anything to do with a firm's performance over the long term, sustainability lies firmly within the context of overall corporate strategy.

8.3.2 Sustainability in Practice

Managers are discovering that the intangible indicators that gauge sustainability can also be indicators of efficacy – that is, how well a company is *run*. From the management of corporate liabilities to new market ventures, a sustainable business strategy can improve all segments of corporate activity. Indeed, some argue that environmental management in particular is a good proxy for gauging overall management capabilities at both the strategic and operational levels. Matthew Kiernan, founder of investment advisory firm Innovest, believes it is a robust metaphor because the environmental problem is more multifaceted than most, as it touches all aspects of a firm's operations from product design to finance, and also holds implications for a wide range of stakeholders from the government to investors to community citizens (Kiernan, 2000). Perhaps more significantly, environmental problems are relatively "young" issues; addressing them demands strategic foresight, superior

---

[16]Our analysis of actual nonfinancial performance as correlated with market value revealed the following value drivers: Innovation, Quality, Customer, Management, Alliances, Technology, Brand, Employee, Environment. Multiple, statistically independent measures are used as inputs for each driver in order to ensure a robust model.

[17]Cap Gemini Ernst & Young Center for Business Innovation, *Measures That Matter* study. Survey of 300 sell-side analysts, 275 buy-side analysts, as well as interviews with portfolio managers. Summaries of these studies are available at http://www.cbi.cgey.com/.

execution, and organizational agility, factors that cannot be fully identified on a balance sheet.

If a large part of a company's equity valuation is riding on nonfinancial data, it would seem sensible to mine these data. As managers seek revealing and reliable performance measures of these factors, they can make use of information that their companies already have available – not only data that they are required to report, but data that are collected for one purpose that may be used to glean information on other dependent issues. Proactively managing and interpreting these figures will better allow companies to be in control of the measures that matter in their market and industry. Indeed, companies may not have a choice: as data gathering and management systems get more sophisticated, *monitoring* performance will by necessity be more tightly linked to *improving* performance. For example, emissions data are required for some regulated toxics, but this same information can also help to establish materials use goals in a production process and communicate progress to interested stakeholders. In many ways, the metrics for sustainability and market performance are strategically linked.

8.3.3   Driving Innovation

Traditionally, environmental compliance and social welfare expenditures have been viewed as costs that correlate negatively with returns. However, recent studies suggest that there are several opportunities for competitive advantage and increased profits to be gained by strategic sustainability initiatives (Reinhardt, 1999).

This reasoning reflects a shift from viewing business expenditures in a static world, to viewing them in a dynamic one based on innovation. Claas van der Linde and Michael Porter (1995) argue that in a static model, firms have already made their cost-minimizing choices and therefore any imperative to spend in the name of the environment or social responsibility "inevitably raises costs and will tend to reduce the market share of domestic companies on global markets." A static world, they say, falsely assumes that through profit-seeking, companies are already pursuing all profitable innovations. However, a dynamic (i.e., real) world is shaped by the stimulation and development of innovations. Porter and van der Linde argue that "net compliance costs [for environmental regulations] are overestimated by assuming away innovation benefits" (they also cite numerous examples in support of this).

In other words, regulation and market pressures have helped spark innovations that have eventually improved process efficiencies, tapped new markets, streamlined production and materials use, and led to many other benefits beyond reduced pollution.[18] For example, Fortune 500 company Ashland thinks of their hazardous waste management program not as a necessary evil, but as a value-added customer service. Ashland Distribution Company offers a one-source, "closed-loop" process to not

---

[18]A debate against Porter's claim of "the innovation-stimulating effect of regulation" casts doubt on the assumption that regulation would offer the possibility of a "free lunch," in part because "there is considerable doubt as to whether regulators would know more about these better methods of production than firm managers." For this reference and an extended discussion, please see Jaffe *et al.*, (1995).

only supply chemicals, plastics, and other materials, but also to manage hazardous and on-hazardous waste streams for customers. Ashland's Environmental Services group leverages their in-house expertise in handling Ashland's own chemical businesses into a value-added service for other customers. They offer a range of processing and treatment options, compliance assurance, and other services throughout North America. Ashland believes they have decreased customers' costs of ownership for purchased chemicals, and benefited from improved customer satisfaction and from revenue from the environmental management services (Fiksel et al., 2004b).

Many continuous improvements at companies are beyond compliance yet "make good business sense" – such as reducing greenhouse gas emissions, improving energy efficiency, and widening the recycling program to all materials and discharges. As such, mere environmental compliance may be considered a "lagging indicator." Forward-looking indicators that measure a company's ability to make efficient use of materials and energy (e.g., throughput data or energy use per consumer product unit), as well as any downstream revenues, can therefore be a measure of innovative ability, effective capital utilization, and value creation.

8.3.4 Communicating Transparency, Transparently

Effective communication can be the linchpin of corporate reputations; negative impacts can be dramatic when stakeholders are not given the information or ability to make an informed choice. Transparency has become a critical business issue. The Sarbanes–Oxley Act is the legislative incarnation of the spotlight that investors, consumers, and employees now shine on the financial statements of a company.[19]

Indeed, companies may pay a price for *not* managing the disclosure of their information, given the ease with which consumers and regulators can now access information on corporate practices. When the Toxic Release Inventory (TRI), a U.S. EPA database of waste management activities, was first disclosed, shares of publicly traded companies reporting data markedly declined in the short term.[20] The implication is that investors updated their expectations of future returns for high TRI companies. This feedback from the market prompted change. The firms with the largest decline in market value subsequently responded by reducing emissions more than their industry peers.[21]

---

[19]The SEC adopted the Sarbanes–Oxley Act in 2002, requiring CEOs and CFOs to certify the accuracy of their public reporting and to file federal forms in a timely manner so as to improve the accuracy of information that is available to investors.

[20]"Stockholders in firms reporting TRI pollution figures experienced negative, statistically significant abnormal returns upon the first release of the information" (Hamilton, 1995).

[21]In Khanna et al. (1998) it is found that the negative stockmarket returns following TRI disclosure had "a significant negative impact on subsequent on-site toxic releases and a significant positive impact on wastes transferred off site, but their impact on total toxic wastes generated by these firms is negligible." The discrepancy with the Konar and Cohen (1997) findings is that the firm samples are different; Konar and Cohen analyze 130 publicly traded firms from several industries, while Khanna et al. analyze 40 firms from the chemical industry. In any event, there was a response by the firms in their treatment of waste, upon disclosure of TRI data.

Companies stepping up to this demand for information disclose not only credible financial statements, but also their policies and procedures concerning the environment, their community, and governance. One recent study shows a relationship between companies that disclose more detailed information about their governance, and higher shareholder return (Sibson Consulting, 2003). Although this correlation is not conclusive, it does underscore the validity of transparency in governance as a value driver. Indicators of transparency include whether or not a company discloses its governance policies and procedures, degree of stakeholder engagement, timeliness of communications, and quality and depth of environmental/sustainability reporting.

8.3.5   Creating Differentiation

Some companies face the challenges of operating in a "commodity" industry. This makes differentiation particularly difficult, and companies typically compete through operational efficiency gains and productive capacity. Yet, a company's reputation for environmental and social responsibility can have an important impact on strategic issues, and sustainability can be a differentiator. While the primary negotiating levers for most businesses are based on economics, concern for the environment can, for example, provide access to capital and global markets. Some host governments may even demand adherence to sustainable development principles as a price of entry. There is evidence of increasing regulatory pressures, particularly in Europe. On 14 April 2003, *The New York Times* reported that "the European Union is adopting environmental and consumer protection legislation that will go further in regulating corporate behavior than almost anything the United States government has enacted in decades." However, it has become clear that being assertive makes good business sense. In the words of Gary Pfeiffer, Chief Financial Officer of DuPont Corporation (Fiksel *et al.*, 2004a): "Every corporation is under intense pressure to create ever-increasing shareholder value. Enhancing environmental and social performance are enormous business opportunities to do just that."

Measurable indicators of sustainability reputation include a company's regulatory compliance record, third-party recognition and awards, participation in environmental/sustainability consortia, and extent of community development and philanthropic support.

8.3.6   Managing Risk

Proactive investing in environmental measures beyond that required by law can be good for the bottom line, if for no other reason than to limit downside risk (Reinhardt, 1999). Damages and hefty litigation fees are incentive enough to manage proactively the risk of environmental, social, and public relations disasters. Often, this is relegated to an Environment, Health and Safety office that is far removed from the hub of corporate activity. However, if pursuing sustainable business strategies can increase a firm's expected value, it would be sensible to infer that integrating sustainability considerations into other kinds of risk management will lead to better decision-making. Forest Reinhardt argues that "environmental problems are best analyzed as business problems ... the basic tasks do not change when the word 'environmental' is included in the proposition" (Reinhardt, 1999). For example, more efficient use of materials is one environmental goal that can also help to stem product obsolescence, a strategic concern of paramount importance. For example, one can fight obsolescence in the consumer products industry by brute-force fast and frequent product introductions. However, another way may be to extend the product life itself – designing parts and materials for ease of upgrades rather than disposal. The bottom-line benefits can run from customer loyalty or even customer lock-in with a service relationship over a longer product lifetime, to lower disposal costs, particularly in countries that mandate product take-back.[22]

Measures related to sustainable risk management would include accrued environmental liabilities, fines, warnings and penalties, as well as product take-back programs whether in compliance with or in addition to regulatory initiatives.

8.3.7 Enhancing Growth and Expansion

A company that is expanding its worldwide operations would be wise to tie sustainability considerations to its strategic management of social, political, and economic factors. Some famous episodes in the public eye – Shell's conflict with the Ogoni people and allegations about Nike's labor practices, for example – demonstrate that sustainable operations are an opportunity to avoid or reduce future costs. Adopting a sustainability mindset could also lead to increased access to capital. The primary negotiation lever is likely to be focused on price, profit, and economics, but concern for sustainability could certainly be a differentiator *ceteris paribus*. Some host governments may even demand adherence to sustainable development principles as a price of entry.

The monitoring and disclosure of information important to stakeholders in any given region of operation, the involvement of local action groups at the level of investment in local development, and measures of political and economic risks, would be leading indicators of this kind of risk and growth management. Early measurement and reporting of leading indicators of sustainability initiatives also helps build better relationships with stakeholders, especially at the local level. Sarah Severn, Director of Corporate Sustainable Development at Nike, states that proactively listening to the stakeholders that voice their concerns has helped Nike discern exactly where their business has an impact. As a result, Nike can prioritize and address multiple issues from water use to organic cotton to climate change to human rights – and create initiatives to improve their performance along each of these objectives.

---

[22]Laws in Europe, Japan, and in some U.S. localities require some manufacturers to take back their products from consumers or to set up easily accessible collection systems for disposal, reuse, or recycling.

8.3.8 A Company-Level Sustainability Model

Most of the intangibles and indicators referred to in this sustainability model should resonate with common sense, but do not have to rely solely upon it. By mining what they already know, managers can obtain a more comprehensive view of how to encourage sustainable growth. For any given business, a host of qualitative evidence, quantitative measures, and the particulars of its industry context can be mined to create a hypothetical model of the relevant sustainability drivers (see Section 6.2.2 for a discussion of indicators used in the Value Creation Index). Surveys of internal and external stakeholders as well as existing operational data can then be used to identify, for each driver, groups of measurable sustainability indicators that cut across all the business's functions (procurement, supplier relations, produce design, and so on). That might, for example, include such indicators as R&D investment, worker-safety statistics, accrued environmental liabilities, and the effectiveness of risk prevention.

The resultant company-specific model can be tested empirically. The impact of indicators on drivers and, in turn, on overall sustainability can be quantified, and changes in both indicators and drivers can be mapped to such performance factors as stock price, earnings, and market share.

Identifying leading indicators of sustainability may never be a perfect science. However, financial indicators will be more helpful when combined with intangible indicators as we move from last year's year-end figures to what a company's value proposition is *now*. Performance measurement that is incomplete and imperfect is better than measurement that is disconnected from business objectives. Competitive advantage will likely go to those who avoid getting caught managing what they cannot measure. A long-term strategic view of sustainability may seem to contradict the urgency of immediate results for *this* quarter, but it is precisely the availability of both financial and intangible performance information – and

---

BOX 8.2 KEY INTANGIBLES FOR CHEMICALS, ENERGY, AND UTILITIES INDUSTRIES

The Value Creation Index (see Section 6.2.2, Intangibles and Sustainability) quantifies nonfinancial corporate performance and links key intangible factors to a company's valuation in the market. Using a mix of publicly available and proprietary data, multiple metrics for each intangible were collected and scrutinized statistically to discern their relationship with real – not perceived (based on opinion surveys, and so on) – market value.[i]

The index also weighs each intangible according to its impact on market value in order to identify the value drivers that are most significant in a given industry. The indicators that measure each intangible, and the key intangibles themselves, are industry specific. For example, the "environment" intangible value driver may includes regulatory compliance metrics such as number of Occupational Safety and Health Administration violations and worker safety statistics.

In 2002, Cap Gemini Ernst & Young developed VCI models for the energy, utilities, and chemicals industries. About 70 of the largest United States-based companies are included in the data set after filtering for a minimum market capitalization in each industry, and for availability of data. The models ended up explaining at least 70 percent and up to 90 percent of the variability in market value in an industry.

The intangible value drivers, along with the ones of greatest significance for each of the three industries, are ranked in Table 8.3.[ii]

TABLE 8.3. Key Intangibles for Chemicals, Energy, and Utilities Industries

Driver Rank	Chemicals	Integrated Oil and Gas	Electric Utilities
1	Innovation*	Innovation*	Social responsibility/ Environment*
2	Alliances*	Brand*	Human capital*
3	Leadership*	Human capital*	Quality of product
4	Social responsibility/ Environment*	Quality of product	Size
5	Quality of product	Leadership	Alliances
6	Business to business exchanges	Alliances	Leadership
7	Technology	Technology	Innovation
8	Human capital	Social responsibility/ Environment	Technology

*Indicates the intangibles that are highly statistically correlated with market value.

[i]Three major Cap Gemini Ernst & Young studies contributed to the VCI: "Measures That Matter," a survey of more than 300 sell-side analysts and 275 buy-side investors carried out in 1996, determined that nonfinancial performance accounts for up to 35 percent of institutional investors' portfolio allocation decisions; "Success Factors in the IPO Transformation Process," a study conducted first in 1998 and again in 2001, which showed that a new company's value proposition is highly correlated with its intangibles offer, including such factors as managerial effectiveness, intellectual capital, and organizational capabilities; "Decisions That Matter," a 2001 survey, identified the critical nonfinancial drivers of long-term economic value according to senior managers and indicated that deficient measurement of these factors correlated with weaker financial results. This underscored that internal performance measures are not currently aligned with corporate strategies. These studies and more details on Cap Gemini Ernst & Young's body of work on intangibles valuation is available in Low and Kalafut (2000).

[ii]For a fuller explanation of the VCI energy, utilities and chemicals industries' models, please see C. B. Cobb, "Measuring What Matters: Value Creation Indices for the Energy, Utilities and Chemical Industries," Perspectives on Business Innovation, Cap Gemini Ernst & Young Center for Business Innovation.

the ability to take action based on its interpretation – that can give decision-makers a more comprehensive understanding of what is important for performance over the long term.

8.4   DUPONT: GROWING SUSTAINABLY

DICKSEN TANZIL
BRIDGES to Sustainability

DAWN G RITTENHOUSE
DuPont

BETH R BELOFF
BRIDGES to Sustainability

Manufacturing companies have incorporated sustainable development into their business practices to varying degrees. Many, perhaps most, have concentrated on cost and risk reductions through initiatives such as pollution prevention and eco-efficiency. Some, however, have begun to emphasize the value-generation aspect, using a sustainability frame of mind to create new opportunities and markets. To learn about the latter, this chapter considers the case study of DuPont, a science company that has the rare trait of having sustainable development at the core of its business strategy. The chapter will review how sustainability is implemented in DuPont's management, research and development (R&D), and operations, as well as the results and challenges that the company faces to date.

Founded in 1802, the Delaware-based DuPont is the oldest company on the Fortune 500 list. It is a diversified global company with $27 billion of revenues in 2003, employing about 81,000 employees in over 70 countries. DuPont styles itself as a "science company," offering products and services founded on chemistry, material science, biology, and other sciences. The company is currently organized into five business groupings or "growth platforms," namely Electronic & Communication Technologies, Performance Materials (including polymers), Coatings & Color Technologies, Safety & Protection, and Agriculture & Nutrition. A sixth business group, Textiles & Interiors, was divested in April 2004.

8.4.1   Sustainability as a Growth Strategy

Incorporating sustainability into a company's business strategy can be daunting. To be sure, it is an inherently complex concept. One may easily come up with an endless list of economic, environmental, and societal needs and concerns that should be addressed. Thus, in constructing a sustainability strategy, it is imperative that a company understands what sustainability means specifically to it and recognizes the most relevant aspects of sustainability.

As an innovation-based science and technology company, it is DuPont's business to deliver products and services that improve quality of life – making people's life better, safer, and healthier. Furthermore, in an economically viable enterprise, growth is essential. This led DuPont to craft the term "sustainable growth," defined as creating shareholder and societal value while reducing its environmental footprint through the value chain. This definition provides a decision-making framework for DuPont that emphasizes the value creation aspect of sustainability. Shareholder and societal value can be simultaneously created when the business is finding solutions to society's needs for better quality of life. At the same time, the framework recognizes a broader scope in considering health and environmental footprint (including injuries, incidents, emissions, use of depletable resources, and so on). By striving to reduce the footprint through the entire value chain, rather than limited to the company's boundaries, DuPont aims to create an added value in helping customers to reduce their own footprints.

In 2000, sustainable growth was adopted as DuPont's "official mission." Since then, it has been regarded as the company's core strategy and direction. The sustainable growth strategy is further organized into three strategic pathways (Holliday, 2001):

- *Integrated science*: broadening its technological basis, integrating biology and other sciences with its long-established strength in chemistry and material science. Biology, especially, is critical in developing sustainable solutions, such as more efficient synthesis using biological catalysts and new products from renewable biomass feedstocks.
- *Knowledge intensity*: adding more knowledge content to products and services. This aims at creating more value through less material-intensive means, capitalizing on the company's experience, expertise, and brand equity. One way to achieve this is by marketing products along with services such as implementation assistance and training.
- *Productivity improvement*: implemented primarily with Six sigma, a rigorous data-driven methodology for reducing defects well to the per-million level. Six sigma ensures quality and customer satisfaction while reducing wastes and environmental impacts.

A metric used by DuPont to "measure" sustainable growth is the "shareholder value added per pound of production." Pound of production is used in the denominator as a proxy for footprint, as impacts typically increase in proportion to production mass. As a quantitative measure, this metric may be overly sensitive to market fluctuations and thus not a good tool for measuring year-to-year progress towards sustainable growth. However, it remains useful at the strategic and planning level. Integrated science, knowledge intensity, and productivity improvement all lead to higher shareholder value per pound. In recent years, DuPont has divested from businesses that generate low shareholder value per pound, such as commodity polymers and oil and gas. At the same time, it has acquired and invested in knowledge-intensive businesses such as electronics, biotechnology, and nutrition.

DuPont's corporate vision, mission, and strategy for sustainable growth, summarized in Table 8.4, form a basic comprehensive framework for its business approach.

TABLE 8.4. DuPont's Framework for Sustainable Growth

Corporate vision

    To be the world's most dynamic science company, creating sustainable solutions essential to a better, safer, healthier life for people everywhere.

Corporate mission

    Sustainable growth – to create shareholder and societal value while decreasing the environmental footprint along the value chains in which DuPont operates.

Strategy for sustainable growth

    1. Integrated science;
    2. Knowledge intensity;
    3. Productivity improvement.

The three-pronged strategy described above leads to greater shareholder value per footprint. However, the greater value advantage of sustainable growth still lies in the simultaneous pursuit of *both* shareholder and stakeholder value. This model is captured schematically in Figure 8.17. Stakeholder engagement – meeting people's expectations and finding essential solutions to their needs – remains the strategic driver for reaching more people through new business models, markets, and partnerships. Aligned with the company's own self-interest, this strategy is aimed at bringing benefits of the company's technology and knowledge to more people worldwide. Extensive stakeholder participation is implemented through corporate advisory panels (such as for biotechnology and health issues) as well as thought leaders and stakeholder representatives from around the world. It is part of DuPont's strategy to put the company's leaders out to interact with thought leaders and the community to better inform the company's strategies.

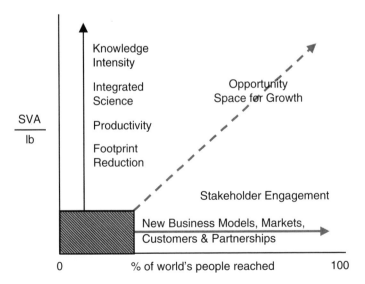

Figure 8.17. Sustainable growth as a business approach.

8.4.2 DuPont's Journey

The integration of sustainable growth as DuPont's global business is brought about by a number of factors, most crucially public expectation, leadership vision, and the its own corporate values. Public concerns drove DuPont to pollution prevention, such as preventing hazardous discharges to river, even in the early 19th century. Engineers were assigned full time to pollution prevention activities as early as in 1938. With the rise of modern environmentalism in the United States in the 1960s and the proliferation of environmental laws and regulation that follows, DuPont established a corporate Environmental Quality Committee in 1966 to ensure compliance and to search for ways to tackle the economic problems associated with them. In the 1970s, DuPont was faced with public pressure as the world's largest producer of chlorofluorocarbons (CFCs). With emerging scientific findings on the role of CFCs in destroying Earth's protective ozone layer, DuPont went beyond compliance and led the industry's turnaround in phasing out CFC production and developing alternatives. This earned DuPont a position to work on the issue with different groups, including NGOs, industry, academia, and government.

Nevertheless, transforming a large global company like DuPont requires vision and commitment from the top leadership. In 1989, Edgar Woolard became the company's CEO. Having witnessed the business challenges associated with the CFCs as well as external events such as the catastrophic Union Carbide incidence in Bhopal and the start of the Toxic Release Inventory (TRI) communication, he recognized that it was no longer sufficient to be in-compliance and it was in the DuPont's interest to be right where the *society* wanted it to be. In his first public speech as a CEO, Woolard coined the term "corporate environmentalism," a proactive concept that goes beyond compliance to meet the public expectations of DuPont. This was institutionalized in a new corporate policy, adopted in 1994, that set the target of reducing DuPont's waste and emissions to zero – captured in the company's mindset-changing motto "The Goal is Zero!"

The transformation continued when Charles (Chad) Holliday became the company's CEO in 1997. Having served as President of DuPont Asia for seven years, Holliday witnessed the rapid economic growth in the region, including its unintended consequences in terms of environmental quality and economic disparity. Corporate environmentalism became no longer sufficient, as there was the need for products and services that create better quality of life for all. Holliday believed that sustainability was a necessary part of the solution, especially in the global growth of DuPont in Asia and elsewhere. Discussions between DuPont's executives and academics were initiated in 1998 to explore how sustainable development could be incorporated into DuPont's business strategy without sacrificing growth, which led to the term "sustainable growth." At the same time, DuPont began to assume a leadership role in the field of sustainable development, with Holliday serving as chair of the World Business Council for Sustainable Development (WBCSD) for 2000–2001.

DuPont views sustainable growth as an outgrowth of its corporate values: safety and health, high ethical standards, environmental stewardship, and treating people with respect and dignity. These are parts of what is driving DuPont's journey towards sustainable growth. The most important driver for sustainable growth at DuPont, however, is innovation – DuPont's growth engine. Sustainable growth brings a set of new challenges to DuPont, such as reducing customers' impacts, new solutions for energy use, and safer and healthier food. These are the drivers for the new ways and solutions that keep DuPont from competing in the same old way, thus allowing it to differentiate itself in the marketplace.

In 2002, DuPont celebrated its 200th anniversary. Its past 100 years were marked by its transformation from a company known for its explosives to a diversified chemical company. Today, DuPont is in the midst of another remake, into an integrated science company. It realizes that only by aligning itself with society's needs can a corporation like DuPont continues its successes into its next century.

8.4.3   Realizing the Transformation

Business integration is the key to DuPont's transformation model, depicted in Figure 8.18. In this model, a change is founded on the CEO's commitment and leadership that is "felt" throughout the business structure. For sustainable growth, this is articulated by the corporate vision and mission (see Table 8.4). Business integration requires sustainable growth to be well understood by employees, especially the business leaders. To do this, senior leaders are enrolled in forums that cover various issues of sustainability, such as the Prince of Wales' Business & Environment Programme, Keystone Leadership Forum, and Sustainable Enterprise Academy. Internally, there are "Stewardship to Sustainability" and other training programs

Figure 8.18. DuPont's model for business transformation.

attended by business managers, sales managers, as well as R&D leaders. Equipped with the right people and expertise, sustainable growth can be incorporated in every business decision-making process through expert- and question-driven approaches. This is done, for example, in evaluating opportunities and risks in acquisitions and joint ventures. Sustainable growth is further promoted throughout the businesses through metrics, goals, and rewards programs. The integration of sustainable growth into the company's management, R&D, and operational systems is the topic of this section.

8.4.4 Managing Sustainable Growth

Consistent with the business integration model, the responsibility for sustainable growth is not delegated to any single entity, but to the company as a whole. At the corporate level, DuPont's sustainable growth is coordinated by both the Sustainable Growth Council and the Environmental Policy Committee. The Sustainable Growth Council looks over strategic matters for the company. It is headed by the CEO and involves senior leaders (i.e., corporate vice presidents) representing DuPont's businesses and functions. Equally important is the Environmental Policy Committee, a committee of the Board of Directors. Despite its name, this Committee is charged with determining not only the company's environmental policy, but also those related with safety, health, and other aspects of sustainable growth.

At the operational level, rather than adding a separate layer of management, sustainable growth is matrixed into the entire system with a network of functional experts. This network is coordinated by a sustainable growth champion, and was served by the corporate vice president for safety, health, and environment. This function, however, has been taken over by the newly created post of Chief Sustainability Officer of DuPont.

8.4.5 Sustainable Growth in R&D

As a science company, R&D innovations has contributed to a substantial portion of DuPont's revenue growth and is a critical part of the corporate strategy. By finding new ways to meet society's needs, sustainability is an important driver for innovation. Sustainable growth considerations are incorporated into DuPont's R&D through a process called Stewardship and Sustainable Growth Assessment. This question- and list-driven self-assessment process is applied not only to DuPont's own innovations, but also to products and technologies acquired externally through acquisitions and majority-owned joint ventures before they become fully commercialized. For illustration, Table 8.5 lists some of the questions considered in the concept phase (or early phase) of R&D, intended to screen the sustainability of the product or technology to both the company and the intended customers. More detailed analysis of risks in terms of human and ecosystem toxicity, exposure, occupational health and safety, and so on, is performed at later stages of development. The assessment also considers questions related to business strategy, such as whether to market the innovations as products or services. The stewardship and

TABLE 8.5. Some Sustainability Considerations in the Concept Phase of DuPont's R&D Process

- Is this product more or less toxic than what it would replace? What about the processing of raw materials used to manufacture it? List any issues of concerns.
- Are emissions from the product higher than similar products?
- Would this process use more or less energy?
- Would this process be more or less efficient in the use of water and materials?
- Is risk management an issue or can the material be created by an inherently safe process?
- How does it fit into the intended customers' SHE and sustainability goals and policies?
  - Would it reduce the customers' emissions or wastes?
  - Is the new product more durable or longer lasting?
  - Can this product relieve customers of regulatory burdens?
  - Can and should the product be refurbished, reused, or recycled? What additional infrastructure is called for? Should the product contain recycled materials?
  - Are there any hazards on disposal (e.g., landfill or incineration)?
  - What is the ultimate fate of the product?

Source: DuPont SHE Guideline S19X, Stewardship and Sustainable Growth Assessment for Research and Development Projects.
SHE, Safety, Health, and Environment.

sustainability review continues after commercialization, based on two-, three-, or four-year cycles depending on the assessed level of potential hazards.

DuPont also employs a set of "sustainable growth metrics" that are considered at least qualitatively to compare the innovations with competing alternatives including the products or processes being replaced. These metrics include energy use, material efficiency, water use, toxic chemical dispersion issues, global warming potential, ozone depletion potential, and acid rain contribution. These footprint metrics are considered for not only processes within DuPont, but also the product lifecycle. When appropriate, a formal lifecycle assessment (LCA) is performed. Lifecycle metrics that are lower than competing products and services reflect increased value to the customer. Results of the lifecycle assessment may be used in formulating business strategy for the new product or service.

The emphasis on innovation and integrated science is further strengthened through collaborations. DuPont has, in recent years, established substantial alliances with various research entities around the world. The foremost example of this is the $38 million DuPont-led industry, academia, and government consortium for the development of renewable energy and feedstock technologies, known as the Integrated Corn-Based Bioproducts Refinery (ICBR) project.

8.4.6 Sustainable Growth in Operations

Sustainable growth must also be translated into clear targets, responsibility, and recognition. DuPont business units are responsible for meeting objectives and overall goals for the organization, such as 10 percent energy from renewable sources, reduction in air carcinogens and greenhouse gas emissions, and so on. How this happens is up to each business, recognizing that the ways to minimize footprint and grow sustainably vary across businesses. In addition to the conventional Safety, Health, and Environment (SHE) audits, every year businesses have to submit current performance and programs that will improve performance over time. The annual Corporate Environmental Plan allows the corporation to evaluate all the plans to assure their cost-effectiveness in meeting corporate goals. As part of the Strategy Review in 2003, each business was required to complete an 8-Parameter Sustainable Growth Self-Assessment (Fig. 8.19) in discussion with the Office of the Chief Executive. It was not expected that each business would be impacted or possess all eight of the attributes in the self-assessment. In fact, the company rates itself right in the middle in all eight attributes. The goal is not to grade a business, but to generate a clear idea of a business's current level of sustainability, primary strengths and weaknesses, and where the most business value can be gained.

The use of recognition programs is another important tool. The annual Sustainable Growth Excellence Awards Program is chaired each year by a corporate vice president. It is organized in a pyramid scheme where each business and region runs its own process and the best are sent for consideration at the corporate level. Winners are chosen by a selection panel that includes internal participants and external judges representing NGOs, government, academia, media, and other companies. Such external engagement is one of the more valuable aspect of this program, as it provides a channel for soliciting external inputs. The awards honor teams and individuals from DuPont's businesses worldwide have made significant contributions towards DuPont becoming a sustainable growth company. Twelve global winners are recognized annually at an award banquet hosted by the CEO, and each award includes a grant donation to safety, health, environmental, or educational initiative chosen by the winner.

Factor	Assessment Result – Level of Sustainability & Trajectory	Comments
Economic Growth/ Profitability	(1)—(2)—(3)—(4)—(5)	earnings and revenue growth, cash generation and margin/ROIC
Customer or Market Pull	(1)—(2)—(3)—(4)—(5)	interest and suction for "green"/"sustainable" products, product attributes and services like biodegradable, bio-derived, recycle content recyclability, chlorine-free, etc.
Absence of Environmental and Energy Footprint Issues	(1)—(2)—(3)—(4)—(5)	are TRI, waste, emissions, energy significant issues for the business/corporation; are there sensitive issues like persistence, bio-accumulation, toxicity, climate impacts, underground injection wells?
Renewability	(1)—(2)—(3)—(4)—(5)	use of renewable feedstocks and/or energy; products that are recyclable, recycled, reusable, use recycle content
Knowledge Intensity	(1)—(2)—(3)—(4)—(5)	degree to which value is created using technology, know-how, services versus materials and energy. (SVA/LB)
Value as Perceived by Society	(1)—(2)—(3)—(4)—(5)	do the products and services meet important societal needs (as viewed by civil society)? Are there sensitive issues like gmos, climate impacts, persistence, etc.
Participation in Emerging Economies	(1)—(2)—(3)—(4)—(5)	degree to which products and services reach/are designed to reach beyond developed economy markets and top of the pyramid customers.
Stakeholder Engagement	(1)—(2)—(3)—(4)—(5)	degree to which the business engages and utilizes a broad and diverse set of stakeholders to guide their thinking and business strategies

where 5 is highest level of sustainability

Key Actions
- ...
- ...

Figure 8.19. DuPont's 8-Parameter Sustainable Growth Self-Assessment.

DuPont also issues a separate SHE and sustainable growth progress report. The first safety, health, and environment report was issued in 1992. In 1996, the company decided that the key audience for the report was their own employees, so they reduced the size to eight pages and sent it to each employee's home, with more detailed information made available on the company's website. DuPont began reporting consistent with the Global Reporting Initiative (GRI) format in 2003, with the reports posted for the public on its website. By including detailed information on the web, the reporting is aimed at stimulating dialog with employees and community stakeholders.

8.4.7   Results and Challenges

DuPont has shown some impressive footprint reductions since the development of their original goals and the adoption of its corporate environmental policy in 1994 and sustainable growth in 2000. Compared to the 1990 baseline, hazardous waste and air carcinogens were reduced globally by 44 and 89 percent, respectively, by 2002. Total global energy use was reduced by 9 percent, which has saved over $2 billion in energy costs since 1990. Reduction in energy use, along with reduction in chlorine and N_2O emissions, contributed to 67 percent reduction in greenhouse gases over the same period. All these were achieved while production mass was increased by 30 percent. In manufacturing operations, Six sigma is driving DuPont toward greater efficiencies, resulting in the reduction of defects and elimination of wastes. In the Titanium Technologies business, for example, a quick survey indicated that 60 percent of the completed Six sigma projects could demonstrate footprint reductions.

The goal of sourcing 10 percent of the global energy use from renewables, however, has proven to be challenging to achieve. Baseline of approximately 2 percent renewable energy (primarily hydropower) is currently obtained via supplier agreements. A number of projects are currently under way to develop new renewable energy sources for the company. Wind energy is being tested in a few locations, and it was shown to be economically competitive when supported with adequate government subsidies. The most cost-effective options come from the generation of steam from landfill gas and waste biomass. Proximity to and availability of these sources, however, are key (Mongan, 2003).

In transforming itself into integrated science company, DuPont has divested a number of resource-intensive businesses such as energy (Conoco) and nylon and polyester fibers. It has also acquired new knowledge-intensive ventures in biotechnology (primarily agriculture and nutrition), specialty polymers, and others. DuPont is also expanding its comprehensive consulting services. Its Safety Resources business, for example, combines DuPont's strengths in protective apparel products (Kevlar®, Nomex®, Tyvek®, and so on) with DuPont's knowledge and expertise in operational safety by offering comprehensive services that include training and implementation assistance.

Through product stewardship, DuPont differentiates itself in the marketplace by providing opportunities for customers to reduce their footprints. It helps customers to eliminate disposal and emission issues, such as by offering agricultural products in water-soluble packages, and lightweight polymer materials and improved coatings for the automotive industry. It also provides a market for recycled materials, such as Tyvek® from post-consumer, high-density polyethylene (i.e., milk jugs). Recycling programs for some DuPont products have also been successfully implemented, such as the post-consumer recycling program for Tyvek® banners in Malaysia, which produced an important competitive advantage and 6 percent growth in DuPont's banner and sign segment in that country.

DuPont's successes have been externally recognized. DuPont was awarded, for example, the Best Environmental Practice by *Financial Times* in its 2000 Global Energy awards. It is recognized in socially responsible investing (SRI) indices such as the FTSE4Good Index (since 2001 inception), Dow Jones Sustainability Indexes (DJSI, since 1999 inception), and Innovest Strategic Value. In fact, it is acknowledged as the 2003 and 2004 Chemical Market Sector Leader by DJSI and rated number one in Innovest's 2002 EcoValue Model rating among 37 chemical companies. Furthermore, DuPont's collaborative research with the U.S. Department of Energy's National Renewable Energy Laboratory (NREL) on the development of biomass feedstock manufacturing of 1,3-propanediol, the key building block of DuPont's Sorona® polymer, earned it the 2003 Presidential Green Chemistry Award.

Nevertheless, sustainable growth continues as a journey of transformation for DuPont. It needs to continuously navigate around emerging industry issues such as "chemical trespass" (unwanted exposure to chemicals in products) and children's health issues. These concerns go beyond manufacturing waste and emissions to product impacts on people. Recent issues concerning the EPA reporting requirement on perfluorooctanic acid (PFOA) used in the manufacture of Teflon® illustrate the challenges that DuPont continues to face. While the safety of Teflon® products were never questioned in the legal dispute, it resulted in a public relation crisis in China where rumors regarding their safety led to department stores pulling Teflon®-coated products from their shelves. Although the rumor was founded largely on misinformation, the business challenge it generated was real.

On a broader view, acceptability of new technologies remains an issue that must also be anticipatively managed. DuPont, and the industry as a whole, still needs to find a good way to dialog and respond to the society's concerns around risks and benefits of new technologies. It needs to figure out what is and is not acceptable and how the debate can be framed. While DuPont is a company based on science and technology, it needs to be cautious not to force innovations faster than the readiness and acceptance levels of society.

8.4.8 The Business Case

For DuPont, the business case of sustainability lies in it being a growth engine. Sustainability is the driver of innovations, both technical and in delivery to the market. Finding solutions to global societal needs, such as access to clean energy, safe food, shelter, and information, presents new business opportunities for DuPont and the industry.

Sustainable growth also enhances DuPont's reputation. A large number of employees are interested in sustainable growth training programs, especially on how to make the world a better place and sharing the values of the community. Anecdotal evidence suggests that hiring of employees and employee satisfaction may have benefited from it. DuPont's reputation also opens up access to emerging economies. The DuPont crop protection business, for example, was invited to enter Vietnam due, in part, to the company's good reputation. Furthermore, DuPont believes that there are a growing number of business consumers interested in sustainability and that its progress in operating responsibly is one aspect that helps DuPont be a preferred supplier.

Sustainability strengthens the company's top line and bottom line. Innovation will continue to be the primary driver in the future. Cost reductions and improvements in operational efficiency are also important, as are better community relations. In the end, sustainable growth is about finding opportunities to create value for shareholders and society, while reducing the company's footprint. For the industry, depending on where on the journey the company is, getting in line with societal expectations is the place to start. To have the right to operate and grow, a company needs to be in line with society's expectations.

8.5   BUSINESS VALUE FROM SUSTAINABLE DEVELOPMENT AT SHELL

MARK WADE

Shell International Limited

Joe Machado

Shell Chemical LP

Business is facing unprecedented risk, scrutiny, and challenge in the face of dramatic change sweeping the world. Nowhere has this been more clearly demonstrated in the recent past than at Shell[23] itself. The need to recategorize oil and gas reserves in the first months of 2004 has had wide-ranging consequences, denting public trust and damaging the Group's reputation.

It will take a considerable effort to recover. But that recovery is in the sights of many employees, and central to that is a firmly held commitment to sustainable development. Only by continuing to focus on achieving a strong economic, environmental, and social performance can the expectations of shareholders, customers, and society at large be met, licences to operate retained, and the damage to the Group's reputation repaired.

The journey, which has delivered a deep-rooted commitment to sustainable development, has been long and challenging. It has reached to the core of Shell's values and drawn widely on society's expectations. It has engaged the time and emotions of senior leaders, and impacted the people, business models, governance,

---

[23]The expressions "Shell" and "Group" refers to the companies of the Royal Dutch/Shell Group of companies. Each of the companies that make up the Royal Dutch/Shell Group of companies is an independent entity and has its own identity.

and behaviors of a global organization. This article provides an insight into that journey, demonstrates how sustainable development is becoming "the way we do business" and focuses on some of the specific impacts on the Chemicals Business.[24]

8.5.1 The Lessons of the 1990s

Shell has always been a values-driven company. Its core values of honesty, integrity, and respect for people are the foundations of Shell's General Business Principles, first codified in 1976. Having set the framework, Shell prided itself on operating to what it felt were the highest business standards. It came as a shock in 1995 when two events – the disposal of the Brent Spar and human rights issues in Nigeria – catapulted Shell into the international headlines and called into question its behavior. The Group found itself ill-equipped to deal with the public criticism and damage to its reputation and was badly shaken by these events.

There was an urgent desire by senior leaders to find out what had gone wrong and why. So, in 1996, Shell embarked on one of the largest stakeholder engagement exercises ever conducted in the energy sector. It included NGOs, government officials, academics, labor representatives, and community leaders in 14 countries, who were engaged in a series of 20 or so roundtable discussions. Gradually, through the process Shell learned to "bite its tongue," stop defending, and listen.

The roundtables were followed by a quantitative survey of 7500 members of the public in 10 countries, 1300 opinion formers in 25 countries, and 600 Shell employees. The results made salutary reading. They showed that a significant minority had serious concerns over what they perceived as Shell's lack of regard for the environment and human rights. As Managing Director Mark Moody-Stuart said at the time "we looked in the mirror and neither liked nor recognised what we saw."

What was clear from the evidence was that having values is not enough. The way values are expressed, as behavior, needs to be constantly refreshed to reflect changing public expectations.

In today's world, there are growing calls for companies to move beyond being responsible to shareholders for financial performance and become accountable to stakeholders for their wider economic, environmental, and societal impacts. The risk is that if customers believe that a company is not behaving ethically they are more likely to vote with their wallets and make their protest known by boycotting products or services. At a time of unprecedented scrutiny for companies, made easier by the CNN/Internet world, there is simply "no hiding place."

A second key learning was that the world was moving from a "trust me" to a "show me" world. People are less willing than they were to take the assurances of authorities such as government, scientists, and companies on trust. There is an increasing call for corporations to show what it is they are doing. And in the absence of trust – something that characterizes the modern world – there is a demand for independent verification of what is being shown.

---

[24]The Group is organized into five Businesses. Businesses are portfolios of activities operating according to common objectives and strategies. One such Business is the Chemicals Business.

Against this backdrop Shell saw there was a growing worry in society about the long-term social and environmental "sustainability" of today's economic develop-ment. The powerful forces of globalization, liberalization, and technology have brought rapid change and advancement. But the speed and scale of the impacts are posing far-reaching questions. They are also exposing dilemmas and threats as they push the boundaries of natural and man-made systems. People are concerned about climate change, erosion of ecosystems, issues of health, and extremes of poverty and wealth.

With the global population set to rise by 50 percent and global energy demand expected to double or even triple by 2050, there is a real sense of urgency and alarm at the prospects for the planet and its future generations. Shell recognized more clearly than before the implications of this to the future success of the Group. It also saw its wider responsibilities to the societies it serves. There was a clear imperative to respond.

8.5.2   Responding to the Challenge

Shell's first response to these findings was the acceptance by senior management that they were important. The second was to face up to the fact that this would require some significant changes to the way the business operated.

Clearly, that required getting a few basics in order. The first, in 1997, was the updating of the Shell General Business Principles to include specific commitments to support fundamental human rights and to contribute to sustainable development. At the same time the health, safety and environment (HSE) commitment, policy and procedure was greatly strengthened. Strict governance and clear processes were put in place to ensure adoption and compliance with these measures worldwide.

A global issues identification and management system was developed and a Social Responsibility Committee at Board level was established to review the over-all social and HSE performance and policies of the Group. The Business Principles were translated into 51 languages covering 99 percent of Shell people. Minimum environmental expectations were laid down, based on norms in OECD countries, defining the standards of performance to be applied wherever Shell operates. Group targets were set for reducing greenhouse gas emissions to 10 percent below 1990 levels by 2003, and for ending venting of gas by 2003, and continuous flaring by 2008. Shell Renewables was set up to grow a viable business in solar, wind, and other forms of alternative energy.

After making these moves people said "great intentions – now show us the per-formance!" This led to the first Shell Report "Profits and Principles – does there have to be a choice?" published in 1998. Over night it broke the mold on Shell's corporate reporting. It sought to be an honest, transparent, and verified account of Shell's environmental, social, and ethical performance worldwide. It surprised many people, including Shell's fiercest critics, with its frankness in discussing the most sensitive issues facing the Group such as climate change, operating in politically sensitive areas, and dealing with the legacies of past industrial activity. It acknow-ledged where mistakes had been made and lessons learnt.

The decision to produce the Shell Report was evidence of a greater commitment to openness and transparency – an acknowledgement of the "show me" world. That meant a willingness to engage with groups that had not traditionally been informed or involved in business decisions. That in turn would help ensure that the values of the company continued to be sensitive and responsive to public expectations. With time this would go a step further and begin to open up an "involve me" world.

These moves were part of a wider strategy of ensuring Shell was fit and competitive for the 21st century. It was the beginning of a journey, one based on a core conviction that business had an essential role to play in finding new ways of meeting present and future needs that were more socially and environmentally sustainable.

8.5.3 A Different Way of Doing Business

After getting the foundations in place, the next step was to understand more clearly what this meant in practical terms and then to integrate sustainable development thinking across the Group in a structured and consistent way. This was a real challenge given the diversity of Shell's operations (spanning five Businesses) and the range of environments in which it works.

It was clear that this would only be successful if all those who worked for and with Shell understood what sustainable development meant to Shell and how they could contribute. A difficulty was that there was very little understanding of what such a commitment meant. There had been a long tradition of social investment by Shell in the communities in which it worked, but this had tended to be philanthropic and seen as an optional extra.

What was now being established was a different way of doing business. This would entail integrating environmental and social aspects into the mainstream of the Group's activities along with economic considerations. It also required the balancing of short-term priorities with longer-term needs, and a willingness to engage the outside world in the decision-making process. This meant asking those impacted by Shell's activities what they thought at the beginning of new activities rather than telling them what had been decided at the end.

There was undoubtedly significant skepticism at the outset on the need to make a commitment to sustainable development. "Why us now? Most of our competitors seem to be doing well if not better without it" was a common retort. And "What about the shareholder? Companies are not charities. Surely the prime responsibility of any corporation is to its shareholders?" This sentiment was driven by the widely held perception that although the moral case was clear, such a commitment might mean additional burdens on the business and therefore make it less effective.

People would need to be convinced of the value of this approach. There was a need to demonstrate that the best way of serving shareholders is, paradoxically, not to focus exclusively on the shareholder. There was a need to make the case that the best way of maximizing long-term total shareholder returns is by taking a wider perspective of the expectations placed on a company – and thus avoiding risks that destroy value – and grasping future opportunities that create value.

8.5.4   The Business Case for Sustainable Development

At the heart of the effort to convince the organization that this was a sensible way to proceed was the need to show how a commitment to sustainable development could add value to the business – to present the business case for sustainable development (SD) in the same way as for any other element of the Group's activity.

That business case has seven key components:

- Attracting and motivating employees;
- Reducing costs through efficient use of materials and energy;
- Reducing financial risks;
- Steering the portfolio for the future;
- Influencing product and service innovation;
- Attracting more loyal customers;
- Enhancing our reputation.

As these components are so fundamental to the internal understanding and acceptance of SD, it is worth providing further insight into each one.

Attracting and motivating employees. Most people want to work for a company whose values and concerns are aligned with their own. As competition grows for talent, this can provide a real competitive advantage.

Reducing costs through efficient use of materials and energy. This spurs innovation and helps reduce costs through eco-efficiency. It means finding cost-effective ways to produce more with less material, energy, water, and waste, and turning waste into saleable products.

Reducing financial risks. This is concerned with reducing risk by understanding what stakeholders see as responsible behavior and meeting those expectations. Gaining speedy approvals from government for major multimillion dollar projects, for example, and gaining a "license to operate" from local communities can save substantial amounts of time and money. Demonstrating the ability to manage risk is attractive to financial institutions, leading to lower cost of capital and improved confidence among investors.

Steering the portfolio for the future. Anticipating new markets, driven by societal and customer desires for a cleaner, safer more sustainable world, can give strategic direction to the evolution of product portfolios and supply chain relationships to match.

Influencing product and service innovation. Sustainable development provides a lens that can be used to understand the wider impacts of a business and foresee future impacts. Understanding changes in customers' lifestyles, values, and their sometimes conflicting priorities gives a direction to product and service innovation and differentiation.

Attracting more loyal customers. It is essential that suppliers consistently understand and meet their customers' needs. This is about anticipating how those needs will change, but also about giving customers confidence that their suppliers will do their part in looking after the value chains in which both parties are involved.

Enhancing our reputation. By being seen as a credible, good corporate citizen whose performance matches its words, the company becomes first choice for customers, staff, investors, suppliers, partners, and communities in which it operates. Having a good reputation is a clear advantage in a business climate where it is increasingly clear that failure to safeguard reputation has long-term financial consequences. It not only secures a "license to operate," but more crucially a "license to grow."

8.5.5 Embedding Sustainable Development into the Organization

In the early stages, Shell's understanding of how contributing to sustainable development could be a "win–win" was limited to a few who had had the opportunity to explore the concept more deeply. It was understood that it would be wrong to try to prescribe from the center how the sustainable development approach should be applied in detail in each and every operation. Instead, a guide was developed called the Sustainable Development Management Framework. This described the case for sustainable development, what it meant for Shell and provided practical guidance on how to apply its concepts. It was intentionally distributed by the businesses, not from the corporate center, to emphasize that its implementation was to be owned and managed by the businesses. Local adaptation and exploration was encouraged and emerging best practise shared.

At the same time as the framework was being rolled out, the Chemicals Business was putting in place a new operating model that would define its core business activities and provide a backdrop for decision-making (Fig. 8.20). It was agreed that SD would be placed at the heart of this model, emphasizing that SD is "the way Shell chemicals companies do business."

As the development of sustainable thinking in the Chemicals Business grew, a further step was taken to move from a high-level commitment to SD to embedding a practical action plan. This plan was built around four strategic priorities:

- *Reducing our footprint* is concerned with reducing emissions, waste, energy, and other resources. Examples include an "Energize" programme that advises manufacturing facilities on ways to reduce energy usage and the sale of waste CO_2 to manufacturers of carbonated drinks.
- *Sustaining value chains* is about finding ways to make the value chains in which Shell chemicals companies participate more sustainable. Examples include the development of cool wash detergents, and the PETFix technology to recycle waste PET drinks bottles into building materials.
- *Building social capital* is about engaging with external stakeholders, such as local communities, and supporting employees. Examples include a job creation program at Berre, where site facilities were offered for use by start-up companies to create new jobs, and the promotion of diversity among employees.

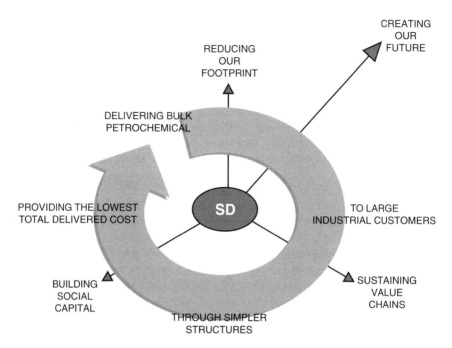

Figure 8.20. New operating model for the Chemicals Business.

- *Creating our future* reflects a commitment to innovation and to delivering solutions that respond to society's evolving needs. Examples include a new technology to produce monopropylene glycol, with higher efficiency and less water, using CO_2 as a starting material rather than propylene, and searching for a new process for making organic carbonate that would avoid the need to use phosgene in polycarbonate manufacture.

The Chemicals Business and other Group Businesses still have much to do, but good progress has been made in a number of areas including integrating sustainable development considerations into the business decision-making process. For example, any new investment proposal has to include the cost of carbon in the economics. This drives investment in the most energy-efficient technologies, improves competitiveness, and prepares for a time when there will be a financial penalty for carbon use. This is in addition to a requirement that all project proposals must have an assessment of the environmental, health, and social implications in submissions for capital allocation.

The hard wiring is reinforced in internal communications, training, and development of Shell people. There is a particular recognition of the new skills required by leaders within the organization, many of whom have engineering and technical backgrounds and who have had little experience of, for example, engaging with stakeholders. A better understanding of the social and environmental factors may mean not pursuing the most obvious technical option if it has too high an environmental or social cost. The re-routing of pipelines to avoid sensitive areas is a case in point.

That makes it essential to provide learning opportunities to help managers deal with what are often complex dilemmas. That exposure is provided both within the organization but also with the help of external providers. Recognizing that sustainable development is still a relatively new concept for business, Shell supports a number of academic institutions in their provision of leadership development and research into approaches to sustainable development within business contexts.

The embedding process is reinforced by appraisal and reward systems. Since 1997, contributing to sustainable development has counted in the remuneration of Shell's most senior executives and in the appraisal of Group Business. Similar reward schemes are adopted across Shell's operating companies, providing very powerful incentives to operate in a sustainable way.

8.5.6 Measuring and Communicating Performance

The Shell Report, by reporting environmental and social performance, also underlined the way the old adage that "what gets measured gets done" is as true for integrating sustainable development as it is for any other business need. This focus on measurement meant that a set of key performance indicators was developed to measure and drive progress in critical areas of economic, environmental, and social performance. They serve as the logical basis for the setting of targets and milestones and for developing standards of reporting and verification for the Shell Report. The indicators include quantitative measures such as Return on Average Capital Employed, greenhouse gas emissions, and diversity in the organization. They also include qualitative measures such as stakeholder views on the acceptability of environmental performance, and wider measures of reputation.

Specific SD performance goals have also been set for the Chemicals Business that focus on continuous performance over a five-year period between 2002 and 2007. The goals center on key aspects of environmental and social performance and are linked to major programmes of activity across the Shell chemicals companies. The goals are:

- *energy efficiency* improved by 7.5 percent;
- critical *air emissions* reduced by 25 percent;
- *spills* reduced by 50 percent;
- initial *health risk assessments* completed for all products;
- *social performance* improved, as measured in community surveys;
- alignment of *business processes* with SD increased by 40 percent.

Progress against the goals is shared annually with all employees, and corrective actions agreed as required.

8.5.7 Engaging with Stakeholders

One of the key changes that a commitment to SD has delivered is an acceptance that it is necessary to involve external stakeholders in dialog and the decision-making process right from the start of a project. That involvement varies from specific engagement on the particular implications of a new project, to work on overall policy themes such as human rights.

Some of this work has led to partnerships with international organizations on major issues such as biodiversity or climate change to ensure that Shell has access to the widest expertise and is in touch with the latest thinking in these areas.

On biodiversity, Shell is working with partners ranging from the IUCN to UNESCO on policy development, and with the Smithsonian Institution on the effect of oil operations on the local environment in Gabon. In addition, it works with industry partners to develop and share best practice.

Informed by these relationships, Shell developed a biodiversity policy, which applies across the Group. It led it to announce, in August 2003, a commitment not to explore for oil and gas in Natural World Heritage sites. This was welcomed by many conservation groups and helps to create a situation where Shell is seen as being a legitimate part of the overall debate on where it is acceptable for extractive industries to operate.

On human rights, Shell has established a regular dialog with groups such as Amnesty International and Pax Christi. This helps to inform Shell's approach and to develop understanding of the challenges and dilemmas faced by the business when operating in sensitive countries. A range of training materials has also been developed on issues including child labor, bribery, and corruption, and working with indigenous people. All of these help to provide practical support and guidance to staff on the ground so that they can take appropriate action to deal with local conditions and uphold the highest standards of behavior.

8.5.8 Working with Others

This is a part of a wider ongoing debate with international NGOs and forums such as the International Chamber of Commerce, the European Roundtable, and the World Business Council on Sustainable Development (WBCSD), where Shell is a leading player. The WBCSD has undertaken a considerable amount of work on many of the key issues such as energy and climate, biodiversity, and corporate accountability and reporting. By drawing together best practice, this helps individual businesses keep up with the leading edge of thinking and gives business a stronger role in the debate by providing a consistent and authoritative voice.

What is emerging from this continuing debate is that there is an increasing recognition that business is part of the solution to sustainable development questions. This was acknowledged at the 2002 World Summit on Sustainable Development in Johannesburg, where business embarked on a range of partnerships that will have a very practical benefit for significant social and environmental problems. These include poverty alleviation through economic development and energy provision.

As Kofi Annan, Secretary General of the United Nations, put it at the Business Day of the Summit: "Business leaders now understand that companies can only succeed long-term if they put values at the heart of their business and address social and environmental issues in a sustainable way". He added, "the corporate sector need not wait for governments to take decisions. It is only by mobilising the private sector that we can make progress."

8.5.9 Contributing to Sustainable Development and the Use of Finite Resources

There are, of course, some who believe that business and the oil and gas business in particular cannot act in a sustainable way. A common challenge to Shell is "How can you claim to be committed to sustainable development when you go on exploring for oil and gas? That simply doesn't add up!" To answer that, one needs to go back to where the concept of sustainable development came from and its purpose.

It was the Brundtland Commission of the United Nations in 1987 that defined sustainable development in inspiring but very open-ended terms as "development that meets the needs of the present without compromising the ability of future generations to meet their own needs." Crucially, it acknowledged that economic development was necessary to generate the wealth for cleaner, more efficient technologies, and social improvement. Going back to pre-industrial times was not the answer to the world's problems.

As "economic engines," companies clearly have a role to play. That is why Shell's commitment is couched in terms of contributing to this wider movement of sustainable development. Shell did not pretend that it could become sustainable in the sense some people ascribe to the term as in only using renewable or recyclable resources, and *only* supplying renewable forms of energy to its customers. This is simply unrealistic with hydrocarbons set to remain the mainstay of the energy scene for at least another three decades. But what Shell felt it could do is continually improve its performance across the three dimensions of sustainable development: economic, environmental, and social, and use this approach to give purpose and impetus to its strategic direction as a progressive energy company.

Shell saw that it could contribute in three ways using the fourth dimension of sustainable development – time. This perspective provides a readily understood framework for action for people who ask "What can I do on Monday morning?"

The first way is to look back in time and "learn the lessons of the past." This can involve bringing standards up to date, cleaning up legacies, and retrofitting sustainable development thinking to older operations. It is the least glamorous way of contributing. It is a cost and a distraction from current operations. Nevertheless, it is an essential starting point and payback is achieved through reduced risk of liabilities, and a protected reputation.

The second way is in the present – doing things better now. This generally involves incremental improvements to how things are done such as making products or processes cleaner and safer or in managing major projects in a more informed way. An example of the former is the development of a new process that allows some of the waste water produced in the manufacture of styrene monomer and propylene oxide (a combined process) to be recycled back into the production process. Applying this process at a plant in Singapore – where fresh water is a very precious resource – has reduced fresh water use by 9 percent.

An example of the latter is the development of the CSPC-Nanhai petrochemical project in China. This development between Shell and its Chinese partners is, with an investment of $4.3 billion, the largest ever Sino-foreign joint venture. Some 2.3 million tonnes of product will be produced in the plant each year, feeding local demand from manufacturers of telephones, TV sets, computers, cameras, car components, and so much more.

From the outset, the project presented a number of social and environmental challenges and the joint venture, in which Shell has a 50 percent share, has worked closely with the local government to meet them. In particular, great care has been taken to ensure that the resettlement of households has been carried out to international standards, with a range of livelihood restoration initiatives put in place. Resettlement has been monitored by external observers and publicly reported. It is hoped that the processes developed at Nanhai, and the lessons learned, will provide a valuable model that can be applied to other projects in China.

The third way is to look forward in time and redesign for the future. This is the most inspiring and far-reaching way of contributing. It involves "thinking outside the box" and seeking answers to fundamental questions of sustainability. This can lead to new products and services, new supply chain relationships, and portfolio evolution. For Shell this has led to looking at how to radically reduce the impact of the production and use of oil and gas and helping the world shift towards a low-carbon energy system. In practical terms this means growing Shell's natural gas business, while working to make solar and wind power competitive and supporting the development of an infrastructure for hydrogen fuels. Other developments include new fuels from crops (biofuels), and finding ways of capturing greenhouse gases from fossils fuels cheaply.

8.5.10   Conclusion

Shell believes there is a compelling business case for sustainable development. It has learnt from experience – no more so than in the first half of 2004 – that companies need to embrace the challenges of heightened expectations and increasing demands for openness and transparency. They need to do this if they are to secure the long-term growth and success of their businesses. It is also clear that this commitment needs to be maintained in both good and bad times.

The challenge of achieving a sustainable world is huge and daunting, but there is no alternative. No one group will have all the answers, and partnerships will be even more essential in ensuring that the full range of perspectives and views are received and pursued. The organizations that succeed in today's fast-changing world will be those that have strong core values but that are committed to developing and refining those values; they will be those that are humble enough to accept that they do not have all the answers; they will be prepared to be transparent about their mistakes, and work with others, even their critics, to improve the way they operate. Companies that learn to do so will be those that are best placed to avoid the risks and seize the commercial opportunities ahead.

8.6 SUSTAINABLE DEVELOPMENT: AN INTEGRAL PART OF BASF'S CORPORATE VALUES

ERNST SCHWANHOLD

BASF Aktiengesellschaft

BASF is the world's leading chemical company. We offer our customers a range of high-performance products, including chemicals, plastics, performance products, agricultural products, and fine chemicals, as well as crude oil and natural gas. BASF manufactures and sells more than 8000 products and has grown into a very successful company since its birth in 1865. BASF operates production facilities in 41 countries and maintains contact with customers in more than 170 countries. The BASF Group encompasses BASF Aktiengesellschaft with its parent plant in Ludwigshafen, Germany, as well as over 160 subsidiaries and affiliates. In 2003, BASF had about 87,000 employees worldwide.

Responsibility towards society and the environment is a key element of BASF's corporate philosophy. This philosophy is based on the principles of Sustainable Development. Sustainable Development is anchored into our "Values and Principles" and BASF commits itself to the principles of the United Nation's Global Compact initiative. For BASF, Sustainable Development means being committed to helping meet the needs of society today without compromising the ability of future generations to meet their own needs. It also includes an open dialog within the company, with its business partners and neighbors, and with society. The principles of Sustainable Development are important for BASF to operate successfully in a global competitive environment. The economic importance of Sustainable Development is acknowledged by financial markets.

Implementing Sustainable Development is a challenging task for a transnational company. Economic, environmental, and social conditions vary from country to country. This means that on one hand we need to develop strategies and tools that can be applied worldwide, and on the other hand these strategies and tools must be flexible enough to conform to the requirements of a given country – something that is particularly challenging in the social context. It was essential to establish suitable management structures to support the implementation of Sustainable Development worldwide. Examples for BASF's implementation include Responsible Care®, long-term global environmental goals, the "Verbund" structure, and innovations.

In the following, the implementation and practical handling of sustainable development within BASF is presented in more detail. Furthermore, an important instrument, the eco-efficiency analysis, is described. BASF was one of the first chemical companies to develop this method for use in its business activities.

8.6.1   Values and Principles

Our commitment to sustainability is integrated into the "Values and Principles of the BASF Group." It is a core component of the conduct that is binding for all our employees.

As the world's leading chemical company, it is our mission to benefit our customers, our shareholders, our company, our employees, and the countries in which we operate. BASF's values describe the orientation and the manner in which we want to reach our goals. The first of six values states: "Ongoing profitable performance in the sense of Sustainable Development is the basic requirement for all our activities. We are committed to the interests of our customers, shareholders and employees and assume a responsibility towards society."

8.6.2   Our Commitment to the Global Compact

The Global Compact initiative was established by UN Secretary-General Kofi Annan in 2000. As a founding member, BASF has committed itself to promoting and implementing the Global Compact's nine principles. These principles formulated by the United Nations extend from the demand to put human rights principles into practice, through respect for trade union rights to the encouragement and spread of environmentally friendly technologies. These are principles on which sustainable development is built.

8.6.3   Management Structures

BASF's Board of Executive Directors established the Sustainability Council in June 2001. The Sustainability Council (Fig. 8.21) is chaired by a Board member and ensures that BASF Group policy is in accord with the principle of sustainability. The key task of this Council, whose members include seven division presidents in addition to the Board Member, is to draw up strategies for the three aspects of Sustainable Development: the economy, the environment, and society.

Figure 8.21. Organization of sustainability management in the BASF Group.

To implement strategic decisions, the Sustainability Council is supported by the International Steering Committee Sustainability. Its ten members, senior executives from various regions and disciplines, reflect the range and diversity of the BASF Group and the significance of the topic. They plan and oversee strategy implementation. A number of task-specific project teams report to this Committee and draw up specific measures.

The Sustainability Center acts as an interface between the Sustainability Council and the International Steering Committee. It coordinates various internal projects and teams and is responsible for BASF Group communication on sustainability.

8.6.4 Examples for the Implementation of Sustainable Development within BASF

8.6.4.1 The Responsible Care® Initiative. Continuous improvement in the areas of environmental protection, health and safety is the goal of Responsible Care®, the chemical industry's global voluntary initiative. BASF joined this initiative at the very beginning and is committed to its goals and principles. By establishing the global Competence Center Responsible Care in 2001, BASF's Board of Executive Directors has once more endorsed and strengthened the worldwide implementation of Responsible Care®. Experts in a truly global network have been entrusted with the task of setting goals, implementing measures, managing processes, monitoring progress, and ensuring compliance with our standards in all key sectors of Responsible Care®:

- Environmental protection;
- Product stewardship;
- Occupational safety and occupational health;
- Process safety and emergency response;
- Distribution safety; and
- Dialog.

The BASF Group's Responsible Care® initiative consists of many elements, two of which are described in more detail below.

Site Audits. Regular site audits carried out on behalf of the Board of Executive Directors are an important means of answering questions regarding the success in implementing Responsible Care® goals. These audits are essential for an assessment of the actual improvement with respect to the Responsible Care® goals. They also contribute to the identification of areas with room for further improvement.

When auditing a site or a plant, we rely on clearly defined criteria to produce an environmental and safety profile. Separate audits are conducted to determine the occupational health conditions at a facility. A site's performance profile covers both "hard facts" such as accident numbers and emission levels, as well as "soft facts" such as effective organizational structures, defined responsibilities, and a culture of safety or safe practices. In addition to the environmental and safety profile, an audit generates a risk profile for a site, based on the materials and processes used and the facility's immediate surroundings. A risk matrix helps illustrate the results and shows whether additional measures are needed.

An Example: Responsible Care in Raw Material Purchasing. In 2002, BASF bought more than 10,000 different raw materials from about 5000 different suppliers worldwide. Along with price, quality, and reliability, compliance with specific environmental, health and safety standards is a precondition for our business partnerships with raw materials suppliers. Our purchasers negotiate terms, but also evaluate the risk associated with each product and supplier. For this purpose, we have developed a safety matrix (Fig. 8.22). According to this matrix, all raw materials purchased are classified into one of three hazard categories according to their environmental, toxicological, and safety properties: A (safe), B (harmful), or C (e.g., toxic). For example, sodium chloride belongs to category A. Ethanol is classified as highly flammable, therefore it is part of category B. Methanol, a highly toxic type of alcohol, belongs to category C.

Suppliers and potential suppliers are classified with respect to expected or known compliance with environmental and safety standards. In a first approach, this classification is performed according to their location in OECD countries or non-OECD countries, since the risk of noncompliance with environmental and safety standards is expected to be higher in non-OECD countries (Kranz and Sargasser, 2003). Products/producers assigned a C3 rating represent a potentially high risk and are therefore subjected to a particularly careful check. This means that BASF employees visit the supplier and carry out an EHS assessment to determine whether the supplier's plant operates according to Responsible Care standards. A company can only be included in our list of suppliers if its facilities meet our requirements.

In the case where a supplier does not meet our EHS requirements, BASF offers its support and the development of a joint action plan. If the company implements the improvements, it will remain or become listed as a qualified supplier. BASF favors

Product risk	Supplier risk		
	1	2	3
A	safe area I		
B		safe area II	low risk
C			potentially high risk

Figure 8.22. Safety matrix of raw material purchasing.

long-term business relationships with its suppliers. Therefore, suppliers' compliance with international standards is a clear benefit for them. The procedure is part of our risk management and we are convinced it represents a competitive advantage for BASF.

8.6.4.2 Long-Term Global Environmental Goals. We have already successfully reduced our emissions to air and water by continuously increasing the efficiency of our production processes and energy generation. For example, between 1990 and 2002, BASF Group worldwide reduced greenhouse gas emissions by 38 percent in absolute terms. In the same period, production increased by 45 percent. As a result, we more than halved greenhouse gas emissions per metric ton of sales product (61 percent reduction).

However, we want to do even better. In order to ensure transparency of the process of improvement, we have published long-term global environmental goals that we want to achieve. By 2012, BASF plans to cut emissions of greenhouse gases by 10 percent per metric ton of sales product and emissions of air pollutants from its chemical plants by 40 percent compared with 2002. We will also reduce emissions of both organic substances and nitrogen to water by 60 percent and heavy metal emissions by 30 percent.

BASF additionally plans to significantly improve its safety record by 2012. The goals are 80 percent fewer lost-time accidents compared with 2002, and 70 percent fewer transportation accidents compared with the value for 2003.

By 2008, BASF intends to extend its data on chemical substances even further to include the relevant information on all substances handled worldwide in volumes exceeding one metric ton per year.

8.6.4.3 The Verbund. The "Verbund" is an integral part of BASF's corporate philosophy. It was the original idea of BASF's founder, Friedrich Engelhorn, to link each production facility to other plants so that the products, byproducts, and waste streams from one plant could serve as feed in the next. Production plants in this Production Verbund are connected through an intricate network of piping that provides an environmentally friendly method of transporting raw materials quickly and safely. This way, BASF created efficient value-adding chains, starting from basic chemicals and extending to higher value products like coatings and crop protection products.

In the Energy Verbund, excess heat from chemical reactions in production plants is converted into steam. This is fed into the steam grid and thereby becomes available to other plants. Thanks to the Energy Verbund, we could reduce the use of fossil fuels at our main production site in Ludwigshafen by 49 percent compared with the mid-1970s, while production increased by 45 percent in the same period.

This Verbund structure successfully combines environmental protection with economic advantages. Verbund, however, has taken on a much wider meaning for BASF. It also encompasses the Procurement Verbund, Knowledge Verbund, Verbund with customers, and so on.

8.6.4.4 Innovation and Sustainable Development: Example "Three-Liter-House." With innovations based on chemistry, we are helping to put the principles of sustainability into practice. Innovations from BASF make our customers' products and systems as well as their entire production processes environmentally sounder, and help save resources.

As an example, BASF is contributing to sustainable urban development with its innovative "three-liter house" concept for the low-energy modernization of older buildings. Using a number of innovative products, technologies, and systems, we have developed a process for refurbishing old buildings such that the energy demand for heating is reduced to a fraction of its original value, even below the minimum requirements for new buildings. A model project in the Brunck district of Ludwigshafen has already proved the concept's value. A renovated old apartment now requires just three liters of heating oil per square meter per year instead of the 20 to 30 liters needed in unrefurbished older buildings. Optimal thermal insulation is the most important part of the project. Primarily, this was achieved by employing products such as BASF's new insulating material, Neopor®. Further elements are an air exchange system that allows 85 percent of heat to be recovered, special thermal control windows, and an interior plaster that stores heat and ensures a pleasant indoor climate. In addition, a fuel cell provides heat and electricity highly efficiently, with a low level of emissions.

Meanwhile, three years after the launch of the project, the three-liter house has received international acclaim. It serves as a best-practice model for energy-efficient modernization and urban development for the European Union's SUREURO research project into sustainable urban development.

8.6.5 The Eco-Efficiency Analysis

Regarding SD tools, one of BASF's core competencies lies in the field of eco-efficiency analysis. In simple terms, eco-efficiency describes how environmentally friendly and economical a product or process is. BASF, together with an external partner, began to develop the instrument based on the conceptual work of Schaltegger and Sturm (1994) as early as 1996. In the meantime, the method has matured and around 230 products and production processes have been analyzed to date. Half of the eco-efficiency analyses conducted to date have been used for internal strategy and research decisions. The other half of the analyses has been carried out in cooperation with external partners such as customers, NGOs, and governmental institutions. Right from the beginning, the eco-efficiency analysis was discussed in public. This has led to many valuable contributions from external parties and has put the method on a high level of acceptance throughout various industries and stakeholders.

8.6.5.1 The Methodology. The eco-efficiency analysis starts with the definition of a specific customer benefit. The analysis then compares economic and ecological advantages and disadvantages across several product or process solutions, which can fulfill the same function for customers (Saling *et al.*, 2002). This means that products are not compared with one another in overall terms, but rather their application performance such as "painting a square meter of furniture front." The eco-efficiency analysis focuses on each phase of a product's lifecycle "from cradle to grave," beginning with the extraction of raw materials from the Earth and ending with recycling or waste treatment after use. The basis is a lifecycle analysis according to standard ISO 14040 and the following.

In this way, the environmental impact of the production, the usage behavior of the consumers, and the various possibilities for reuse and disposal are analyzed. Additionally, a comprehensive economic assessment is performed, including all costs incurred in manufacturing or use of a product. The economic analysis and the overall environmental impact are then combined to evaluate the eco-efficiency. Thus, all relevant decision factors are analyzed with specific customer benefits always being the focus of attention.

Eco-efficient solutions to the problems are those that provide a better customer benefit from a cost and environmental point of view.

8.6.5.2 The Indicators.

The representation of a multiplicity of individual results from the actual lifecycle assessment is frequently opaque and difficult to interpret. To improve the interpretation of results, BASF has developed a method that combines the ecological and economic parameters, plotting them as a single point in a coordinate system. The environmental impact is described with reference to six categories, and each of these categories covers a large number of detailed individual criteria. The results of all environmental impact categories are combined by weighting.

Also, the economic data are compiled over the entire lifecycle. The total costs are normalized with respect to the average of all alternatives. This helps in identifying cost drivers and areas offering potential for cost reductions. The data on relative costs and environmental impacts are used to construct a diagram, the so-called eco-efficiency portfolio, which clearly shows the strengths and weaknesses of each product or process.

8.6.5.3 Further Enhancing the Eco-Efficiency Analysis by Including Social Aspects.

Based on the principles of the BASF eco-efficiency analysis on the one hand, and a product-related specification of the social sustainability dimension on the other, a social lifecycle assessment procedure has been developed. The new integrated instrument, the so-called SEEbalance compares the costs, environmental impact, and social effects of different product or process alternatives. Socio-eco-efficient solutions combine a good environmental performance with high social benefit and at the same time low costs for the end customer. Social indicators for the evaluation of products and processes are developed. The indicator classification is based on the "business management stakeholder approach" (Schmidt *et al.*, 2004). The results of the social, ecological, and economic assessments are combined in the three-dimensional portfolio, the so-called SEEcube®. The instrument described here is to be regarded as "work in progress," that is, it needs to be constantly and critically reviewed and adjusted to the ongoing developments in society and discussions in politics, society, and science.

8.6.5.4 *Eco-Efficiency and the Global Compact.* As part of its active involvement in the Global Compact, BASF started in 2002 a project together with UNIDO (United Nations Industrial Development Organization) and UNEP (United Nations Environmental Program). The goal of this partnership is to investigate and improve the eco-efficiency of small- and medium-sized enterprises (SMEs) in developing countries. The partnership facilitates the transfer of know-how, contributing to sustainable industrial development.

In the first step of the partnership, BASF provided its eco-efficiency expertise (see Section 6.1.3 for detailed discussion of BASF eco-efficiency analysis) to help Moroccan textile dye works operate in a more efficient and environmentally friendly way. Based on the eco-efficiency analysis and the experience of many years with products and processes in the textile sector, BASF developed a software package, which was given to Moroccan companies free of charge. The software tool substantially simplifies the compilation of an eco-efficiency analysis. It uses key technical data to calculate how the manufacturing process can be improved.

Such a software tool can be expanded to other industrial sectors and other countries. Combined with consulting services of the UNEP/UNIDO National Cleaner Production Centers, it enables SMEs in developing countries to perform eco-efficiency analyses.

Applying the results of these analyses, scarce resources can be conserved and the burden on the environment reduced. This enables SMEs to more easily adhere to international standards (Bethke, 2003).

8.6.6 Dialog: An Important Part of Sustainability

As stated in our Values and Principles, BASF is committed to an open dialog within the company, with its business partners and neighbors, and with society. For BASF, dialog means providing insights, accounting for our activities, and actively seeking debate with those who have other opinions. Transparency and openness are important instruments to establish trust.

For the year 2000, BASF published for the first time a report about Social Responsibility. This completed our traditional Annual Report and the Environment, Safety and Health Report to a comprehensive Sustainable Development reporting. The independent verification of our reports ensures the validity of the information we publish. Following the verification of the Social Responsibility 2001 Report, the Environment, Health and Safety 2002 Report was also verified by the independent accountancy firm Deloitte & Touche.

8.6.6.1 *Crises Communication.* BASF goes to great lengths to ensure that the facilities at its worldwide sites are as safe as possible, often putting procedures in place that exceed the statutory regulations. It is of course impossible, even though we employ the most sophisticated measures, to totally rule out an accident or plant malfunction. In such a case we consider it our duty to inform our employees, neighbors, the authorities, and the general public swiftly, honestly, and comprehensively about any potential risks. Our comprehensive crisis management system at the Ludwigshafen site therefore includes as an integral part a team of the Corporate Communications Department on permanent duty. The task of this team is to inform the public and the workforce around the clock and throughout the entire year about emergencies through news releases, flyers, the Web, and a telephone hotline.

8.6.6.2 Community Advisory Panels: Institutionalized Dialog. A permanent, close, and institutionalized dialog is the goal of the community advisory panels (CAPs) that BASF has set up at many sites, either alone or together with other companies. Representatives from BASF meet with interested neighbors and discuss topics that include emissions, plant safety, and procedures in case of plant malfunctions. Based on the experiences of existing CAPs, the "Dialogue" expert group has drawn up standards for CAPs. They ensure that CAPs adhere to certain features that distinguish them from other forms of contact and interaction with our neighbors: membership should reflect the diversity of the community's interest groups, meetings should be regular, and topics of mutual interest should be discussed with the aim of guaranteeing transparency.

8.6.7 Financial Markets Honor Sustainable Development

Increasingly, investors are diversifying their portfolios by investing in companies that set industry-wide best practices with regard to sustainability. Reasons are:

- *Low risk.* Companies applying the principles of sustainable development systematically assess future economic, social, and ecological risks. Strategies are developed early to minimize such risks.
- *Shareholder value.* Business in accordance with sustainable development is attractive to investors because it aims to increase long-term shareholder value.
- *Success.* Sustainability is considered a catalyst for enlightened and disciplined management, and, thus, a crucial success factor.

Which companies operate according to the principles of sustainability? The answer can be found in special share indices – the so-called sustainability indices. In 2003, BASF was again included in the Dow Jones Sustainability Index for the third year in succession. This index includes only 10 percent of chemical companies represented in the Dow Jones Global Index. BASF shares are also listed in the FTSE4Good index and are included in the portfolios of various sustainability funds such as Storebrand's Principle fund.

To be included and to remain listed in these sustainability indices, BASF has to undergo an annual detailed assessment by a rating agency. The agency's analysts scrutinize to what extent we have integrated sustainability principles into our business processes and how we implement them using appropriate tools. This critical external investigation is an important supplement to our internal control mechanisms. Inclusion in sustainability indices is important for BASF because it shows our progress towards a sustainable company.

8.6.8 Conclusions

BASF is pursuing a strategy of increasing and sustaining its value through growth and innovation. We want to achieve steady, high profits in a manner that maintains and extends possibilities for further growth. We are convinced that we will best succeed in this by combining our business growth with environmental protection and social stability. Sustainable development therefore is an integral part of BASF's corporate values.

8.7 THE GSK APPROACH TO SUSTAINABLE DEVELOPMENT

DAVID J C CONSTABLE, ALAN CURZONS, AILSA DUNCAN, CONCEPCION JIMÉNEZ-GONZÁLEZ AND VIRGINIA L CUNNINGHAM

GlaxoSmithKline, USA

8.7.1 Introduction

GlaxoSmithKline (GSK) is a research-based firm that is among the world's largest transnational Pharmaceutical and Consumer Healthcare companies. It is dedicated to the development of medicines and consumer healthcare products that are used to treat a broad range of diseases across multiple therapeutic areas and support a better quality of life. GSK is a market leader in four major therapeutic areas: anti-infectives, central nervous system (CNS), respiratory, and gastro-intestinal/ metabolic, and has a growing portfolio of oncology products. In addition, it is a leader in the development and manufacture of vaccines. The company's commanding position in genomics/genetics and new discovery technologies has resulted in one of the best early phase pipelines in the industry. GSK products are manufactured in 99 different facilities in 39 countries and marketed throughout the world to a diverse population of customers.

As a result of the 2001 merger of Glaxo Wellcome and SmithKline Beecham, a comprehensive EHS framework was extensively revised to reflect best practices in both legacy organizations, while positioning GSK for future challenges. This framework provides the conceptual and practical underpinnings for moving GSK towards Sustainability in Environment, Health and Safety and is comprised of the GSK EHS Vision, Policy and Global Standards. The 63 standards, including high-level standards on Sustainable Development and Product Stewardship, were approved by the GSK Board of Directors and therefore have the highest level of approval possible. The structure for the GSK Global EHS Standards is shown in Figure 8.23. The Standards and their accompanying Guidelines, EHS Business Processes, and Guidance have been augmented by technical and training manuals, information,

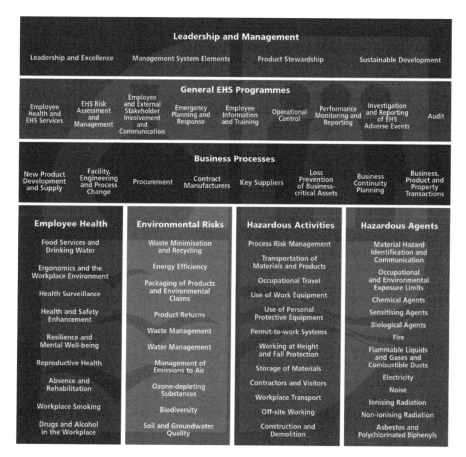

Figure 8.23. Structure of Global EHS Standards.

corporate intranet sites, methodologies, and tools that can be seamlessly integrated into existing business processes.

Following on from the development and implementation of the EHS framework, Corporate EHS (CEHS) recognized the need to lead and sustain movement towards more sustainable business practices. Beginning in 2003, CEHS embarked on an extensive planning exercise to develop the GSK Plan for Excellence. The GSK Plan for Excellence provides a road map to the Corporation for achieving certain aspirations by 2010. These aspirations include: Optimal Health and Sustainable Business, EHS Integral to New Product Development, EHS adds Value to Product Commercialization, Acknowledgement as an EHS Leader, and Most Efficient and Sustainable Processes. Our performance towards these aspirations will be measured by the achievement of global metrics and targets that are linked to our strategic objectives and the continuous improvement process. The first set of targets for 2005 is shown in Figure 8.24.

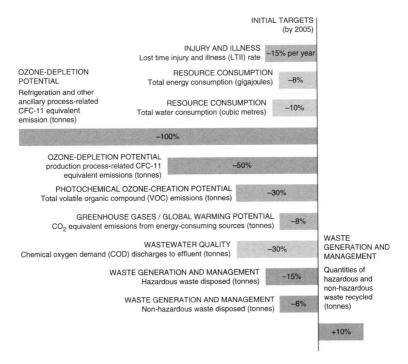

Figure 8.24. GSK 2005 targets.

The remainder of this section is a description of GSK's efforts to move towards environmental sustainability.

8.7.2 Sustainability Strategy

Sustainable Development is generally understood as not impairing the ability of future generations to enjoy the same or a better standard of living than what is currently enjoyed in developed countries. Environmental sustainability means we aspire to replace our use of nonrenewable or scarce materials with materials that are renewable and more abundant, while reducing our dependence on fossil fuels. We also aspire to take from the environment no more than we return, leaving ecosystems in a healthy state. This requires us to take only raw materials that can be relatively quickly renewed by naturally occurring Earth systems processes and leave only waste that serves as raw materials for industrial or Earth systems processes.

While GSK aspires to be a sustainable company, we recognize that moving from aspiration to reality will take many years. Initially, we will work toward enhancing our environmental sustainability while delivering new products faster and better. Environmental sustainability may be achieved at GSK in three steps.

- The first step is to decrease the materials and energy intensity of our products through improvements in production efficiency. Improvements in our primary and secondary manufacturing process efficiencies should decrease the mass and energy we use while decreasing the amount that is wasted within the boundaries of GSK operations.

- The second step is to increase the use of renewable raw materials throughout the product lifecycle. This represents a shift in historic approach and will require partnerships with suppliers to find alternative raw materials for our processes that can be sustainably produced from renewable resources.
- The third step is to balance consumption and waste generation so that our raw materials are taken from the planet in a sustainable manner and our waste materials serve as nutrients and feedstocks to other industrial or Earth systems processes.

These three steps will require work over many years. GSK is beginning the first step by determining how to optimally measure production and operational efficiency. A series of publications is available that describes GSK's efforts to develop appropriate metrics for its R&D chemical development activities (Constable *et al.*, 2001, 2002).

Our R&D organization is also working to improve the effectiveness and efficiency of current production processes that are developed and handed over to manufacturing while it works on replacing current batch processing technology with continuous, just-in-time, right-sized processing technology. In the design and development of new synthetic chemical processes, scientists seek to use fewer materials and less of them, while substituting materials with the best environment, health and safety profile. This is not only the lowest cost approach, but also results in the greenest processes. Some of the tools that have been developed to assist this effort are described below.

In addition, some GSK production processes address the second step noted above through the use of microorganisms as a sustainable resource to produce raw materials, enzymes, or advanced intermediates that either make up or make our medicines. Harnessing biological processes to make complex materials and active pharmaceutical ingredients is clearly an extremely desirable goal for the Pharmaceutical Industry, but the promise remains largely unrealized. The third step towards more sustainable business practices will be taken in the future as the business evolves and opportunities become more available and are better understood.

Our key strategies for moving towards sustainability in EHS are highlighted in Table 8.6. First among these is senior management involvement, providing high-level support for moving the corporation towards more sustainable practices through its R&D and environment, health and safety activities. As an integral part of its business operations, EHS issues are included on the agendas of the Audit Committee of the Board of Directors, the Corporate Executive Team, and other senior management teams.

In providing a core EHS resource group, the Corporate Environment, Health and Safety (CEHS) Department recognizes the critical importance of bridging the gap between the aspirations of sustainable development and the practical integration

TABLE 8.6. Strategies in Moving Towards Sustainability in Environment, Health, and Safety at GlaxoSmithKline

- *Senior management involvement.* High-level support for business activities that move the corporation towards more sustainable practices.
- *Support for a core resource group.* Corporate support for a central resource of EHS specialists who lead and support sustainability initiatives, including traditional EHS activities.
- *A systematic and structured EHS framework.* Detailed and comprehensive framework of documents and program available to all operations for the enhancement and improvement of EHS performance.
- *Product stewardship.* Corporate recognition of its responsibility to identify and mitigate potential adverse EHS impacts through the products' lifecycle.
- *Management systems.* Comprehensive and global EHS management system that is aligned with internationally recognized standards and includes routine auditing at the operations.
- *Integration into the business.* Explicit integration of EHS standards into existing business practices, including new product development, facility, procurement, key suppliers, property transactions, etc.
- *Operational excellence and continuous improvement.* Alignment of GSK's operational excellence (OE) program, which combines Lean Manufacturing and Six Sigma, with EHS activities' goals and targets.
- *Managing key risks.* Managing short-term, traditional EHS risks as well as longer-term, externally driven risks as part of GSK Plan for Excellence.
- *Target setting.* New and revised targets produced every five years to drive improvement and demonstrate commitment to improvement.
- *Stakeholder involvement.* Inputs from internal and external stakeholders, which are critical to continuous improvement and continued success of the company.

and implementation of these principles into the business. Thus, in addition to teams focused on auditing and regulatory issues, CEHS has various other teams dedicated to:

- *EHS hazard assessment and communication*, which carries out comprehensive material hazard testing and assessment to produce best-in-class safety data sheets;
- *EHS product stewardship*, which promotes the early identification of EHS issues and product-based risk assessment, mitigation, and management during product R&D, and provides new product support during technical transfer from R&D to manufacturing; and
- *EHS global information and reporting*, which is responsible for EHS data collection from GSK operations (including third-party manufacturing), internal and external reporting, and communication of technical guidance, training, targets, and company goals.

These teams provide a versatile and flexible structure for supporting the business.

GSK's comprehensive EHS system also includes a central technical documentation that promotes consistency and best practice while relieving individual operations of the burden for developing their own materials. Examples of technical support tools are product stewardship guides that supplement and enhance information in Safety Data Sheets; detailed information about process safety, environmental controls, ergonomics, and other issues; training information on technical topics; as well as intranet R&D tools discussed later in this section.

Product Stewardship, while widely recognized and accepted in the chemical industry, is not generally embraced by the pharmaceutical industry. Product Stewardship efforts by GSK represent industry leading practice and demonstrate its commitment to Corporate Social Responsibility. GSK's pharmaceutical products are carefully designed to create a biological change in patients and as a result have potential EHS risks and impacts throughout their lifecycle, that is, from R&D and manufacturing through to patient use and disposal. Product stewardship provides a systematic way for GSK to identify product or process risks early and proactively mitigate and manage them. EHS Product stewardship encompasses the assessment of the health, safety (excluding patient safety), and environmental risks created during all stages of the product's lifecycle, from discovery and development; through manufacturing, marketing, distribution, and use; to recycling or final disposal. In particular, we also strive to apply product stewardship principles at the key decision stages in R&D and to our contract manufacturers and key suppliers. Integration of product stewardship into business activities protects people and the environment, enhances compliance with regulations, and avoids interruption of supply.

As summarized in Table 8.6, other strategies to move towards sustainability in EHS include management systems, integration into the business, operational excellence and continuous improvement, managing risks, target setting, and stakeholder involvement. Managing key risks to the corporation is in fact an early part of the Plan for Excellence briefly described above. To that end, CEHS has identified eight key risks and is organizing to address each area in a systematic, cross-functional, multiyear effort. These GSK-specific risks include several short-term, traditional EHS risks such as Chemical Agents, Ergonomics, Driver Safety, and Process Safety. The key longer-term and externally driven risks include Energy and Climate Change, Pharmaceuticals in the Environment, Chemicals Policy (e.g., REACH), and Regulatory Tracking.

Input from internal and external stakeholders is also critical to continuous improvement and our continued success. GSK actively seeks stakeholder involvement for its Plan for Excellence and its Corporate Social Responsibility initiatives. In addition, customer feedback is routinely obtained and helps to focus attention on issues that affect the public through the consumption of GSK products. GSK has also published reports of its EHS performance since 1992 and has continued to develop and enhance its reporting to align with many of the features recommended by the Global Reporting Initiative.

8.7.3 Designing Products for Environmental Sustainability

As a result of senior management support over the course of the past eight years, GSK has been able to systematically develop a suite of tools and methodologies to assist the business and R&D in moving towards more sustainable business practices. Given the nature of the Pharmaceutical Industry, where processes are locked in two to three years before a product reaches the market, the only hope of changing what is done is to design sustainability into new product processes within an R&D environment as early as is practical. This is a very difficult and challenging endeavor since principles of sustainability must compete with more traditional measures and demands placed on the pharmaceutical industry by the regulatory agencies governing new product approval. Given the very strict regulatory oversight encountered in the pharmaceutical industry and the pressure to provide clinical trial supplies for products that will never enter the market, there is a natural and understandable suspicion and aversion to anything that might be construed as either another restriction, or worse, as something that is nice to have, but not really necessary for the survival of the business.

Consequently, we have launched a comprehensive eco-design toolkit that is primarily focused on influencing new product and process development, but may also be of help for product transfer or redesign of processes. The toolkit should assist bench-level GSK scientists and engineers, together with their managers, to bring products to market faster. We believe this is true because the eco-design principles, practices, and guidance that have been provided to the scientists and engineers will help them to design out potential problems early in development. These tools should also help to bring products to market more cost-effectively because the eco-design principles and practices will enable GSK in the longer term to use less material and energy to make products, while ensuring that the material and energy used has fewer EHS impacts. R&D should be able to address potential environment, safety and health issues before a process is handed over to manufacturing where the cost to address EHS issues is considerably higher.

The toolkit is currently composed of five modules, which are discussed below. Each of these modules considers the EHS impacts of materials, processes, and services from the time raw materials are extracted through to the ultimate fate of products and wastes in the environment.

8.7.4   Green Chemistry Guide

Beginning in early 1996, the EHS Product Stewardship Team began to investigate what "green" or "clean" chemistry and/or technology meant in the context of GSK. Given the many different definitions and the general focus on large petrochemical and chemical processes, there was very little information or interest available to help our investigations. We undertook a systematic evaluation of a few GSK chemical processes, and evaluated these processes using a variety of metrics. Work with the American Institute of Chemical Engineers Center for Waste Reduction Technologies helped to narrow the scope and a reduced metrics set, discussed in Section 6.1.2 of this book, was chosen for application to about 250 individual chemical syntheses. From an extended and extensive evaluation of the data, a considerable amount of very useful information was obtained (Curzons *et al.*, 2001) and eventually converted into the GSK Green Chemistry Guide.

Launched in 2000, the GSK Green Chemistry Guide is a comprehensive site that provides guidance to GSK scientists and engineers on applying Green Chemistry concepts that will enable more efficient use of resources, reduce environment, health and safety impacts and minimize costs. It also contains a Technology section that provides both a methodology for ranking (Jimenez-Gonzalez *et al.*, 2001, 2002) and the actual ranking of commonly used technologies against newer alternative technologies. Some of the contents of this module are shown in Figure 8.25.

8.7.5 FLASC©: Fast Lifecycle Assessment for Synthetic Chemistry

While investigating and developing the GSK approach to Green Chemistry and technology, a multiyear investigation into the lifecycle inventory (LCI) of a typical active pharmaceutical ingredient (API) was undertaken. This was necessarily an extended investigation given that there was a tremendous lack of data available for those materials generally found in the supply chain of an API. In addition, data in most commercial software is generally not that transparent or modular; that is, one cannot deconvolute a typical LCI and dissect the supply chain to determine where the greatest impacts arise or with which materials the greatest impacts are associated.

In collaboration with researchers at North Carolina State University, a modular, *ab initio* approach based on chemical engineering design principles was developed

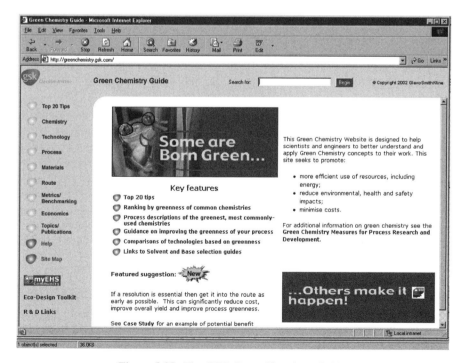

Figure 8.25. The GSK Green Chemistry Guide.

(Jimenez-Gonzalez, 2000, 2003a). This methodology is transparent, well documented, and extremely flexible. Modular Gate-to-Gate (GTG) LCIs for materials can be used to construct chemical trees from an advanced material back to the cradle, and the inventories "rolled-up" to obtain a Cradle-to-Gate (CTG) LCI. Its utility was demonstrated for a typical, commercially successful GSK API (Curzons *et al.*, 2001). These GTG LCI data were also available for additional analysis that led to the development of the FLASC© tool (Curzons *et al.*, 1999).

Very briefly, FLASC©is a web-based tool and methodology that delivers fast lifecycle assessments of potential chemical synthetic routes or manufacturing processes used to make GSK APIs or intermediates. It also provides guidance about which materials have the greatest lifecycle environmental impacts and allows a GSK scientist or engineer to benchmark between existing or proposed routes or processes. Route or process assessment is initiated by supplying the FLASC© site with the quantity (kg/kg final product) of all materials used in the process via a simple, automated spreadsheet upload procedure. FLASC© has been developed to accept process and material information routinely generated by GSK R&D scientists using existing software systems or reports and formats. Information and guidance from the FLASC©site should, over time, help GSK move away from materials and processes that have significant lifecycle impacts.

8.7.6 Materials Guides

8.7.6.1 Solvent Selection. The GSK Solvent Selection Guide (SSG) was created in response to initial lifecycle inventory/assessment work that clearly showed solvents make a considerable contribution to the overall lifecycle environmental impact of GSK. The SSG, containing detailed information on 45 solvents, was launched in 1998 as a paper/EXCEL based version (Curzons *et al.*, 1999) and was rapidly followed by a corporate intranet site. The Guide was unique in combining a systematic assessment, review and ranking of solvents from an Environment, Health and Safety perspective and presenting information in a simple, easy to use and understand format. Underpinning the guide was a comprehensive evaluation of solvents from first principles.

By the time of the merger, continuing work on solvents and on the lifecycle of solvents afforded the opportunity to revise the guide. The new guide contains a revised set of solvents (some solvents whose performance was not acceptable were removed), a revised health scoring methodology that incorporates recent advances in harmonizing risk phrases, and detailed lifecycle information (Jimenez-Gonzalez *et al.*, 2003b) for each solvent. The original scoring methodology was also revised to include a solvent lifecycle score for each solvent.

An extensively revised and expanded intranet site was launched in July 2003 and is shown in Figure 8.26. This site contains:

- A comparison and ranking of 47 solvents according to environmental waste profile, environmental impact, safety profile, health impact, and lifecycle profile;

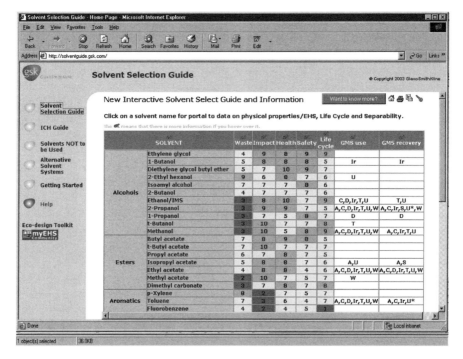

Figure 8.26. The GSK Solvent Selection Guide.

- A comparison of International Conference on Harmonization (ICH) guidelines on allowable concentrations of solvents in active pharmaceutical ingredients against EHS characteristics of solvents;

- Information on solvents that should be avoided;

- Information on boiling point and azeotrope formation to assist in the selection of separable co-solvents; and

- Detailed information on selected physical properties, safety, health, and environmental issues.

Future work on solvents is focusing on the interaction of solvents and chemistry and greater emphasis on replacement solvents.

8.7.6.2 Base Selection. Following on the success of the solvent selection guide and in response to specific requests from a number of synthetic chemists, a similar exercise was undertaken to evaluate chemical bases. Bases are used extensively in many synthetic chemical procedures and are often used in stoichiometric quantities. Given their use and properties, their EHS performance is understandably of great concern.

The approach taken to evaluating bases for the Base Selection Guide (BSG) is similar to that taken for the SSG, but is simpler. As can be seen in Figure 8.27,

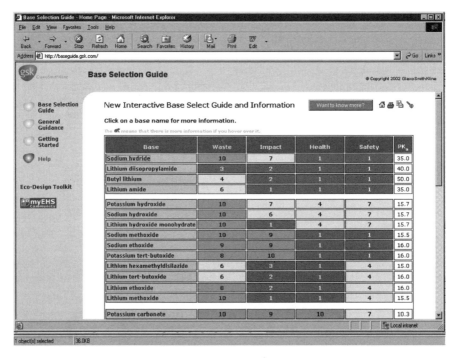

Figure 8.27. The GSK Base Selection Guide.

the BSG contains a ranking of 42 chemical bases according to their environmental waste profile, environmental impact, safety profile, and health impact. As with the SSG, the user is directed to a readily recognizable summary table format that functions as a departure point for more detailed information on each base. In addition, there is general guidance on the use of bases. This site was updated and revised in late 2002.

8.7.6.3 Green Packaging Guide. GSK has worked since the early 1990s to influence packaging decisions, especially in its Consumer Healthcare business. Since the mid-1990s, GSK has had a corporate intranet site dedicated to improving the EHS and lifecycle performance of its packaging. An extensively revised site was launched in 1998, and again, in 2003. An additional major enhancement to the site in 2003 was the incorporation of a new tool known as WRAP – Wizard for the Assessment of Packaging (Duncan and Curzons, 2004). WRAP is an interactive tool that allows users to assess the greenness of their packaging, benchmark against different packaging choices, and provides guidance on improving packaging performance. General guidance for the Pharmaceuticals and Consumer Healthcare businesses in the intranet site includes a business process for the selection of packaging, a module on packaging legislation throughout the world, and a module describing new packaging materials.

8.7.6.4   Sustainability Opportunities. As a transnational pharmaceutical company, there are several sustainability business opportunities available to GSK. These include:

- *New disease targets.* For example, GSK already has a significant investment in the development of medicines for diseases of the developing world. These initiatives include the development of vaccines, antidiarrheals, antimicrobials, and antimalarials. There remains a plethora of diseases afflicting huge segments of the worlds' population that have not been addressed, and there are clear opportunities for companies that are able to overcome equally clear challenges.
- *Green building design.* There are significant opportunities to reduce our overall environmental footprint through building designs that include consideration of materials, energy use, location, community impact, and so on.
- *Changes in sales and marketing.* There are currently very large sales and marketing forces traveling throughout huge areas to educate and inform physicians about GSK products. Opportunities for accomplishing the same tasks through advanced telecommunications and Internet applications hold some promise for reducing the impact while maintaining or increasing the effectiveness of the sales force. These same techniques hold additional promise for reaching currently unreached or underserved people groups throughout the world.
- *Targeted API delivery.* Advances in genomics and proteomics, nanotechnology, and other areas hold significant promise for delivering the right drug in the right amount at the right time as opposed to the heavy-hammer (particularly in oncology), "one-size-fits-all" approach of modern pharmaceutics. This opportunity is, however, confronted by significant challenges and is clearly a long-term opportunity.

8.7.6.5   Sustainability Challenges. There are significant technological, societal, and political challenges facing implementation of principles of sustainable development in the pharmaceutical industry. These include, but are not limited, to the following:

- *Healthcare infrastructure.* In many parts of the world, there is a very poor or nonexistent healthcare infrastructure. This exacerbates current initiatives for treating diseases such as HIV and prevents the development of additional medicines. A lack of healthcare infrastructure usually is highly correlated with political instability, and this makes the conduct of normal business very difficult.
- *Patent protection.* Fair patent laws for medicines and the processes used to make them need to be harmonized throughout the world. Adequate patent protection inhibits the discovery and development of new drugs, particularly in those parts of the world that are currently underserved.
- *Technology.* Significant effort must be undertaken to develop more flexible, efficient, and effective processes to make APIs. A paradigmatic shift is required for the industry to move from batch to continuous processing, but this must be accompanied by a mechanism that makes the switch economically viable.

There is significant capital investment in place for a typical multipurpose chemical plant that is fully functional and will continue to be so. This is also locked in place by regulatory agencies with changes extremely difficult to implement.

- *Chemistry.* Efforts to demonstrate sustainable chemical synthesis must be undertaken at a scale that is meaningful to the industry.
- *Materials.* The building blocks for APIs are currently derived almost exclusively from petrochemical sources. New materials, including biologically derived materials, must be developed.

8.7.7   Conclusions

GSK's sustainability initiatives begin with the selection of potential new drug candidates that have more favorable EHS characteristics, and move into process research and development where green chemistry and green technology concepts are incorporated. In addition, there are inherent safety, pollution prevention, waste minimization, and green packaging programs that are integral parts of manufacturing operations. Finally, there are initiatives to ensure responsible use and disposal of GSK products and wastes. GSK business units are enabled, by real-time access to expert data, information, knowledge, and key experts, to take sustainable design principles and make them integral to the business.

While there are implicit social and economic aspects to the GSK programs described in this chapter, it must be understood that explicit "triple bottom line" discussions within the corporation are not conducted at this time. However, there is considerable investment by GSK throughout the corporation in initiatives that would be considered to be moving towards more sustainable business practices. In addition, GSK continues to partner and work with a variety of organizations as part of GSK's Corporate Social Responsibility initiatives, which clearly falls under the banner of sustainable development. Examples include GSK's initiative to eliminate elephantiasis fillariasis, a crippling disease that potentially affects at least one billion of the worlds' people or GSK's leading efforts to make anti-HIV medicines more accessible to those populations most affected by HIV. In addition, in light of recent corporate scandals affecting other industries, the corporation has taken a number of steps to make its financial accounting and general business practices more transparent.

As compelling and significant as these initiatives are, opportunities to pursue Sustainable Development as a business opportunity in the context of a developed world, transnational pharmaceuticals company continues to be difficult and must for the present remain in the future. However, these business opportunities are not out of the realm of the possible and they are achievable.

8.7.8   Acknowledgments

The authors wish to acknowledge that the GSK achievements in the area of Sustainable Development are the result of the efforts of all of the staff of GSK

CEHS, and many others within the corporation, without whose contributions these initiatives would not be progressing. In particular, we wish to especially recognize the significant contributions of James R. Hagan, Nancy English, Steve Bailey, and Jon Wilson, whose leadership has made our progress possible.

8.8 MOVING 3M TOWARD SUSTAINABILITY: THE BUSINESS CASE FOR SUSTAINABLE DEVELOPMENT

KEITH J MILLER

3M

The concept of sustainability – meeting today's needs while preserving the ability of future generations to meet their needs – is ancient. Along with other elements of society, commercial interests have long understood this premise, although their commitment was sometimes influenced by short-term business demands. Sustainability was considered a benign but noncritical business objective – good, but only insofar as it did not hurt profitability.

Today, progress toward environmental, social, and economic sustainability is viewed as an absolute requirement for the success of multinational corporations. And, increasingly, it has an impact on the performance of businesses that operate in more local markets. In fact, companies that cannot demonstrate progress toward sustainability often find themselves at a competitive disadvantage: The expectations of shareholders, customers and other stakeholders, as well as regulatory demands, are beginning to make sustainability the price of entry into many markets.

The most successful corporations have seized upon this business reality. They recognize that sustainability increases market appeal and promotes operational excellence (and therefore profitability). In addition, attention to sustainability makes them more likely to anticipate and address environmental, economic, and social issues.

Because of its culture and values (Box 8.2), 3M for many years has been adopting the precepts of sustainability. Implicit in these values is the expectation that the company will act in a way that preserves the trust of its constituencies. 3M is committed to the highest ethical standards and to transparency whenever possible (that is, when transparency does not compromise confidentiality or intellectual property).

When 3M established its Environmental Policy in 1975, it was one of the first companies to do so (Box 8.3). In that same year, it introduced its Pollution Prevention Pays program, a pioneering effort in its recognition of the business case for environmental performance and in its reporting on environmental progress. These statements of its progress were later followed by explicit goals for solid waste and releases to air and water.

In recent decades, the company has sought continual improvement in its environmental performance, and it has detailed its goals and its efforts to achieve them. It has expanded its reporting to increase transparency. And, through its recently revised Environmental, Health and Safety (EHS) Management System, it has made sustainability a required part of each business' strategic planning.

BOX 8.3   3M CORPORATE ENVIRONMENTAL POLICY

3M will continue to recognize and exercise its responsibility to:

- Solve its own environmental pollution and conservation problems;
- Prevent pollution at the source wherever and whenever possible;
- Develop products that will have a minimal effect on the environment;
- Conserve natural resources through the use of reclamation and other appropriate methods;
- Assure that its facilities and products meet and sustain the regulations of all federal, state, and local environmental agencies;
- Assist, wherever possible, governmental agencies and other official organizations engaged in environmental activities.

—Adopted by 3M Board of Directors, February 10, 1975

BOX 8.4   CORPORATE VALUES AND SUSTAINABILITY

3M's sustainability policies and practices are directly linked to our four fundamental corporate values:

- Satisfying our customers with superior quality and value;
- Providing investors an attractive return through sustained, high-quality growth;
- Respecting our social and physical environment;
- Being a company that employees are proud to be part of.

Today, employees and shareholders increasingly view sustainability as a primary consideration in the company's key functions – product development, manufacturing, human resource activities and community involvement – because it is the right thing and because it makes business sense.

8.8.1   Transition to Sustainability

While there are many factors supporting the business case for sustainability at 3M, the main drivers for sustainability are operational excellence, organic growth, and risk management. Each of them is linked to corporate business initiatives.

Improving operations has an immediate impact on the bottom line through cost reductions *and* supports sustainability. Our Pollution Prevention Pays (3P) program has reduced waste and resulted in operating cost savings. Linked with our corporate Six Sigma initiative, improving process yields results in eliminating pollution at the source and lowering operating costs.

Organic growth at 3M means growth through the sales of more and new, innovative products. In recent years, an emphasis has been on products that help customers achieve their goals for sustainability. Through our Design for Six Sigma (DFSS) process and linking with our 3M Acceleration initiative, these sustainability efforts are aimed at creating a competitive advantage.

Risk management benefits present a convincing business case for sustainability alone. Companies need to anticipate and address potential problems before they become liabilities. Companies that do not manage their risks have seen their value drop significantly.

While operational excellence, growth, and risk management are all drivers for sustainability, the discipline required by a pursuit of sustainability frequently spills over into other aspects of a company's operations. Indeed, recent studies suggest that companies that manage their EHS systems well tend to have higher long-term returns than other companies. Innovest (www.innovestgroup.com) cites studies by researchers at Duke University and ICF Kaiser that show environmentally superior companies financially outperform environmentally inferior companies. WestLB AG recently revisited their study "More gain than pain – Sustainability pays off" in October 2003. According to the updated results of the study, "sustainability is an independent return-driving factor that can have a positive impact on shareholder value." In studying the performance of sustainability investments, the Dow Jones Sustainability Index achieved a risk-adjusted outperformance of 3.1 percent positive alpha. They conclude, "it pays off to be good."

8.8.2 Organizing the Move Toward Sustainability

8.8.2.1 Sustainability and Operational Excellence. Since the earliest days of the Industrial Revolution, successful manufacturers have understood that waste (in the form of unused raw materials and defects) is costly. In the last half of the 20th century, the definition of waste grew to encompass energy inefficiencies and emissions. An early proponent of this broader view was then 3M President William L. McKnight, who in 1948 labeled as costly waste anything that went "up the smokestack" rather than into products. He urged the creation of products and manufacturing technologies that would conserve energy and raw materials.

McKnight also espoused an unusual management philosophy for the times: he believed in encouraging individual initiative, risk-taking, and the freedom to fail – qualities that promoted employee growth and satisfaction as well as the development of diverse businesses and technologies.

3M's operations became more sophisticated – and productive – with the introduction of the Pollution Prevention Pays (3P) program in 1975. As the name suggests, the program was founded on two assumptions: that pollution should be prevented rather than treated after the fact, and that such an approach made economic sense. The results of the program have borne out these assumptions. To date, the 3P program has implemented over 5200 projects, most of them stemming from employee suggestions. Tallying just first-year results from these projects, 3M calculates around $950 million in operating cost reductions and the prevention of more than one million tons of pollutants.

3M has continued to expand and refine the 3P program. In 2001, for example, the company provided more opportunities for participation by our research and development, logistics, transportation, and packaging employees, with the addition of new award categories and criteria.

The connection between operational excellence and sustainability has been reinforced since 2001, when 3M adopted Six Sigma methodology as a corporate initiative. In some cases, projects that have been designed and implemented to achieve Six Sigma goals, that is, to improve productivity and reduce defects, have also improved the sustainability of our products, human resource practices, and manufacturing processes. For example, Six Sigma projects to improve product yield have frequently resulted in reduced waste; most of these Six Sigma projects also qualify as 3P projects. In addition, all new products must go through a Life Cycle Management review that is part of 3M's Design for Six Sigma (DFSS) process. By addressing defects such as inherent process waste during the manufacturing phase of the lifecycle, these reviews improve the resource intensity and cost to produce the product.

The benefits of moving toward sustainability are not dependent on country-specific regulation. To the contrary, their impact on productivity and profitability are similar around the world. For this reason, 3M facilities worldwide are held to the same environmental standards that the company applies to its facilities in the United States (except where local standards are higher). For the same reason, all 3M manufacturing facilities making products for transnational markets are required to become ISO 14001 certified.

8.8.3 Sustainability and Growth

Early on, the most obvious benefit of sustainability – at least in terms of business success – was its emphasis on cost-reduction through better management of resources. In recent years, though, another benefit has arisen: products that help customers achieve their own goals for sustainability are more likely to succeed. In other words, sustainability efforts can create a competitive advantage.

At 3M, product development teams use the company's EHS Management System to design products that capitalize on this advantage. 3M's EHS system is multidisciplinary and multilevel, involving employees at virtually every level of the company. Top managers set EHS goals, establish the measurement systems, and hold employees accountable. Middle managers provide the necessary resources and manage the business unit planning process (including goal-setting and measurement systems). Supervisors and employees in other capacities – such as line manufacturing operations, labs, offices, and sales and marketing operations – all help to ensure that the company is reducing its environmental footprint and protecting the health and safety of employees.

In some cases, the advantage of a sustainable product is specific and readily observable by the customer, such as a Post-it™ Recycled Paper Notes, made with 100 percent recycled paper fiber (20 percent post-consumer waste). 3M paint replacement films, which allow automobile and aircraft manufacturers to reduce their use of paints and solvents, are an example of an industrial application where the customer is striving to achieve its own sustainability goals.

3M products that have increased energy efficiency have been particularly successful with customers and end-users. 3M window insulator kits for homes and its solar-reflective films for cars and buildings are well-known examples. Less obvious are 3M's Brightness Enhancement Films, which greatly reduce the light needed to illuminate electronic displays and therefore allow laptop computers, PDAs, and cell phones to run almost twice as long on a single battery charge. Electronics manufacturers have quickly grasped the power of energy efficiency in the marketplace: because no consumer wants the device with half the efficiency of the competition, virtually all these devices now use these films.

The obvious appeal of energy efficiency has been one of the forces behind 3M's significant investment in fuel cell technology. As the world moves inevitably toward a hydrogen economy, 3M intends to be a leader. The company has applied a broad range of technological strengths to the development and manufacture of the membrane electrode assembly that is at the core of any fuel cell system. Today, 3M is capable of manufacturing large quantities of this key component and is positioned to expand its manufacturing capacity as demand increases.

In other instances, the benefit in the marketplace is more general, and reflects customers' perceptions of 3M as a responsible manufacturer. 3M earns this reputation (and reinforces it) through its use of Life Cycle Management (LCM), its reporting, and its careful attention to environmental marketing claims.

With 500 new products introduced each year, 3M has a continuous flow of opportunities for improving its environmental, health and safety performance. To make the most of these opportunities, 3M requires that all new products undergo LCM review prior to commercialization. Business units also conduct LCM reviews for existing products on a prioritized basis. In these reviews, product development teams address the environmental, health and safety opportunities, and issues present at each stage of a product's life – from development and manufacturing, through distribution and customer use, to disposal.

A Chemical Data Management System (CDMS) and other material tracking systems allow 3M to track the procurement and use of chemicals and other materials that may contain chemicals throughout the product lifecycle. The CDMS is also used to generate Material Safety Data Sheets (MSDSs), which help assure product safety.

3M continues to monitor the environmental performance of its facilities, but LCM's product focus emphasizes the business unit's role in sustainability, as well as personal accountability and leadership roles within the laboratory and marketing functions.

3M's sustainability efforts have been given additional prominence in the company's New Product Introduction framework, which relies on Design for Six Sigma (DFSS) tools to elicit customer preferences for sustainability and other product characteristics. Lifecycle management has been incorporated into each stage of the New Product Introduction framework and into the associated DFSS process.

Growth opportunities are also enhanced by a corporation's embrace of its social responsibility. First, community involvement through volunteerism, community affairs activities, and corporate giving (directly or through a foundation) has an invaluable impact on the reputation of the company and its brands. In addition, such involvement correlates strongly with employee satisfaction and retention; it also plays a role in attracting new talent. Finally, companies that are known for the contributions to the surrounding community – and their respect for the needs and preferences of their neighbors – invariably fare better when they need to appear before local bodies of government. Clearly, a company's desire to expand must be matched with the community's willingness to support such growth.

8.8.4   Sustainability and Risk Management

By focusing on sustainability, a company can anticipate and deal with potential problems before they become serious liabilities. The ability to anticipate and manage such risks is invariably rewarded by investors, customers, and employees. Conversely, companies that do not manage risks have seen their value (stock price) plummet. Indeed, the risk reduction and risk management benefits alone constitute a convincing business case for pursuing sustainable development.

Two examples help illustrate this benefit.

- *Global sourcing.* Multinational companies (and many local ones) source raw materials and manufactured goods from suppliers located around the world. For economic or cultural reasons, not all of these suppliers use products or processes that are sustainable – a situation that can compromise the ethics of purchasers and can reduce the vendor's ability to provide a consistent supply of high-quality products. Because it considers a supplier's sustainability when it begins developing business partnerships, 3M is more likely to build relationships with responsible and reliable vendors.

- *Global climate change.* Because of its attention to sustainability, 3M began to assess and respond to global climate change over six years ago. It created a cross-functional task force that inventoried the company's greenhouse gas (GHG) emissions and proposed strategies for significantly reducing them. These strategies include major investments in new abatement technologies, the inclusion of global climate change impacts in LCM reviews, and process changes to reduce GHGs. Because these strategies were developed in advance of regulation or sudden shifts in marketplace demands, 3M has been able to implement them in a deliberate, well-planned fashion that has minimized the disruption to employees, customers, and shareholders. 3M has received third-party verification of its GHG inventory, which was developed using the World Resource Institute's/World Business Council for Sustainable Development's GHG Protocol. The company plans to include additional greenhouse gas emissions data, including expected reductions, in its sustainability reporting.

Such benefits are usually the result of a systematic approach for facilities and business units to identify and manage their risks. The 3M EHS Management System, described above, provides an example of how such an approach can be applied. This system imposes a formal process for identifying risks, prioritizing them, and then addressing them through action plans. An EHS Scorecard tracks progress toward targets; senior management regularly reviews this progress and charts the course for further EHS improvements.

8.8.4.1 External Communications: Reporting and Marketing Claims. Reporting offers additional opportunities to reinforce 3M's commitment to sustainability and enhance its reputation. In the mid-1970s, 3M's reports on its environmental progress (issued as part of the 3P program) were among the first by any major corporation. Since then, the company's reporting has been explicit about its goals and its success in meeting those goals.

In 2002, the company also published its first Sustainability Report (available online at 3M.com), which documents ethical and employment policies, community involvement, and contributions to economic health, in addition to environmental programs and results. This report was prepared using the Global Reporting Initiative's (GRI) June 2000 Sustainability Reporting Guidelines. In addition, it follows the overarching principles detailed in the GRI Working Paper, April 16, 2001, "Overarching Principles for Providing Independent Assurance on Sustainability Reports."

Although 3M recognizes the competitive advantage of sustainable products and processes, the company is very careful about marketing claims that refer to such qualities or benefits. Before such claims are allowed, they must be approved by the 3M Environmental Marketing Claims Review Committee, which consists of legal, marketing, technical, and other experts. Approval is based on technical accuracy and clarity. Broad environmental claims, such as "environmentally friendly," are prohibited because they are ambiguous and impossible to document. Rigorous attention to these claims helps assure 3M customers that they can rely on their accuracy.

8.8.5 Benefits/Opportunities

The benefits to implementing sustainability practices include improved operational efficiency, increased sales growth, reduction of EHS risks, enhanced corporate/ brand reputation, and employee retention and talent attraction. Opportunities arise from new sustainable product innovation that allows access to new markets and provides a competitive advantage.

One example of this type of product innovation is the 3M Novec 1230 Fire Protection Fluid that was developed using a balanced lifecycle approach. Until the early 1990s, halon was used for critical fire protection applications (e.g., communications and electronic equipment). With the production of halons banned because of high ozone depletion potential and most of the halon replacement chemistries (primarily hydrofluorocarbons, HFCs) having high global warming potentials and problems with toxicity, 3M developed a new technology platform using 3M's LCM process as a key business decision-making tool. This novel chemistry has a short atmospheric lifetime, a global warming potential of 1 and is low in acute toxicity. The reinvention of the business model to include EHS as a driver using the LCM process has changed the basis of competition and supported 3M's fire protection leadership position.

The 3P program, aimed at eliminating pollution at the source, has improved operational efficiency through the many projects that have reduced waste and/or energy consumption and saved money (nearly $1 billion in first year savings alone). Teamed with our 3M Six Sigma initiative, the number of 3P projects has been increasing as yield improvement projects increase (as improving yield means less waste and cost savings).

The implementation of our EHS Management System has required each facility, business unit, and subsidiary to understand and address their EHS issues through an EHS planning process. This process, when integrated into the business strategic planning process, provides better management of business EHS risks and identifies new business opportunities.

Additionally, all of the business processes will only be successful if they are supported by strong corporate values and uncompromising ethics. Social responsibility and ethical business practices must be as strong as the other leg of sustainability. 3M has developed Leadership Attributes that includes living 3M values. Each employee's performance is rated based on these attributes in addition to their performance against specific objectives. It is taught within the 3M culture, our values, and business conduct to "do the right thing."

8.8.6   Challenges

A focus on sustainability brings undeniable benefits for companies, but it also presents risks and challenges that companies must manage if they hope to be successful.

The most obvious is balance. Sustainability encompasses environmental, social and economic goals that can conflict. Companies need to achieve a level of economic health – so that they can provide employment and wealth to the current generation – without compromising the opportunities of subsequent generations. During tough economic times, companies are challenged to make investments in their communities and in the technology needed to improve environmental performance. Conversely, the demand for social change and environmental improvement must be balanced against companies' needs to achieve adequate financial results to survive.

Companies face another challenge in finding the right balance between transparency and confidentiality. Individuals and organizations have an almost unquenchable thirst for data on companies' activities. While most companies acknowledge their responsibility to provide a reasonable description of their practices and progress toward environmental, social, and economic goals, they must also hold some information confidential. In some instances, the release of information on product composition or unpatented manufacturing processes would have grave economic consequences. The dilemma for companies is that they must maintain a reputation for sustainability and transparency while resisting calls for additional information.

A more subtle issue is the concern that sustainability can stifle innovation. More than most employees, researchers and product developers are intimately aware of any potentially detrimental impacts – again, environmental, social, or economic – from a new product or process. These impacts cannot be ignored, but they must not be allowed to quash a project too early, since alternatives can often be found if other technical and marketing issues can be resolved. Furthermore, even if alternatives are not found, experience teaches that potential negative impacts can often be managed so that the risk is minimized.

As real as these challenges are, they are offset by the opportunities presented by the pursuit of sustainability. Much has been made of so-called "disruptive technologies" – the advances rewrite the basis of competition and force abrupt changes on markets and the companies that participate in them. Whenever a disruptive technological change occurs, companies that have pushed the advance, and others that have the capacity for rapid adaptation, enjoy dramatic successes. Slow-moving, backward-looking companies falter.

Sustainability could be considered a "disruptive perception" that causes similar consequences. As customers and end-users (as well as regulators) become even more attuned to the desirability of products and processes that reflect the principles of sustainability, markets will change. Companies that can seize the opportunities presented by this shift – either because they are leading the change or because they can respond quickly to it – will thrive in this new environment.

8.8.7 The Business Case for Sustainability: The Bottom Line

Certainly, the world is full of companies that appear to disdain the principles of sustainability. Indeed, in some countries, these companies seem to prosper over the short term. There is no question, however, that these companies hold the seed of their own demise. They are, themselves, not sustainable.

When companies do a superior job of managing their environmental, health and safety systems and addressing their social responsibility, they enjoy a range of competitive advantages over those that do not. By embracing sustainability, these companies avoid the risks inherent in unsustainable practices. They also operate more efficiently, are more productive, and are more likely to succeed in the market. Little wonder that they have higher long-term returns and more robust prospects than other companies.

The most obvious signs of this embrace of sustainability are the systems, business processes and programs that companies institute and support. For this "superstructure" to function successfully, however, it must be firmly rooted in core values and principles of business conduct that have been articulated by management and are lived by all employees. We have seen too many examples of corporate failures in the recent past where the systems and processes were in place, but values and ethical business practices were lacking. Values and business conduct are the foundation for sustainability and long-term business success.

**8.9   ACHIEVING BUSINESS VALUE: THE INVESTMENT
COMMUNITY PERSPECTIVE ON THE IMPORTANCE OF
INCLUDING ENVIRONMENTAL AND SOCIAL ASPECTS
IN VALUATIONS**

DON REED, CFA

ECOS Corporation

8.9.1   Introduction

The mainstream investment community has made little explicit reference to sustainability issues. Despite this fact, the topic is not really new. There are plenty of examples of investors thinking about the business implications of environmental and social issues in the chemical industry.

Each year Dow Chemical brings the financial analysts who follow the company to Midland for a conference. In the past, this event has featured sessions on how Dow was better than other companies in the industry in every area including the environment. While it is not necessarily clear that analysts cared that this was the case, at least they heard the message.

To understand the view of institutional investors towards sustainability in the chemicals sector and capture a bit of the flavor, this section looks at how environmental and social issues are becoming increasingly seen as business issues by both companies and investors. Then, I address how chemical companies have value at stake on sustainability as well as some generalizations about how sustainability leaders in the industry have addressed the value of their efforts in this area. I then explore how mainstream investors are approaching these issues in the industry, beginning with the context, then going on to what tools they have to integrate an understanding of these issues into their work. Finally, I look at the emerging sustainability issues in the industry, how investors might be approaching these issues in years to come, and what opportunities for growth in the industry are based on sustainability as an insight.

8.9.2   Environmental and Social Issues as Business Issues

While the proposition that environmental and social issues are business issues still strikes some as a novel perspective, increasing numbers of business leaders and investors understand that this is the case. Social and environmental issues can be seen as market forces. The following framework, shown in Figure 8.28, helps explain this in the same terms an analyst would use to understand any market force and how it affects the prospects of a business.

Like other market forces, the business impact of social and environmental market forces is either increased or decreased by other trends in society and business. Over the past few decades, a familiar set of trends (e.g., globalization, connectedness, rise of civil society, and so on) have come together in such a way as to significantly magnify the business impact of social and environmental market forces. Essentially,

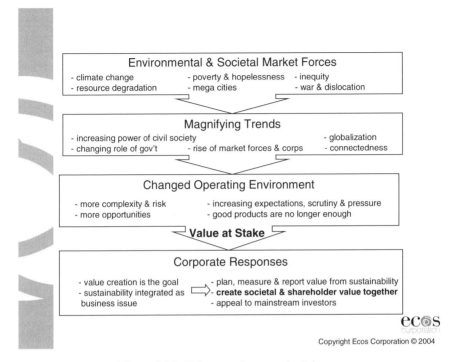

Figure 8.28. Value at stake on societal forces.

there are more people with greater expectations and much more information about corporate actions affecting society and the environment, with much greater ability to organize globally to act on those issues in their roles as voters, citizens, employees, investors, and consumers.

These magnified market forces have changed and continue to change the operating environment for companies. Most good business leaders understand this, whether or not they can put their finger on the causes. They know that many aspects of their business are subject to tipping points in public attitudes and that there is no option to mask poor performance against societal expectations with good public relations. All this makes business more complex. As these market forces have changed the operating environment for companies, one can fairly say that businesses have value at stake on social and environmental issues.

Good business strategy requires responses to changes in the operating environment. That is clearly the case here. While the corporate strategic responses vary by industry and company, our experience is that the successful ones have some common attributes. Rather than focusing on improving social and environmental performance, they focus on creating shareholder value and societal value together. Value creation remains the driving goal. They serve to integrate sustainability issues into business processes rather than concentrate responsibility for them in separate functional groups.

Because these strategic responses focus on delivering shareholder value, they have the potential to appeal to mainstream investors. From the investors' standpoint, the impact of this framework is potentially profound. Most mainstream investors have historically thought of social and environmental issues as being nonfinancial, and therefore outside their scope of responsibility. In fact, their fiduciary duty actually required them to ignore these issues. However, if social and environmental issues can have a material impact on financial performance and business strategy, then investors have a fiduciary duty to examine those issues, understand how the company has value at stake and how their strategy will affect value. In one giant change, the consideration of sustainability issues goes from being prohibited to required.

There are clear signs that exactly this process is happening. For example, Hermes is a large asset manager in the United Kingdom that started out managing the pension fund assets of BT (then British Telecom). Today, Hermes has £44 billion under management. Hermes has recently required the leaders of firms held in portfolio to explain how their companies have "value at stake" on SEE (social, environmental and ethical issues) (Monks, 2001).

This concept of having "value at stake" on social and environmental issues is broad and more powerful than the more traditional view that the main financial import of environmental issues is as potential liabilities. There have been significant efforts made to improve corporate disclosure of material environmental liabilities,[25] and the increased emphasis on the identification and disclosure of risks under the Sarbanes–Oxley Act should drive this further. Value at stake includes these potential environmental liabilities as well as the opportunities and competitive advantage that can come from the right environmental strategies.

8.9.3　"Value at Stake": Environmental and Social Issues in the Chemical Industry

So how does this look in the chemical industry? More so than in many other industries, investors generally accept that social and environmental issues can be financially significant in the chemical industry, but within traditional parameters. Most chemical industry analysts lack training in how to analyze the financial impact of social and environmental considerations in the industry. It is not an ordinary or systematic part of what they consider. There are no accepted standards for when these issues are and are not important and there is no accepted means of integrating and understanding social and environmental issues into financial analysis.

Analysts typically operate from specific examples that demonstrate the case they believe to be happening. They were not particularly surprised that Union Carbide was terminally damaged by the Bhopal tragedy, but they do not have a way of analyzing all the other instances in which a chemical company may have its right to operate impinged despite having complied with the letter of the law. Likewise,

---

[25]See the work of Corporate Sunshine Working Group (http://www.corporatesunshine.org/) and Michelle Chan-Fishel (http://www.environmental-finance.com/2003/0309sep/disclsre.htm).

analysts may understand that a specific emission regulation is coming down the track and that it will affect the industry, but they typically do not have the capacity to analyze how companies are differently positioned for the regulation based on the details of their environmental capabilities.

Empirical research is one of the ways to understand how investors view how the companies have value at stake on sustainability issues. Here, I will briefly discuss two types of research that are directly applicable to the chemical industry.[26]

A number of research papers have shown how investors react negatively to the disclosure of data from the Toxics Release Inventory when a company is a leading emitter (Konar and Cohen, 1997). This is not particularly surprising; investors typically react negatively to news events they believe have some dark connotation. Environmental news is no different from other types of news. While it is true that investors dislike bad environmental news, this research does not show how long this effect lasts.

While there are lots of general empirical studies about the statistical relationship between environmental and financial performance, most of this research looks across all the companies in the market rather than focusing on specific industries. Most financial researchers and portfolio managers evaluate companies relative to their industry peers and the major trends in the industry. One of the few bodies of research to examine the specific issues involved in financial analysis of environmental strategies in the chemical industry was done by Pamela Cohen Kalafut and Jonathan Low at the now defunct Ernst & Young Center for Entrepreneurship (see Chapter 6, Section 6.2.2). Karina Funk describes this work, which ranks the different drivers of shareholder value in the global chemical sector.[27] Of eight drivers tested, Social Responsibility/Environment was the fourth behind Innovation, Alliances, and Leadership, all of which were highly statistically correlated to the market value of the companies in the industry.

Taken together, this combination of intuition, examples, and research build a clear case that the chemical industry has material value at stake from environmental issues.

8.9.4 Examples of How Sustainability Leaders in the Chemical Industry View Value

It would be presumptuous to attempt to describe the many different ways in which chemical companies have created significant shareholder value through environmental strategies.[28] My purpose here is to offer a few examples that illustrate the

[26]There are a number of summaries of this research. For example, see http://www.figsnet.org, Dinah Koehler, "Capital Markets and Corporate Environmental Performance: Research in the United States," http://opim.wharton.upenn.edu/risk/downloads/03-36-DK.pdf, and Donald Reed, "Green Shareholder Value, Hype or Hit?" World Resources Institute, 1998. http://www.getf.org/file/toolmanager/O16F3346.pdf

[27]"Sustainability and Performance," *MIT Sloan Management Review*, Winter 2003 and C. B. Cobb, "Measuring What Matters: Value Creation Indices in the Utilities and Chemical Industries," CBEY Center for Business Innovation.

[28]For more on approaches to measuring the shareholder value created by corporate sustainability strategies, see Reed (2001).

potential and the breadth of approaches used at DuPont and Dow, two companies that have often appeared at the top of sustainability ratings of the chemical industry.

DuPont has a long-standing reputation for leadership on environmental issues. One of the hallmarks of that has been the degree to which the company has integrated the concept of creating value through sustainability into key business systems.

Perhaps the best example of this is the widespread use of a metric that relates value and environmental footprint, best known as shareholder value added per pound of material produced (a suitable proxy for environmental footprint in many bulk chemical businesses).[29] Perhaps the most powerful aspect of DuPont's use of this metric is the range of applications to which it has applied the metric. These include:

- Measuring overall progress at producing more value per unit of environmental footprint over time in a particular business and across the company;
- Comparing businesses based on their relative value per footprint as one criteria for additional investment; and
- Applying the metric to understanding "knowledge intensity" in determining the best long-term portfolio of businesses.

The on-the-ground impact of the use of this metric has been to drive the idea and practice that many production processes can be improved to reduce environmental impact while improving profitability. A great deal of the value DuPont has created through environmental insight has been by profitably reducing the negative environmental footprint.

Despite the important role this metric has played, DuPont does not routinely assess the financial value of sustainability initiatives as an ordinary part of making operating or product development decisions.

While profitably reducing environmental footprint has become a systematic source of value, DuPont has achieved less success applying the concept of sustainable growth to the innovation and new product development processes. DuPont does have examples of new products for which a major portion of the customer value results from the elimination of environmental problems for the customer. Examples of this include:

- Cyrel FAST, an offering that eliminates the need for solvents in the process of making the plates used in certain printing processes. Because solvents require particular handling, specialty "job shops" or printers primarily owned the earlier equipment and produced the plates. With the new equipment and no need to handle solvents, graphic design businesses and others became a whole new market for the system;

---

[29]For a more detailed treatment of DuPont's use of this metric, see Holliday (2001).

- Solution Dyed Nylon (now a product of Invista, which DuPont sold to Koch Industries in 2004). Historically, DuPont's nylon customers purchased raw product and dyed it themselves. More recently, DuPont has offered these customers nylon that is dyed to their specifications in the manufacturing process. This has a number of benefits to the customer including the steadfastness and consistency of the colors. For the environment, this process means more of the overall dying process happens in a centralized location with high environmental controls. From the DuPont business perspective, this adds more value to their customers and raises the cost of switching to another material supplier.

While these examples indicate the potential, DuPont has yet to make such new offerings developed around environmental benefits a systematic product of technology-led innovation rather than happy byproducts of other innovations.

Dow has a long history of viewing social and environmental issues as a key part of their business. The company has also carried out specific work on how to embed value creation through sustainability into key business decision-making.

Among Dow's best-known efforts is its work with stakeholders on the value to be obtained from process changes that would reduce local emissions. This work involved collaborating with local community groups near its facilities, a national environmental group, Natural Resource Defense Council (NRDC), and engineering consultants to explore specific process changes.[30] The process yielded several projects that met the criteria of both the community and the company.

Dow has also had some success using environmental benefits as a source of innovation in new product development. These include:

- Stalk Board, an alternative to wood fiber board that is made from wheat stalk and does not use formaldehyde;
- Dow's joint venture with Cargill produces PLA, a biodegradable substitute for plastic that is based on a renewable crop rather than a fossil fuel.

While these innovations are worthy of note, it is not clear that Dow has developed the ability to systematically apply an insight into social and environmental problems in developing new products.

Some argue that this lack of a systematic approach to applying sustainability insights to business decision-making was evident in Dow's acquisition of Union Carbide. It appears that Dow took a legalistic view of the risks associated with acquiring the company closely associated with the 1984 tragedy in Bhopal, India. While it is true that Union Carbide reached settlements with strictly limited ongoing legal liability, this did not change the perception held by many that the company did not make those who suffered whole. In a report prepared on behalf of a group of socially responsible investors, Innovest made the case that the risks associated

---

[30]For more on this collaboration, see http://www.nrdc.org/water/pollution/ndow.asp and http://www.nrdc.org/water/pollution/msri/msriinx.asp

with Bhopal and a host of other developments at Dow justified downgrading the company's environmental rating significantly (Innovest, 2004).

Overall, the chemical industry has approached sustainability as a way to preserve their right to operate in the face of significant environmental challenges. That is obviously an important source of value. While there are a number of examples of applying sustainability to improve margins, there are fewer examples of applying the concept to drive revenue growth.

The majority of sustainability investments to date in the chemical industry have played the role of insurance to guarantee current right to operate until legacy issues pass their statute of limitations (if there is such a thing with such legacies). Had the chemical companies not acted to preserve their right to operate, the value lost could have been enormous, even though we do not normally think of the value of a dodged bullet.

8.9.5 The Prevailing View of Mainstream Investors on Sustainability in the Chemical Industry

While it is fairly hard to generalize, it is safe to say that most mainstream investors in the United States have not focused on these issues as being significant drivers of value overall. While there are a few examples to the contrary around specific legislative initiatives, they tend to confirm the general impression.

Most financial analysts generally accept that environmental regulations may matter a great deal at specific times in some industries, but that is fairly different from the notion that companies typically have some value at stake on social and environmental issues.

Part of the issue is one of communication from the companies themselves. Analysts and investors depend a great deal on communications from the company to know what issues are financially material. In a public forum hosted by the New York Society of Security Analysts, a chemical industry sell-side analyst said, "We rely on the company to tell us when these issues are material."

The other attitude found frequently among financial analysts is the belief that voluntary programs have effectively reduced many of the risks associated with social and environmental issues in the chemical industry. There are obviously ways in which this is true of programs such as Responsible Care®, but it is by no means true of all the risks and opportunities of all chemical companies at all times.

The overall financial context of the chemical industry gives investors plenty of other issues upon which to focus. The chemical industry is a capital intensive, cyclical business with mature products typically sold into commodity markets. It often does not earn its cost of capital (Lev, 2001a). This has led to a number of sales of mature assets in the industry to financial buyers (Firn, 2004). Overall, cost cutting reigns as the principle financial strategy while the capacity for innovation and growth is modest.

There are early signs that mainstream financial analysts are beginning to develop their own understanding of how to approach social and environmental issues. A number of mainstream investment banks have published investment research for their institutional clients on the implications of climate change. UBS and Dresdner

Klienwort Wasserstein both published reports on climate change and the Pan-European utility sector. WestLB has published a report on the scope of financial value at risk in the economy overall and by specific industries. UBS, Credit Suisse First Boston, and Citigroup have published reports on the implications of carbon trading. Goldman Sachs has published an environmental and social index for the global oil and gas industry.[31]

There are also indications of growing demand in the United States for institutional asset management services that integrate social and environmental issues into asset management. Several prominent leaders of major public pension funds in the United States have publicly expressed concern that global warming posed long-term economic risks that threatened the value of retirement funds. These state officials have formed the Investors Network on Climate Risk around an action plan calling for action by the Securities and Exchange Commission and companies on disclosure of climate change risks and of investors to consider climate change in their decision-making and proxy voting.[32]

Early actions on this agenda include:

- CalPERS making initial allocations to invest $200 million in cleantech venture capital and $500 million in publicly traded equities that use environmental criteria in the asset management;
- These investors and others sponsored and supported a number of proxy resolutions on disclosure of climate change risk, some of which resulted in direct votes of shareholders while others led to negotiated agreements to meet the general terms of the resolution.

8.9.6 Sustainability Research, Ratings, and Indexes

The fact that ratings of corporate sustainability performance exist and are marketed to investors has been one of the most public manifestations of action where sustainability issues meet capital markets. Obviously, there is an appetite for this research and for investment products that are based on it. Approximately three billion euros are invested based on the Dow Jones Sustainability Index.[33]

There are, however, relatively few signs that this has moved from being a niche activity to being part of the mainstream world of investment. For example, only three of 35 SRI research organizations evaluated by SustainAbility in *Values for Money* are found to "currently analyze the link between social/environmental issues and material impacts on investment value drivers" (Beloe *et al.*, 2004). These are SAM Research, Innovest, and CoreRatings. These have the greatest potential to cross over into more widespread use among mainstream investors.

---

[31]See http://unepfi.net/stocks/oilandgas_gsachs.pdf.
[32]See http://www.incr.com/.
[33]See http://www.sustainability-indexes.com/djsi_pdf/news/PressReleases/DJSI_PressRelease_040902_Review.pdf. Baruch Lev, "Grey Matters: CFO's Third Annual Knowledge Capital Scorecard," *CFO Magazine*, April 2001b.

Mainstream investors face big challenges in trying to apply research about corporate sustainability into their existing investment process. They can use the indexes or the ratings themselves as a part of the quantitative investment approach, but most investors use more traditional methods of stock picking and it is not obvious how to apply sustainability research to that process.

One of the leading solutions to this problem is to think of the sustainability capability of companies as being one of several "nonfinancial" factors or intangible assets of the company. The whole area of valuing intangible assets is a rapidly emerging issue in investing overall. There is a high degree of acceptance that these assets are a huge and growing portion of the market value of companies, but there is little agreement about how to value these assets. Forty to fifty percent of market capitalization of the S&P 500 is from intangible assets and the figure is much higher for some companies (Lev, 2001b).

One of the proposed approaches to the general problem of valuing intangible assets includes explicit reference to social and environmental performance. Jonathan Low, Pamela Cohen Kalafut and others formerly with Cap Gemini Ernst & Young's, Center for Business Innovation have pioneered a means of measuring the financial value of intangibles such as brand equity, leadership, reputation, and innovation. Their Value Creation Index includes Environment among these value drivers.[34] While this topic does not consistently rate among the top intangible assets in all industries in its effect on market value, it is significant enough to include in the model and is statistically significant in the chemical industry. Other aspects of sustainability also contribute to the value of intangible assets. For example, diversity and employee relations are key components of human capital and environmental performance is a key part of brand equity.[35]

Overall, these various attempts to measure sustainability performance and relate it to shareholder value reflect the understanding that the ability to handle the complexity of sustainability reflects good management and that good management performs financially.

Sustainability research and ratings remain a niche market, but they have made some significant progress in capturing initial sales, including some very demanding mainstream investors. Several have also successfully forged alliances with big-time mainstream institutions, such as SAM's work with Dow Jones on the Dow Jones Sustainability Index and ABP's ownership interest in Innovest after having become a customer. Perhaps most importantly, several of these research and rating providers have successfully made the transition from merely gathering what data happened to be available to seeking out the right information to answer the right questions about how these factors do or do not connect to value.

While the strides forward for corporate sustainability research have been impressive, the reality remains that only a small minority of institutional investors use this research. There is little consensus about how this analysis should be

---

[34]See www.invisibleadvantage.com and http://www.gemi.org/docs/conf2002/graphics/gemi2002/Session8/GEMI_JonLow.ppt.

[35]See *Clear Advantage: Building Shareholder Value*, GEMI, 2004, available at http://www.gemi.org/GEMI%20Clear%20Advantage.pdf.

done, with at least two primary differences. The first is between using standardized questionnaires to gather information, which has resulted in significant "questionnaire fatigue," and relying on analyst contact with the company. The second goes straight to the heart of what is in fact being measured. We have moved beyond just gathering publicly available data on measures of environmental performance (e.g., TRI), but some ratings put the emphasis on the existence of the right sustainability policies while others explicitly seek to make a judgment about ability of management to handle risks related to social and environmental issues.

8.9.7 Key Issues for Analysts of the Chemical Industry Now and in the Future

From an analysis of the published financial research on the chemical industry, the sustainability issues of concern in the mainstream investment world revolve first around exposure to "hot-button" issues such as asbestos and genetic modification. There are some indications that this concern may extend to issues that have not yet fully blossomed, such as endocrine disruptors. There is also lingering concern about the potential for European Union rules that, if adopted, would essentially shift the burden of proof from showing that a chemical caused harm to having to demonstrate the safety of a number of chemicals in current use. Among those analyzing the sustainability of the chemical industry, the primary task appears to be to evaluate the companies' ability to deal with complex issues of safety, emission reductions, community relations, and so on.

Looking around the corner, there are a number of sustainability issues that could be financially material in the chemical sector. The research on "body burden," or what trace (and otherwise) substances exist in humans as a result of environmental exposure is perhaps the most troubling in that little is known about how this occurs, especially among populations that are not in proximity to known sources. Look for the level of concern on this issue to rise and for potential societal and policy pushback fuelled by the uncertainty. "Body burden" could serve as a gateway issue to a wide range of actions by civil society on a whole host of specific substances and pathways of exposure. In particular, results that pertain to "body burden" in children have a huge emotional resonance beyond the scope of most issues of "pollution."

If the EU were to adopt the precautionary principle versus demonstrated harm as the standard for regulating the chemical industry, there would be natural pressure for this to spread to other regions, with potential material financial impact on the industry. Even if the EU proposal is watered down or rejected, this issue is not going away. The trend is clearly for activists to force a debate about the safety of chemicals currently in use and there are a number of different mechanisms through which that could take place.

8.9.8 What the Future May Hold for Investors

What does the long term hold for investors in the chemical industry? Given the clear nature of how the industry has value at stake on social and environmental issues and the fiduciary duty of those acting on behalf of most institutional investors to explore all issues that materially affect value, it is not difficult to imagine a setting in which a thorough examination of these issues was ordinary investment practice. This analysis would take the form of integrating social and environmental issues into existing financial analysis as a normal part of discharging one's fiduciary duty. The precursors of this are already observable among investors such as Hermes, mentioned above.

The climate change actions led by some large public pension funds and state treasuries in the United States could well be the beginning of a sea change. If their actions are successful financially, this approach could expand to a much larger universe of investors across a more expansive range of sustainability issues.

This could fuel a rapid expansion of the product offerings based on integrating social and environmental issues into investment decisions. As noted above, this is already a growing segment, but off a relatively small base. Overall, several of the "best-in-class" funds have sufficiently long track records of good performance to remove some of the standard barriers to attracting institutional investment.

There is also a maturation of the analytical techniques investors can use to integrate sustainability considerations into their established investment practices. While there are many possible alternatives, the work of the World Resources Institute on a technique built around conventional tools of financial analysis is quite promising (Repetto and Austin, 2000; Austin and Sauer, 2002; Austin et al., 2003).

It is instructive to note related changes elsewhere in the financial services sector that appear to confirm the trends in the investment industry. The Equator Principles in commercial banking are the first industry-wide effort to establish standards for how to apply an understanding of social and environmental issues to providing credit. Likewise, action by some leaders in the reinsurance industry such as Swiss Re and Munich Re on the risks associated with climate change may well foretell a broader pattern in considering these issues in insurance underwriting.

8.9.9   What Could Be Next for Chemical Companies

The next big horizon for creating value through sustainability in the chemical industry is systematically using sustainability as a value-creating insight in new product development. The chemical industry has the potential to lead the way to solutions to a number of environmental and social dilemmas. Among the innovations in the industry with the ability to add substantial societal value are:

- Nanotechnology materials and catalysts;
- Coatings without solvents;
- Advanced membranes;
- Electro-conductive chemicals.

As an industry, there is a large potential to create additional shareholder value by strategically engaging stakeholders and forming partnerships with commercial and noncommercial partners who have the ability to drive the demand for sustainability attributes and reduce risks.

While the industry has accomplished a great deal by using sustainability as an insight to improve operations by driving out costs and reducing risks, there is certainly room to create more shareholder value through this avenue. To effectively unlock this, companies will most likely need to develop the capacity to systematically measure the value created by these types of sustainability efforts to put profitable footprint reduction on the same sort of routine platform as quality control through Six Sigma.

To truly unlock the potential value to shareholders, chemical companies will not only have to expand their ability to create value through sustainability, they will also need to measure it at the corporate level. This is a necessary precursor to being able to explain to the investment community how sustainability is creating value. The end game would be to be able to treat these efforts as they would other strategic initiatives by explaining how they will create value, how they will measure that value, and what results to expect in the future.

8.10 INVESTMENT ANALYSIS: DOW AND BHOPAL, INDIA

Marc Brammer
Innovest Strategic Value Advisors

[Editor's note: The following Investment Analysis of Dow by Innovest Strategic Value Advisors is intended only as an illustration of potential investment community concerns arising from decisions that fall within the CSR/sustainability realm. While Dow Chemical is the subject of the analysis, it is not the intention of the editors to recognize all of the positive developments that have been undertaken by Dow and that are referenced elsewhere in this book.]

> *Companies that don't meet their responsibilities to all their constituencies will have a difficult time. Responsible customers won't want to buy their products ... Enlightened communities won't want them as neighbors, and wise investors won't entrust them with their economic futures.*
> —Dow's Chairman and CEO, William Stavropoulos, The Business of Business Managing Corporate Social Responsibility: What Business Leaders are Saying and
> Doing 2002–2007

This chapter details the effects of the Bhopal, India, chemical disaster on Dow Chemical, the world's largest commodity chemicals business, after it merged with the Union Carbide Corporation (UCC) in 2001. In the late 1990s, Dow was beginning to gain a reputation for environmental excellence and innovation, both in facilities management and product development. However, the merger with Union Carbide brought with it a number of high-profile liabilities such as asbestos cases and the Bhopal disaster. Through the addition of UCC's portfolio of products, the merger entrenched the company's commitment to developing organochlorines, a dangerous class of chemicals. In addition, many long-term issues such as Agent

Orange manufacturing and dioxin contamination at the company's main plant in Midland, MI, began to re-emerge as major causes of concern. Dow's handling of the Bhopal issue after the UCC merger provides a good case study of corporate responsibility in the age of globalization and mega-mergers.

8.10.1   The Meaning of Bhopal for Investors

The 1984 Bhopal disaster, involving a catastrophic failure at a Union Carbide pesticide manufacturing plant, represents a serious problem for Dow with the potential to damage the company's reputation. Over 14,000 deaths, and 50,000+ permanent injuries have been attributed to the event and its aftermath by Indian government officials. It has been the source of ongoing legal battles in both India and the United States. This case study covers the existing and potential financial and reputational impacts for Dow related to the Bhopal disaster, as well as a general overview of the ongoing events and controversies.

While Dow claims that it has no responsibility for the incident and that it is a tragic event of the past, investors should be concerned about current developments and the possible financial ramifications that could result. Key issues include:

- *Union Carbide's status in India.* Dow's subsidiary, Union Carbide, has been listed as an "absconder from justice" by the Chief Bhopal Magistrate for failing to appear before the court on criminal charges relating to the disaster. Likewise, an extradition order has been issued by the Indian government[36] for Warren Anderson, former CEO of Union Carbide at the time of the accident, so that he may face criminal charges as well. Efforts are under way to summon Dow to deliver Union Carbide to appear in the criminal case and to have Dow clean up the Bhopal site (Singh, 2002). While Union Carbide no longer has a presence in India, Dow does. Its current holdings[37] registered in India[38] include:
  - Dow Chemical (India) Holdings Private Limited held by Dow Chemical Pacific;
  - Dow Chemical International Pvt. Ltd.;

---

[36]*The Indian Express* (July 16, 2003), "Government Finally Acts on Anderson's Extradition;" *The Hindu* (March 15, 2003). "Centre Blamed for Delay in Anderson's Extradition;" *Frontline Magazine* (August 2–15, 2003). "Request for Anderson's Extradition," Vol. 20, Issue 16.

[37]When asked about the size of its holdings in India, the company stated that the above companies were "minor holdings" and that it did not disclose information regarding the valuation of individual subsidiaries.

[38]Dow Chemical India Holdings Pvt. Ltd. (Registration #11-113550). Registered in Mumbai, Maharashtra; Dow Chemical Intl. Pvt. Ltd. (Registration #11-1135512) Registered in Mumbai, Maharashtra; Dow Polymers Pvt. Ltd. (Registration #04-27447) Registered in Ahmedabad, Gujarat; DE Nocil Crop Protection Ltd. (Registration #11-83566) Registered in Mumbai, Maharashtra; Dow Chemical International Ltd., USA (Registration #833). Registered in New Delhi; Dow Corning India P Ltd; Dow Corning Singapore Pte Ltd, Singapore (Registration #1299) Registered in New Delhi; Dow Elanco B.V., Netherlands (Registration #1236) Registered in New Delhi; Anabond Essex India Pvt Ltd. Registered in Chennai, Tamilnadu.; NEW DELHI: Registrar of Companies, 2nd Floor, B Block, Paryavaran Bhavan CGO Complex, Lodi Road, New Delhi 110 003; MUMBAI: Registrar of Companies, Maharashtra, Everest 100, Marine Drive, Mumbai 400 002, Tel: +91 22 22812639 fax: +91 22 22811977.

- Anabond Essex India Private Limited held by the Mortell Company (50 percent);
- Dow Polymers Pvt. Ltd;
- DE Nocil Crop Protection Ltd (Joint venture);
- Dow Corning India P Ltd Dow Corning Singapore Pvt. Ltd, Singapore.

- *India's increased economic importance.* India is a growing player in the globalized chemical industry and opportunity costs could be sizable for Dow if it is constrained in such a large and growing market by the "substantial legal risk" that Dow says exists regarding Bhopal. The Indian economy was projected to grow by 7.2 percent in 2004.[39] It boasts the world's 12th-largest chemical industry in terms of production, which is valued at Rs 1200 Billion ($26 bn) and has been growing at twice the rate of Asia's overall chemical market since 1998, and in the late 1990s India became a net exporter of chemicals.[40] Agribusiness is an important consumer of Dow's products and India's agricultural sector is projected to grow 13.8 percent in 2004 (Shah, 2003). Dow currently reports that it has 12 percent of revenues from Asia, while only 2 percent of fixed assets there (Fig. 8.29). This implies that for Dow to increase its market share in India it may have to acquire more fixed assets in the country and therefore increase its exposure to potential liabilities from Bhopal.

- *Bhopal as a long-term issue.* Many serious points of contention remain between the survivors and Dow despite the funds from the civil case ($470 million). An estimated Rs 1500 crores (approx. $340 million) remains undistributed in the Bhopal compensation fund.[41] The balance of the amount remaining in the fund is committed to compensation of victims, and cannot be used for the many other needs of the community, such as the public health and economic impacts resulting from the disaster, or for remediation of the contamination left behind by Union Carbide. Ongoing issues include remediation of water pollution alleged to be from the site, cleanup of the site itself,[42] and further social support for widows, orphans, and the many thousands who are unable to work as a result of injuries. Given this, it seems likely that legal challenges and controversies will continue regardless of the outcomes of individual cases.

- *Potential for divestment.* Bhopal creates the real potential for escalating exclusions against Dow Chemical by money managers who run portfolios screened for social responsibility. In 2003 a study showed that there is now over $2.18 trillion under SRI management.[43] Corporate directors can no longer ignore this level of investment. The issue is gaining a higher profile and will likely begin to affect investment decisions among this expanding group of investors.

---

[39]*Indian Business Insight* (January, 2004) "Sustained Growth: Indian Economy to Post 7.2% Growth in 2003–2004."

[40]*Indian Business Insight* (November 2003) "The Chemical Industry: An Indian Perspective," vol. 38; issue 11, pp. 44, 46.

[41]*The Hindu* (March 22, 2004) "Bhopal Tragedy: Payment Sought from Compensation Fund Balance."

[42]*Indian Express* (October 20, 2002) "MP [Madhya Pradesh] Wants Dow to Clean Up Mess."

[43]Social Investment Forum.

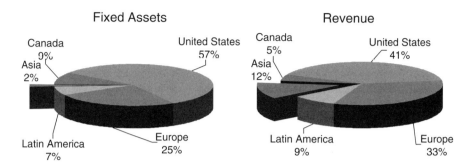

Figure 8.29. Breakdown of Dow's fixed assets and revenue by region. (*Source*: William S. Stravropoulos, Chairman and CEO, Dow Chemical, Presentation to Morgan Stanley Conference, February 24, 2004.)

- *Regulatory trends.* Recent events in India involving pesticides (including Dow product: chlorpyrifos) in soft drinks has led to demands for greater regulatory oversight of water quality and other aspects of environmental quality. India's economic growth and combined with renewed concern over pollution may provide the necessary funds and motivation to institute more effective environmental cleanup and monitoring laws, which would affect both the status of the Bhopal site and Dow's operations in general.

For investors to put Bhopal into perspective it is important to recognize one important factor – Bhopal is not just a synonym for industrial disaster, it is a leading test case for what is alleged to be wrong with the lack of corporate social accountability in the globalized economy. The actions of Union Carbide, and now Dow, are perceived worldwide as those of foreign investors more concerned about damage control than about the hundreds of thousands of human lives their operations have destroyed.

When Dow acquired Union Carbide's shares in February of 2001, it stated that "the company conducted an exhaustive assessment to ensure that there was absolutely *no outstanding liability* in relation to Bhopal. There was none; the company that Dow acquired retained absolutely no responsibility for either the tragedy or for the Bhopal site."[44]

However, Dow states in its 2002 Global Public Report:

> We respect that for some people, responsibility for the Bhopal tragedy continues to be an unresolved issue. This doesn't change the facts that the Government of India, through the Settlement agreement, has full authority and responsibility over issues arising from the tragedy and that, upon acquiring Union Carbide, Dow inherited no responsibility. Still, some people would have us take action to resolve their concerns. *But, we*

---

[44]Statement of The Dow Chemical Company Regarding the Bhopal Tragedy, available at http://www.dow.com/environment/debate/d15.html.

are aware of potentially significant legal risks associated with such actions and we will not compromise our obligation to protect our shareholder interests. (Author emphasis.)

The management's reporting of the Bhopal case presents a number of apparent omissions, inaccuracies, and contradictions that are misleading to investors. It implies that resolving outstanding concerns would raise significant legal risks. This amplifies the reality that the management is aware of outstanding risks related to Bhopal that are not being reported to shareholders, such as the fact that Union Carbide is considered an "absconder from justice" by the Bhopal magistrate.

8.10.2   The Bhopal Disaster

On the night of December 2–3, 1984, water leaked into a tank containing methyl isocyanate (MIC) at the Bhopal, India, pesticide plant of Union Carbide India, Ltd. The resulting runaway reaction vaporized an estimated 30 to 40 tons of MIC, releasing a massive toxic cloud of methyl isocyanate gas.[45] The cloud drifted across Bhopal, a city of some 900,000 residents at that time. The estimates of casualties vary somewhat depending on the source. However, the dense slums that sprawled right up to the wall surrounding the plant meant that the Bhopal accident would have the highest death and injury rate of any industrial disaster yet recorded. The initial death toll was officially placed at 2500, but other estimates based on the sale of shrouds and cremation wood begin at 7000 (Table 8.7) (Mukerjee, 1995). The company that owned and ran the site, Union Carbide India Ltd., was a closely held company (50.9 percent) owned by the U.S. parent company, Union Carbide.

Sixteen years later, in February 2001, Dow Chemical acquired Union Carbide as a wholly owned subsidiary, purchasing 100 percent of its stock and consolidating its balance sheet. Dow Chemical management has gone to lengths to say that when it acquired Union Carbide it thoroughly investigated the matter and did not acquire with it any remaining liabilities with the stock purchase of the company. At the 2003 annual shareholder meeting, for instance, Dow Chemical CEO William S. Stavropoulos described Bhopal as a tragic bygone that is all but resolved in the courts, and for which the company is helpless to take any actions.

As a result of the acquisition, however, Dow has become an international target of protest and media scrutiny. The management's public denials of liability and responsibility are fueling the protest movement. Those denials may also prove to be misleading to investors in light of ongoing efforts to hold Union Carbide liable

---

[45]In addition, some reports suggest that hydrogen cyanide was also present. When methyl isocyanate is pyrolyzed at temperatures above 427°C, hydrogen cyanide is formed as a breakdown product. Union Carbide denied that cyanide gas was present, likely because the health effects of cyanide gas are more well known than methyl isocyanate and the company wished to reduce any potential liability. Doctors treating victims found that treatments used for addressing cyanide exposure were effective and eyewitnesses reported the small of "rotten almonds." Cyanide has an almond-like odor. See Lueck (1984), Ashford (1984), Mukerjee (1995), and, for a description of cyanide, see Center for Disease Control, Agency for Toxic Substances and Disease Registry; ToxFAQs[TM] for Cyanide (Cianuro) September 1997; CAS# 74-90-8, 143-33-9, 151-50-8, 592-01-8, 544-92-3, 506-61-6, 460-19-5, 506-77-4.

TABLE 8.7. Bhopal Disaster Casualty Figures: Multiple Estimates

	Initial Deaths	Initial Injuries	Permanent Disabilities	Long-term Mortality
Indian Government – ICMR	2500+		50,000	14,400+
Congressional Research Service	2000+	100,000	50,000	
Claims filed with Indian Government			600,000	16,000
NGO estimates	7000–8000+			15,000–20,000

(*Sources*: Indian Center for Medical Research – ICMR, the U.S. Congressional Research Service, Lapierre D, Moro J., 2001[46])

for criminal charges and environmental remediation and to link Dow with UCC in the Indian courts.

8.10.3   History of the Bhopal Case

In May of 1982, Union Carbide inspected its pesticide plant in Bhopal, India, and found 10 major deficiencies (Lueck, 1984; Ashford, 1984). The facility had been losing money for two years already and in the four years before the accident it lost a total of $4.5 million and the company was looking to sell it (Lueck, 1984; Ashford, 1984).

Charges in the pending criminal litigation assert that Union Carbide was guilty of "culpable homicide not amounting to murder." Many issues of fault have been alleged, some of which place blame with the parent corporation rather than its Indian subsidiary. It is alleged that dangers found in safety audits of the Bhopal plant were not remedied, even though remedial measures for similar dangers were undertaken at Union Carbide's MIC plant in West Virginia.[47] According to allegations made in various arenas regarding the events of December 2, 1984, water (that was being used for washing the lines) entered the tank containing MIC through leaking valves. The refrigeration unit, which should have kept the MIC close to zero degrees centigrade, was shut off by company officials to save on electricity bills.[48] The entrance of water to the tank, full of MIC at ambient temperature, triggered off

[46]Bhopal People's Health and Documentation clinic for the Sambhavana Trust; The Bhopal Gas Tragedy: 1984; New Dehli, India: Bhopal People's Health and Documentation Clinic for the Sambhavana Trust, 1998 November.

[47]The *Wall Street Journal* reported that several weeks before the Bhopal disaster, Union Carbide changed procedures at its Institute West VA plant, on which the Bhopal plant design was based, to prevent a similar incident. Rep. Henry Waxman (D. CA.) disclosed that a Union Carbide safety team in September had warned of a potential "runaway reaction" of methyl isocyanate at Institute and the EPA had reported that there were 28 small leaks of the chemical in 1980 at the facility (Winslow, 1985).

[48]The *New York Times* (January 27, 1985) conducted an extensive investigation involving more than 100 interviews in India and the United States about the gas leak. Its findings indicated that the plant design was from Union Carbide in the United States and that many of the safety features of the plant were inoperable at the time of the accident, in some cases to save on costs ("Design Flaws Contribute to Gas Leaks").

an exothermic runaway reaction and consequently the release of the lethal gas mixture. The safety systems, which, in any event were not designed adequately to control such a runaway situation, were awaiting repair. To avoid having the community be unduly alarmed due to occasional leaks, the siren in the factory had been switched off (Morehouse and Subramaniam, 1986; Everest, 1985).

In addition, there are allegations that the parent company had underinvested in the technologies at the plant. According to representatives of Bhopal survivors, documents obtained through discovery in cases filed against Union Carbide show that in the early 1970s, in response to the Indian government's efforts to promote import substitution through the dilution of foreign equity (The Foreign Exchange Regulatory Act – FERA), the company reduced its investment in the Bhopal plant from $28 million to $20.6 million. Despite reducing the total cost of investment in line with FERA, the company retained its 51 percent share in the UCIL subsidiary by back-integrating the equity formulation for the plant. Under FERA, this reformulation required the transfer of additional technology not currently available in India.[49] In order to prevent the dilution of its ownership, Union Carbide transferred substandard technology, which had only a limited test run because it lowered costs in addition to meeting, on paper at least, the requirements of FERA.

According to Himanshu Rajan Sharma – the attorney representing the Bhopal plaintiffs against Dow in the District Court for the Southern District of New York – these documents also show that the pesticide production system at Bhopal had only a trial run and that Warren Anderson, then CEO, and the management committee of the Union Carbide Corporation made the key decision to transfer inferior technology to Bhopal. This would be especially significant as FERA stipulated that the technology transfer had to involve "state-of-the-art" technology. Comparisons of the Bhopal plant design and a plant owned by Union Carbide in West Virginia producing the same "Sevin" pesticide at the time allegedly show design differences between the plants.

In short, faced with losing majority ownership under FERA, it has been alleged that Union Carbide made the overall plant cheaper to build by importing inferior technology, thus putting Bhopal at risk from the plant's inception.

Both civil and criminal cases were brought against Union Carbide and other parties, including its Indian subsidiary, Union Carbide India Ltd. The civil case filed by the Indian government on behalf of the survivors was settled for $470 million in 1989. While by the terms of the Indian Bhopal Gas Disaster Relief Act the settlement resolves all claims of survivors for injuries resulting from the disaster, according to survivors' organizations today, the amount of the settlement was based on estimations that have proved far too low, whether quantifying the dead, the injured, or the property lost. It also never accounted for future medical claims. As a result,

---

[49]Foreign Exchange Regulation Act, 1973. Details from an interview with a U.S.-based attorney who has brought the New York litigation against Union Carbide regarding Bhopal in "Proof from Carbide Itself ": Interview with Himanshu Rajan Sharma, U.S.-based Attorney, in *Frontline Magazine*, 19(26), December 21, 2002–January 3, 2003 (published in India, The Hindu Group).

say the survivors' organizations, the $470 million dollars has proved to be inadequate even to satisfy the claims of the acknowledged victims of the disaster. The balance of the amount (approx. $340 million[50]) remaining in the fund is committed to compensation of victims, and cannot be used for the many other needs of the community – not the public health and economic devastation resulting from the disaster, and not for remediation of the contamination left behind by Union Carbide.

In November 1994, Union Carbide sold its interest in Union Carbide India Limited (renamed Eveready Industries India Ltd, or EIIL) to MacLeod Russell (India) Ltd. of Calcutta. As a consequence of that sale, Union Carbide asserted that it retained no interest in or liability for the Bhopal site.[51] However, the Chief Judicial Magistrate for Bhopal has held that Union Carbide Corp.'s transfer of shares was not bona fide because it was done to evade potential liabilities arising out of the ongoing criminal case in Bhopal. As a result, $74 million in Union Carbide assets were attached pending appearance of Union Carbide in the criminal case.[52]

Union Carbide has never filed an appearance in the criminal case, and asserts that the courts of India lack jurisdiction to make the company appear as a defendant.

8.10.4 After the Dow Chemical Acquisition of Union Carbide

Many investors may erroneously take at face value the management's statements that the company acquired no liability for Bhopal and the company cites a number of issues and developments to back up this claim.[53] However, Union Carbide remains a named accused in the criminal case in India, placing Dow Chemical assets in India at risk for potential attachment in the ongoing case. In addition, the Indian government is reportedly processing a request to sue Dow to engage in remediation of the site.

8.10.5 Union Carbide's Pending Criminal Charges in Bhopal

A criminal case was filed against 12 defendants, including Union Carbide, alleging the crime of "culpable homicide not amounting to murder."[54] As part of the Indian government's 1989 settlement, outstanding criminal liabilities against Union Carbide Corporation and other accused were quashed. However, India's Supreme Court reviewed this component of its order and reinstated criminal charges against all accused in 1991.[55]

[50]From *The Hindu* (March 22, 2004) "Bhopal Tragedy: Payment Sought from Compensation Fund Balance."
[51]EIIL took exclusive possession of the land under lease from the government of Madhya Pradesh.
[52]These assets have since been reallocated towards construction of a hospital for Bhopal. However, this asset transfer has not obviated the potential liability of Union Carbide in the criminal case.
[53]See http://www.dow.com/environment/debate/d15.html.
[54]Some of the Indian defendants have had their charges reduced to lesser charges of criminal negligence in the intervening years. However, the charges against Union Carbide and former CEO Warren Anderson remain as culpable homicide.
[55]Paragraph 214 of Supreme Court judgment of October 3, 1991, in (1991) 4 SCC 584 Union Carbide Corporation and Ors versus Union of India and Ors: "The Court sets aside the quashing of criminal proceedings against UCC and others, and reinstates the proceedings."

Indian courts are based on a British system of jurisprudence, which means that penalties of restitution can be levied in criminal cases. In the case of the crimes that Union Carbide is charged with, there is no upper limit for the penalty amount. According to Indian lawyer S. Muralidhar, the penalty amount is left to the discretion of the Court, and usually depends upon the magnitude of the crime and the ability of the criminal to pay.

In 1992, the Chief Judicial Magistrate ordered that Union Carbide Corporation be proclaimed an absconder for repeatedly failing to honor summons to face trial. Thus, in India, Union Carbide is considered a criminal fugitive.

Dow continues to have a presence in India in the form of several subsidiaries and at least two joint ventures with assets and businesses (see above). The continuing presence of these assets, regardless of their size, may give the Indian courts some leverage to attempt to enforce the cooperation of the parent corporation Dow Chemical in producing its subsidiary Union Carbide in court.[56]

8.10.6   Toxic Wastes and Contaminated Groundwater

Union Carbide operated waste evaporation ponds on a 14 hectare (35 acres) plot 400 meters from the factory.[57] These ponds received toxic effluents. Such effluents were also discharged through an open drain flowing nearby. Several tons of obsolete pesticides and process wastes lie strewn around the factory site.

The environmental contamination of the site – much of it created prior to the chemical disaster – has rendered the city a dangerous place to live. Thousands of tons of toxic wastes, including obsolete pesticides such as the persistent and bio-accumulative poison HCH and persistent metals such as mercury, have been abandoned at the factory site (Brown, 2002). Mercury levels in some areas are 6 million times the background values. The groundwater carries high loads of heavy metals, persistent chemicals and solvents, and chlorinated chemicals. Although a portion of the residents have access to overhead tanks of clean water, many of the nearly 20,000 people living in the vicinity are routinely exposed to these chemicals in their drinking water from local wells (Labunska *et al.*, 1999).

Numerous studies, including by Union Carbide itself, acknowledge the presence of contamination and toxic wastes on and off site. A 1999 Greenpeace study confirmed the presence of persistent organic pollutant chemicals, such as DDT

[56]In addition, on February 26, 2004, the Bhopal court was petitioned by a group of Bhopal survivors (the Bhopal Group for Information and Action) to issue a summons to Dow Chemical to produce Union Carbide for the criminal case. In an initial hearing, the government prosecutors (CBI) took the position that such a procedure is unavailable under Indian law; however, at this writing the court had not yet decided for or against the application. Recent developments in courts in India and the United States appear to be directed to further involving Dow Chemical or its subsidiary in issues of site contamination and remediation. In January 2005, the Chief Judicial Magistrate's court issued a summons to Dow; however, this has not been transmitted to Dow by Ministry of External Affairs. The Dow management has repeatedly asserted or implied that there are no criminal cases pending against Union Carbide over Bhopal, that the company faces "absol-utely" no liability associated with Bhopal, and that the only defendants associated with Bhopal are individuals or companies in India.

[57]Bano *et al.* versus Union Carbide Corp. & Warren Anderson, United States District Court Southern District of New York, Civil Action No. 99, Civ. 11329 (JFK).

and Lindane (as contamination) and hexachlorocyclohexane (as stockpiles) among others (Labunska *et al.*, 1999).

Although the factory was returned to the Madhya Pradesh state government in 1998, the condition in which the property was returned may violate the lease agreement. Members of the Indian government and survivor organizations assert that Union Carbide was bound under the terms of their site lease to return the land to the Indian government in usable and habitable condition. Instead, they say, the company undertook some cosmetic remediation work, which community residents assert did nothing to diminish the danger of contamination to local aquifers, before requesting local government departments to take back the lease. Although the government took the site back from Carbide, the Madhya Pradesh government is reportedly working to sue Dow as current owner of Union Carbide, asserting a failure of the company to abide by the lease terms.[58]

Dow Chemical's management has declined any responsibility for clean-up of the toxic wastes, stating, among other reasons, that the 1989 compensation settlement covers all liabilities. However, the Supreme Court order states that the settlement amount "... shall be made to the Union of India as claimant and for the benefit of all victims of the Bhopal Gas disaster under the Bhopal Gas Leak Disaster (Registration and Processing of Claims) Scheme, 1985 ..." The Bhopal Gas Leak Disaster Act, 1985, which describes the framework of the abovementioned scheme, clearly refers to four kinds of claims that can be addressed using the settlement money. All these are claims "... arising out of, or connected with, the disaster."

The toxic wastes, obsolete stockpiles, and groundwater contamination were a result of the routine operation of the factory rather than primarily arising out of or connected with the disaster. Therefore, it appears that the liability for clean-up of the toxic wastes and remediation of contaminated groundwater were not necessarily extinguished by the 1989 settlement.

The courts of India have formally adopted a "polluter pays" principle, which requires that where possible, environmental contamination costs will be charged to the entities causing pollution. The Indian Supreme court has adopted this as a binding principle of law:

> ... once the activity carried on is hazardous or inherently dangerous, the person carrying on such activity is liable to make good the loss caused to any other person by his activity irrespective of the fact whether he took reasonable care while carrying on his activity. The rule is premised upon the very nature of the activity carried on.[59]

Consequently, polluting industries are "absolutely liable to compensate for the harm caused by them to villagers in the affected area, to the soil and to the underground water and, hence, they are bound to take all necessary measures to remove sludge and other pollutants lying in the affected areas."

In the case of Vellore Citizens Forum, the Indian Supreme court clarified further that the "Polluter Pays Principle" as interpreted by this Court means that the absolute

---

[58] *Indian Express* (October 20, 2002) "MP [Madhya Pradesh] Wants Dow to Clean Up Mess."
[59] Indian Council for Enviro-Legal Action versus Union of India [(1996) 3 SCC 212: JT (1996) 2 SC 196 at p. 246, para 65].

liability for harm to the environment extends not only to compensate the victims of pollution but also the cost of restoring the environmental degradation. Remediation of the damaged environment is part of the process of "Sustainable Development" and as such the polluter is liable to pay the cost to the individual sufferers as well as the cost of reversing the damaged ecology.[60]

By the Polluter Pays principle it may be expected that Union Carbide may ultimately be found to be at least one of the responsible parties.

In addition to activities in India, a class action suit brought by survivors against Union Carbide regarding environmental remediation is currently pending on appeal in New York, after having been resolved in favor of the company in the district court. The suit also seeks additional compensation for the long-term health damages that the plaintiffs assert to have not been addressed in the original Indian government settlement. An Amicus Brief filed by 18 Congressmen on behalf of the victims in response to the court's ruling stated, "There is strong support in Congress for holding those responsible for this horrific tragedy accountable for their actions. It is unacceptable to allow an American company not only to exploit international borders and legal jurisdictions but also the ability to evade civil and criminal liability for environmental pollution and abuses committed overseas."[61]

In March 2004, the Second Circuit Court of Appeals for New York issued an order allowing litigation by Indian property owners near the Bhopal plant to proceed against Union Carbide in the New York District Court for property damages, including environmental remediation of their properties. The case that is being allowed relates to contamination that is not from the gas disaster, but rather from the operation and abandonment of the plant site.

In addition, the Circuit Court left an opening for the District Court to allow litigation regarding remediation of the Bhopal plant site itself. The district court had denied the case for remediation of contamination of the UCIL site itself primarily because it would require cooperation of the government of Madyha Pradesh, which currently holds ownership of the site. The appeals court allowed that if the Madyha Pradesh (MP) government were to intervene in the New York litigation, that aspect of the case could also be revived. The MP government has been looking for a mechanism by which to hold Union Carbide Corp. and Dow Chemical Co. responsible for remediation costs. Given that Union Carbide Corp. has denied the Indian government the ability to assert jurisdiction over the company in India, cooperation by the MP government in the New York case may be forthcoming, thus potentially reviving a large case for remediation of the plant site. It should also be noted that the recent Circuit Court decision provides an opening for personal injury cases related to the contamination to be pursued in the New York litigation, provided that they are not barred by the statute of limitations.[62]

---

[60]Vellore Citizens Welfare Forum versus Union of India, WP 914/1991 (1996.08.28).

[61]Office of Frank Pallone, 6th Dist. of NJ; Press Release: Pallone, "17 Colleagues Call on Dow Chemical to Provide Reparations to the Victims of Bhopal Disaster in Letter to Company's CEO: Jersey Congressman Announces Plans To File Amicus Brief In Support of Bhopal Survivors"; July 21, 2003.

[62]United States Court of Appeals for the Second Circuit, Docket no. 03-7416 Decision March 17, 2004. Bano et al. versus Union Carbide and Warren Anderson.

The decision also upheld the plaintiffs' right to seek medical monitoring from UCC for injuries and symptoms associated with groundwater contamination. UCC may be held liable for funding a medical monitoring program for the benefit of the exposed population of 20,000 persons.

8.10.7 Impact on Dow Chemical's Reputation

Aside from direct financial liabilities, investors should recognize the international significance and historical resonance of the Bhopal chemical disaster, and therefore its powerful effect on corporate reputation. Since the purchase of Union Carbide in 2001, Dow has been subjected to escalating public scrutiny and demands for action refocused from Union Carbide to Dow Chemical as the new parent corporation. These include:

- In 2001 and 2002, survivors of the Bhopal disaster and their representatives met with the management of Dow regarding its completed acquisition of Union Carbide. The discussion ended inconclusively after Dow Chemical's CEO was replaced.
- In a letter to Dow management of March 11, 2002, the survivors wrote in follow-up to conversations with the management requesting that the company address the needs of survivors that remain substantially unmet, including clean-up of the site and groundwater, medical monitoring of survivors, further economic compensation and economic rehabilitation for those whose livelihoods are impaired by injury, and social support for widows and orphaned children.
- Survivors appeared at the 2003 Dow Chemical shareholder meeting, where CEO William Stavropoulos repeatedly stated that there was nothing the company could do to answer the survivors' pleas for help – that the company had neither liability nor responsibility for the prior disaster nor its continuing after effects.
- In December 2003, the 19th anniversary of the Bhopal disaster, protests erupted at Dow facilities worldwide. This included the first organized student protest of Dow Chemical since Vietnam on 25 American campuses. A total of 65 activities worldwide protested against Dow calling for "Justice for Bhopal."
- A large coalition of organizations met in Bhopal in January 2004 and announced an escalating campaign against Dow in the coming months, to culminate in December 2004 with the year's 20th anniversary of the Bhopal disaster.[63]
- On January 19, 2004, more than 500 people, including Bhopal survivors and representatives from Dow-impacted communities in Vietnam (Agent Orange) and Saginaw, Michigan, demonstrated outside the Dow Chemical India headquarters in Chembur, Mumbai. An eight-member delegation, including an Agent Orange victim and a former parliamentarian from Belgium, presented a memorandum to Dow Chief Ravi Muthukrishnan. Among

[63]International Campaign for Justice in Bhopal, Activists mount global challenge to Dow, Press Release, January 16, 2004.

other things, the memo urged Dow to present itself to the court in the ongoing criminal proceedings in Bhopal.

- Eighteen members of Congress sent a letter to Dow management on July 18, 2003, urging the company to provide medical rehabilitation and economic reparations for the victims of the tragedy, clean up contamination in and around the former factory site in Bhopal, provide alternative supplies of fresh water to the affected communities, and ensure that the Union Carbide Corporation appears before the Chief Judicial Magistrate's court in Bhopal where it faces criminal charges of culpable homicide. Similar declarations have also been proffered by 50+ parliamentarians in the United Kingdom.

- In April 2004, Rashida Bee and Champa Devi Shukla, two survivors of the gas disaster and leaders of ICJB, were honored with the Goldman Environmental Prize, known widely as the "Nobel Prize for the Environment". The award brought widespread international recognition of their campaign, and came with a $125,000 "no strings attached" cash award. Rashida and Champa selflessly donated the entire sum to form the Bhopal Ki Chingari Trust, which will be used to provide jobs to woman gas victims, medical treatment of disabled children, and to fund an annual award to honor people fighting against polluting companies.

- In December 2004, the 20th Anniversary of the disaster, there was massive press coverage and NGO activity – over 400 events worldwide focused on Dow, Union Carbide and Bhopal. Amnesty International joined the International Campaign for Justice In Bhopal and released "Clouds of Injustice-Bhopal Disaster 20 years on" which, address the many issues with the Bhopal case through a human rights lens. So far, four campuses have organized "Divest from Dow" campaigns built on the company's role in Bhopal. On March 11, 2005 the University of California student assembly called on the University to divest and refuse donations from Dow Chemical. Dow is a major donor to the University of California, Berkeley, with cumulative donations totaling $4.3 million as of October 2003. The resolution is the third of its kind in the United States.

- On April 15, 2005 more than 1500 Amnesty International student members from 10 states throughout the Northeast descended on the Indian Consulate in New York to demand action from the Indian government on Bhopal. The Association for India's Development participated in solidarity actions and events at various Indian government offices in the United States that same day. These actions indicate that Dow's inability to resolve the Bhopal issues is threatening its license to operate.

In light of these developments, it is reasonable for investors analyzing the situation to determine that the Bhopal controversy will not go away on its own and indeed will always be a black mark on the company's record. Increased attention to Bhopal heightens the real potential that money managers who run portfolios that incorporate environmental and social analysis will screen out Dow stock. There is over $2.18

trillion under socially responsible investment management.[64] The issue is gaining a higher profile as a result of continuing inaction by Union Carbide and Dow, and will likely affect investment decisions among this group of investors.

To learn more about these and other similar issues facing Dow Chemical, see "Dow Chemical: Risks for Investors" by Marc Brammer, Sr. Analyst, Innovest Strategic Value Advisors (available at www.innovestgroup.com).

8.11 SCIENTIFIC, POLITICAL, AND INVESTOR DRIVERS OF CHEMICAL INDUSTRY SUSTAINABILITY: AN NGO PERSPECTIVE

RICHARD LIROFF
WWF

The saga of brominated flame retardants offers a cautionary tale for the chemical industry.[65] These chemicals, developed in the early 1970s, are used in a wide range of consumer products, such as furniture, foam, and plastic casings of electronic devices. In 1998, Swedish scientists reviewing archived human breast milk samples discovered that certain flame retardant chemicals (polybrominated diphenyl ethers, or PBDEs) had doubled in concentration in Swedish breast milk about every five years over the preceding twenty. This was a source of concern, as studies of laboratory animals had shown that PBDEs disrupt thyroid hormones. Such disruption yields neurobehavioral effects similar to those of PCBs (polychlorinated biphenyls), whose manufacture the United States banned in 1976.

The Swedish research led to additional research in the United States that revealed similarly rising levels. These data ultimately prompted the European Union in 2003 to outlaw two forms (penta- and octa-) of PBDEs by 2004. In August 2003, the State of California took similar action, requiring a phase-out by 2008.

On September 23, 2003, the U.S. daily newspaper *USA Today* published a top-of-page 1 story entitled "Flame Retardant Found in Breast Milk." The story cited a spokesperson for industry's Bromine Science and Environmental Forum as saying that human effects cannot be extrapolated from rodents. A page 1 story in the *Wall Street Journal* in October 2003 traced PBDEs' scientific and regulatory history and quoted a U.S. EPA official as saying that more research was needed before national regulatory action could be taken. Nevertheless, a few weeks later, the main U.S. manufacturer of PBDEs announced an agreement with EPA to voluntarily phase out production of the two major PBDEs of concern. Since then, new scientific research has emerged suggesting problems with a third PBDE (deca-) that had originally been thought to be less of a risk than penta- and octa-.[66]

[64]Social Investment Forum.

[65]The first two paragraphs of this essay are based on "As Flame Retardant Builds Up in Humans, Debate Over a Ban," *Wall Street Journal*, October 8, 2003, p. A-1.

[66]"New Research Challenges Assumptions About Popular Flame Retardant," *Environmental Science and Technology Science News* (November 6, 2003). Available at http://pubs.acs.org/subscribe/journals/esthag-w/2003/nov/science/kb_flame.html, accessed November 7, 2003. See also Hites (2004).

This cautionary tale captures well the new and renewed challenges the chemical industry faces as companies enter the 21st century. Such pressures, elaborated below, increase the urgency of companies developing new business models for long-term sustainability. The chemical industry's long-term financial health could benefit substantially by its systematically reviewing the toxicity of its products and refreshing its product lines with less toxic alternatives. In some cases, this may prompt corporate exits from certain product lines to limit reputational risks and reduce legal liabilities. Conversely, corporate recognition of the need to reduce product toxicity could drive new research investments yielding more environmentally friendly products.

8.11.1   New Scientific Understanding of Chemicals' Health Effects

As discussed elsewhere in this book, scientists increasingly understand that it is not only the dose that makes the poison, but also the timing of the dose. There is also rising appreciation for the special vulnerability of certain subpopulations, for example, infants, children, and the developing fetus. Awareness is growing that exposures in the womb can contribute to irreversible health effects, including compromised learning abilities and enhanced susceptibility to certain cancers. Even though there continues to be scientific debate about some of these findings, the markets for chemicals previously regarded as safe or insufficiently tested for these effects are likely to be affected; reputations and financial bottom lines will be placed at risk.

8.11.2   Increased Biomonitoring Programs

The chemical industry has long expressed concern that governments' methods for risk estimation, based on modeling and conservative assumptions, overstate risks from chemicals, and that it would be desirable to have real measurements of human exposures. Recent events serve as a reminder to be careful about what you wish for. For example, a 2001 report from the Centers for Disease Control and Prevention highlighted unexpectedly high levels of phthalates, ingredients in many personal care products, in American women of child-bearing age. Some phthalates have been shown in laboratory animal studies to cause testicular harm. The CDC report contributed to a campaign by public health organizations to encourage companies to remove these chemicals from their products.[67]

Indoor studies of homes underscore that common consumer goods can be a source of hazardous chemicals. For example, a study by the Silent Spring Institute identified 67 suspected endocrine-disrupting chemicals in air and dust samples taken from 120 homes on Cape Cod, Massachusetts.[68] Growing interest in monitoring breast milk and sampling other bodily fluids can only heighten public recognition of the intrusiveness and ubiquity of chemical contaminants. With respect to breast milk, even though it is generally recognized that "breast is best" and contamination

---

[67]See www.notyoopretty.org, accessed February 17, 2004.
[68]"Are U.S. Homes a Haven for Toxins?" *Environmental Science and Technology*, November 1, 2003, p. 407A.

of breast milk by chemicals does not justify a switch to baby formula, parents of nursing babies may increasingly ask what steps can be taken to reduce flame retardants and other chemicals in breast milk.

8.11.3   Media and Public Attention to Chemical Exposures and Disease Incidence

During the past several decades, media attention to chemical exposure issues and environmental health has ebbed and flowed. Media interest in chemical issues appears to be growing again, as suggested by the front page newspaper coverage in September and October 2003 addressing flame retardant breast milk in the United States. This drumbeat of media attention continued later in 2003. DuPont's production and use of a fluorinated chemical in Teflon production was featured on the U.S. ABC television network's *20:20* news magazine program in mid-November 2003, and the CBS network's news magazine program *60 Minutes* focused in mid-December on allegations of links between chemicals used in production of IBM's computer chips and the incidence of cancer among IBM workers. In January 2004, scientists' findings of elevated levels of PCBs in farm-grown salmon, attributable to the use of contaminated feed, drew worldwide attention (Hites *et al.*, 2004).

8.11.4   Changing Demand From "Chemical Choosers," a/k/a Consumer Goods Companies

Growing public recognition of toxic hazards from common consumer goods, rising management awareness of hazards to product reputations, and concomitant management realization that "going green" can be good for the bottom line, is prompting companies that formulate products from commodity and specialty chemicals to press for substitutions of less toxic alternatives. For example, S.C. Johnson and Son, Inc. launched its trade-marked Greenlist process in 2001, through which it systematically reviews the toxicity of chemicals used in its product lines and, working with its suppliers, seeks to marry excellent product performance with reduced toxicity.[69] The Greenlist process moves the company beyond its previous successes in eco-efficiency – reducing waste and lowering packaging requirements – to eco-effectiveness – using better materials to get a job done.[70] While this might entail more research, it may also simply mean encouraging an existing supplier to substitute one off-the-shelf chemical for another. S.C. Johnson ties managers' compensation to their success in reducing product toxicity. The company has also realized cost savings. For example, in reformulating floor wax emulsion, the company is eliminating the most hazardous chemicals that it labels "restricted use materials (RUMs)," and is cutting cost without sacrificing performance. It

---

[69]This description of Greenlist in the text derives principally from materials found at http://www.scjohnson.com/community/greenlist.asp, accessed September 23, 2003.

[70]The concept of "eco-effectiveness" was developed by William McDonough and Michael Braungart (1998).

eliminated five such ingredients from an all-purpose cleaner, again reducing cost with no sacrifice in performance.

The cosmetics industry is another chemical chooser ripe for change. The European Union amended its Cosmetics Directive in January 2003 to outlaw carcinogens, mutagens, and reproductive toxicants in cosmetics. Member states of the European Union are required to comply by September 2004.[71] Not only will companies marketing cosmetics in Europe need to change formulations, but companies marketing cosmetics in the United States will either need to follow suit or argue to American women that they need not worry about the risks from such chemicals. These and other such choices will flow down the supply chain to chemical producers.

8.11.5 Convergence of Nonprofit Groups' Environmental and Public Health Agendas

When the U.S. Environmental Protection Agency was pulled together from portions of the then Department of Health, Education, and Welfare and other departments in 1970, the foundation was laid for a bifurcated approach to public health issues in the United States. Environmental groups tended to focus their attention on the new EPA. In contrast, public health advocacy groups, groups devoted to specific diseases, and state and local health departments continued to focus their attention on HEW (now the Department of Health and Human Services or HHS) and, within HHS, the National Institutes of Health. The focus on NIH often was on disease treatment and cure, with some attention also to prevention. In the ensuing decades, incidence of various human health disorders has increased, amidst growing suspicions, based on laboratory animal research and limited human epidemiological studies, that exposures to common chemicals may be partly to blame.

Now at the beginning of the 21st century, environmental and public health groups are joining together to address shared concerns. For example, the website of the Breast Cancer Fund in California offers a review of scientific evidence linking chemical exposures to disease incidence.[72] So too does the website of Physicians for Social Responsibility.[73] In New York State, breast cancer activists focusing on avoidable chemical exposures have adopted the slogan "Prevention is the Cure."[74] The Collaborative on Health and the Environment is bringing together health-affected groups with health professionals and researchers and environmental organizations.[75] In 2002, the Center for Children's Health and the Environment at Mount Sinai Medical Center in New York City paid the *New York Times* to publish several display advertisements on chemical exposure and disease linkages.[76] For

---

[71] See http://www.nottoopretty.org/eudecision.htm, accessed February 17, 2004.

[72] www.breastcancerfund.org, accessed February 17, 2004.

[73] www.psr.org, accessed February 17, 2004.

[74] www.preventionisthecure.org, accessed February 17, 2004.

[75] www.protectingourhealth.org, accessed February 17, 2004.

[76] The advertisements and supporting science papers can be found at www.childenvironment.org, accessed February 17, 2004.

example, one was captioned "Johnny can't read, sit still, or stop hitting the neigh-bor's kid. Why?" The sub-head stated "Toxic chemicals can cause learning disabilities." A second advertisement was headed "More kids are getting brain cancer. Why?" The sub-head stated, "Toxic chemicals appear linked to rising rates of some cancers." A third ad was titled "Our most precious natural resource is being threatened. Why?" and the sub-head stated, "Toxic chemicals are being passed on to infants in breast milk."

8.11.6   Tightening Regulations in Europe and at the State and Local Level in the United States

The U.S. federal government during the administration of President George W. Bush is widely recognized as unsympathetic to environmental regulations. Governments elsewhere have stepped into the void with their own regulatory initiatives. California and the European Union's outlawing of brominated flame retardants are prominent examples. Other examples abound, most prominently the current European Union efforts (the so-called REACH initiative) to systematically overhaul chemical management in member states. REACH is based on the idea of "no data–no market" – companies must make public much more data than are available now on the hazards of their products, or face the risk that they will not be able to continue selling them. Chemicals that are persistent, bioaccumulative, and toxic (PBT) and those that are very persistent and very bioaccumulative (vPvB) are the "low hanging fruit" that likely will be the initial regulatory targets. Hormone-disrupting chemicals may also be controlled, as will those that are carcinogens, mutagens, and reproduc-tive toxicants. Companies understandably and justifiably will make the case that certain chemicals have yielded sizeable social benefits (e.g., flame retardants' contributions to reductions in fire deaths), and REACH will allow them to make this case before controls are considered. But those companies that have gone through an exercise like S.C. Johnson's Greenlist evaluation for "eco-effectiveness" may be better positioned to be winners under the REACH process than those companies that have not.

8.11.7   Increasing Investor Concern

Increasingly since 2001, investor groups working together with environmental and public health advocacy groups have been demanding greater accountability in cor-porations' management of toxic chemicals and products containing toxic chemicals. Shareholder resolutions have been directed at chemical producers (e.g., Dow, for its production and discharge of dioxin and other persistent pollutants), at chemical choosers (e.g., Avon, for its use of hormone-disrupting chemicals in its cosmetic products), and retailers (e.g., SuperValu and JCPenney, for selling mercury thermometers). Some shareholder resolutions appear to have encouraged changes in corporate policies, while others have not. In 1999, in exchange for the withdrawal of a resolution proposed by two religious groups and a union, Baxter International signed a memorandum of understanding, agreeing to a timetable for replacing the polyvinyl chloride in containers of intravenous solutions and to work on replacing PVC in all its medical products. Baxter also agreed to ask that various chemical industry groups refrain from using Baxter products in their public advertising campaigns.[77] Amidst a wave of corporate management scandals, some investors believe that how companies manage their chemical challenges may reflect the overall quality of their management and this, in turn, may be reflected in investors' return on investment. Poor quality management may also lead to increased litigation risks and attendant financial liabilities. In a 2002 report for The Rose Foundation, "The Environmental Fiduciary: The Case for Incorporating Environmental Factors into Investment Management Policies," Susannah Blake Goodman and her colleagues show how a corporation's ability to profit from environmental innovations and prepare for future environmental risks and exposures can have a significant impact on corporate earnings potential, cash flow, and growth opportunities.[78]

8.11.8 Insurer and Reinsurer Concern

The Swiss insurance company Swiss Re announced in 2003 that it would take corporations' policies on global warming into account when issuing officer and director liability insurance.[79] Such foresight appears prudent in the wake of the insurance industry funding $160 million of a $600 million settlement of PCB litigation in Anniston, Alabama, in 2003 and in the aftermath of the still unresolved asbestos policy mess in the United States that has bankrupted scores of companies and cost the insurance industry billions of dollars.[80] It is only a matter of time before insurance and reinsurance companies begin to ask much tougher questions than they do now about corporations' manufacture and use of toxic chemicals associated with diseases and disabilities. Chemical producers and users may be able to mount formidable defenses based on scientific uncertainties, but the diversion of corporate resources to legal defenses, and associated reputational and market share risks, suggests that it will be in insurers' and reinsurers' best interest to take searching looks at how corporations are attempting to reduce and avoid toxic liabilities.

The converging pressures outlined above present sizeable risks, but they also represent sizeable opportunities for the chemical industry. Nimble, foresighted companies should acknowledge these risks. Through shrewd product evaluations, research, and asset redeployments, they can select a future path leading to enhanced profitability, a reduced ecological footprint, and a healthier future for all.

[77]The Baxter agreement is available by searching the website of Health Care Without Harm, www. noharm.org, accessed February 17, 2004.
[78]The report is available at www.rosefdn.org, accessed February 17, 2004.
[79]"Global Warming May Cloud Directors' Liability Coverage," *Wall Street Journal*, May 7, 2003, p. C-1.
[80]"Monsanto, Solutia Shell Out on PCBs," *Chemical and Engineering News*, August 25, 2003, p. 10; "Asbestos Wave Surges; Crest Still to Come," *Insurance Journal*, October 8, 2003. Available at www. insurancejournal.com, accessed February 17, 2004.

8.12 THE ROLE OF LEADERSHIP AND CORPORATE GOVERNANCE

DAVID ROBERT TASCHLER

Air Products & Chemicals, Inc.

8.12.1 The Case for Leadership Commitment

Leaders throughout the course of history have been aware of the importance of protecting our world for future generations. Stewardship of our land has been a topic of consideration by society's leaders for thousands of years. The early Western philosophers discussed and debated the concepts of intergenerational philosophy (Constitutional Law Foundation, 2004) from early Greek and Biblical times, emerging in the tenets of the Constitution of the United States. John Locke's theories of intergenerational obligation and stewardship collectively constitute "an early, but remarkably clear articulation of what, today, would be termed a 'sustainability ethic'" (Elliot, 1986, p. 217).

Sustainable development had been espoused in the United States as early as 1789 by President Thomas Jefferson: "Then I say the Earth belongs to each ... generation during its course, fully and in its own right, no generation can contract debts greater than may be paid during the course of its own existence" (Jefferson, 1789, para. 2). As one of the founding fathers of the United States, Mr. Jefferson was one of the first leaders in the United States to recognize the importance of leadership providing a focus on sustainability by taking a public position on the issue. He did so by including the concept of intergenerational stewardship into the Constitution of the United States.

The challenge to contemporary corporate leadership is to internalize the legacy of intergenerational stewardship and discover the role of the corporation in achieving societal sustainability. Leaders of contemporary society and corporations in today's world must carry on that tradition of the great leaders of the past. Corporate executives and directors can no longer ignore the emerging trends and societal pressures to protect our world for future generations. "Our task as responsible and ethical global leaders is to point the path to sustainability" according to William K. Reilly, former Administrator of the United States Environmental Protection Agency and a DuPont director (Reilly, 1999, p. 26). John Elkington succinctly elucidated the importance of corporate attention to sustainability: "To refuse the challenge implied by the triple bottom line is to risk extinction" (Elkington, 1998, p. 2). Chad Holliday, CEO and Board Chair of DuPont, challenged the leaders of industry to recognize that that sustainability is a "business imperative for success in the 21st Century" (*Process Engineering*, 2003, para. 1).

As chemical companies commit to expand their role in assuring a sustainable future for themselves and society, it will be increasingly important to understand what roles leadership and corporate governance play in shaping this commitment. Taschler (2004) analyzed publicly available information on the eight companies in the chemical industry sector of the 2004 Dow Jones Sustainability World Index to understand any contributory patterns between board member characteristics and corporate sustainability performance for the eight companies in the index. This study provides some insight into how the leaders in sustainable development are putting concepts into practice.

8.12.2   Identifying the Leaders

Taschler examined several chemical companies that have achieved recognition as corporate leaders in sustainable development based upon the proprietary assessment methodology developed and validated by Sustainable Asset Management (SAM) Group, (Dow Jones Sustainability Indexes, 2003). The Sustainable Asset Management and Dow Jones Sustainability World Index (DJSI, 2002) offers a methodical and verifiable method of scoring and identifying the leading companies in each of 62 industry groups, including basic materials and chemicals. It utilizes a best-in-class approach to identify the leading sustainability-driven company in these industry groups by requiring all companies to exceed a minimum threshold of sustainability (Baue, 2002). The chemical sector of the 2004 Dow Jones Sustainability World Index included the eight chemical companies listed in Table 8.8.

The boards of directors of these companies were examined in order to determine patterns of commonality in the leadership of sustainability initiatives. A number of findings emerged that may prove valuable to the leaders of other chemical companies who are looking for ways to employ the tenets of sustainable development in their corporate governance strategy.

8.12.3   The Role of Champion

Leadership commitment was found to be a unifying element among the chemical companies in the 2004 DJSI. The personal passion of an executive officer or director of the corporation was found to be a critical element of corporate commitment to sustainable development. Each of the companies had one or more directors, often one of the executive directors, who emerged as the organizational mouthpiece or champion for the company on issues relating to sustainable development. The

TABLE 8.8. Companies in the Chemical Sector of the 2004 DJSI

Company Name	Country of Incorporation
Air Products & Chemicals, Inc.	United States
BASF AG	Germany
Bayer AG	Germany
Royal DSM, N.V.	Netherlands
Degussa AG	Germany
Dow Chemical Company	United States
E. I. DuPont de Nemours & Co.	United States
Praxair, Inc.	United States

champion was suggested by evidence showing higher than average reference to sustainability-related concepts in communications, as well as by personal involvement in key sustainability initiatives and other social causes. For example, *Chief Executive Magazine* points out that Chad Holliday of DuPont "has burnished a reputation for himself and his company as a champion of 'sustainable development'" (Cortese, 2003, p. 22). The presence of multiple champions at a few of the companies, such as Dow, DSM, and DuPont, reflected an even greater influence of the directors on corporate sustainability initiatives.

8.12.4   Engaging the Board

Each of these companies has consciously engaged their board of directors in corporate activities in sustainable development, albeit in different ways. For some, such as BASF, Bayer, Degussa, and DuPont, a board member has direct accountability for sustainability initiatives. At other companies, including Air Products, Dow, DSM, and Praxair, the board plays an oversight role in monitoring the progress and direction of corporate sustainability plans defined by corporate sustainability centers of excellence. This finding leads to the conclusion that board involvement is key to the corporation achieving corporate success in sustainable development.

8.12.5   Experience Necessary

Governance theory suggests that directors may be selected for their ability to provide access to needed resources and information from the external environment (Daily *et al.*, 1999). Board members play the primary role of making information from the external environment available to the company. Directors are valued for their connections and reputations related to sustainable development. One key resource that may be of value to enable a company to become a leader in sustainable development may be the access a director has to others who are knowledgeable about, and committed to, sustainable development in other organizations.

Having a board comprised of individuals with diverse experience and who represent a broad range of stakeholders is important to corporate initiatives in sustainability. The corporations in the chemical sector of the 2004 DJSI valued the presence on their board of individuals with experience in, or commitment to, sustainability-related activities in other aspects of their lives. The companies have retained the services of individuals with demonstrated expertise in education, environmental affairs, and various philanthropic and social causes. The boards of Dow and DuPont are particularly noteworthy for the representation of directors who have received significant public recognition for their contributions to sustainable development.

A commitment to sustainability compels organizations to pursue a course of action involving the recruitment and retention of directors who understand the value of sustainable development to the corporation and society. Criteria for the selection of directors should include evidence of involvement in, or representation of, stakeholder interests. Corporations can and should engage directors who have the background to share experiences from other organizations committed to the tenets of sustainable development. Sources of such talent may be the boards of other companies in other sectors of the DJSI or other recognized organizations such as the World Business Council on Sustainable Development.

8.12.6   Affiliation with Other DJSI Companies and Sustainability Initiatives

In addition to attracting directors with a personal commitment to sustainable development, most of the boards showed the presence of board members from companies in other sectors of the DJSI. Corporations were relying on the resource dependency aspects of the director role (Zahra and Pearce, 1989) by selecting individuals who could link them to others with knowledge and experience in sustainable development. Each of the companies had directors who were either executives or directors at other companies who have demonstrated excellence in sustainability by virtue of their inclusion in one of the other market sectors in the DJSI (Table 8.9).

Seven of the eight companies in the sample population have at least one director who is an executive at another company that is also listed as a 2004 DJSI firm. All of the companies have directors who also serve on the boards of other DJSI companies. These directors may play a role in transferring knowledge and expertise regarding sustainable development between the various companies that director represents. The trend of cross-membership on boards or management teams of other DJSI companies may have a positive influence on other companies for whom those directors may be engaged.

The DJSI is not the sole measure of corporate commitment to sustainable development. The data were reviewed to determine whether any of the directors had an

TABLE 8.9. Number of Board Members Affiliated with Other DJSI Members

Company	Number of Directors	Number of Directors Holding Executive Positions at Other DJSI Companies[a]	Number of Directors on Boards of Other DJSI Companies[b]
Air Products	13	3	1
BASF	30[c]	2	11
Bayer	24	4	8
DSM	12	4	5
Degussa	24	2	8
Dow	14	1	3
DuPont	13	3	5
Praxair	11	0	1

[a,b]Considers September 2003 DJSI listing only. Also includes directors who have retired from the referenced companies.

[c]Includes two members who moved from Executive Board to Supervisory Board in May 2003 replacing two others, all of whom had influence on board actions prior to the datum point for this study.

affiliation with a company committed to other sustainability initiatives. While there are many such initiatives, this group of chemical companies had multiple directors associated with companies involved with the UN Global Reporting Initiative, the UN Global Compact, or the World Business Council on Sustainable Development, as summarized in Table 8.10.

In all cases, the board of each company had at least one member from another company affiliated with a major sustainable development initiative.

8.12.7 Communicating the Commitment

The path towards global societal sustainability must start with the demonstration of individual leadership and commitment. Over 2700 articles, speeches, press releases, and presentations in the public domain were reviewed for evidence of communication content that reflected the director's awareness of, or commitment to, elements of sustainable development. Each item associated with the citation count listed in Table 8.11 reflects an article containing a combined reference to a specific director, as well as some aspect of sustainable development.

Overall, more than 40 percent of the directors had citations that could be related to the tenets of sustainable development. The number of citations is an important indication of the propensity of the directors to speak to the public about issues related to sustainable development. The frequency with which directors refer to sustainable development in their communications is another important indication of personal commitment. The evidence in Table 8.11 suggests that between 29 and 58 percent of board members are actively speaking in terms that reflect a predisposition to sustainable development.

TABLE 8.10. Director Affiliations with Other Sustainability Initiatives

Company	Number of Directors	Directors from Companies Affiliated with:		
		UN Global Reporting Initiative Companies	UN Global Compact Companies	World Business Council on Sustainable Development
Air Products	12	3	0	1
BASF	30[a]	20	13	22
Bayer	24	16	16	15
DSM	12	8	3	7
Degussa	24	1	1	10
Dow	14	6	6	6
DuPont	13	4	2	3
Praxair	11	0	0	1

[a]Includes two members who moved from Executive Board to Supervisory Board in May 2003 replacing two others, all of whom had influence on board actions prior to the datum point for this study.

TABLE 8.11. Directors with Sustainable Development (SD) Citations

Company	Number of Directors	Number of Directors with SD Citations	Percentage of Directors with SD Citations	Number of SD Citations	Number of Total Citations Reviewed
Air Products	13	5	38%	11	157
BASF	28	14	50%	92	458
Bayer	24	12	50%	115	942
DSM	12	7	58%	44	82
Degussa	24	7	29%	26	81
Dow	14	8	57%	108	432
DuPont	13	5	38%	132	461
Praxair	11	4	36%	28	162

8.12.8 Walking the Talk

For those executive directors in the 2004 DJSI chemical sector, sustainable development is a commitment, not window dressing for commercial purposes. Through their personal commitment to stewardship and sustainable development, corporate leaders can help to ensure a secure future for their company and society at large.

Dr. William Stavropoulos, CEO and Board Chair of Dow Chemical, captured the concept of executive commitment in his own words: "Sustainability is a mindset, a leadership philosophy and arguably the only way to conduct business in the twenty-first century ... And, to truly take hold, it demands leadership – commitment, passion and belief – from the very top of an organization" (Stavropoulos, 2001: para 21).

The companies in the chemical sector of the DJSI have committed to sustainability, and some have faced challenges to that commitment in the normal course of business. Executives who have made a personal, public commitment to sustainable development find themselves in a position to be held accountable for their performance by their stakeholders. One of the more public cases involved Bayer AG. One of Bayer's most promising drugs, the anticholesterol medication Lipobay®, was linked to deaths in the United States. Dr. Manfred Schneider, then CEO of Bayer, made the decision to withdraw Lipobay® from the market, recognizing that the action would "... of course have effects on our pharmaceuticals strategy and on the future development of the group. The economic consequences are very serious, but absolute priority for us is the safety and health of our patients" ("Agence France-Presse", 2001: para 5).

The leading sustainability companies in the chemical industry have made a public commitment to sustainable development despite the possibility that this commitment may result in extra scrutiny for those involved and present risks in terms of subsequent commercial and personal outcomes (Hemingway and Maclagan, 2004). This public commitment is a necessary component and antecedent to success in sustainable development.

8.12.9 The Call to Action

An examination of the boards of the companies in the chemical sector of the 2004 Dow Jones Sustainability Index demonstrates the importance of the corporation identifying, selecting, and retaining directors with backgrounds and demonstrated personal commitment to the tenets of sustainable development. While there are sustainability experts within the organization who may be held accountable for execution of sustainable business strategies, the board must be involved in shaping sustainable development initiatives and monitoring corporate progress during the journey along the continuum towards sustainability.

Commitment and corporate performance in sustainability in the chemical industry requires leadership at the corporate governance level. Corporations that wish to take leadership roles in shaping industry's response to societal demands for a more sustainable business approach must engage their board of directors and make a public commitment to drive towards sustainability. They must overcome past disregard for long-term corporate and societal sustainability by engaging their governance bodies in setting a sustainability agenda. Weak leadership and complacent boards of directors must be educated and engaged, or face demands from stakeholders to be replaced by executive champions and committed boards. For a corporation to be recognized as a leader in sustainable development, an executive- or director-level champion must emerge who can engage and involve the board in the definition and implementation of a strategy of sustainability. Sustainability requires alignment between espoused corporate positions and the actions and behaviors of the individuals who comprise the governance body and leadership of the corporation. This combination provides internal and external influences on the strategies and initiatives that support sustainable development. Leadership of sustainability initiatives must come from executives and board members. It is the responsibility and obligation of the corporate directors to ensure the sustainability of their organization and the society in which they operate.

The complex interplay required by industry, government, and society at large will require the leadership of many – not just leaders of businesses, but also governmental, religious, and civil leaders. The chemical industry has a unique role to play because of its impact on economic and environmental issues. It can play a bigger role to enhance societal well-being for a broad array of stakeholders. Sustainable development requires leaders with a holistic systems perspective and who can appreciate and contribute to the resolution of the global issues of sustainable development. It is precisely this global purview that positions the boards of directors of the world's multinational chemical firms as ideal candidates to accept this leadership role.

Executives and directors in the chemical industry must recognize the breadth of their responsibility. "Sustainability is not just about managing the corporation. It is about leading society, industry, and government towards a sustainable future" (Holliday *et al.*, 2002).

REFERENCES

Agence France-Presse, "Bayer to Cut 1,800 Jobs, Close 15 Sites," *Agence France-Presse*, August 9, 2001. Available at factiva™ database.

N. A. Ashford, "Chemical Catastrophes: Steps for Prevention," *New York Times*, December 9, 1984, Business Section, p. 2.

D. Austin and A. Sauer, *Changing Oil: Emerging Environmental Risks and Shareholder Value in the Oil and Gas Industry*, World Resources Institute, 2002. Available at http://pubs.wri.org/pubs_description.cfm?PubID=3719.

D. Austin, N. Rosinski, A. Sauer and C. Le Duc, *Changing Drivers: The Impact of Climate Change on Competitiveness and Value Creation in the Automotive Industry*, World Resources Institute and Sustainable Asset Management (SAM), 2003. Available at http://business.wri.org/pubs_content_text.cfm?ContentID=2255.

W. Baue, "DJSI adds 'Sin' Stocks to Indexes, Calling into Question the Definition of Sustainability," *Social Funds.com News*, December 6, 2002. Available at http://www.social funds.com/news/article.cgi/article983.html.

S. Beloe *et al.*, *Values for Money: Reviewing the Quality of SRI Research*, SustainAbility, London, 2004. Available at http://www.sustainability.com/publications/about/default.asp.

K. Bethke, "Eco-Efficiency for SMEs in the Moroccan Dyeing Industry. A Sustainable Approach to Industrial Development," *Nord-Süd aktuell*, 1, 106–109 (2003).

M. M. Blair and S. M. H. Wallman, *Unseen Wealth: Report of the Brookings Task Force on Intangibles*, Brookings Institution Press, Washington, DC, 2001.

D. Brown, "The Dead Zone," *Guardian Weekend*, September 21, 2002, p. 44.

E. Clark, *The Intangible Economy: Impact and Policy Issues*, Report of the High Level Expert Group on the Intangible Economy, Enterprise Director-General, European Commission, Brussels, October 2000, pp. 6–7.

D. J. Constable, A. D. Curzons, L. M. Freitas dos Santos, G. R. Geen, R. E. Hannah, J. D. Hayler, J. Kitteringham, M. A. McGuire, J. E. Richardson, P. Smith, R. L. Webb and M. Yu, "Green Chemistry Measures for Process Research and Development," *Green Chemistry*, 3, 7–9 (2001).

D. J. C. Constable, A. D. Curzons and V. L. Cunningham, "Metrics to Green Chemistry – Which are the Best?" *Green Chemistry*, 4, 521–527 (2002).

Constitutional Law Foundation, Intergenerational Justice in the United States Constitution. The Stewardship Doctrine: Historical Overview, 2004. Available at http://www.conlaw.org/intergenerational-intro.html.

A. Cortese, "DuPont's Teflon™ Dilemma," *Chief Executive*, November 1, 2003, p. 22. Available at factiva™ database.

Council of Economic Advisors, *Economic Report of the President*, GPO, Washington, DC, 2000, p. 431, Table B-96.

A. D. Curzons, C. Jiménez-González, A. L. Duncan, D. J. C. Constable and V. L. Cunningham, "Fast Lifecycle Assessment of Synthetic Chemistry (FLASCTM) Tool," *Green Chemistry*, submitted.

A. D. Curzons, D. C. Constable and V. L. Cunningham, "Solvent Selection Guide: A Guide to the Integration of Environmental, Health and Safety Criteria into the Selection of Solvents," *Clean Products and Processes*, 1, 82–90 (1999).

A. D. Curzons, D. J. C. Constable, D. N. Mortimer and V. L. Cunningham, "So You Think Your Process is Green, How Do You Know? Using Principles of Sustainability to Determine What is Green – A Corporate Perspective," *Green Chemistry*, 3, 1–6 (2001).

C. M. Daily, J. L. Johnson and D. R. Dalton, "On the Measurements of Board Composition: Poor Consistency and a Serious Mismatch of Theory and Operationalization," *Decision Sciences*, 30(1), 83–106 (1999).

Dow Jones Sustainability Indexes, "DJSI Index Performance," *Sustainability Investment*, September 2002. Available at http://www.sustainability-indexes.com/sustainability/.

Dow Jones Sustainability Indexes, "Dow Jones Sustainability World Indexes Guide, Version 5.0," September 2003. Available at Dow Jones Sustainability Indexes: http://www.sustainability-indexes.com/djsi_pdf/djsi_world/Sectors/DJSI_World_CHM_04.pdf.

A. L. Duncan and A. D. Curzons, "WRAP Up Your Packaging Issues: Development of a Corporate Approach to Assess Packaging," *Clean Techn Environ Policy, 2004*, submitted.

DuPont, "Sustainable Growth 2003 Progress Report," 2003. Available at http://www1.dupont.com/NASApp/dupontglobal/corp/index.jsp?page = /content/US/en_US/social/SHE/usa/us1.html, accessed on August 26, 2004.

J. Elkington, *Cannibals with Forks: The Triple Bottom Line of Sustainability*, New Society Publishers, Gabriola Island, BC, Canada, 1998.

R. Elliot, "Future Generations, Locke's Proviso and Libertarian Justice," *Journal of Applied Philosophy*, 3, 217–227 (1986). Available at http://www.conlaw.org/intergenerational-intro.html.

L. Everest, *Behind the Poison Cloud: Union Carbide's Bhopal Massacre*, Banner Press, Chicago, 1985.

J. Fiksel, J. Low and J. Thomas, "Linking Sustainability to Shareholder Value," *EM Magazine*, June, 19–25 (2004a).

J. Fiksel, K. Funk, P. Kalafut and J. Low, *Clear Advantage: Building Shareholder Value. Environment Value to Investors*, Global Environmental Management Initiative, Washington, DC, February 2004b.

D. Firn, "Private Cash Drives Chemical Actions," *Financial Times*, January 29, 2004.

A. Gillies, "Short Tenures, Big Returns," 2002. Available at www.Forbes.com/2002/04/29/0429sf.html. Accessed on April 29, 2002.

K. T. Hamilton, "Pollution as News: Media and Stock Market Reactions to the Toxics Release Inventory Data," *Journal of Environmental Economics and Management*, 28, 98–113 (1995).

C. A. Hemingway and P. W. Maclagan, "Managers' Personal Values as Drivers of Corporate Social Responsibility," *Journal of Business Ethics*, 50(1), 33–44 (2004).

R. A. Hites *et al.*, "Global Assessment of Organic Contaminants in Farmed Salmon," *Science*, 33, 226–229 (2004).

R. A. Hites, "Polybrominated Diphenyl Ethers in the Environment and in People: A Meta-Analysis of Concentrations," *Environmental Science and Technology*, February 15, 2004, pp. 945–956.

C. Holliday, "Sustainable Growth, the DuPont Way," *Harvard Business Review*, September, 129–134 (2001).

C. O. Holliday Jr., S. Schmidheiny and P. Watts, *Walking the Talk: The Business Case for Sustainability.* Greenleaf, Sheffield, England, 2002.

Innovest, *Dow Chemical: Risks for Investors*, Innovest, New York, April 2004.

A. B. Jaffe, S. R. Peterson, P. R. Portney and R. N. Stavins, "Environmental Regulation and the Competitiveness of U.S. Manufacturing: What Does the Evidence Tell Us?" *Journal of Economic Literature*, 33, 132–163 (1995).

T. Jefferson, *Letter to James Madison*, 1789. Available at http://Lachlan.bluehaz.com.au/lit/jeff03.html.

C. Jiménez-González, S. Kim and M. R. Overcash, "Methodology of Developing Gate-to-Gate Life Cycle Analysis Information." *Int JLCA*, 5(3), 153–159 (2000).

C. Jimenez-Gonzalez, A. D. Curzons, D. J. C. Constable, M. R. Overcash and V. L. Cunningham, "How Do You Select the 'Greenest' Technology? Development of Guidance for the Pharmaceutical Industry," *Clean Products and Processes*, 3, 35–41 (2001).

C. Jimenez-Gonzalez, A. D. Curzons, D. J. C. Constable and V. L. Cunningham, "Life Cycle Assessment of Pharmaceutical Compounds: Cradle-to-Gate LCI/A of a Typical Active Pharmaceutical, a Case-Study," *International Journal for Life Cycle Assessment*, 2003, OnlineFirst. Available at http://dx.doi.org/10.1065/lca2003.11.141.

C. Jimenez-Gonzalez, A. D. Curzons, D. J. C. Constable and V. L. Cunningham, "Expanding GSK's Solvent Selection Guide – Application of Life Cycle Assessment to Enhance Solvent Selections," *Clean Techn Environ Policy*, 7(1), 42–50 (2004).

C. Jimenez-Gonzalez, D. J. C. Constable, A. D. Curzons and V. L. Cunningham, "Developing GSK's Green Technology Guidance: Methodology for Case-Scenario Comparison of Technologies," *Clean Techn Environ Policy*, 4, 44–53 (2002).

M. Khanna, W. M. H. Quimio and D. Bojilova, "Toxic Release Information: A Policy Tool for Environmental Protection," *Journal of Environmental Economics and Management*, 36, 243–266 (1998).

M. Kiernan, Panel discussion on "Sustainability: Social and Environmental Factors in Financial Reporting," held at Cap Gemini Ernst & Young *Measuring the Future* Conference, Cambridge, MA, October 1–3, 2000. Available at http://www.cbi.cgey.com/events/pubconf/2000-10-4/session/breakout/index.html#sustainability.

S. Klinger, C. Hartman, S. Anderson, J. Cavanagh and H. Sklar, *Executive Excess 2001: CEOs Cook the Books, Skewer the Rest of Us*, Ninth Annual CEO Compensation Survey, Institution for Policy Studies and United for a Fair Economy, August 26, 2002, p. 1.

S. Konar and M. A. Cohen, "Information as Regulation: The Effect of Community Right to Know Laws on Toxic Emissions," *Journal of Environmental Economic Management*, 32, 109–124 (1997).

C. Kranz and I. Sagasser, "How Chemical Industry Initiative Contributes to Environmental, Safety and Health Protection in SMEs: An Example from BASF," *UNEP-Magazine "Our Planet"*, accepted, 2003.

I. Labunska, A. Stephenson, K. Brigden, R. Stringer, D. Santillo and P. A. Johnston, "The Bhopal Legacy. Toxic Contaminants at the Former Union Carbide Factory Site, Bhopal, India: 15 Years After the Bhopal Accident," Technical note 04/99, Greenpeace Research Laboratories, Department of Biological Sciences, University of Exeter, UK.

D. Lapierre and J. Moro, *It Was Five Past Midnight in Bhopal*, Full Circle Publishing, New Delhi, India, 2001.

B. Lev, *Intangibles – Management, Measurement, and Reporting*, Brookings Institution Press, Washington, DC, 2001a.

B. Lev, "Grey Matters: CFO's Third Annual Knowledge Capital Scorecard," *CFO Magazine*, April 2001b.

J. Low and P. C. Kalafut, *Invisible Advantage: How Intangibles are Driving Business Performance*, Perseus, Cambridge, MA, 2002.

T. J. Lueck, "1982 Report Cited Safety Problems at Plant in India," *New York Times*, December 11, 1984.

E. Mankin and P. Chakrabarti, "Valuing Adaptability: Financial Markers for Managing Volatility," Perspectives on Business Innovation, Cap Gemini Ernst & Young Center for Business Innovation, November 9, 2002.

W. McDonough and M. Braungart, "The Next Industrial Revolution," *The Atlantic*, October 1998. Available at http://www.theatlantic.com/issues/98oct/industry.htm, accessed February 23, 2004.

E. Mongan, "DuPont's Quest for Renewable Energy," 2003. Available at http://www.brtable.org/pdf/ClimateRESOLVE/12_02_03Mongan_DuPont.ppt, accessed August 26, 2004.

R. Monks, *The New Global Investors: How Shareholders Can Unlock Sustainable Prosperity Worldwide*, Capstone, Oxford, UK, 2001.

W. Morehouse and M. A. Subramanian, *The Bhopal Tragedy: What Really Happened and What it Means for American Workers and Communities at Risk; a Report for the Citizens Commission on Bhopal*, The Council on International and Public Affairs, New York, 1986.

M. Mukerjee, "Persistently Toxic," *Scientific American*, 272(6), 16, 2p, 1c (1995).

M. E. Porter and C. van der Linde, "Toward a New Conception of the Environment – Competitiveness Relationship," *Journal of Economic Perspectives*, 9(4), 97–118 (1995).

Process Engineering, "*CIA Sets Out 'Sustainable Vision' for Chemical Industry*", October 20, 2003. Available at factiva™ database.

D. J. Reed, *Stalking the Elusive Financial Case for Corporate Sustainability*, World Resources Institute, Washington, DC, 2001.

W. K. Reilly, "Private Enterprises and Public Obligations: Achieving sustainable development," *California Management Review*, 41(4), 17–26 (1999).

F. L. Reinhardt, "Bringing the Environment Down to Earth," *Harvard Business Review*, July–August 1999, pp. 149–157.

R. Repetto and D. Austin, *Pure Profit: The Financial Implications of Environmental Performance*, World Resources Institute, 2000. Available at http://www.wri.org/capmarkets/pdf/pureprofit.pdf.

P. Saling, A. Kicherer, B. Dittrich-Krämer, R. Wittlinger, W. Zombik, I. Schmidt, W. Schrott and S. Schmidt, "Eco-Efficiency Analysis by BASF: The Method," *International Journal of Life Cycle Assessment*, 7(4), S. 203–218 (2002).

S. Schaltegger and A. Sturm, *Ökologieorientierte Entscheidungen in Unternehmen*, 2, Auflage, Bern, Stuttgart, Wien, 1994.

I. Schmidt, P. Saling, W. Reuter, M. Meurer, A. Kicherer and C.-O. Gensch, "SEEbalance – Managing Sustainability of Products and Processes with the Socio-Eco-Efficiency Analysis by BASF," *Greener Management International*, accepted 2004.

S. Shah, "Farm Sector Too Reaps Rewards of Reforms," *The Economics Times*, December, 31, 2003.

Sibson Consulting, "A Study of Corporate Governance Disclosure Practices," Segal Company Report, March 2003.

H. Singh, "MP wants Dow to Cleanup Carbide Mess, State to Approach Center for Supreme Court Action," *The Indian Express*, October 20, 2002.

Spencer Stuart, "The 2004 Spencer Stuart Route of the Top Survey of Fortune 700 CEOs." Available at http://www.spencerstuart.com/ArticleViewer.aspx?PageID=10096& ArtID=4074799#gen_ceotenure, accessed September 10, 2004.

W. Stavropoulos "*William S. Stavropoulos Acceptance Speech – Palladium Medal, Societe de Chimie Industrielle*," May 16, 2001. Available at http://www.dow.com/dow_news/ speeches/

D. R. Taschler, "Characteristics of Board of Director Members and Corporate Commitment to Sustainable Development". Available at UMI Proquest Digital Dissertations, Publication No. AAT 3144876, 2004

W. S. Upton, Jr., "Special Report: Business and Financial Reporting, Challenges for the New Economy," Financial Accounting Series, No. 219-A, Financial Accounting Standards Board, Norwalk, CT, April 2001.

R. Winslow, "Union Carbide Moved to Bar Accident at U.S. Plant Before Bhopal Tragedy," *Wall Street Journal*, January 28, 1985.

World Resources Institute, 2000. Available at http://www.wri.org/capmarkets/pdf/ pureprofit.pdf.

S. A. Zahra and J. A Pearce, II, "Boards of Directors and Corporate Financial Performance: A Review and Integrative Model," *Journal of Management*, 15(2), 291–334 (1989).

APPENDIX 1

RESPONSIBLE CARE®
GLOBAL CHARTER

January 2005

 Global Charter

Responsible Care is the global chemical industry's environmental, health and safety (EHS) initiative to drive continuous improvement in performance. It achieves this objective by meeting and going beyond legislative and regulatory compliance, and by adopting cooperative and voluntary initiatives with government and other stakeholders. Responsible Care is both an ethic and a commitment and

Only the English language version is the official document

Transforming Sustainability Strategy into Action: The Chemical Industry, Edited by B. Beloff, M. Lines, and D. Tanzil
Copyright © 2005 John Wiley & Sons, Inc.

a commitment that seeks to build confidence and trust in an industry that is essential to improving living standards and the quality of life.

The Responsible Care Global Charter arose from an examination of chemical industry practices and performance that has evolved since the mid-1980s and was shaped by considering the recommendations of independent stakeholders from around the world. The Charter goes beyond the original elements of Responsible Care since its inception in 1985. It also focuses on new and important challenges facing the chemical industry and global society, including the growing public dialogue over sustainable development, public health issues related to the use of chemical products, the need for greater industry transparency, and the opportunity to achieve greater harmonization and consistency among the national Responsible Care programmes currently implemented. The International Council of Chemical Associations (ICCA), through Responsible Care, will continue to undertake actions consistent with the environmental principles of the United Nations Global Compact.

The Responsible Care Global Charter contains nine key elements. They are:

1. ADOPT GLOBAL RESPONSIBLE CARE CORE PRINCIPLES.

The Global Responsible Care Core Principles commit companies and national associations to work together to:

- Continuously improve the environmental, health and safety knowledge and performance of our technologies, processes and products over their life cycles so as to avoid harm to people and the environment.
- Use resources efficiently and minimise waste.
- Report openly on performance, achievements and shortcomings.
- Listen, engage and work with people to understand and address their concerns and expectations.
- Cooperate with governments and organisations in the development and implementation of effective regulations and standards, and to meet or go beyond them.
- Provide help and advice to foster the responsible management of chemicals by all those who manage and use them along the product chain.

2. IMPLEMENT FUNDAMENTAL FEATURES OF NATIONAL RESPONSIBLE CARE PROGRAMMES.

Each national chemical association establishes and manages its own national Responsible Care programme based on a set of eight common fundamental features. They are:

- Establish and implement a set of Guiding Principles that member companies sign.
- Adopt a title and logo that are consistent with Responsible Care.

- Implement management practices through a series of systems, codes, policies and guidance documents to assist companies to achieve better performance.
- Develop a set of performance indicators against which improvements can be measured.
- Communicate with interested parties inside and outside the membership.
- Share best practices through information networks.
- Encourage all association member companies to commit to and participate in Responsible Care.
- Introduce and apply systematic procedures to verify the implementation of the measurable elements of Responsible Care by member companies.

Industry leaders support the national associations in the fulfillment of these fundamental features. The Charter defines specific commitments consistent with the fundamental features.

3. COMMIT TO ADVANCING SUSTAINABLE DEVELOPMENT.

Responsible Care is a uniquely designed initiative that enables the global chemical industry to make a strong contribution to sustainable development. Through improved performance, expanded economic opportunities, and the development of innovative technologies and other solutions to societal problems, the industry will continue taking practical steps to implement initiatives in support of sustainable development. The industry will expand its dialogue with stakeholders to identify additional opportunities to contribute to sustainable development through Responsible Care.

The chemical industry recognizes the important contribution that can be made through capacity building of the sound management of chemicals to achieve sustainable development goals. The industry will continue to support national and international initiatives to advance these goals.

4. CONTINUOUSLY IMPROVE AND REPORT PERFORMANCE.

Each chemical company that implements Responsible Care is expected to collect and report data for a core set of environmental, health and safety performance measures. Each national association is expected to collect, collate and report this data from its members in each country. The data will also be collated and reported publicly at the international level and be updated every two years at a minimum.

In order to continue to achieve improved performance, each national association that implements Responsible Care will:

- Periodically assess, with the participation of their members, stakeholder expectations for expanded or modified performance reporting or other aspects of performance.

- Commit to providing practical help and support in sharing and adopting best practices to improve environmental, health and safety performance, and other assistance related to Responsible Care implementation needs.

Chemical companies that implement Responsible Care will:

- Adopt a management systems approach to implement their Responsible Care commitments consisting of the internationally accepted elements of Plan-Do-Check-Act.
- Utilize clean and safe-technologies and processes when building new plants or expanding their current facilities around the world.
- Go beyond self-assessment of the implementation of Responsible Care and adopt external verification processes carried out either by associations, government bodies or other external organisations.

5. ENHANCE THE MANAGEMENT OF CHEMICAL PRODUCTS WORLDWIDE – PRODUCT STEWARDSHIP.

Product stewardship issues will increasingly shape the Responsible Care initiative in future years. The ICCA will establish a strengthened global programme to evaluate and manage chemical-related risks and benefits by developing a unified product stewardship management system approach. This approach will be in place by 2006.

National associations, working with their member companies, will commit to this concerted global effort by establishing processes for Responsible Care companies to:

- Re-commit to full implementation of current Responsible Care product stewardship commitments, including all existing codes, guidelines and practices.
- Improve product stewardship performance and increase public awareness of the industry's commitments and results.
- Develop and share best practices through mutual assistance.
- Work in partnership with upstream suppliers and downstream chemical users to collaborate on improved processes for the safe and effective uses of chemicals.
- Encourage and sustain support for education, research and testing approaches that will yield useful information about the risks and benefits of chemicals through such initiatives as the High Production Volume chemical testing program and the Long-range Research Initiative.
- Implement enhanced product stewardship commitments consistent with the ICCA's Global Chemicals Management Policy, and periodically assess product stewardship practices in the light of evolving societal expectations for chemical products.

6.   CHAMPION AND FACILITATE THE EXTENSION OF RESPONSIBLE CARE® ALONG THE CHEMICAL INDUSTRY'S VALUE CHAIN.

Responsible Care® companies and associations commit to promoting the Responsible Care ethic, principles and practices along their own value chains and communicating the importance of the industry's economic and social contributions.

Chemical companies and national associations commit to increase dialogue and transparency with their business partners and other stakeholders and to expand knowledge and understanding of the management of chemicals. They will also work in partnership with national governments, multi-lateral and non-governmental organisations to define mutual assistance priorities and share access to information and expertise.

The global chemical industry will develop and share information and practices across companies consistent with competition law and other legal requirements.

7.   ACTIVELY SUPPORT NATIONAL AND GLOBAL RESPONSIBLE CARE GOVERNANCE PROCESSES.

The chemical industry, through the ICCA, commits to an enhanced, transparent and effective global governance process to ensure accountability in the collective implementation of Responsible Care. The governance process will be implemented by the ICCA and will incorporate such issues as tracking and communicating performance commitments; defining and monitoring the implementation of Responsible Care obligations; supporting national association governance; helping companies and associations to achieve Charter commitments; and establishing a global process for revoking, when necessary, the Responsible Care status of any company or association that fails to meet its commitments.

8.   ADDRESS STAKEHOLDER EXPECTATIONS ABOUT CHEMICAL INDUSTRY ACTIVITIES AND PRODUCTS.

The global chemical industry will extend existing local, national and global dialogue processes to enable the industry to address the concerns and expectations of external stakeholders to aid in the continuing development of Responsible Care.

9.   PROVIDE APPROPRIATE RESOURCES TO EFFECTIVELY IMPLEMENT RESPONSIBLE CARE.

Responsible Care is the signature performance initiative of the chemical industry and will have an increasingly important part to play as a basis for the industry's views in societal and regulatory discussions. Companies participating in Responsible Care® must support and meet the requirements of the national programmes and provide sufficient resources for implementation.

APPENDIX 2

Directory of Standards and CSR-Related Organizations as Compiled by The Future 500

AA 1,000
The Institute of Social and Ethical
AccountAbility, Unit A,
137 Shepherdess Walk,
London, N1 7RQ,
United Kingdom
Tel: +44 (0)20 7549 0400
Fax: +44 (0)20 7253 7440
http://www.accountability.org.uk

The AA 1,000 is designed by the Institute for Social and Ethical Accountability (also known as AccountAbility) located in Britain. The goal of the standard is to secure the quality of social accounting, auditing, and reporting with an emphasis on stakeholder engagement. The standard advocates a corporate reporting process based on engaging with stakeholders, responding to their input, and embedding new practices into their operations.

> *Special Features*: The first global standard for measurement and reporting of social and ethical performance. AA 1,000 is a widely used standard for corporate social reporting in the UK.
> *Users*: Stakeholder groups apply this standard as a tool to audit and rank the quality of social and sustainability reports.

Baldrige
Baldrige Malcolm Baldrige National Quality Award Program,
NIST 100 Bureau Drive,
Stop 1020,
Gaithersburg, MD 20899-1020, United States
Tel: 301 975 2036
Fax: 301 948 3716
http://www.quality.nist.gov

The Malcolm Baldrige National Quality Program was established by the U.S. Congress in 1987. Its annual Quality Award is managed by the National Institute of Standards and Technology. The Award recognizes performance excellence and quality achievement among U.S. manufacturers, service companies, educational institutions, and health care providers.

> *Special Features*: Quality Management. The award criteria provide a similar but more comprehensive system than ISO 9000 series. The Baldrige application and review process is considered the leading audit of corporate quality and health, inside and outside the United States.
>
> *Users*: Used widely by researchers, investors, media, and competitors as the benchmark for performance excellence.

BCCCC
The Center for Corporate Citizenship at Boston College (BCCCC),
Community Involvement Index,
Boston College Center for Corporate Citizenship,
55 Lee Road,
Chestnut Hill, MA 02467-3942, United States
Tel: 617 552 0117
Fax: 617 552 8499
http://www.bc.edu/centers/ccc/index.html

The Center for Corporate Citizenship at Boston College (BCCCC) issues an annual Community Involvement Index. It provides a series of annual snapshots of issues and trends based on surveys of leading U.S. corporations regarding their contribution allocations, management support, budgets, and staffing for community involvement programs.

> *Special Features*: Community programs. This index is the leading resource on community involvement practices that enable both firms and community partners to benchmark their performance.
>
> *Users*: Researchers, corporations, and the media.

BITC
Business in the Community (BITC),
Corporate Responsibility Index,
Business in the Community,
137 Shepherdess Walk,
London N1 7RQ, United Kingdom
Tel: 0870 600 2482
http://www.bitc.org.uk

Business in the Community (BITC) is a UK-based business membership organization, which publishes the Corporate Responsibility Index that annually surveys and ranks the CSR performance of participating companies.

Special Features: Comprehensive CSR management and performance assessment. The BITC index is a unique voluntary CSR initiative, business-led, that engages with companies from all sectors. The index publicly ranks a company's CSR activity, while consolidating information demands made on companies.

Users: More than half of the FTSE 100 companies participated in the survey and the ranking attracts great attention within and outside Britain.

Calvert
Calvert Group Social Criteria,
Calvert Group,
4550 Montgomery Avenue Bethesda,
Maryland 20814, United States
Tel: 800 368 2748
www.calvertgroup.com

Calvert Group is known for offering the largest family of socially screened mutual funds as well as award-winning tax-free investment products. Calvert has approximately $8.5 billion in assets under management. Calvert uses Social Criteria and the analysis of firms' financial performance to identify eligible investment opportunities.

Special Features: Human rights, product safety, environment, community relations, and corporate governance. This is one of the best-known socially responsible investment funds in the United States.

Users: Inclusion in its portfolio is considered a sign of excellent CSR performance by socially responsible investors and other stakeholder groups.

Caux
Caux Round Table Principles for Business,
Caux,
47835-201 Avenue,
Waterville, MN 56096, United States
Tel: 507 362 4916
http://www.cauxroundtable.org

The Caux Round Table, an international network of business leaders, developed Principles for Business in 1994 to advance a world standard in corporate citizenship against which business behavior can be measured.

Special Features: Corporate citizenship, rooted in two basic ethical ideals: kyosei (living together) and human dignity. The Caux standard encompasses CSR practices comprehensively.

Users: One of the leading business-led, peer-to-peer CSR initiatives, with worldwide recognition.

CERES
Coalition of Environmentally Responsible Economies (CERES),
CERES,
99 Chauncy Street, 6th Floor,
Boston, MA 02111, United States
Tel: 617 247 0700
http://www.ceres.org

The CERES Principles were developed by the Coalition of Environmentally Responsible Economies (CERES), a leading coalition of environmental investor and advocacy groups, to establish a set of environmental criteria by which investors and others could assess the environmental performance of companies.

> *Special Features*: Ten-point code of conduct that focuses on environmental awareness and accountability. CERES is a pioneer in setting environmental performance standards. It was originally entitled the Valdez Principles and promoted in response to the Valdez oil spill.
> *Users*: Seventy large companies have endorsed the Principles. Environmental groups consider signing onto the CERES principles an indicator of commitment to excellence in environmental performance.

Domini
Domini Social Investments Social and Environmental Screens,
Domini Social Investments,
P.O. Box 9785 Providence,
RI 02940, United States
Tel: 800 762 6814

Domini Social Investments manages more than $1.5 billion in assets for individual and institutional investors who wish to integrate social and environmental criteria into their investment decisions.

> *Special Features*: Workplace, human rights, environment, and product safety. The Domini 400 Social Index is often considered the leading SRI Index.
> *Users*: Investors, activists, and the media.

DJSI
The Dow Jones Sustainability Index (DJSI),
SAM Indexes,
GmbH Seefeldstrasse,
215 8008 Zurich, Switzerland
Tel. +41 1 395 2828
Fax +41 1 395 2850
www.sustainability-index.com

The Dow Jones Sustainability Index (DJSI) is described as the first global index tracking the financial performance of the leading sustainability-driven companies.

Special Features: The DJSI corporate questionnaire is designed to assess opportunities and risks deriving from economic, environmental, and social activities of companies. Inclusion in this standard is a competitive process; selection is considered a mark of distinction for companies that want investors to see them as sustainability leaders.

Users: Investors and stakeholders

FTSE4Good
FTSE4Good Index Series,
FTSE,
15th Floor,
St Alphage House,
2 Fore Street,
London EC2Y 5DA, United Kingdom
Tel: +44 (0) 20 7448 F17
www.ftse.com/ftse4good/index.jsp

The FTSE4Good Index Series was created by FTSE, a respected global financial index company based in Britain, in response to the increasing interest in SRI. Its inclusion criteria measure the performance of companies that meet globally recognized corporate responsibility standards.

Special Features: Environmental sustainability, stakeholder management, and universal human rights. The visibility and reputation of FTSE4Good provides companies with a powerful vehicle to communicate their CSR achievements. Inclusion is a competitive process and considered a mark of distinction.

Users: Widely used by investors and asset managers, especially in Europe, and, increasingly, in Asia.

GRI
Global Reporting Initiative (GRI),
Sustainability Reporting Guidelines,
GRI,
Keizersgracht 209,
1016DT Amsterdam,
The Netherlands
Tel: +31 20 531 0012
www.globalreporting.org

Global Reporting Initiative (GRI) Sustainability Reporting Guidelines set a globally applicable framework for reporting the economic, environmental, and social dimensions of an organization's activities, products, and services.

Special Features: A comprehensive global standard for sustainability reporting. GRI is the most widely used and internationally recognized standard for corporate sustainability measurement and reporting.

Users: Corporations and stakeholder groups, as a standard with wide support to assess and improve the quality of corporate sustainability reporting.

Goldman
Goldman Sachs Group, Inc.,
Ten Point Plan,
Goldman, Sachs & Co.,
85 Broad Street,
New York, NY 10004, United States
Tel: 212 902 0300
www.gs.com

Goldman Sachs Group, Inc., founded in 1869, is one of the oldest and largest investment banking firms. Its Ten Point Plan for corporate governance was called for by its CEO in June 2002, setting a standard for best practices in the aftermath of the Enron scandal.

Special Features: Governance and disclosure. The Ten Point Plan sets forth best practices based on lessons learned from Enron and other recent corporate accountability scandals. The Ten Point Plan is a bold call-to-action by a respected CEO, its criteria are considered somewhat more stringent than NYSE.

Users: Corporations that wish to demonstrate commitment to manage newly recognized corporate governance risks.

Innovest
Innovest Strategic Value Advisors,
Innovest,
4 Times Square,
3rd Floor,
New York, NY 10036, United States
Tel: 212 421 2000
www.innovestgroup.com

Innovest Strategic Value Advisors was founded in 1995 and has created an analytic platform to assess the risks and value potential in firms' environmental performance.

Special Features: Environmental factors most directly associated with financial performance and competitiveness. Its Industry Benchmark Reports provide direct comparison of companies in more than 50 industries. It is the leading source for the assessment of firms' competitiveness in strategic environmental management.

Users: Investment advisors, mutual fund managers, corporations, insurers, and nonprofit organizations.

ICCR
Interfaith Center on Corporate Responsibility (ICCR),
Principles for Global Corporate Responsibility,
ICCR Room 550 475 Riverside Drive,
New York, NY 10115, United States
Tel: 212 870 2295
Fax: 212 870 2023
www.iccr.org

The Interfaith Center on Corporate Responsibility (ICCR) publishes the Principles for Global Corporate Responsibility: Benchmarks for Measuring Business Performance – a compendium of leading CSR indicators and standards.

Special Features: Workplace and environmental performance criteria. ICCR is the first major standard of corporate social responsibility movement to build a portfolio of holdings based on CSR criteria. The combined portfolio value now exceeds $110 billion.

Users: Religious and socially responsible investors use ICCR to guide investment decisions; media and activist stakeholders use their information on corporate performance.

ICC-BCSD
The International Chamber of Commerce's (ICC),
Business Charter for Sustainable Development (BCSD),
International Chamber of Commerce,
38, Cours Albert 1er,
75008 Paris, France
Tel: +33 1 49 53 29 16
Fax: +33 1 49 53 28 59
www.iccwbo.org

The International Chamber of Commerce's (ICC) Business Charter for Sustainable Development (BCSD) was developed by a group of business executives and sets 16 key principles for environmental management. The principles serve as a foundation upon which companies build their own integrated environmental management systems.

Special Features: Company-driven environmental management, based on 16 principles. The Charter is endorsed by more than 2300 companies worldwide, and is recognized as a complement to environmental management systems.

Users: Major corporations and several industry associations.

ICC-CG
International Chamber of Commerce Corporate Governance Program (ICCCG),
ICC,
38 Cours Albert 1er,
75008 Paris, France
Tel: +33 1 49 53 28 26
Fax: +33 1 49 53 28 59
www.iccwbo.org

The ICC's Corporate Governance Program (ICC-CG) developed a 10-point self-assessment intended to give a simple first-round screening of key corporate governance practices.

Special Features: Basic corporate governance principles for business managers worldwide. It is known for its global reach and has an involvement with a wide range of organizations.

Users: Company managers, board members, and stakeholders.

International Organization for Standardization (ISO)
1, rue de Varembé,
Case postale 56 CH-1211,
Geneva 20, Switzerland
Tel: +41 22 749 01 11
Fax: +41 22 733 34 30
www.iso.ch/iso/en/iso900014000/iso14000/iso14000index.html

ISO includes environmental management systems, environmental auditing, environmental labeling, environmental performance evaluation, and lifecycle assessment. It is a process rather than a performance-based system.

Special Features: ISO is the most widely used environmental management process standard worldwide. ISO 14000 certification is seen as evidence of environmentally responsible management, especially important where regulatory standards may not be consistently enforced.

Users: Thousands of companies apply for the standard or require their suppliers to be certified. Companies use the certification to select business partners and suppliers.

NYSE
NYSE New York Stock Exchange Listing Standards,
NYSE,
20 Broad Street, New York, NY 10005, United States
Tel: (212) 656 5034
www.nyse.com

The New York Stock Exchange (NYSE), the world's leading equities market, revised its Listing Standards in 2002. The new standards put listed companies to a much higher requirement of corporate governance and disclosure practices.

> *Special Features*: Governance and disclosure. The NYSE has nearly 2800 listed companies, with a combined $15.3 trillion total global market capitalization. Its listing requirements are recognized by investors worldwide as the most stringent standards of any marketplace in the world.
>
> *Users*: Corporations, their board members, and investors.

OECD
Organization for Economic Cooperation and Development (OECD),
Principles of Corporate Governance,
OECD 2, rue André Pascal F-75775 Paris,
Cedex 16, France
Tel: +33 1 45 24 82 00
USA Tel: 212 870 2316
www.oecd.org

The OECD Principles of Corporate Governance are intended to assist governments to evaluate and improve the legal, institutional and regulatory framework for corporate governance in their countries.

> *Special Features*: Rights of shareholders, transparency, and disclosures. OECD is an internationally developed and recognized standard for corporate governance.
>
> *Users*: Stock exchanges, investors, corporations, and other parties seeking guidance in the process of developing good corporate governance.

SGN
Smart Growth Network (SGN),
Smart Growth Network,
c/o International City/County Management Association,
777 North Capitol Street, NE Suite 500,
Washington, DC 20002, United States
Tel: (202) 962 3623
www.smartgrowth.org

The Smart Growth Network (SGN) was formed in response to increasing American community concerns about the need for new ways to grow local communities while boosting the economy, protecting the environment, and enhancing community vitality.

> *Special Features*: Community engagement and development, with emphasis on integrative solutions to a mix of community issues, such as traffic, housing, jobs, sprawl, and environment. Smart Growth is gaining attention recently.

It provides an alternative to single-issue focus and engages companies and stakeholders in productive problem-solving.

Users: Companies that seek to build strong community bonds. Numerous states and municipalities have endorsed the Smart Growth approach in their development strategies.

Sullivan
Global Sullivan Principles,
5040 E. Shea Boulevard, Suite 260,
Scottsdale, AZ 85254-4687, United States
Tel: 480 443 1800
Fax: 480 443 1824
http://globalsullivanprinciples.org

The Global Sullivan Principles were developed by Leon Sullivan, theologian and General Motors board member, whose Sullivan Principles for human rights were pivotal in ending racial segregation in South Africa. The Principles provide a framework by which socially responsible companies and organizations can be aligned.

Special Features: Human rights for employees and in communities where companies operate. The Sullivan Global Principles were instrumental in building corporate opposition to South African Apartheid. They were recognized by the UN Secretary General as promoting CSR. More than 100 of the world's leading companies so far have endorsed the Principles.

Users: Corporations, stakeholders, media, to assess corporate and supplier human rights issues and risks.

UN Global Compact
UN Global Compact Principles,
Global Compact Office,
United Nations S-1883,
New York, NY 10017, United States
Tel: 1 917 367 3483
www.unglobalcompact.org/Portal/

UN Global Compact is an initiative proposed by Secretary General Kofi Annan to bring companies together with UN agencies, labor, and civil society to support nine principles in the areas of human rights, labor and the environment.

Special Features: Human rights, labor, and environment. UN Global Compact is a voluntary corporate citizenship program endorsed by more than 1100 corporations worldwide; it is the largest CSR initiative in the world.

Users: Companies, UN agencies, national governments, and civil society groups participate in the initiative.

APPENDIX 3

AUTHOR BIOGRAPHIES

Abraham, Martin received his BS in chemical engineering in 1982 from Rensselaer Polytechnic Institute and his PhD in 1987 from the University of Delaware. After nearly 10 years at the University of Tulsa, he joined the chemical engineering department at the University of Toledo, serving as Associate Dean for Research and Graduate Studies in the College of Engineering from 2000–2004 and currently as Dean of the Graduate School. Dr. Abraham has over 60 refereed publications and nearly 100 technical presentations in the area of chemical reaction engineering. He has served as chair for the Engineering Conferences International sponsored multi-disciplinary conference on "Green Engineering: Defining the Principles," and as Chair of the Industrial and Engineering Chemistry Division of the American Chemical Society. He was named the Outstanding Researcher in the College of Engineering and received the University of Toledo Outstanding Research Award in 1999.

Beaver, Earl R. is Chair of the Institute for Sustainability, Managing Partner of Practical Sustainability, LLC, and Chief Technical Advisor to BRIDGES to Sustainability; he retired from Monsanto after 30 years of service. In his final position there he was responsible for the development of new environmental technology solutions for Monsanto Company's diverse operations, internal and external advocacy for sustainability and pollution prevention, as well as the development of sustainability and eco-efficiency tools. He is a member of the American Chemical Society, a fellow of the American Institute of Chemical Engineers, and a Director of the National Environmental Technology Institute. He has authored many publications and patents.

Bélanger, Jean was the President of the Canadian Chemical Producers' Association from 1979 to 1996. Mr. Bélanger currently serves as Chair of the Advisory Board to the Institute for Chemical Process and Environmental Technology of the National Research Council. In 1996, he was appointed an officer of the Order of Canada. Mr. Bélanger was appointed to the National Round Table on the Environment and the Economy in October 1996. Mr. Bélanger is the Chair of the NRTEE's Ecological Fiscal Reform (EFR) and Energy Task Force.

Berger, Scott is Director of the Center for Chemical Process Safety (CCPS), an Industrial Technology Alliance of the American Institute of Chemical Engineers. Mr. Berger holds SB and SM degrees in Chemical Engineering from the

Massachusetts Institute of Technology. Mr. Berger's 27-year industrial career focused on "greener" and "inherently safer" technologies began at Rohm and Haas, then continued at Owens Corning, where he held positions in Environment, Health and Safety, including Director of Strategic EHS management.

Besly, Michael J. is a Manager in the Sustainability Business Solutions practice of PricewaterhouseCoopers LLP (PwC). This practice specializes in strategic assessment, opportunity and risk evaluation, and the design and implementation of sustainability programs. Michael holds a BSc in International Relations from Southampton University (UK) and an MBA from Boston University, where he graduated Beta Gamma Sigma and with High Honors. He is also an associate member of the American Bar Association and the Environmental Law Institute.

Bigham, Joan is a partner at SmithOBrien, a management consulting firm on corporate responsibility, financial measurement, and executive learning. SmithOBrien is based in Cambridge, Massachusetts.

Marc Brammer is Senior Research Analyst with Innovest Strategic Value Advisors in New York. Mr. Brammer covers companies in the specialty chemicals, industrial manufacturing, and automotive sectors and has researched a number of reports on corporate accountability and environmental policy. He holds a Masters degree in Political Theory from the Graduate Faculty of the New School University.

Constable, David J.C. is Director and Team Leader of the Environmental Product Stewardship Team in Corporate EHS at GlaxoSmithKline. David's Team is responsible for developing and championing programs, systems, tools and methodologies that integrate Sustainability, Life Cycle Inventory/Assessment, Total Cost Assessment, Green Chemistry and Green Technology activities into existing R&D and Global Manufacturing Supply business processes. The Team is also responsible for providing EHS support to New Product Supply Teams. David obtained a PhD in Analytical Chemistry from the University of Connecticut and a BS in Environmental Sciences, Air and Water Pollution, from Slippery Rock University.

Coyne, Karen, MS, CPEA, is a Vice President with CoVeris, Inc., the leading provider of independent corporate performance verification services. She is the current Chairman of the Board of BRIDGES to Sustainability™, a 501(c)(3) corporation fostering practical sustainable development implementation, and Past-President of The Auditing Roundtable, the leading professional association for EHS auditors. Consulting to Fortune 50 and multinational companies for 20 years, Ms. Coyne's practice includes corporate environmental strategy and verification; environmental management systems development, implementation and certification (ISO 14001); environmental performance reporting; and financial market implications of voluntary environmental, health and safety (EHS) activities.

Dorfman, Mark is an environmental scientist working in the public interest sector since 1987 to promote pollution prevention strategies, particularly in the chemical industry. He has authored or co-authored ten sustainability reports ranging from chemical industry case studies to national water pollution report cards. Through a

2002 Chemical Heritage Foundation fellowship, Mr. Dorfman created an introductory lecture on biomimicry and the chemical industry.

Dourson, Michael directs Toxicology Excellence for Risk Assessment (TERA), a nonprofit corporation dedicated to the best use of toxicity data for estimating risk assessment values. Dr. Dourson is a Diplomate of the American Board of Toxicology and served on its Board as President, Vice President and Treasurer. He has published more than 80 papers on risk assessment methods, co-authored over 100 government risk assessment documents, and made over 100 invited presentations. He recently received the Society of Toxicology's Arnold J. Lehman Award. For 15 years, Dr. Dourson held leadership roles in the U.S. Environmental Protection Agency, as chair of EPA's Reference Dose (RfD) Work Group, charter member of the EPA's Risk Assessment Forum, and chief of the group that helped create the Integrated Risk Information System (IRIS) in 1986.

Funk, Karina manages the emerging technologies investment programs at the Massachusetts Renewable Energy Trust, programs that provide venture financing, networking opportunities, and other resources for entrepreneurs in the state. Prior to joining the Trust, Karina spearheaded research and consulting at Cap Gemini Ernst & Young, focusing on intangibles valuation and its links with environmental management. Karina's work at the intersection of technology, finance, and the environment have led to private sector and government roles in the United States, Europe, and Latin America, as well as numerous publications and speaking engagements on the business case for sustainability. Karina holds Masters degrees at MIT as a National Science Foundation Fellow, and a post-graduate degree in Technology and Management from the Ecole Polytechnique in France.

Gable, Cate is President of Axioun Communications International (www.axioun. com) and serves as Director of Product Stewardship for the Future 500 (www. future500.org). Gable is the author of *Strategic Action Planning NOW!* (St. Lucie Press, 2000) and teaches "Strategic Action Planning and Innovation for the New Century" at the top business college in Paris, France, Ecoles des Hautes Etudes Commerciales. Gable is an expert facilitator in the new field of multistakeholder dialog.

Gadagbui, Bernard joined Toxicology Excellence for Risk Assessment (TERA) in 2004 as a toxicologist after a combined eight-year toxicologist position at the University of Florida and the Bureau of Pesticides of the Florida Department of Agriculture and Consumer Services. Dr. Gadagbui has a wealth of research and teaching experience in environmental toxicology, experience in evaluating human health risks posed by chemicals including pesticides and in developing health-related toxicity values for chemicals, and is familiar with state and federal pesticide regulations and risk assessment.

Geiser, Kenneth is Professor of Work Environment and Director of the Lowell Center for Sustainable Production at the University of Massachusetts Lowell. Dr. Geiser is one of the authors of the Massachusetts Toxics Use Reduction Act and served as Director of the Massachusetts Toxics Use Reduction Institute from its founding in 1990 to 2003. His research and writing focus on pollution prevention and cleaner production, toxic chemicals management, international chemicals policy, safer technologies, and green chemistry and, in 2001, he completed a book, *Materials Matter: Towards a Sustainable Materials Policy*, published by MIT Press. As a recognized expert on environmental and occupational health policy, he has served on various advisory committees for the U.S. Environmental Protection Agency, and the United Nations Environment Program.

Gillen, Art has over 30 years experience in environmental regulatory and legislative support and environmental management systems within the chemical industry. After 25 years with Union Carbide and BASF Corporation in their corporate groups, he has spent the last five years providing environmental consulting support to private industry, EPA and the U.S. Air Force. He recently worked with the American Chemistry Council on its pilot program combining the requirements of Responsible Care® with ISO 14001. He has been associated with SOCMA's environmental activities for nearly 20 years. He has a BE and ME in Chemical Engineering from Stevens Institute of Technology.

Greener, Catherine, Principal and Leader of Rocky Mountain Institute's Commercial and Industrial Team, has more than 20 years of experience in industrial quality management, lean manufacturing, operations and manufacturing sustainability. Ms. Greener holds a degree in Industrial Engineering from Northwestern University and a Masters in Business Administration from University of Michigan.

Haber, Lynne has more than 12 years of experience in applying risk assessment methods in evaluating the toxicity, toxicokinetics, and mode of action of chemicals. Her current interests are in the application of mechanistic information in risk assessment and in methods for extending the dose–response curve to low doses. Other current work includes research on children's risk issues, consideration of mode of action in cancer risk assessment, incorporating data on polymorphisms into risk assessment, and development of scientifically based occupational exposure limits.

Hileman, Doug is a Director in the Sustainability Business Solutions practice of PricewaterhouseCoopers LLP (PwC). He is also the primary Western Region contact for Sustainability, including environmental, health and safety, corporate social responsibility, reporting and communications, and supply chain issues. Doug has 28 years of experience in the field, including nine in industry for chemical and energy companies. Doug is a Qualified Environmental professional, a registered Professional Engineer (California, Ohio), and a Nevada Registered Environmental Manager. Doug has a BS in Chemical Engineering, a BS in Mathematics (both from Rose-Hulman Institute of Technology) and an MBA (Cleveland State University).

Heine, Lauren joined GreenBlue as Director of Applied Science after four years with Zero Waste Alliance (ZWA), where she was Director of Green Chemistry and Engineering. Earlier, Lauren was a Fellow with the American Association for the Advancement of Science in the Industrial Chemicals Branch of the U.S. EPA.

At EPA she worked with the Green Chemistry Program, promoting R&D of products and processes with benign human and environmental health impacts. She was senior editor for two American Chemical Society Symposium Series books, *Green Chemical Syntheses* and *Processes and Green Engineering*. As Director of Applied Science, she guides the development of technical tools and approaches that help organizations integrate principles of Green Chemistry and Engineering into their operations to become more sustainable. She has a doctorate in Civil and Environmental Engineering from Duke University.

Kalafut, Pamela Cohen and **Jonathan Low** are Partners in Predictiv LLC, a consulting firm that specializes in performance management, organizational design, and the measurement of intangibles like leadership. They have over ten years experience working with companies that range from the *Fortune* 500 to start-up enterprises. In the public sector, they have worked with the U.S. Defense Department, the Environmental Protection Agency, and the Department of Commerce. They are the authors of numerous studies evaluating the impact of intangibles on financial and operational outcomes. Their work has appeared in the *Wall Street Journal, Forbes, The New York Times*, and they have appeared on ABC, CBS, MSNBC, and CNN. They have testified before the Securities and Exchange Commission, the Federal Reserve Bank of New York, the European Commission, and the Organization of Economic Cooperation and Development. Jon and Pam are the authors of *Invisible Advantage: How Intangibles Are Driving Business Performance*, published by Perseus Press in 2002.

Larson, Andrea conducts research and teaches as a member of the faculty at the Darden Graduate School of Business Administration. Her work focuses on corporate entrepreneurial innovation at the nexus of business and natural systems. A forthcoming book and wide range of curriculum materials highlight cost, profit, and market advantage opportunities for firms adopting sustainability principles.

Laurin, Lise founded EarthShift to focus on supporting businesses that help the Earth. EarthShift supports and sells lifecycle assessment (LCA) and total cost assessment (TCA) software, and provides LCA and TCA consulting in association with Sylvatica. Prior to EarthShift, Lise consulted with a number of high-tech and environmental organizations, providing market research and marketing strategy. Lise uses skills developed through 20 years in manufacturing firms within the semiconductor industry. She began her professional life as a process engineer at Intel. She holds a BS in Physics from Yale University.

Liroff, Richard A., a political scientist, is Senior Fellow in the Toxics Program at World Wildlife Fund in Washington, DC. Dr. Liroff has published numerous books and reports on a wide range of environmental policy issues. Currently he is developing a benchmarking framework that can be used by corporate managers and investors for rating corporate management of toxic chemicals in consumer products. The views expressed here are his own and not necessarily those of World Wildlife Fund.

Low, Jonathan Predictiv (see Kalafut and Low).

Machado, Joe is currently the Manager of Technology Strategy and Innovation for Shell Chemicals. Prior to that, Joe was Manager of Sustainable Development for Shell Chemicals for four years, leading the effort to implement sustainable development across Shell chemicals companies' global petrochemical operations and business units. Joe has a PhD in Polymer Science and Engineering and a BS in Plastics Engineering. He joined Shell in 1988 in Houston, Texas, working in polymer product research. Since then, he has worked in technology, manufacturing, business development, and sustainable development for Shell in a number of locations in the United States, United Kingdom, and Belgium. Joe holds over 25 patents, has written numerous articles, and lectured at many universities both on technical subjects and on sustainable development, and the challenges and opportunities it holds for business.

Mathis, Mitch is Environmental Economist at the Houston Advanced Research Center (HARC), where he specializes in natural resource and environmental policy and Latin American geography. He established and leads HARC's Valuing Nature in Texas program, and has worked extensively on issues concerning water, development, and the environment. Mitch holds a doctorate in economics from The University of Texas at Austin.

Meyer, Stephanie is a Principal of Stratos, a Canadian-based sustainability consultancy. She works with a wide range of private and public sector clients to improve their sustainability performance through enhanced performance measurement, management and reporting practices. She was project head for Stratos's 2001 and 2003 benchmark surveys of corporate sustainabiity reporting in Canada: Stepping Forward and Beyond Confidence.

Miller, Keith J. leads the Environmental Initiatives and Sustainability Group within the Environmental Technology and Safety Services (ET&SS) organization at 3M. He is responsible for 3M environmental initiatives including Environmental Targets for 2005 (ET'05), Pollution Prevention Pays (3P), and the Global Climate Change Program. He also leads sustainability activities including sustainability reporting, metrics and goals development/implementation and strategic planning.

Norris, Greg founded and directs Sylvatica, a lifecycle assessment (LCA) research consulting firm. Norris co-developed the Total Cost Assessment program, TCAce. He is Program Manager for the United Nations' Environment Program's (UNEP) global Life Cycle Initiative. He teaches graduate courses on LCA and Industrial Ecology at the Harvard School of Public Health. He consults to UNEP, Federal and state agencies in the United States, and the private and nonprofit sectors worldwide. Norris is founder and executive director of New Earth, a foundation for grassroots sustainable development. Norris is Adjunct Research Professor at the University of New Hampshire, a Program Associate at Kansas State University, and an editor of the *International Journal of Life Cycle Assessment*.

Ochsenkuehn, Rainer is a project manager at First Environment, Inc. His work includes consulting, auditing, and training regarding ISO 14001, RC14001, and

RCMS. He is a highly regarded RAB-certified Lead Auditor performing certification audits for ISO 14001 and RC14001. He has developed and performed training courses and presentations for ACC regarding RC14001 and RCMS, as well as numerous public and inhouse workshops on EHSMS implementation and auditing. He is principal author of the *Introduction to ISO 14000 Handbook* and other papers on management system inplementation and integration.

Pojasek, Robert is president of Pojasek & Associates, a management consulting firm specializing in process improvement and performance measurement. He is also an adjunct professor at Harvard University, where he teaches a distance-learning course on the path to sustainable development.

Poltorzycki, Stephen is the founder and President of The Boston Environmental Group, a firm dedicated to providing excellence in environmental, health and safety ("EHS") consulting. The focus of his work is to assist clients with sustainable development and EHS strategy, organizational and management challenges in maximizing the business value of their EHS activities. For these clients he has developed strategies and processes to create and deliver business value, evaluated and designed organizational frameworks, and assessed, designed, and implemented management systems to assure compliance, manage risk, and obtain competitive advantage. He is the author of *Creating Environmental Business Value*, published by Crisp Publications in 1998, as well as a number of articles in various publications on the subjects of environmental management and sustainable development.

Reed, Don is a Principal with Ecos Corporation, a corporate advisory that works with companies to reduce business risks and open up new opportunities to lower costs, increase capital efficiency, and enhance growth. As a Chartered Financial Analyst, he brings a rigorous financial perspective to demonstrating and measuring the relationship between sustainability and shareholder value. Prior to joining Ecos, he led the World Resources Institute's corporate engagement program, served as Director of Research at The Carson Group (a capital markets advisory), and co-founded Excelsior Capital Corporation.

Rittenhouse, Dawn is Director, Sustainable Development, for the DuPont Company. Dawn joined DuPont in 1980 and has held positions in Technical Service, Sales, Marketing, and Product Management within the Packaging and Industrial Polymers business and Crop Protection businesses. In late 1997, she began working in the corporate organization to assist DuPont businesses integrate sustainability strategies into strategy and business management processes. She leads DuPont's efforts at the World Business Council for Sustainable Development and with the UN Global Compact. She also manages the corporate recognition program for Sustainable Growth Excellence. In 2001 and 2002, Dawn served as co-chair of the GEMI working group that developed the SD Planner™. She also co-chaired the WBCSD working groups on Innovation and Technology and Sustainability through the Market. Dawn has a double major in Chemistry and Economics from Duke University.

Robèrt, Karl-Henrik, is one of Sweden's foremost cancer scientists who, in 1989, initiated the organization "The Natural Step." In 1982, Dr. Robèrt became a Professor of Internal Medicine. In 1984, he won the Swedish Hematological Association Research Award. From 1985 to 1993 he headed the Division of Clinical Hematology and Oncology and the Department of Medicine, Huddinge Hospital. He was Editor of *Reviews in Oncology* from 1987 to 1993. In 1995, Dr. Robèrt was appointed Professor of Resource Theory at the University of Gothenburg, Sweden. He has written many publications on the environment and sustainability, including: "Det Naturliga Steget," a booklet distributed to all households and schools in Sweden, "From the Big Bang to Sustainable Societies" with K.-E. Eriksson in *Reviews in Oncology*, and "Socio-ecological principles for sustainability," with Holmberg and Eriksson, published in *Practical Applications of Ecological Economics*.

Savitz, Andy is a Partner in the Environmental and Sustainability Services group at PricewaterhouseCoopers LLP (PwC). He assists companies to manage risks and identify opportunities related to economic, environmental, and social activities – the "triple bottom line" of sustainability. Andy specializes in strategic assessment, risk evaluation, and the design and implementation of sustainability programs. As a former senior environmental regulator, Andy also provides traditional environmental services, including development of compliance-focused environmental management systems and litigation support. Andy graduated Phi Beta Kappa, from the Johns Hopkins University, attended New College, Oxford, where he was a Rhodes Scholar, and graduated from Georgetown University Law Center, where he was an editor of the Law Review.

Scarbrough, Paul is an Associate Professor of Accounting and Finance at Brock University, St. Catharines, Ontario, Canada and a SmithOBrien consultant. His research interests are cost reduction and the effect of corporate culture on the use of cost reduction methods. Prior appointments include Boston University and Bentley College. International teaching experience includes Waseda University (Tokyo) and Vaxjo University (Sweden).

Schoen, Elizabeth C. Girardi is Senior Director EHS Strategic Partnerships & Planning of Pfizer Inc. Liz is responsible for setting strategy, establishing programs and systems for all EHS, and for ensuring the success of internal and external partnerships.

Schwanhold, Ernst is Head of the competence center Environment, Safety & Energy at BASF Aktiengesellschaft. He was born in Osnabrück, Germany, in 1948. After training as a laboratory assistant, he studied chemical engineering in Paderborn and graduated as an engineer.

Shanley, Agnes is currently Editor-in-Chief of *Pharmaceutical Manufacturing* magazine, published by Putman Media. She has specialized in covering the pharmaceutical and chemical industries as well as environmental and business issues for companies including McGraw-Hill and *Chemical Week*, and has also worked in the competitive intelligence and market research field. She has a BA from Barnard College and a degree in life sciences from the City University of New York.

Sherman, Dave is a founding partner of Sustainable Value Partners, a strategy consulting firm that helps companies deliver business value from social, environmental, and economic leadership. He has more than 25 years of experience in strategy consulting and process industries. He is currently a Mandel Fellow at the Weatherhead School of Management at Case Western Reserve University where he is earning an Executive Doctorate. He holds an MBA from UC Berkeley and a BS in Chemical Engineering from Purdue University.

Sigman, Richard is a Principal Administrator for the Organisation for Economic Co-operation and Development (OECD) in Paris. Since joining in 1993, he has been responsible for OECD's work on risk management, new chemicals, pesticides, biocides, environment and trade, environmental outlooks, and liaison with newly emerging economies. From 1990 to 1992, he worked for the U.S. Chemical Manufacturers Association (now called the American Chemistry Council), as director of the Air Program. Prior to joining CMA, Mr. Sigman worked for the Executive Office of the U.S. President, and the Environmental Protection Agency's Office of Toxic Substances.

Smith, Neil is co-founder and managing partner at SmithOBrien in Boston, where he leads the firm's corporate responsibility auditing and financial measurement capabilities and shareholder resolution advisory service. Mr. Smith has extensive experience in helping corporations apply their core values and code of practice as the foundation for organizational change and growth. He has authored journal articles and books on corporate responsibility, including "Shopping for Safer Boat Care." He is an adjunct faculty member at Boston's College's Carroll School of Management, a guest lecturer on corporate responsibility at the Harvard University School of Public Health, and a consultant for the Center for Corporate Citizenship at Boston College.

Spitz, Peter has spent most of his career in the chemical industry, first as a group leader and executive at Esso Engineering and Scientific Design Company, and later as head of Chem Systems, a respected global management consulting firm. He recently founded Chemical Advisory Partners. He is the author of two books on the chemical industry.

Taschler, David R., PE is a business manager at Air Products and Chemicals, Inc. in Allentown, PA and an adjunct professor in the Engineering Division at Lafayette College in Easton, PA. He conducted his doctoral research on the "Characteristics of Board of Director Members and Corporate Commitment to Sustainable Development" at the University of Phoenix. Dr. Taschler is a former Management Verifier for the American Chemistry Council's Responsible Care program and is a member of the AIChE Sustainability Division and the Academy of Management.

Thorpe, Beverley is International Director of Clean Production Action, a nonprofit resource for advocacy groups working towards sustainable chemicals policy and clean product design. As a consultant she has also worked internationally with governments and industry to promote sustainable materials strategy.

Tickner, Joel is Research Assistant Professor in the Department of Work Environment at the School of Health and Environment, University of Massachusetts Lowell. His scholarly interests include development of innovative scientific methods and policies to implement a precautionary and preventive approach to decision-making under uncertainty. His teaching and research focus on regulatory science and policy, risk assessment, and cleaner production.

Underwood, Joanna D. is the President of INFORM, Inc., a national, nonprofit environmental research organization, which she founded in 1974 to identify practical and innovative solutions to some of this country's most complex environmental and health-related challenges. INFORM's research is widely used by government, business, and environmental leaders around the world. Ms. Underwood is a graduate of Bryn Mawr College and holds an Honorary Doctor of Science Degree from Wheaton College. In 2000, she was named by *The Earth Times* as one of the 100 most influential voices in the global environmental movement. She is listed in *Who's Who in America* and *Who's Who of American Women.*

Verrico, Brad is President of Verrico Associates, LLC. He has consulted with over 130 chemical and petrochemical companies about Responsible Care®, has visited over 250 facilities and interviewed over 8500 employees, from CEOs to plant operations staff. Mr. Verrico has trained over 250 people in RC14001 and RCMS. He personally developed much of the RC14001 technical specification and guidance documents currently being used today. He also helped to develop the technical specification and guidance documents for the ACC's Responsible Care Management System (RCMS), the "non-ISO" option for Responsible Care certification. He is a Certified Professional Environmental, Health, and Safety Auditor (CPEA) from the Board of Environmental, Health, and Safety Auditing Certification (BEAC).

Vigon, Bruce W. is a Research Leader in Battelle's Chemical and Environmental Technologies group. He has over 28 years of experience conducting successful research and development programs for his clients, including analysis and management of diverse issues associated with products and processes in the chemical, defense, and energy industry sectors. Mr. Vigon has led Battelle's efforts in the development and application of methods for integrated lifecycle assessment (LCA) of products and materials. Mr. Vigon's educational background combines a strong science capability (BS, Chemistry, University of Illinois, and MS, Water Chemistry, University of Wisconsin) with related competencies in engineering, economics, resource management and statistics (MS, Water Resources, University of Wisconsin).

Wade, Mark joined Shell in 1979 as a research biochemist. Since then he has served in a wide variety of roles for Chemicals and Shell International. In 1997 he moved to the Corporate Centre to become a founder member of Shell's Sustainable Development Group as head of Sustainable Development Strategy, Policy and Reporting. In early 2003 he moved into Leadership Development to lead the sustainable development learning programme. Mark currently serves as Shell's Liaison Delegate to the World Business Council for Sustainable Development (WBCSD) and is Chairman of the Business Network of the European Academy of Business in Society (EABIS).

Waage, Sissel launched and directs the R&D Program at The Natural Step. She develops strategy and implements research and strategic dialog projects. In addition, Sissel works with the Services Group advising Fortune 500 companies on integration of sustainability into strategy, operations, and reporting. Previously, she worked with Sustainable Northwest and the World Wildlife Fund's (WWF) East and Southern Africa Program. Her work has been published in numerous journals, and she edited the book, *Ants, Galileo, and Gandhi: Designing the Future of Business through Nature, Genius, and Compassion* (Greenleaf Publications, 2003). Sissel has a PhD and MS from the University of California, Berkeley, in the Department of Environmental Science, Policy and Management, and BA degree magna cum laude from Amherst College. She studied at University of Oslo, Norway, as a Fulbright Scholar.

Weinrach, Jeff has been a senior examiner and is currently Lead Judge for *Quality New Mexico*, as well as a judge and senior examiner for the *Green Zia Environmental Excellence Program*. Dr. Weinrach has 13 years of experience in pollution prevention, waste minimization, and related environmental quality programs. He has provided leadership to the Pollution Prevention Program Office at Los Alamos National Laboratory (LANL) and has worked to bring innovative approaches to LANL pollution prevention and waste minimization activities. He has worked with the U.S. DOE, U.S. EPA, U.S. Department of Commerce, U.S. Department of Defense, the state of New Mexico, and the University of New Mexico (UNM) to integrate ideas and projects. He has served as co-Editor-in-Chief of *International Journal of Environmentally Conscious Design and Manufacturing* and has served on the Editorial Board of the *Encyclopedia of Life Support Systems*. He has served as General Chairman of the *International Congress of Environmentally Conscious Design and Manufacturing* and Chairman of the New Mexico Pollution Prevention Advisory Council. He writes a regular column for *Environmental Quality Management* entitled "The Business of EQM."

Wiesner, Mark R. is the Director of the Environmental and Energy Systems Institute at Rice University, where he holds appointments as Professor in the Departments of Civil and Environmental Engineering, and Chemical Engineering. His recent research addresses the applications of emerging nanomaterials to membrane science and water treatment and an examination of the fate and transport of nanomaterials in the environment.

Yosie, Terry is Vice-President, Responsible Care, American Chemistry Council. He has served in senior management positions at the U.S. Environmental Protection Agency, the petroleum industry, and management consulting. He has been a member of the National Academy of Sciences Board on Environmental Studies and Toxicology and is the author of over 50 publications.

INDEX

513